# 8th Edoardo Amaldi Conference on Gravitational Waves 2009

AA002775

**Journal of Physics: Conference Series Volume 228**

## New York, New York, USA
## 21-26 June 2009

**ISBN: 978-1-61738-798-2**
**ISSN: 1742-6588**

**Printed from e-media with permission by:**

Curran Associates, Inc.
57 Morehouse Lane
Red Hook, NY 12571

**Some format issues inherent in the e-media version may also appear in this print version.**

Copyright© (2009) by the Institute of Physics
All rights reserved.

Printed by Curran Associates, Inc. (2010)

For permission requests, please contact the Institute of Physics
at the address below.

Institute of Physics
Dirac House, Temple Back
Bristol BS1 6BE UK

Phone: 44 1 17 929 7481
Fax: 44 1 17 920 0979

techtracking@iop.org

**Additional copies of this publication are available from:**

Curran Associates, Inc.
57 Morehouse Lane
Red Hook, NY 12571 USA
Phone: 845-758-0400
Fax: 845-758-2634
Email: curran@proceedings.com
Web: www.proceedings.com

# TABLE OF CONTENTS

# PREFACE

At Amaldi7 in Sydney in 2007 the Gravitational Wave International Committee (GWIC), which oversees the Amaldi meetings, decided to hold the 8th Edoardo Amaldi Conference on Gravitational Waves at Columbia University in the City of New York. With this decision, Amaldi returned to North America after a decade.

The previous two years have seen many advances in the field of gravitational wave detection. By the summer of 2009 the km-scale ground based interferometric detectors in the US and Europe were preparing for a second long-term scientific run as a worldwide detector network. The advanced or second generation detectors had well-developed plans and were ready for the production phase or started construction. The European-American space mission, LISA Pathfinder, was progressing towards deployment in the foreseeable future and it is expected to pave the ground towards gravitational wave detection in the milliHertz regime with LISA. Plans were developed for an additional gravitational wave detector in Australia and in Japan (in this case underground) to extend the worldwide network of detectors for the advanced detector era. Japanese colleagues also presented plans for a space mission, DECIGO, that would bridge the gap between the LISA and ground-based interferometer frequency range.

Compared to previous Amaldi meetings, Amaldi8 had new elements representing emerging trends in the field. For example, with the inclusion of pulsar timing collaborations to the GWIC, gravitational wave detection using pulsar timing arrays was recognized as one of the prominent directions in the field and was represented at Amaldi8 as a separate session. By 2009, searches for gravitational waves based on external triggers received from electromagnetic observations were already producing significant scientific results and plans existed for pointing telescopes by utilizing gravitational wave trigger events. Such multimessenger approaches to gravitational wave detection also received special attention at the meeting.

For the first time in the history of Amaldi conferences, plenary and contributed sessions were held to transfer ideas and experience gained with gravitational wave science inspired education and outreach projects. Additionally, Columbia University faculties working in frontier fields, which currently may not have direct connection to the field of gravitational wave science, gave enthralling presentations in the form of a 'wake-up' lecture series.

The meeting also facilitated the exchange of scientific results and new ideas among all members of gravitational wave experiment collaborations and the gravitational wave theory community. Additionally, future directions in gravitational wave detection were discussed in a special session dedicated to the Gravitational Wave International Committee Roadmap.

A highly entertaining and inspiring public talk titled "Songs from Space: Black Holes and the Big Bang in Audio" was given by Janna Levin and it attracted over 300 young and old science enthusiasts from and around New York City.

This special issue of *Classical and Quantum Gravity* (Volume 27, Number 8, 2010) is published as the proceedings of Amaldi8. It contains the overview articles by invited plenary speakers, and some of the highlights of the meeting as selected by session chairs and organizers. Other Amaldi8 talks and posters appear in the refereed issue of the electronic *Journal of Physics: Conference Series*. This issue of CQG and the JPCS issue are electronically linked.

The conference organizers gratefully acknowledge the financial support of Columbia University in the City of New York, the International Union of Pure and Applied Physics and the United States National Science Foundation.

We would like to thank the Gravitational Wave International Committee for the scientific oversight, the Local Organizing Committee for the program selection, and the session chairs for the thoughtful delivery of conference goals and enabling many lively discussions that are much needed for the development of our frontier field. We would also like to thank the hard work of Columbia University Physics Department administrators for handling some of the practical aspects of the organization. Finally, a team of students (Jonathan Berliner, Rutu Das, David Fierroz, Alyssa Miller, David Murphy and Keith Redwine) provided assistance behind the scene. Their hard work, practical ideas, and dedication are greatly appreciated. We are also grateful for Keith Redwine for his help in organizing the JPCS proceedings articles.

Finally, we would like to thank the over 310 participants for their interesting and lively contributions that ultimately made the conference a success.

**Zsuzsa Marka** Local Organizing Committee
**Szabolcs Marka** Chair, Local Organizing Committee
Columbia Astrophysics Laboratory and Physics Department, Columbia University
Guest Editors

Opening Ceremony - Zsuzsa Marka (Local Organizing Committee) introduces first speaker Eugenio Coccia

Eugenio Coccia - lecture on the life of Edoardo Amaldi

Plenary session - Rana Adhikari

Columbia University students at the reception desk

Conference Break

Conference Break

Conference Break

Conference Break

Janna Levin - Public Lecture

Janna Levin - Public Lecture

Conference Dinner - American Museum of Natural History

Conference Dinner - American Museum of Natural History

Conference Dinner - American Museum of Natural History

Parallel Session

Poster Session

Poster Session

Poster Session

Confernce Break

And the winner is... - Szabi Marka (Chair of Local Organizing Committee) and Elisabetta Cesarini, winner of the Best Student Poster Prize

All papers published in this volume of *Journal of Physics: Conference Series* have been peer reviewed through processes administered by the proceedings Editors. Reviews were conducted by expert referees to the professional and scientific standards expected of a proceedings journal published by IOP Publishing.

# Astronomy's New Messengers: A traveling exhibit to reach out to a young adult audience

**Marco Cavaglià[1], Martin Hendry[2], Szabolcs Márka[3], David H. Reitze[4], Keith Riles[5]**

[1] The University of Mississippi, University, MS 38677, USA
[2] University of Glasgow, Glasgow, G12 8QQ, United Kingdom
[3] Columbia University, New York, NY 10027, USA
[4] University of Florida, Gainesville, FL 32611, USA
[5] University of Michigan, Ann Arbor, MI 48109, USA

E-mail: marco.cavaglia@ligo.org, szabolcs.marka@ligo.org, martin.hendry@ligo.org, david.reitze@ligo.org, keith.riles@ligo.org

**Abstract.** The Laser Interferometer Gravitational-wave Observatory exhibit *Astronomy's New Messengers: Listening to the Universe with Gravitational Waves* is traveling to colleges, universities, museums and other public institutions throughout the United States. In 2010, an extended version of this exhibit will appear in a New York City venue that is accessible to a large and diverse cross section of the general public. *Astronomy's New Messengers* primarily communicates with an adolescent and young adult audience, potentially inspiring them into the field of science. Acknowledging that this audience is traditionally a difficult one to attract, the exhibit publicly announces itself in a charismatic fashion to reach its principal goals of broadening the community of people interested in science and encouraging interest in science among young people.

## 1. Introduction

As a frontier physics effort, a core mission of the Laser Interferometer Gravitational-wave Observatory (LIGO) [1] Scientific Collaboration is to inspire interest in astronomy and fundamental science among students and to educate the broader community. *Astronomy's New Messengers: Listening to the Universe with Gravitational Waves* is an innovative outreach project whose aim is to inform the general public of the physics of gravitational waves and the science of the LIGO detectors. Funded by the National Science Foundation [2] through grant NSF-0852870 from the Informal Science Education program, the EPSCOR program, and the Office of Multidisciplinary Activities, and managed by the LIGO Scientific Collaboration, *Astronomy's New Messengers* consists of an approximately 200 sq. ft. touring exhibit which will be displayed at junior colleges, universities, museums, and other public institutions through the nation.

## 2. Learning objectives and target audience

The learning objectives of *Astronomy's New Messengers* are to increase interest in, and understanding of, gravitational-wave astronomy and LIGO science. The opportunity to discover the beauty of the cosmos should not be limited by age, culture or abode. The LIGO exhibit

© 2010 IOP Publishing Ltd

strongly reflects this viewpoint and reaches out to citizens from diverse socio-economic groups, different areas of the nation, and underserved groups.

The target audience of *Astronomy's New Messengers* is an adolescent and young adult population of students, college age visitors, and informal learners, i.e. people who typically may have some general awareness of astronomy and its long and rich cultural heritage, but will have essentially no understanding of the fundamental nature of gravitational-wave astronomy and its relevance to basic physics. To attract this difficult audience, the exhibit includes possibilities for age-appropriate self exploration and a messaging that allows visitors to first discover their own excitement in the material, and then delve deeper into its complexity. The exhibit conveys to its visitors three main concepts:

- The LIGO scientific endeavor is motivated by the same desire for exploration, the curiosity for the unknown and the awe of nature which motivated Galileo four hundred years ago. In this respect, science and art are two facets of the same human quest for beauty and truth.
- The distant Universe can be studied by non-electromagnetic means, i.e. through the detection of gravitational waves. A tumultuous Universe is awash with gravitational waves, but at the moment these phenomena are entirely unexplored. Many prospective gravitational wave sources, for example black holes, have no corresponding electromagnetic signature. Thus mapping the gravitational-wave sky will provide an understanding of the Universe in a way that electromagnetic observations cannot. As a new field of astrophysics, it is quite likely that gravitational wave observations will uncover new classes of sources not anticipated in our current thinking.
- Giant, new non-conventional "telescopes" are needed to detect the gravitational-wave spectrum. The cutting-edge technology of these telescopes, called interferometers, is pushing back the frontiers of many scientific fields. The wonder of the LIGO interferometers lie in the wide range of distance scales the instruments are exploring: to probe unimaginably violent events at large cosmological distances, infinitesimally small movements must be detected. Achieving this degree of sensitivity requires a remarkable combination of technological innovations in vacuum technology, precision lasers, and advanced optical and mechanical systems.

To maximize the impact of the exhibit, and achieve these goals, the design of *Astronomy's New Messengers* reproduces the physics and technology of the actual LIGO instrument in an eye-catching and entertaining way.

## 3. Design

The exhibit's introductory area presents a general overview of the LIGO detectors and their science, while inviting visitors to step inside, explore and find out more. Text panels and a large LCD screen with a looping high-quality video produced by Milde Marketing [3] for the International Year of Astronomy [4] cornerstone project "100 Hours of Astronomy" deliver key informational points (Figure 1, left.)

The main area of the exhibit explains how gravitational waves are generated by cataclysmic events in the distant universe, how LIGO "listens" for these events, and demonstrates how scientists decode their signals. Three interactive components engage visitors in discovering how LIGO operates and understanding some of the foundations of gravitational wave astronomy: a working and interactive laser interferometer, a grid-patterned rubber sheet to illustrate the curvature of space-time, and the black hole hunter game [5] kiosk. A second LCD screen with looping video compiled of clips from the NSF movie *Einstein's Messengers* [6] and several panels with photographs and diagrams further encourage visitors' interest in LIGO and the deep universe.

Journal of Physics: Conference Series **228** (2010) 012001 doi:10.1088/1742-6596/228/1/012001

**Figure 1.** Visitors at the 2009 World Science Festival [9] browsing the exhibit graphic panels (left). K-12 after-school students at the University Museum, Oxford, Mississippi, simulating curved space-time with the rubber sheet interactive (right. Photo credit: Elizabeth Herren.)

The interferometer model, constructed by the Ann Arbor Hands-On Museum [7] in collaboration with the Michigan LIGO group, intuitively shows visitors how an interferometer operates. A fringe pattern is projected on a screen, illustrating the concept of light interference and the effect of environmental noise. By tapping the interferometer case, visitors can disturb the fringe pattern. A photodiode at the output port of the interferometer measures the fringe variation and speakers produce a sound for an ultimate multi-sensory experience. Each key element of the interferometer is called-out and explained on a LCD screen. The concepts of space-time and of gravity as space-time curvature are illustrated with the rubber sheet interactive. Visitors can set a heavy steel ball on the sheet, which stretches the grid-patterned sheet around it. Rolling a second, smaller ball into the depression formed around the larger one, visitors can visually understand how space-time warps around massive objects and affects the motion of bodies in space. (Figure 1, right.) The purpose of the black hole hunter game, developed by Cardiff University, is to give visitors an opportunity to do (in a figurative sense) what LIGO scientists do, i.e. look for gravitational signals in noisy streams of data. Simulated gravitational-waves are translated into sound clips. Through a GUI interface on a computer screen, visitors try to detect recorded sounds of black hole events buried in different static noise clips. (Figure 2.)

Crucial to the design, integration and implementation of these elements was the collaboration between the LIGO Education and Public Outreach group and professional exhibit designers. *Astronomy's New Messengers* was designed by Lee H. Skolnick Architecture + Design Partnership [8], a multi-disciplinary firm of architects, designers, and educators providing award-winning architecture and exhibit design services.

### A larger exhibit blending art and science

*Astronomy's New Messengers* premiered at the Street Fair of the World Science Festival [9] in New York City in June 2009. After a month-long stay at the Adler Planetarium in Chicago, the traveling LIGO exhibit will be on display in a network of museums, science museums, public libraries and academic institutions across a cluster of southern states from August 2009 to February 2010.

Phase II of the project envisages the design and construction of a large-scale, interactive exhibit, which will draw upon elements of the touring exhibit (and upon lessons learned from the former's evaluation). The large-scale exhibit will blend LIGO science with high-concept artwork through incorporation of an interpretive 3-D lighting display. This large-scale exhibit

**Figure 2.** Playing with the black hole hunter game. The interferometer kiosk is in the background. Photo credit: Elizabeth Herren.

is scheduled to debut at a prominent public location at the 2010 World Science Festival in New York City, alongside a signature event on the theme of gravitational-wave astronomy and featuring high-profile keynote speakers and artists. After the 2010 Festival, the exhibit will be transferred to a permanent home in a public institution or science museum.

The large-scale exhibit will incorporate elements from the touring exhibit, but on a grander scale – e.g. using larger flatscreen displays and poster boards and including a larger table-top interactive interferometer. A key additional element of the large-scale exhibit will be the blending of the table-top interferometer with a major, high-concept work of art in the form of an interactive three-dimensional lighting display. This will allow visitors to be exposed to the science at the basis of LIGO research in an exciting and visually stunning manner. To highlight and illustrate the grand scale of gravitational-wave astrophysical sources, the output of the interferometer will be linked directly to a large-scale three-dimensional artwork suspended from the ceiling, symbolizing the Universe above the Earth. This artwork will be easily visible from afar and its distinctive appearance will be instrumental in drawing visitors towards the exhibit. Changes in the interferometer's fringes will be translated and displayed on a spatial field of strands of diodes, creating a dazzling show of light and sound in real time, with each and every visitor who disturbs the interferometer contributing to the patterns of light, color and sound of the "sky" above them, representing the everlasting link between Earth and the Universe. Leni Schwendinger Light Projects [10], a NYC-based art studio which specializes in creating lighting environments for architectural and public spaces all over the world, will design and realize the artistic component of the exhibit.

**Acknowledgments**
The authors gratefully acknowledge the support of the United States National Science Foundation for the construction and operation of the LIGO Laboratory and the Science and Technology Facilities Council of the United Kingdom, the Max-Planck-Society, and the State of Niedersachsen/Germany for support of the construction and operation of the GEO600 detector. The authors also gratefully acknowledge the support of the research by these agencies and by the Australian Research Council, the Council of Scientific and Industrial Research of India, the

Istituto Nazionale di Fisica Nucleare of Italy, the Spanish Ministerio de Educación y Ciencia, the Conselleria d'Economia, Hisenda i Innovació of the Govern de les Illes Balears, the Royal Society, the Scottish Funding Council, the Scottish Universities Physics Alliance, The National Aeronautics and Space Administration, the Carnegie Trust, the Leverhulme Trust, the David and Lucile Packard Foundation, the Research Corporation, and the Alfred P. Sloan Foundation.

## References

[1] http://www.ligo.org
[2] http://nsf.gov
[3] http://www.milde-marketing.de
[4] http://www.astronomy2009.org
[5] http://www.blackholehunter.org
[6] http://einsteinsmessengers.org
[7] http://www.aahom.org/
[8] http://www.skolnick.com
[9] http://www.worldsciencefestival.com
[10] http://www.lightprojectsltd.com

Journal of Physics: Conference Series **228** (2010) 012002 doi:10.1088/1742-6596/228/1/012002

# LIGO-Virgo searches for gravitational waves from coalescing binaries: a status update

**Anand S. Sengupta for the LIGO Scientific Collaboration and the Virgo Collaboration**

LIGO Laboratory, California Institute of Technology, Pasadena CA 91125, USA.

E-mail: `sengupta@ligo.caltech.edu`

**Abstract.** Coalescing compact binaries of neutron stars and/or black holes are considered as one of the most promising sources for Earth based gravitational wave detectors. The LIGO-Virgo joint collaboration's Compact Binary Coalescence (CBC) group is searching for gravitational waves emitted by these astrophysical systems by matched filtering the data against theoretically modeled template waveforms. A variety of waveform template families are employed depending on the mass range probed by the search and the stage of the inspiral phase targeted: restricted post-Newtonian for systems having total mass less than $35M_{\odot}$, numerical relativity inspired complete inspiral-merger-ringdown waveforms for more massive systems up to $100M_{\odot}$ and ringdown templates for modeling perturbed black holes up to $500M_{\odot}$. We give a status update on CBC group's current efforts and upcoming plans in detecting signatures of astrophysical gravitational waves.

## 1. Introduction

Gravitational waves emitted by compact binaries of neutron stars and/or black holes in the last few minutes before merger lie within the sensitive bandwidth of present ground-based interferometric detectors like LIGO [1] and Virgo [2] and are considered likely candidates for first detection in these instruments. This is partly motivated by the fact that the phase and amplitude evolution of waveforms from such sources are well modeled analytically in the inspiral phase and ringdown phase when the excited final black hole emits gravitational waves and loses energy [3, 4]. This allows the construction of templates for use in the application of matched-filtering techniques [5] to filter the noisy data. The intermediate merger phase is not well modeled but can be obtained by numerical relativity [6]. These have, in turn, inspired a whole generation of template waveforms which bridge the inspiral, merger and ringdown parts by various approximations [7, 8].

The three LIGO detectors at Livingston and Hanford operated at their design sensitivity in a continuous data acquisition mode between November 2005 and September 2007 in their fifth science run (S5), collecting a full year of triple coincident data. The first run of the Virgo detectors (VSR1) coincided with the last 6 months of S5 at a sensitivity comparable to the LIGO instruments above 400 Hz. The aim of this article is to provide a snap-shot of the current activities of the CBC group and present our plans for the near future.

© 2010 IOP Publishing Ltd

## 2. Data processing

At the core of the CBC group's data analysis pipeline [9] is the matched filtering engine that correlates data from detectors against theoretical template waveforms expected from the sources. A discrete lattice of templates spanning the intrinsic parameter space of the binary is constructed such that the maximum loss in signal-to-noise ratio (SNR) is no more than a prescribed limit. Filtering the data $x$ against a template $h(t;\mu_i)$ yields the SNR $\rho$ given by

$$\rho = \frac{\langle x, h\rangle}{\sqrt{\langle h, h\rangle}}, \tag{1}$$

where $\langle x, h\rangle$ denotes the noise-weighted scalar product of the data and the template. For each template, triggers that have SNR greater than a prescribed threshold are retained. The templates are typically parametrized by the component masses of the coalescing binary (or, for ringdowns, the mass and spin of the final blackhole). These triggers are tagged by the template parameters and coalescence time, which can be later used for clustering [10] and determining coincidence [11]. Although the trigger generation by matched filtering is at the heart of the CBC pipelines, other important steps including: determining detection efficiency and accidental background estimation; vetoing glitchy times; determining coincidence across different instruments and following up interesting coincident triggers; all form important parts of the full data analysis pipeline and are used both for GW-triggered and externally triggered (eg, GRB events) searches.

## 3. Status of current activities of the CBC group

The template families are chosen to optimize the sensitivity of the searches to probe the part of the signal parameter space accessible by the ground-based detectors. The current activities of the group include: the analysis of coincident LIGO S5 and Virgo VSR1 data for low mass coalescing binaries (systems with total mass between 2 and $35\mathrm{M}_\odot$); analysis of the S5 data for high-mass binaries (systems with total mass $25-100\mathrm{M}_\odot$) using inspiral-merger-ringdown (IMR) waveforms; those for ringdown of perturbed black holes using exponentially damped sinusoidal waveforms; and analysis of S5 data around times triggered by short, hard gamma-ray bursts.

(a) The 'S5 low-mass search' covers the $2-35\mathrm{M}_\odot$ region of the total-mass parameter space employing frequency domain restricted post-Newtonian (PN) template waveforms accurate to 2PN order in phase evolution up to the innermost stable circular orbit (ISCO). Data is cross-correlated against such template banks consisting of 5-7 thousand templates to provide adequate coverage via semi-automated data analysis pipeline. Historically, it was divided into three separate searches to analyze different subsets of the S5 data: (a) The S5 first calendar year search [12] - sensitive to CBC as far as 150 Mpc for systems with total mass $\sim 25\mathrm{M}_\odot$ at an SNR of 8. No gravitational waves were detected by this search. Improved upper limits on the rate of coalescence for neutron star and black hole binary systems were established. (b) The S5 12-18 month analysis covering 186 days beyond the first calendar year [13]. No gravitational wave candidate events were reported in this search which yielded a 90%-confidence rate upper limit of CBC of neutron stars at $1.4\times10^{-2}$ per year per $L_{10}$ where the latter is $10^{10}$ times the blue solar luminosity. (c) The last six months of S5 data was analyzed in coincidence with VSR1 allowing the first joint LIGO-Virgo data analysis. This work is nearing completion and publication. The upper limits are cumulative (i.e, they include the results from all previous analyses) and are combined in a Bayesian manner.

(b) The 'S5 high-mass search' uses time-domain numerical-relativity tuned effective one-body (EOBNR) [8] template waveforms to cover the total mass space between $25-100\mathrm{M}_\odot$ with some overlap with the low mass search. The frequency domain templates used in the low mass search are not suitable in this case as a significant portion of the SNR comes from

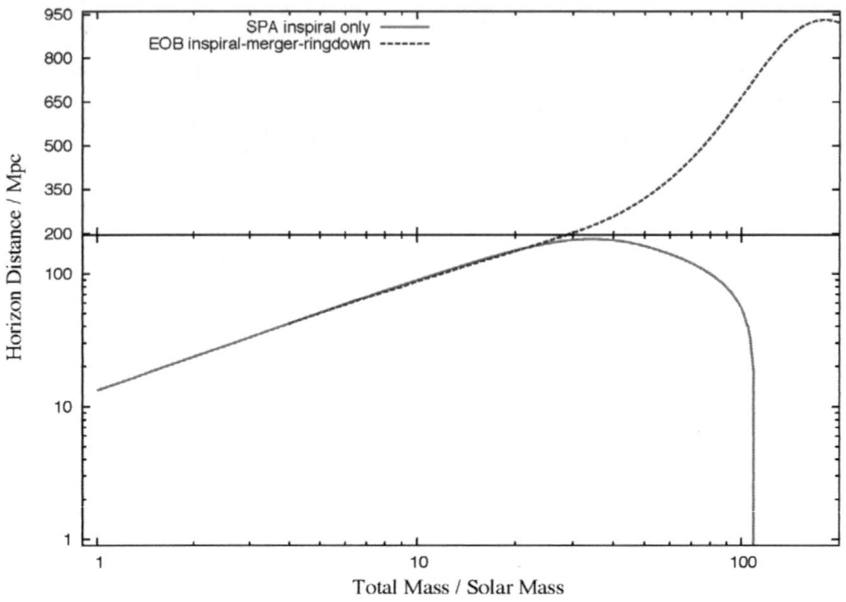


**Figure 1.** Horizon distance (distance at which an optimally located and oriented binary would be detected in the LIGO detectors with SNR of 8) for restricted 2PN (red trace) and EOBNR inspiral-merger-ringdown waveforms tuned to numerical relativity; for equal-mass binaries where the symmetric mass ratio $\eta = 1/4$. To convert the horizon distance for other mass ratio, we need to multiply by $\sqrt{4\eta}$. Note that the high-mass search using time-domain EOBNR waveforms greatly extends the reach of the search for systems with total mass above $\sim 25M_{\odot}$. Initial LIGO design noise power spectral density is assumed.

the merger and ringdown phase of the coalescence for these massive systems and are not captured by the post-Newtonian template waveforms. The high-mass search compliments the low-mass search by extending the boundaries of the parameter space explored and also the volume reach of the search. At present, spin effects are not taken into account in the template bank used for this search. We hope to take spin effects into account by running spinning waveforms through to evaluate efficiencies. Filtering, coincidence and signal-based veto techniques used in this search are similar to the low-mass search as is the technique of 'effective SNR' ranking of the coincident triggers. The S5 data is being analysed on a month by month basis. New structured query language (SQL)-based post-processing tools developed are being tried out on this search by the CBC group for the first time.

(c) Recently, the CBC group completed the first ringdown search [3] using the fourth LIGO science data (S4) spanning black holes with masses between $10 - 500M_{\odot}$ with spins ranging from $0 - 0.994$. The waveform templates are modeled as damped sinusoids spanning this two-dimensional parameter space. It is currently gearing up to analyze the S5 data with an eye to determining coincidence with the triggers from the S5 high-mass search. Although the fundamental data analysis pipeline as outlined in S4 paper has been retained for S5, a new coincidence algorithm has now been deployed which lowers the false alarm rate and effectively makes the search more sensitive. Tuning of the search pipeline is in progress using complete IMR and ringdown-only waveforms.

(d) Coalescence of two neutron stars or a neutron star and black hole are considered as likely progenitors of short gamma-ray bursts (GRBs). The CBC group has analyzed the science data around 22 such GRBs that occurred over the duration of S5/VSR1 to look for gravitational-wave counterparts, employing restricted 2PN templates with total mass between $2 - 35\mathrm{M}_{\odot}$. Although the actual number of short hard GRBs was larger, only 22 had enough science data available across multiple instruments to guarantee at least a two-interferometer coincidence. In this search both time and directional coincidence for triggers are required, enabling a lowering of SNR thresholds to enhance the sensitivity of the search. An earlier analysis of GRB 070201 [14] (whose sky position error box overlaps the M31 Andromeda galaxy) was repeated in view of the subsequent improvement to the data analysis pipeline. The earlier analysis carried out by the CBC group lead to a null result excluding a CBC progenitor in M31 ( at 99% confidence).

Apart from these searches, the CBC group is also developing techniques to follow-up interesting candidate events with a view to developing detection confidence. These include using Markov-chain Monte-Carlo methods [15] to estimate the parameters of detection candidates. Other activities include: developing better IMR waveforms that are faithful to the numerical waveforms over a wider range of parameters; including spin as a parameter in our detection templates - especially for high-mass end of the parameter space; detector characterization to understand the source of coincident glitches; and preparing the ground for upcoming, more sensitive science runs.

## 4. New data analysis techniques on the horizon

Since the beginning of S5, the basic data analysis pipeline has seen several improvements to its core functionality and many new features have been added. In this section, we highlight some of the most important ones. A key area of improvement has been the addition of numerical relativity inspired waveforms to the LSC and Virgo collaboration's arsenal of existing templates: time-domain EOB waveforms calculated up to the light ring, calibrated to numerical simulations and then matched to the quasi-normal modes of the ringdown at the end, have extended the effective bandwidth of the high-mass search thereby increasing the distance reach and sensitivity of the high-mass search; see Figure 1. Phenomenological IMR waveforms [7, 16] that include spins (albeit restricted to non-precessing aligned/anti-aligned configurations) now allow us to not only probe significantly larger parts of the CBC parameter space but also to inject such signals into the data stream for more reliable estimates of the efficiency of our pipeline in detecting gravitational waves from CBCs. Several new analysis technologies are on the horizon: hardware-accelerated signal processing where the computer's graphical processing units (GPU) are used to parallelize the fast-Fourier transform calculations have been tested and found to make our search pipeline an order of magnitude faster. Multi-variate statistical classification tools are being developed to bring in machine-learning algorithms that are able to classify our putative events and better distinguish them from background [17]. New ranking statistics based on likelihood ratios have been found to perform better than traditional 'effective SNR' statistic and have been deployed in some of the searches. The coherent inspiral pipeline has matured to the point where it can be run in a follow-up mode for interesting candidates that pass the stringent mass and time coincidence tests. Efforts are also on towards a Bayesian analysis of data from arbitrary network of detectors for the extraction of parameters of a likely signal in the data [18]. The full posterior probability density function for parameter estimation (using a wide range of template waveforms) can be computed, allowing estimation of likely parameters and the range of their uncertainty. These new techniques are at various stages of development and stem from a common desire of the group to work towards building confidence in our first detection.

## 5. Future plans: S6, VSR2 and beyond

Following S5 and VSR1 data taking, the detectors are undergoing scheduled upgrades which will improve the sensitivity in these instruments and the volume of space that these 'enhanced' detectors can probe. Data taking with the enhanced detectors (LIGO's S6 and Virgo's VSR2) began in June 2009 and will continue for approximately 1.5 years. At that point, the detectors will undergo more major upgrades. Advanced LIGO is scheduled to be fully commissioned in 2015. In these instruments, the improvement in sensitivity is expected to be a factor of 10 over initial LIGO and a 1000 fold increase in the event rates, making the detection of gravitational waves routine. S6/VSR2 data is being used as a test bed for several data analysis techniques developed by the CBC group in the recent past. A major paradigm shift has been to focus on rapid turn around of the results by analyzing a week's data at a time. Low latency pipelines being developed by the group have been deployed and are being extensively tested. Many of the post-processing tasks and follow-up of interesting triggers from the main pipeline are being automated: among them, rapid sky localization and parameter estimation; and improved background estimation are being explored. The latter is motivated by a chance for electromagnetic followup by quick identification and localization of a transient corresponding to an interesting gravitational-wave candidates. This is also establishing the protocol for external collaborations with other ground and space based telescopes and the exciting opportunity to 'see' a gravitational wave trigger in different wavelengths.

## Acknowledgments

I would like to thank Szabi Marka, Zzusa Marka and other organisers of the 8th E. Amaldi conference in New York City. I also thank Ajith Parameswaran, Craig Robinson and Alan Weinstein for reading and editing the manuscript, comments and feedback.

## References

[1] B. P. Abbott *et al.*, Reports on Progress in Physics **72**, 076901 (25pp) (2009).
[2] F. Acernese *et al.*, Class. Quant. Grav. **23**, S635 (2006).
[3] B. P. Abbott *et al.*, Physical Review D **80**, 062001 (2009).
[4] A. Buonanno, B. R. Iyer, E. Ochsner, Y. Pan, and B. S. Sathyaprakash, Physical Review D (Particles, Fields, Gravitation, and Cosmology) **80**, 084043 (2009).
[5] B. S. Sathyaprakash and S. V. Dhurandhar, Phys. Rev D **44**, 3819 (1991).
[6] F. Pretorius, Phys. Rev. Lett. **95**, 121101 (2005).
[7] P. Ajith, M. Hannam, S. Husa, Y. Chen, B. Bruegmann, N. Dorband, D. Mueller, F. Ohme, D. Pollney, C. Reisswig, L. Santamaria, and J. Seiler, ArXiv e-prints (2009).
[8] A. Buonanno, Y. Pan, J. G. Baker, J. Centrella, B. J. Kelly, S. T. McWilliams, and J. R. van Meter, Physical Review D **76**, 104049 (2007).
[9] B. Allen, W. G. Anderson, P. R. Brady, D. A. Brown, and J. D. E. Creighton, FINDCHIRP: an algorithm for detection of gravitational waves from inspiraling compact binaries, 2005.
[10] A. S. Sengupta, J. A. Gupchup, and C. A. K. Robinson, (2008), in Preparation.
[11] C. A. K. Robinson, B. S. Sathyaprakash, and A. S. Sengupta, Physical Review D **78**, 062002 (2008).
[12] B. P. Abbott *et al.*, Phys. Rev. **D79**, 122001 (2009).
[13] B. P. Abbott *et al.*, Phys. Rev. **D80**, 047101 (2009).
[14] LIGO Scientific Collaboration and K. Hurley, The Astrophysical Journal **681**, 1419 (2008).
[15] M. van der Sluys, I. Mandel, V. Raymond, V. Kalogera, C. Roever, and N. Christensen, CLASS.QUANT.GRAV **204010**, (2009).
[16] P. Ajith, S. Babak, Y. Chen, M. Hewitson, B. Krishnan, A. M. Sintes, J. T. Whelan, B. Brügmann, P. Diener, N. Dorband, J. Gonzalez, M. Hannam, S. Husa, D. Pollney, L. Rezzolla, L. Santamaría, U. Sperhake, and J. Thornburg, Physical Review D **77**, 104017 (2008).
[17] T. Miller, A. S. Sengupta, K. Hodge, and A. Weinstein, LIGO Technical Document (LIGO-T0900454) from 2009 SURF program (2009).
[18] J. Veitch and A. Vecchio, Classical and Quantum Gravity **25**, 184010 (2008).

8th Edoardo Amaldi Conference on Gravitational Waves IOP Publishing
Journal of Physics: Conference Series **228** (2010) 012003 doi:10.1088/1742-6596/228/1/012003

# Recovering a spinning inspiralling compact binary waveform immersed in LIGO-like noise with spinning templates

**László Veréb**[1,2], **Zoltán Keresztes**[1,2*], **Péter Raffai**[3†], **Szabolcs Mészáros**[4], **László Á. Gergely**[1,2,5‡]

[1] Department of Theoretical Physics, University of Szeged, Tisza Lajos krt 84-86, Szeged 6720, Hungary
[2] Department of Experimental Physics, University of Szeged, Dóm tér 9, Szeged 6720, Hungary
[3] Institute of Physics, Loránd Eötvös University, Pázmany Péter s 1/A, Budapest 1117, Hungary
[4] Department of Optics and Quantum Electronics, University of Szeged, Dóm tér 9, Szeged 6720, Hungary
[5] Institute for Advanced Study, Collegium Budapest, Szentháromság u 2, Budapest 1014, Hungary

E-mail: *`zkeresztes@titan.physx.u-szeged.hu` †`praffai@bolyai.elte.hu` ‡`gergely@physx.u-szeged.hu`

**Abstract.** We investigate the recovery chances of highly spinning waveforms immersed in LIGO S5-like noise by performing a matched filtering with $10^6$ randomly chosen spinning waveforms generated with the LAL package. While the masses of the compact binary are reasonably well recovered (slightly overestimated), the same does not hold true for the spins. We show the best fit matches both in the time-domain and the frequency-domain. These encompass some of the spinning characteristics of the signal, but far less than what would be required to identify the astrophysical parameters of the system. An improvement of the matching method is necessary, though may be difficult due to the noisy signal.

We report on an attempt to recover a gravitational wave signal $h_{injection}$ immersed in LIGO S5-like noise by a matched filtering with $10^6$ randomly chosen templates generated with spinning waveforms [1]. We applied the following process to generate a LIGO S5-like noise sample of N elements: (i) cut out a stretch of 2N elements from an arbitrary sample of real LIGO S5 noise, (ii) cut the stretch into half, resulting in two stretches of N elements, (iii) Fourier-transfom both stretches with an FFT method, resulting in two samples of N complex elements, (iv) combine the magnitude values of the first spectrum with the phase values of the second spectrum, resulting in a complex spectrum of N elements, (v) apply an inverse FFT on the combined spectrum, and use the real part of the resulting array as our LIGO S5-like noise sample. This method has the advantage of preserving the basic spectral and statistical properties of the original LIGO data, while discarding any real gravitational wave signatures that might be present within the original data stretch.

The matched filtering follows the method described in Ref. [2]. Namely, we calculate the

© 2010 IOP Publishing Ltd

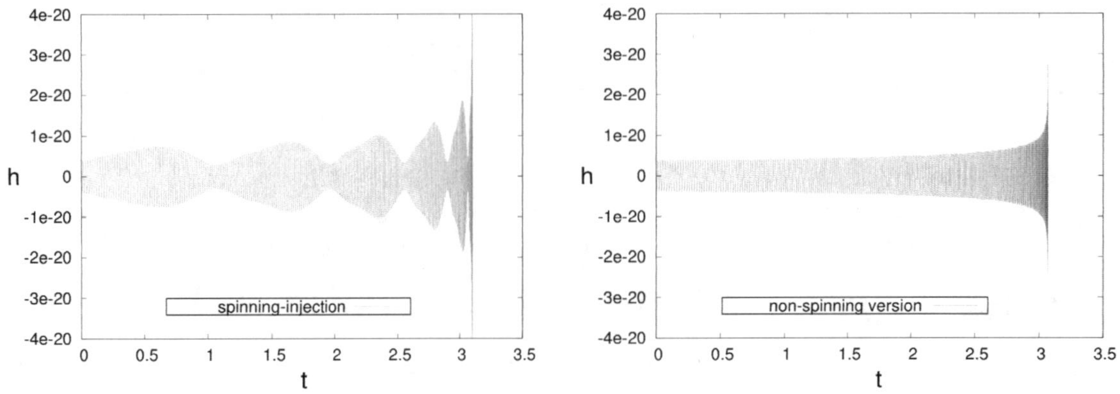


**Figure 1.** The injected signal (red) and its spinn-less counterpart (blue).

*overlap* between the noisy injection $h_{n,i}$ and the spinning templates $h_{template}$ as

$$O\left[h_{n,i}, h_{template}\right] = \frac{\langle h_{n,i}|h_{template}\rangle}{\sqrt{\langle h_{n,i}|h_{n,i}\rangle \langle h_{template}|h_{template}\rangle}} \,, \tag{1}$$

where

$$\langle h_1|h_2\rangle = 4Re\int_{f_{\min}}^{f_{\max}} \frac{\tilde{h}_1\left(f\right)\tilde{h}_2^*\left(f\right)}{S_n\left(f\right)}df \,, \tag{2}$$

with $\tilde{h}_i\left(f\right)$ the Fourier transform of $h_i\left(t\right)$ and $S_n\left(f\right)$ the power spectral density of the noise. We choose $f_{\min} = 50\text{Hz}$ and $f_{\max} = 600\text{Hz}$, which lies in the best sensitivity of the LIGO detectors, also the post-Newtonian prediction for the waveform can be considered reasonably accurate. The most lengthy templates in this domain were found to last for approximately 3 sec.

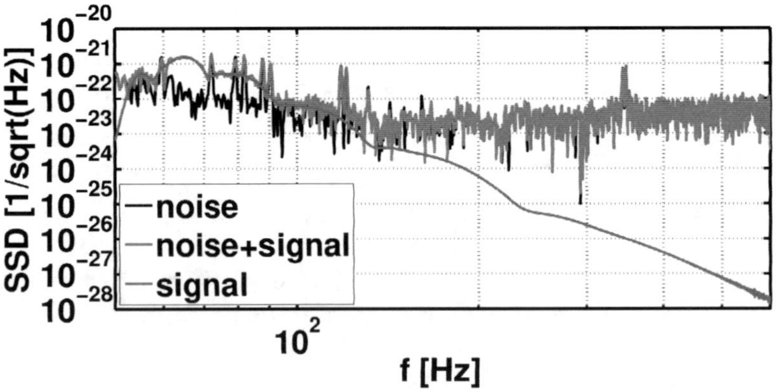


**Figure 2.** The strain spectral density of the signal (red), noise (grey) and injected signal + noise (magenta).

On theoretical grounds, an infinitely long data series would be required for exact determination of the power spectrum. In order to achieve a stability of the power spectrum of the order of 1%, we would like to have at least 100 periods of all frequencies to be included in the power spectrum, therefore a minimal length of the templates of $100 \times (1/50\text{Hz}) = 2$ sec was imposed. As we know where the injection commences in the noise time sequence, we simplify

**Table 1.** The parameters of the injection and 5 best fit templates.

| name | $m_1(M_\odot)$ | $m_2(M_\odot)$ | $\chi_1$ | $\chi_2$ | $\cos\beta_1$ | $\varphi_1$ | $\cos\beta_2$ | $\varphi_2$ |
|---|---|---|---|---|---|---|---|---|
| injection | 3.553 | 3.358 | 0.983 | 0.902 | 0.984 | 1.109 | 0.978 | 0.957 |
| 560884.wave | 4.805 | 3.336 | 0.230 | 0.293 | −0.599 | 1.327 | −0.621 | 0.006 |
| 467403.wave | 4.092 | 3.405 | 0.868 | 0.891 | −0.520 | 1.275 | 0.363 | 3.925 |
| 313940.wave | 3.569 | 3.540 | 0.753 | 0.413 | 0.306 | 1.458 | −0.092 | 2.679 |
| 93938.wave | 5.012 | 3.152 | 0.280 | 0.424 | 0.947 | 2.593 | 0.888 | 2.340 |
| 80521.wave | 4.278 | 4.360 | 0.736 | 0.475 | −0.620 | 5.959 | 0.945 | 3.290 |

our analysis by comparing with templates beginning at the same instant $t_0$. Further, we also assume identical antenna functions of $\sqrt{2}/2$ both for the injection and templates. Moreover, all parameters of the injection and templates other then the masses $m_i$ and spins (given by their dimensionless magnitude $\chi_i$, polar and azimuthal angles $\beta_i$, $\varphi_i$, taken in the inertial system with the line of sight on the $z$-axis) were kept fixed. These parameters were the distance of the binary $d = 1$Mpc, and the polar angle of the orbital angular momentum $\Theta = 1.4299$. The initial phase $\Phi_S = 0$.

We picked up the injection signal as one of the longests available (see Fig 1). The presence of a high spin induces a strong amplitude modulation, but also a less visible frequency modulation. For comparison, we have also represented on Fig 1 the corresponding non-spinning signal, with the same masses, but the spins switched off.

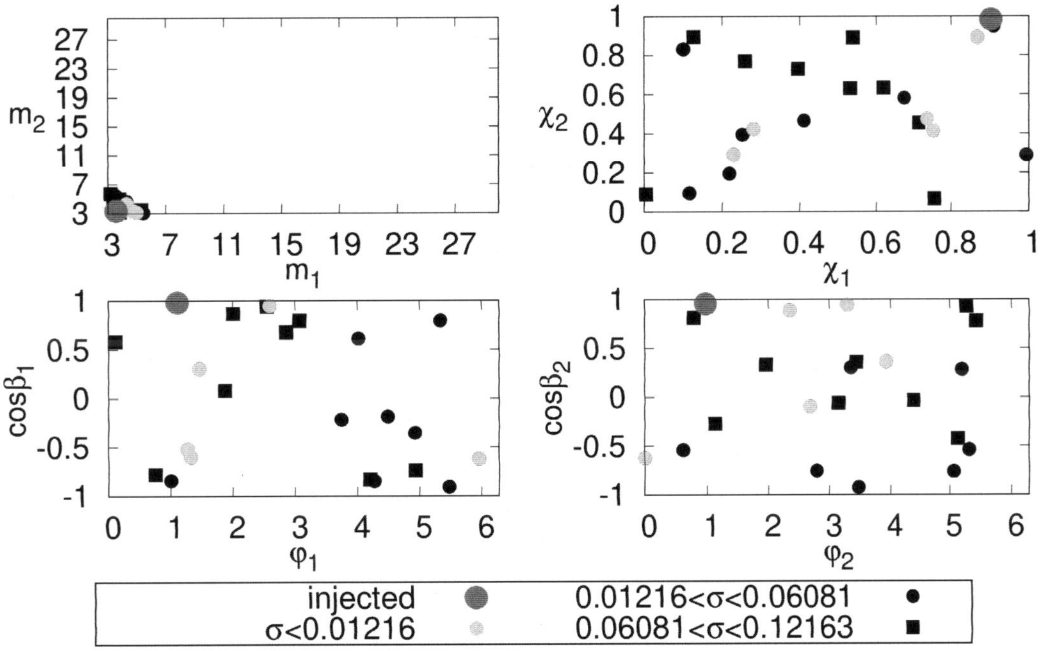


**Figure 3.** The injection and the best fit matches from $10^6$ templates. The mass is roughly recovered, though sligtly overestimated, however the information contained in the spins is not.

The spinning signal is injected in the middle of a finite stretch of LIGO-like data with double the length of the signal. The noise has the basic statistical characteristics (e.g. amplitude distribution, spectral compatibility) of the LIGO 4 km detectors during the 5th science run. The segment to be analyzed is filtered with a zero-phase $50-600$ Hz bandpass filter, and a zero-phase notch filter mitigating the effect of narrow insensitivity peaks of detector spectrum. After removing one-fourth of the data segment from both ends due to filter transients, this leaves half the length of the initial noise sequence including the signal for further processing. The strain spectral density, defined as the square root of the periodogram, computed under MatLab [3], is represented on Fig 2 for the signal, noise and noisy signal.

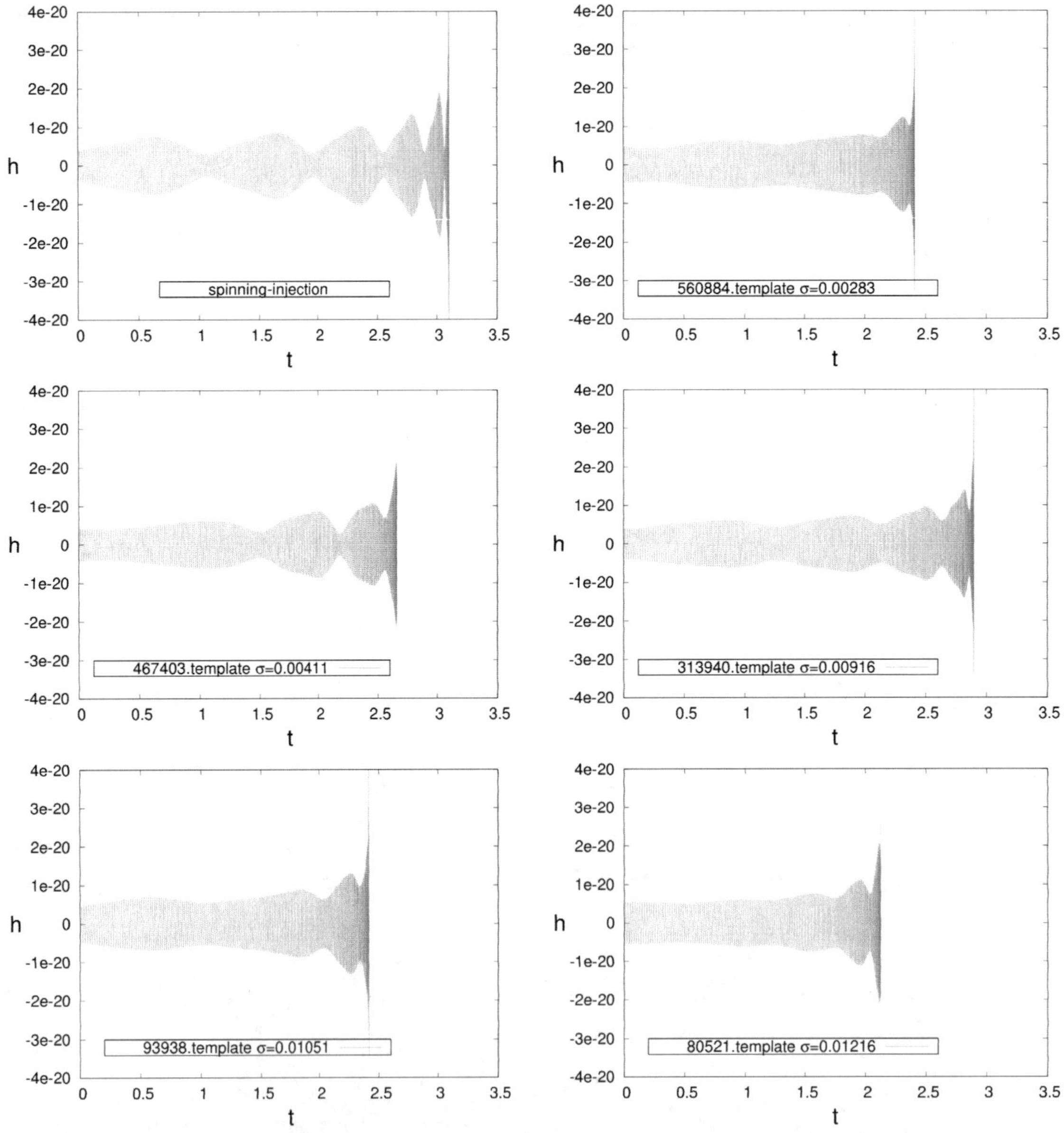


**Figure 4.** The injected signal and the five best matches. Their serial numbers from among the $10^6$ randomly generated templates is also indicated.

The overlap between the noisy injection and the injected signal itself was found to be $O\left[h_{n,i}, h_{injection}\right] = 0.83426$. Therefore we define an auxiliary quantity

$$\sigma = \left|1 - \frac{O\left[h_{n,i}, h_{template}\right]}{O\left[h_{n,i}, h_{injection}\right]}\right| \tag{3}$$

for characterizing the closeness of the overlap of a template with the noisy injection to $O\left[h_{n,i}, h_{injection}\right]$. The parameters of a number of $> 20$ templates with the lowest values of $\sigma$ are represented on Fig 3. While it is comforting to realize that (although slightly overestimated) the masses are reasonably well recovered, the spins are not. There is a considerable scatter in both their magnitudes and orientations.

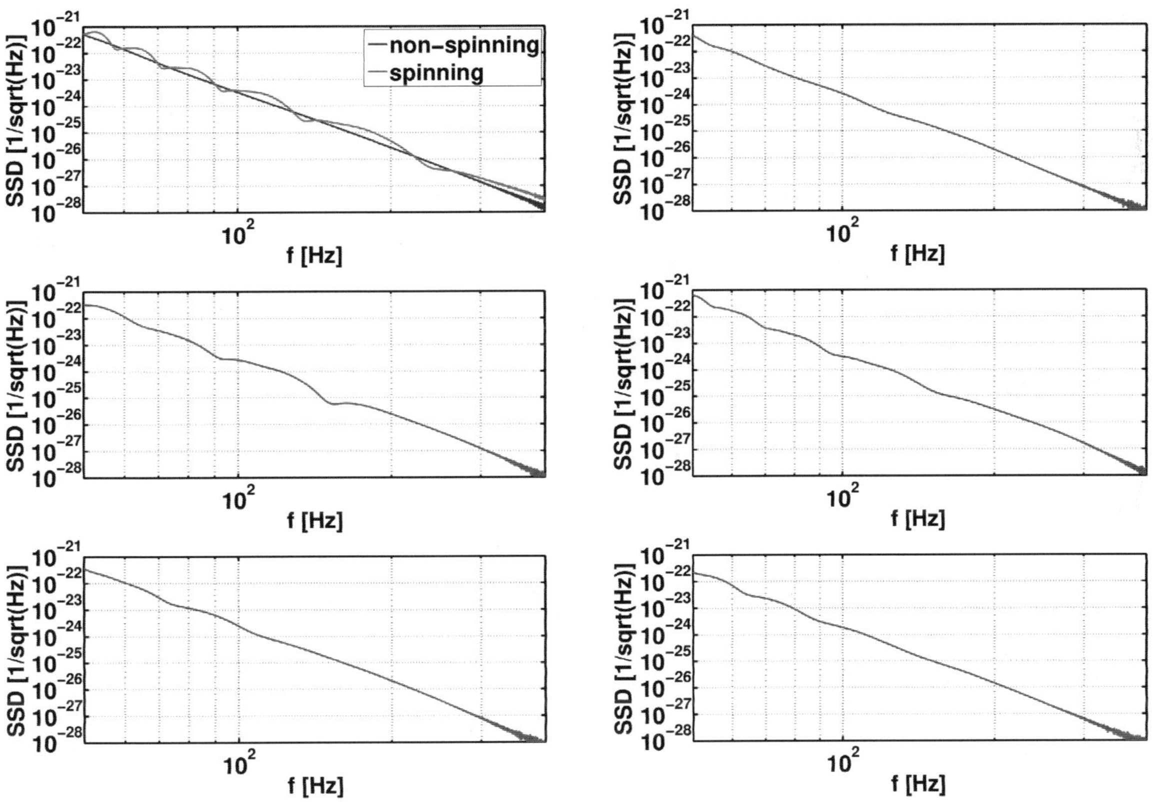


**Figure 5.** The strain spectral density of the injected signal (red) and its spin-less counterpart (blue) and the five best matches (green).

We picked up the 5 best fits, showed their parameters in Table 1 and represented the respective waveforms on Fig 4; also their strain spectral density on Fig 5. For comparison, the injected signal is also shown.

We conclude that even with all the simplifying assumptions adopted, (a) either the number of $10^6$ templates employed was still insufficient or (b) the 6 additional parameters introduced by the spins cause severe degeneracies, which cannot be resolved by this method. In this context we note that the use of a Markov-chain Monte-Carlo technique [4], [5] may turn useful in the recovery of spinning signals. We also plan to investigate how the use of Advanced LIGO-like data would change these conclusions.

*Acknowledgements*: We acknowledge interactions with Gyula Szabó and István Szapudi in the early stages of this work and discussions with Vivien Raymond on the suitable frequency range. This work was supported by the Polányi and Sun Programs of the Hungarian National Office for Research and Technology (NKTH). ZK and LÁG were further supported by the Hungarian Scientific Research Fund (OTKA) grant no. 69036. LÁG is grateful to the organizers of the Amaldi8 meeting for support.

## References

[1] https://www.lsc-group.phys.uwm.edu/daswg/projects/lal/nightly/docs/html/LALSTPNWaveformTest_8c-source.html
[2] Vaishnav B, Hinder I, Herrmann F, Shoemaker D 2007 *Phys. Rev.* D **76** 084020
[3] http://www.mathworks.com/access/helpdesk/help/toolbox/signal/index.html?/access/helpdesk/help/toolbox/signal/periodogram.html
[4] van der Sluys M, Raymond V, Mandel I, Roever C, Christensen N, Kalogera V, Meyer R, Vecchio A 2008 *Class.Quant.Grav.* **25** 184011
[5] Raymond V, van der Sluys MV, Mandel I, Kalogera V, Roever C, Christensen N, 2009 *Class.Quant.Grav.* **26** 114007

# Hierarchical Hough all-sky search for periodic gravitational waves in LIGO S5 data

**Llucia Sancho de la Jordana for the LIGO Scientific Collaboration and the Virgo Collaboration**

Departament de Física, Universitat de les Illes Balears, Cra. Valldemossa Km. 7.5, E-07122 Palma de Mallorca, Spain

E-mail: llucia.sancho@uib.es

**Abstract.** We describe a new pipeline used to analyze the data from the fifth science run (S5) of the LIGO detectors to search for continuous gravitational waves from isolated spinning neutron stars. The method employed is based on the Hough transform, which is a semi-coherent, computationally efficient, and robust pattern recognition technique. The Hough transform is used to find signals in the time-frequency plane of the data whose frequency evolution fits the pattern produced by the Doppler shift imposed on the signal by the Earth's motion and the pulsar's spin-down during the observation period. The main differences with respect to previous Hough all-sky searches are described. These differences include the use of a two-step hierarchical Hough search, analysis of coincidences among the candidates produced in the first and second year of S5, and veto strategies based on a $\chi^2$ test.

## 1. Introduction

Spinning neutron stars are the most promising sources of continuous gravitational wave signals for ground-based interferometers such as GEO600 [1], LIGO [2] and VIRGO [3]. Using data from the different science runs, there have been two kinds of searches for gravitational waves from pulsars: i) targeted searches [4, 5, 6, 7, 8] for periodic gravitational radiation from pulsars whose parameters (sky-position and frequency evolution) are known through radio observations, and ii) searches for pulsars yet unobserved by radio telescopes [9, 10, 11, 12, 13, 14]. For this second kind of search the optimal method based on a coherent integration over the full observation time is computationally prohibitive for a wide-parameter search. Therefore all-sky searches require the use of alternative techniques. These can be semi-coherent techniques [15, 16, 17, 18, 19, 11], that are less sensitive for the same observation time but are computationally inexpensive, or hierarchical approaches. Hierarchical approaches combine both coherent and semi-coherent techniques, optimizing the sensitivity of a search for a given computational power [20, 19, 21, 22, 13]. Some of these semi-coherent methods operate on successive Short Fourier Transforms (SFTs) of the measured strain [23] searching for cumulative excess power from hypothetical periodic gravitational wave signal. All these methods take into account the Doppler modulation of the detected frequency due to the Earth's rotational and orbital motion with respect to the Solar System Barycenter, and the time derivative of the frequency intrinsic to the source. The Hough transform [10, 11, 15, 16] is an example of such a method.

© 2010 IOP Publishing Ltd

Two flavors of the Hough transform have been developed and employed for different searches. The 'standard Hough' [10, 15], computes the cumulative excess power as a sum of binary zeroes and ones, where a SFT contributes a one if and only if the power exceeds a normalized power threshold, and the 'weighted Hough' [11, 17] in which the contribution of each SFT is weighted according to the noise and detector antenna pattern to maximize the signal-to-noise ratio. Using the 'weighted Hough' has two notable benefits: i) the gain in sensitivity, and ii) that this method can be used to analyze data from multiple detectors, since the different weights automatically take into account the different sensitivities.

This paper describes the pipeline employed to analyze the data from the fifth science run of the LIGO detectors to search for periodic gravitational waves using a hierarchical Hough transform, emphasizing the changes with respect to previous Hough searches [10, 11]. The paper is organized as follows: In section 2 we briefly summarize the basic principles of the Hough transform and its statistics. In section 3 we give an overview of the search pipeline and comment on the data employed and the parameter space of the search. In section 4 we describe two main features of our two-step hierarchical search, and we conclude in section 5.

## 2. The Hough transform

The starting point for the Hough transform is a set of $N$ SFTs, short stretches of Fourier transformed data which are digitized by setting a threshold $\rho_{th}$ on the normalized power $\rho_k$ in the different frequency bins:

$$n_k^{(i)} = \begin{cases} 1 & \text{if} \quad \rho_k^{(i)} \geq \rho_{th} \\ 0 & \text{if} \quad \rho_k^{(i)} < \rho_{th} \end{cases}, \tag{1}$$

where

$$\rho_k = \frac{2|\tilde{x}_k|^2}{T_{\text{coh}} S_n(f_k)}. \tag{2}$$

Here $\tilde{x}_k$ is the value of the Fourier transform in the $k^{th}$ frequency bin corresponding to a frequency $f_k$, $T_{\text{coh}}$ is the time baseline of the SFT (in our case $T_{\text{coh}} = 1800$ s), and $S_n(f_k)$ is the single sided power spectral density of the detector noise at the frequency $f_k$.

For a given template, the Hough number count is a weighted sum of the binary zeroes and ones, along the corresponding time-frequency pattern

$$n = \sum_{i=0}^{N-1} w_k^{(i)} n_k^{(i)}, \tag{3}$$

where the weights are defined as

$$w_k^{(i)} \propto \frac{1}{S_k^{(i)}} \left\{ \left(F_{+1/2}^{(i)}\right)^2 + \left(F_{\times 1/2}^{(i)}\right)^2 \right\}, \tag{4}$$

being $F_{+1/2}^{(i)}$ and $F_{\times 1/2}^{(i)}$ the values of the beam pattern functions at the mid point of the $i^{th}$ SFT. Thus, we take a binary count $n_k^{(i)}$ to have greater weight if SFT $i$ has a lower noise floor or if, in the time-interval corresponding to this SFT, the beam pattern functions are larger for a particular point in the sky.

The weight normalization is chosen according to

$$\sum_{i=0}^{N-1} w_k^{(i)} = N. \tag{5}$$

With this normalization the Hough number count $n$ lies within the range $[0, N]$. Note that the sensitivity of the search is governed by the ratios of the different weights, not by the choice of overall scale. The robustness of the Hough transform method in the presence of strong transient disturbances is not compromised by using weights because each SFT contributes at most $w_i$ (which is of order unity) to the final number count.

The natural detection statistic is not the Hough number count $n$, but the *significance* of a number count, defined by:

$$s = \frac{n - \bar{n}}{\sigma}, \tag{6}$$

where $\bar{n}$ and $\sigma$ are the expected mean and standard deviation for pure noise. It is proved [15] that the mean number count in the absence of signal is $\bar{n} = Np$, and the standard deviation would be $\sigma = \sqrt{||\mathbf{w}||^2 p(1-p)}$, where $||\mathbf{w}||^2 = \sum_{i=0}^{N-1} w_i^2$, and $p$ is the probability that a frequency bin is selected in the absence of a signal. Values of $s$ can be compared directly across different templates characterized by different weight distributions and $\sigma$ values. Furthermore, in the case of Gaussian noise, the relation between the significance and the false alarm probability $\alpha$ is given by:

$$\alpha = 0.5\,\mathrm{erfc}(s/\sqrt{2})\,. \tag{7}$$

Setting a threshold on the significance would then identify interesting candidates. We refer the reader to [11] for further details.

## 3. Description of the pipeline

This paper presents a new pipeline used to analyze the data from the LIGO detectors to search for continuous gravitational waves from isolated spinning neutron stars. The LIGO detector network consists of two interferometers in Hanford Washington, one 4-km and another 2-km (H1 and H2) and a 4-km interferometer in Livingston Louisiana (L1). The search described here is currently carried out over the entire sky using the data produced during LIGO's fifth science run (S5) that started on November 4, 2005 and ended on October 1, 2007, at initial LIGO's design sensitivity. Data from each of the three LIGO interferometers is used to perform the search.

The method employed here is based on the Hough transform which we use to find signals in the time-frequency plane of the data whose time-frequency evolution fits the pattern produced by the Doppler shift imposed on the signal by the Earths motion and the pulsars spin-down during the observation period. The starting point is a collection of SFTs generated directly from the calibrated data stream, using 30-minute intervals of data for which the interferometer is operating in what is known as science mode. With this requirement, we search 32295 SFTs from the first year of S5 (11402 from H1, 12195 from H2 and 8698 from L1) and 35401 SFTs from the second year (12590 from H1, 12178 from H2 and 10633 from L1).

The search is performed in the frequency range 50 − 1000 Hz and with the frequency's time derivative in the range $-8.9 \times 10^{-10}$ Hz s$^{-1}$ to zero, being those values limited by the computational cost of the search. We use a uniform grid spacing equal to the size of a SFT frequency bin, $\delta f = 1/T_{\rm coh} = 5.556 \times 10^{-4}$ Hz. The resolution $\delta\dot{f}$ is given by the smallest value of $\dot{f}$ for which the intrinsic signal frequency does not drift by more than a frequency bin during the total observation time $T_{\rm obs}$: $\delta\dot{f} = \delta f / T_{\rm obs} \sim 1.8 \times 10^{-11}$ Hz s$^{-1}$. This yields 51 spin-down values for each frequency. $\delta\dot{f}$ is fixed to the same value for the search on the first and the second year of S5 data. The sky resolution, $\delta\theta$, is frequency dependent, with the number of templates increasing with frequency, as given by Eq.(4.14) of Ref. [15]. This yields a resolution of about $9.3 \times 10^{-3}$ rad at 300 Hz, which corresponds to $\sim 1.5 \times 10^5$ sky locations for the whole sky at that frequency.

The key difference from previous searches is that, starting from 30 min SFTs, we perform a multi-interferometer search analyzing separately the two years of the S5 run, and we study coincidences among the source candidates produced by the first and second years of data. This is inspired by a similar coincidence search using VIRGO data [24]. Furthermore, we use a $\chi^2$ test adapted to the Hough transform searches, as described in [25], to veto potential candidates.

The approach used to analyze each year of data is based on a two-step hierarchical search for continuous signals from isolated neutron stars as described below. In both steps, the weighted Hough transform is used to find signals whose frequency evolution fits the pattern produced by the Doppler shift and the spin-down in the time-frequency plane of the data. The search is done by splitting the frequency range in 0.25 Hz bands and using the SFTs from multiple interferometers.

In the first stage, we break up the sky into smaller patches with frequency dependent size. The size of the sky-patches ranges from $\sim 0.4$ rad $\times$ 0.4 rad at 50 Hz to $\sim 0.07$ rad $\times$ 0.07 rad at 1 kHz. Ideally, to obtain the maximum increase in sensitivity, we should calculate the weights, based both on the noise and the antenna pattern, for each sky-location. In practice, we calculate the weights just once for the center of each sky-patch. In this first stage, we perform the Hough multi-interferometer search using the traditional *look up table* approach, that enormously reduces the computational cost. For each SFT, the *look up table* pre-computes all the sky locations compatible with a given frequency and spin-down (and this is valid for a certain frequency range). This avoids having to determine these quantities for each individual point in parameter space (see [15] for a detailed description). But limitations on the memory of the machines constrain the volume of data (i.e., the number of SFTs) that can be analyzed at once and the parameter space (e.g., size and resolution of the sky-patches and number of spin-down values) we can search over. For this reason, in this first stage, the Hough transform is not applied using all the available SFTs, but selecting the best 15000 SFTs (according to the noise floor and the beam pattern functions) for each frequency band and sky-patch. A top-list keeping the best 1000 candidates is produced for each 0.25 Hz band.

In a second stage, we recompute the Hough significance of each candidate in the top-list using the complete set of available SFTs from all the interferometers, and at the same time we reduce the mismatch of the template, i.e., we calculate the number count and the corresponding significance without the roundings introduced by the *look up table* approach and by recomputing the weights for the precise sky location and not for the center of the corresponding patch, as previously done. For each candidate, we also compute the $\chi^2$ value that will help in vetoing the resulting candidates.

## 4. A two-step hierarchical Hough search

In this section we describe in more detail the two main features of the two-step hierarchical Hough search, i.e., the selection of the best SFT data and the comparison of the significance values produced in both steps.

### *4.1. Selection of the best SFT data*

The first of the two main changes to the Hough code consists in selecting the best SFTs based on the weights given by Eq. (4). In this way, when doing the all-sky search for each 0.25 Hz band, for each sky-patch, we will keep the 15000 SFTs that have lower noise and that are more sensitive at that sky location.

Figure 1 shows the percentage of SFTs from each of the three LIGO detectors (H1, H2 and L1) that have been selected when doing the Hough search on a 0.25 Hz band at a frequency of 420 Hz for the S5 first year of data. At this frequency, the number of sky-patches is 426 with a size of $\sim 0.17$ rad $\times$ 0.17 rad. The figure shows the percentage of SFTs from each detector on each sky-patch. In this figure we can see that, for this particular frequency, the detector that contributes the most at almost all the sky locations is H1. The maximum contribution of H1,

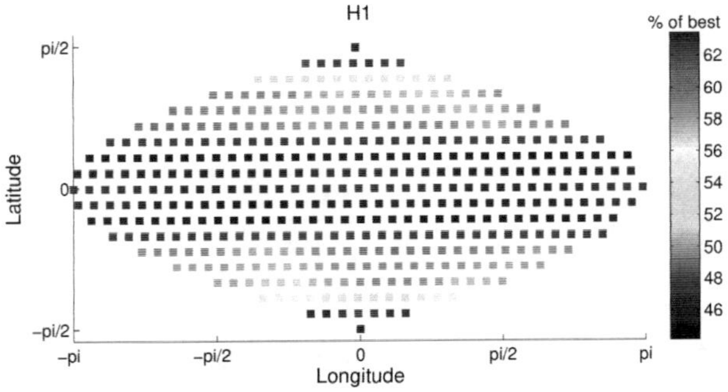


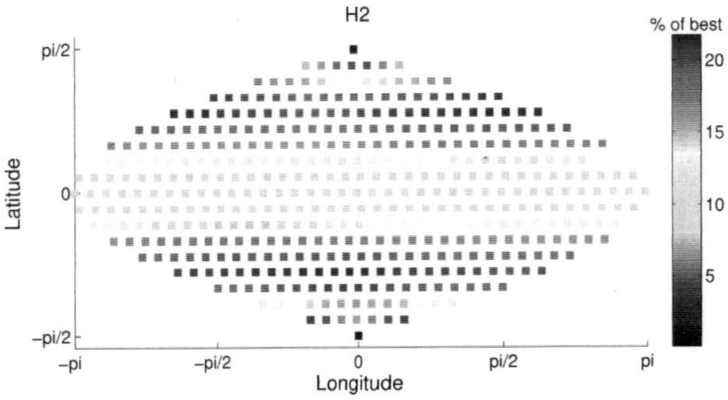


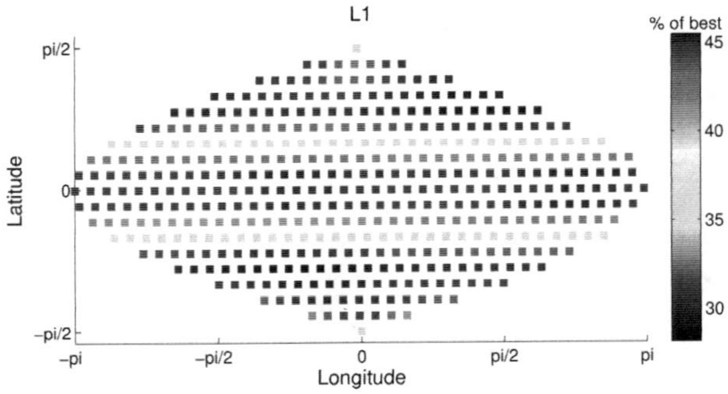


**Figure 1.** Percentage of SFTs that each detector has contributed at each particular sky-patch when selecting the best 15000 SFTs based on the Hough weights. These figures correspond to an all-sky search performed on a 0.25 Hz band at 420 Hz for the S5 first year of data.

about 64%, is at the poles, L1 has it maximum contribution, about 46%, around the equator and H2 contributes at most 23% of the SFTs in the region in between.

The selection of SFTs has been done based on the total weights, that depend both on noise floor and antenna pattern. If this selection had been based only upon the weights due to the noise floor, the H2 detector would not have contributed at all in this first stage. By having the weights antenna pattern dependent, H2 has a certain contribution in some sky locations where it is more sensitive. If the noise floor ratio between the different interferometers would be constant for different frequencies, these figures would have been independent of the chosen frequency.

### *4.2. Comparison of the significance values*

In the first stage of the hierarchical search, the Hough significance is computed using the *look up table* approach to enormously reduce the computational cost. The code uses the selected 15000 best SFTs to compute the Hough significance and make a list of interesting candidates.

In the second stage of the search, for each candidate stored in the top-list produced for each 0.25 Hz band, the code recomputes the Hough significance without any rounding, and using all the SFTs from the three LIGO detectors. In this stage, the significance is computed directly from the number count that is obtained from the digitized time-frequency plane by summing weighted binary zeros or ones along the expected path of the frequency of a hypothetical periodic gravitational wave signal, taking into account the Doppler shift and the spin-down.

Figure 2 shows how this recomputation affects the previous value of the significance in the case where an injected pulsar was present and in the case of a frequency band free of spectral disturbances. To illustrate the effect, we have used a 'quiet band' at 205.25 Hz and a band with an injected pulsar at 575 Hz. In both bands we perform an all-sky search and compare the significances computed on the first and second stages. The 'old significance' corresponds to the one computed in the first stage, using the *look up table* approach and using only the best 15000 SFTs. The 'new significance' corresponds to the one computed in the second stage using all the SFTs and without the roundings introduced by the *look up table* approach. In the same

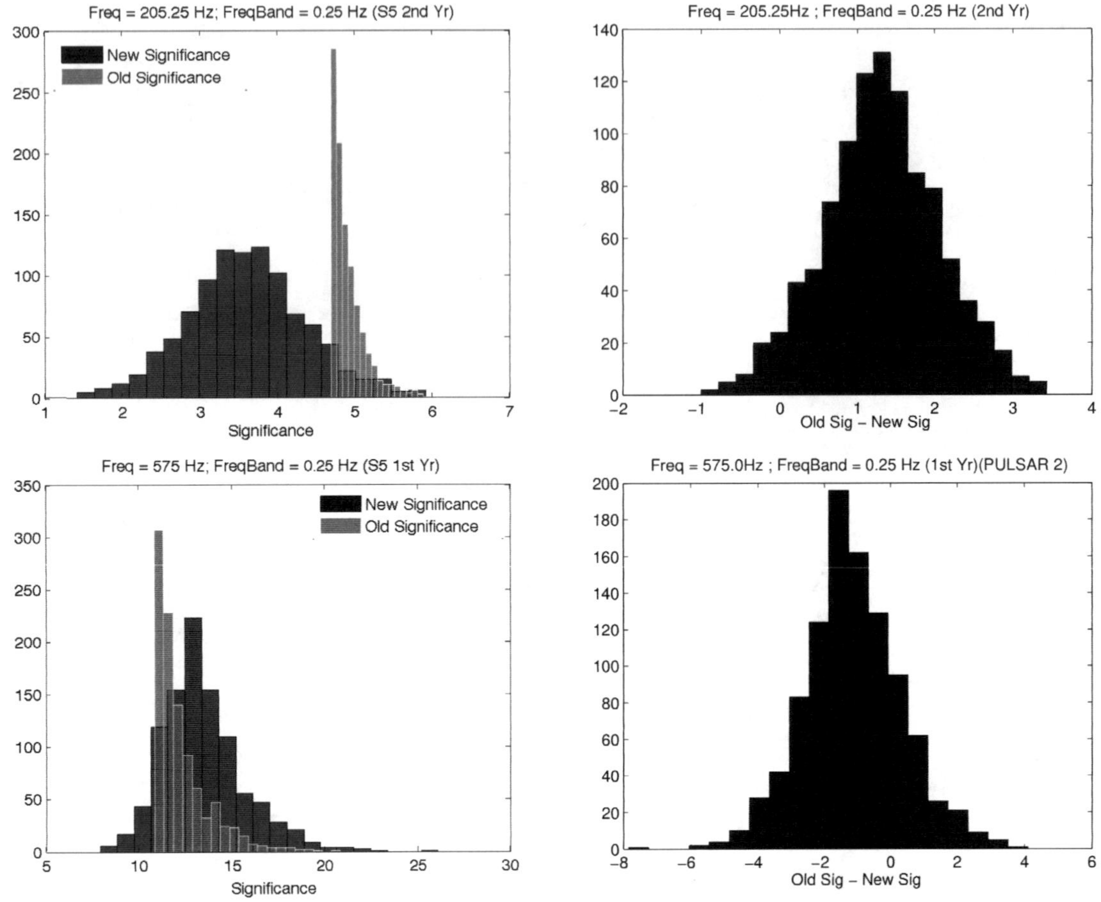


**Figure 2.** Comparison of the distribution of the significance values when calculated in the first stage using the *look up table* approach ('old significance') and after recomputing its value using all the SFTs and without any rounding ('new significance'). These results correspond to an all-sky search on a 0.25 Hz band at a frequency of 205.25 Hz for the S5 2nd year of data (top panels), where no relevant spectral disturbances were present, and at 575 Hz for the 1st year of data (bottom panels), containing a hardware injected pulsar.

figure we also provide the histograms of the differences between the 'old' and 'new' significances, showing that in the case where we have the hardware injected signal, the values of the 'new' significance are, in most cases, higher than the 'old' values. While in the case of the 'quiet' band, the 'new significance' becomes smaller than the 'old' one in the majority of the cases.

## 5. Conclusions and future work

In this paper, we have presented the improved Hough search pipeline, which is being used to analyze data from the fifth science run of the LIGO interferometers, describing in detail the two main features of the two-step hierarchical search, i.e., the selection of SFT data and recomputation of the significance values. We have shown how these improvements perform on the data, showing examples at some particular frequencies, either with interesting artifacts, such as hardware injected pulsars, or others where we expect approximately Gaussian noise. Work is in progress to analyze the coincidences among the candidates produced in the first and second year of S5 for the full parameter space, and to compute astrophysical upper limits using the search pipeline presented in this paper.

**Acknowledgments**
The authors gratefully acknowledge the support of the United States National Science Foundation for the construction and operation of the LIGO Laboratory, the Science and Technology Facilities Council of the United Kingdom, the Max-Planck-Society, and the State of Niedersachsen/Germany for support of the construction and operation of the GEO600 detector, and the Italian Istituto Nazionale di Fisica Nucleare and the French Centre National de la Recherche Scientique for the construction and operation of the Virgo detector. The authors also gratefully acknowledge the support of the research by these agencies and by the Australian Research Council, the Council of Scientic and Industrial Research of India, the Istituto Nazionale di Fisica Nucleare of Italy, the Spanish Ministerio de Educacion y Ciencia, the Conselleria dEconomia Hisenda i Innovacio of the Govern de les Illes Balears, the Foundation for Fundamental Research on Matter supported by the Netherlands Organisation for Scientic Research, the Royal Society, the Scottish Funding Council, the Polish Ministry of Science and Higher Education, the FOCUS Programme of Foundation for Polish Science, the Scottish Universities Physics Alliance, The National Aeronautics and Space Administration, the Carnegie Trust, the Leverhulme Trust, the David and Lucile Packard Foundation, the Research Corporation, and the Alfred P. Sloan Foundation. LIGO Document No. LIGO-P0900212

## 6. References

[1] Luck H *et al.* 2006 *Class. Quant. Grav.* **23** S71–S78
[2] Abbott B *et al.* (The LIGO Scientific Collaboration) 2009 *Rept. Prog. Phys.* **72** 076901 (*Preprint* 0711.3041)
[3] Acernese F *et al.* 2008 *Class. Quant. Grav.* **25** 114045
[4] Abbott B *et al.* (The LIGO Scientific Collaboration) 2004 *Phys. Rev.* **D69** 082004 (*Preprint* gr-qc/0308050)
[5] Abbott B *et al.* (The LIGO Scientific Collaboration) 2005 *Phys. Rev. Lett.* **94** 181103 (*Preprint* gr-qc/0410007)
[6] Abbott B *et al.* (The LIGO Scientific Collaboration) 2007 *Phys. Rev.* **D76** 042001 (*Preprint* gr-qc/0702039)
[7] Abbott B *et al.* (The LIGO Scientific Collaboration) 2008 *Astrophys. J.* **683** L45–L50 (*Preprint* 0805.4758)
[8] Abbott B *et al.* (The LIGO Scientific Collaboration and The Virgo Collaboration) 2009 (*Preprint* 0909.3583)
[9] Abbott B *et al.* (The LIGO Scientific Collaboration) 2007 *Phys. Rev.* **D76** 082001 (*Preprint* gr-qc/0605028)
[10] Abbott B *et al.* (The LIGO Scientific Collaboration) 2005 *Phys. Rev.* **D72** 102004 (*Preprint* gr-qc/0508065)
[11] Abbott B *et al.* (The LIGO Scientific Collaboration) 2008 *Phys. Rev.* **D77** 022001 (*Preprint* 0708.3818)
[12] Abbott B *et al.* (The LIGO Scientific Collaboration) 2009 *Phys. Rev.* **D79** 022001 (*Preprint* 0804.1747)
[13] Abbott B P *et al.* (The LIGO Scientific Collaboration) 2009 *Phys. Rev.* **D80** 042003 (*Preprint* 0905.1705)
[14] Abbott B *et al.* (The LIGO Scientific Collaboration) 2009 *Phys. Rev. Lett.* **102** 111102 (*Preprint* 0810.0283)
[15] Krishnan B *et al.* 2004 *Phys. Rev.* **D70** 082001 (*Preprint* gr-qc/0407001)
[16] Palomba C, Astone P and Frasca S 2005 *Class. Quant. Grav.* **22** S1255–S1264
[17] Sintes A M and Krishnan B 2006 *J. Phys. Conf. Ser.* **32** 206–211 (*Preprint* gr-qc/0601081)
[18] Antonucci F *et al.* 2008 *Class. Quant. Grav.* **25** 184015 (*Preprint* 0807.5065)
[19] Brady P R and Creighton T 2000 *Phys. Rev.* **D61** 082001 (*Preprint* gr-qc/9812014)
[20] Papa M A, Schutz B F and Sintes A M 2000 *ICTP Lecture Notes Series Vol. III, edited by V. Ferrari, J. C. Miller and L. Rezzolla (Italy)* 431–442 (*Preprint* gr-qc/0011034)
[21] Frasca S, Astone P and Palomba C 2005 *Class. Quant. Grav.* **22** S1013–S1019
[22] Cutler C, Gholami I and Krishnan B 2005 *Phys. Rev.* **D72** 042004 (*Preprint* gr-qc/0505082)
[23] Astone P, Frasca S and Palomba C 2005 *Class. Quant. Grav.* **22** S1197–S1210
[24] Acernese F *et al.* 2007 *Class. Quant. Grav.* **24** S491–S499
[25] Sancho de la Jordana L and Sintes A M 2008 *Class. Quant. Grav.* **25** 184014 (*Preprint* 0804.1007)

# Searching for gravitational waves from pulsars in binary systems: an all-sky search

**S. van der Putten[1], H. J. Bulten[1], J. F. J. van den Brand[1], M. Holtrop[2]**

[1]Nikhef, Sciencepark 105, 1098 XG Amsterdam, The Netherlands
[2]Department of Physics, University of New Hampshire, Durham, N.H. 03824, USA

E-mail: `sputten@nikhef.nl`

**Abstract.** Non axisymmetrical neutron stars emit continuous gravitational waves. The periodicity of the system allows for the improvement of the signal-to-noise ratio by integrating the signal over time. Many of the known neutron stars are in binary systems and their orbital movement will Doppler-shift the signal. When performing an all sky-search for such systems the use of exact binary motion templates become unfeasible. The Doppler shifts causes the power in the signal to be spread out over a large enough frequency band so that the gain of integrating the signal over the observation time is negligible. Here, the principles of a new analysis method for LIGO and Virgo data, called the polynomial search, will be presented together with first tests on simulated data.

## 1. Introduction

Gravitational wave detectors like LIGO and Virgo have reached their design sensitivity and direct detection of gravitational waves may become possible in the near future. Non axial-symmetric rotating neutron stars, or pulsars (when they emit radio or X-ray pulses), are systems that emit gravitational waves. The strain wave amplitude from such a system is calculated to be [1]

$$h_0 = \frac{16\pi^2 G}{c^4}\frac{\epsilon I_{zz} f^2}{r} \approx 4.2\times 10^{-27}\left(\frac{I_{zz}}{10^{38}\text{kg m}^2}\right)\left(\frac{10\text{kpc}}{r}\right)\left(\frac{f}{100\text{Hz}}\right)^2\left(\frac{\epsilon}{10^{-6}}\right), \qquad (1)$$

where $\epsilon$ represents the ellipticity of the neutron star, $I_{zz}$ the $z$-component of the moment of inertia, $f$ the rotation frequency of the neutron star, $G$ and $c$ are the gravitational constant and the speed of light, respectively.

Neutron stars can be found as either isolated objects or in a binary system. In the following we discuss a new data analysis method termed the polynomial search, that enables an all-sky search for gravitational waves from neutron stars in binary systems.

## 2. Motivation

Most of the current all-sky searches [2, 3, 4, 5, 6, 7] focus on isolated neutron stars. From the ATNF pulsar catalog [8] it appears that the majority of the known pulsars with rotational frequencies above 10 Hz are in binary systems (see Fig. 1). The corresponding gravitational wave amplitude scales with $f^2$ (Eq. 1). However, the Keplerian orbit of the pulsar will introduce

© 2010 IOP Publishing Ltd

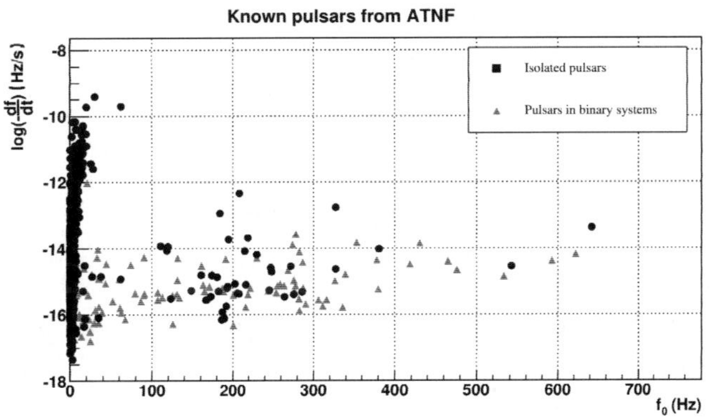


**Figure 1.** *The spin versus the rotational frequency the down for isolated pulsars (black squares) and for pulsars in binary systems (red triangles). The data are taken from the ATNF pulsar catalog (June 2009) [8].*

a Doppler shift to the signal in addition to the Doppler shift from the detector motion. These Doppler shifts cause the signal power to be spread out in frequency.

In order to take full advantage of the long integration times, templates must be matched with the data and they must be coherent with the signal over as long a time as possible. When attempting an all-sky coherent search for neutron stars in binary systems the number of required signal templates was found to be so large that such a search would be unfeasible [9].

Another strategy which is used in all-sky searches for isolated neutron stars is the Hough transform search [4]. This method uses shorter integration times so that the signal stays in a single frequency bin. When applying this principle to neutron stars in binary systems [9] one finds that the maximum coherent time for a signal $T_{coh}$ to stay in one frequency bin can be written as

$$T_{coh} \leq 131.6 \left( \frac{(1+q)^{1/3}}{\sqrt{q}} \right) \left( \frac{P}{1\text{ day}} \right)^{2/3} \left( \frac{m_{NS}}{1.4\ M_{\odot}} \right)^{-1/6} \left( \frac{f}{1\text{ kHz}} \right)^{-1/2} \text{sec}, \quad (2)$$

where $q = m_2/m_{NS}$ represents the ratio of the masses of the two objects, $P$ the period of the binary, $m_{NS}$ the mass of the neutron star and $f$ the gravitational wave frequency. For an example binary system with parameters $m_2 = m_{NS} = 1.4\ M_{\odot}$, $P = 2$ h and $f_0 = 1$ kHz we find $T_{coh} \approx 20$s. This coherent time is too short to gain any significant signal to noise ratio by continuous observation.

The polynomial search is a hierarchical search that employs a limited number of filters and aims to increase the possible coherent time by more than an order of magnitude compared to the Hough transform search. Next, the first step of the hierarchical pipeline will be presented.

## 3. The polynomial search method

The Doppler shifts induced by the Kepler orbit of the neutron star will spread the signal over many frequency bins. The parameter space can be represented by the $12 + s$ dimensional vector

$$\vec{\lambda} = (P, a_p, e, \omega_0, h_0, \phi_0, \tau, f_0, \alpha_{ns}, \delta_{ns}, i, \psi, f^{(1)}, ..., f^{(s)}),$$

where $P$ represents the period of the binary, $a_p$ the projected semi major axis, $e$ the eccentricity of the orbit, $\omega_0$ the longitude of the periastron, $\tau$ the time of periastron passage, $h_0$ the gravitational wave amplitude the signal has at the detector, $\phi_0$ the initial phase of the gravitational wave, $f_0$

the rotation frequency of the neutron star, $\alpha_{ns}$ and $\delta_{ns}$ the right ascension and declination of the source, $i$ the angle between the rotation and propagation axis, $\psi$ the polarization angle of the gravitational wave and $f^{(1)}$ to $f^{(s)}$ the spin down parameters.

Instead of attempting to create templates in the $12 + s$ dimensional parameter space, the polynomial search employs the empirical phase model given by

$$\Phi(t;\vec{\xi}) = 2\pi\left(\phi_0 + f_0 t + \frac{\alpha}{2}t^2 + \frac{\beta}{6}t^3\right), \tag{3}$$

where the vector $\vec{\xi} = (\phi_0, f_0, \alpha, \beta)$ lives in a four dimensional filter parameter space. This phase model has the advantage of a lower dimensionality of the parameter space and is a priori not dependent on any (astro)physical assumptions. However, the model will only hold as long as the coherent time is short enough or the Keplerian parameters are small enough in such a way that the phase from Eq. (3) does not deviate more than a certain value from the real signal. The filters

$$F(t;\vec{\xi}) = \sin\left(\Phi(t;\vec{\xi})\right) \tag{4}$$

are to be applied to the data in stretches of coherent time $T$. The maximal values for $\alpha$ and $\beta$ can be calculated from the Kepler orbits assuming that the dominant contribution in the frequency dependence on time arises from the Doppler shifts. The values are calculated for a specific binary by taking the maximal values for the first and second derivatives for $\alpha_{max}$ and $\beta_{max}$. [1] For the previously introduced example binary, the values are $\alpha_{max} = 1.3$ mHz/s and $\beta_{max} = 1.1\ \mu\text{Hz/s}^2$. Thus when applying the filters described in Eq. (4) and stepping in parameter space up to these maxima, all binary systems will be covered by these filters if the binary parameters lead to Doppler shifts smaller than the extreme case. The step size has been chosen by performing Monte-Carlo simulations assuming that the higher order corrections are small. The stepsize was chosen by computing the maximal distance of the true parameters to the nearest point in parameter space which results in a phase difference of $\pi/2$ for the complete filter. The details of these simulations will be presented in a future publication. The maximal coherent time, defined as the maximal time that the filter stays in phase within 1 radian with the original signal, can be calculated to be $T = 500\ s$ for this binary. Thus the polynomial search will gain a factor of 25 in coherent time over a Hough search.

The polynomial search is performed by stepping in the parameter space and creating a collection of filters. The filters are then applied to the data by calculating the correlation with the data $D(t)$ and each filter $F(t-\tau)$ where $\tau$ represents the lag parameter which can be interpreted as the time shift of the filter. This correlation $C$ as a function of lag $\tau$ and filter with parameters $\vec{\xi}$ is computed in the frequency domain for computational efficiency which, when taking the noise level of the detector into account can be written as

$$C(\tau;\vec{\xi}) = \mathcal{F}^{-1}\left[\frac{\mathcal{F}\left[D(t)\right]\mathcal{F}\left[F(t;\vec{\xi})\right]^*}{\sqrt{2P_F S_n(\omega)}}\right], \tag{5}$$

where the operator $\mathcal{F}[]$ represents the operation of taking the Fourier transform, $*$ denotes the complex conjugate, $D(t)$ the data from the detector, $F(t;\vec{\xi})$ the filter with parameters $\vec{\xi}$, $P_F$ the spectral power in the filter and $S_n(\omega')$ the single-sided power spectral density of the data, which is used to estimate the noise level of the detector. The resulting statistic can be considered Gaussian distributed with unit variance[2].

[1] In practice, the choice for $\alpha_{max}$ and $\beta_{max}$ will determine the maximum span of the binary parameter space one would wish to search.

[2] In this case, the correlation $C$ can also be interpreted as a signal to noise ratio as can be found with the Neyman-Pearson criterion

Eq. (5) gives the correlation for each filter with the data, shifted with lag parameter $\tau$. The phase as a function of the lag can be written as

$$\Phi(t-\tau) = 2\pi \left[ \underbrace{\left(\phi_0 - f_0\tau + \frac{\alpha}{2}\tau^2 - \frac{\beta}{6}\tau^3\right)}_{\phi(\tau)} + \underbrace{\left(f_0 - \alpha\tau + \frac{\beta}{2}\tau^2\right)}_{f_0(\tau)} t + \frac{1}{2}\underbrace{(\alpha - \beta\tau)}_{\alpha(\tau)} t^2 + \frac{1}{6}\beta t^3 \right]. \quad (6)$$

From Eq. (6) it can be seen that the lag parameter creates an interdependency of the filter parameters. This property can be exploited to avoid generating filters for each point in the parameter space which will reduce the amount of CPU time and memory needed. In order to use the lag parameter as a time shift it must be possible to shift the filter within the chunk of data between 0 and $T$ s. In our implementation the filters are zero-padded for $t > T/2$ and only the lags $0 < \tau < T/2$ are considered.

The filters are produced in a frequency band $\Delta f$ and are compared to the data by shifting the spectrum of the filter which is equivalent to heterodyning the data. When discetizing the filter $N = \Delta f T$ values of the correlation $C(\tau, \vec{\xi})$ for the lags between 0 and $T/2$ are obtained, where Eq. (6) shows the dependency of the parameters on the lag. In order to reduce the number of correlations to process, only the maximum correlation and corresponding values for $\tau$, $f_0$, $\alpha$ and $\beta$ are recorded[3] for each applied filter $F(t; \vec{\xi})$.

Since the correlation is normalized to the power spectral density, a threshold in terms of the estimated noise level of the detector can be set. When applying such a threshold to the collection of maxima the resulting numbers are called *hits*. These hits can be displayed by plotting the frequency as a function of time,

$$f(t-\tau) = \frac{1}{2\pi}\frac{d\Phi(t-\tau; \vec{\xi}_{hit})}{dt}, \quad (7)$$

where $\vec{\xi}_{hit}$ represents the parameters of the filters for which the correlation exceeds the threshold.

## 4. Sensitivity

The probability that, given a threshold and in absence of a signal, at least one filter will give a hit is called the false alarm probability. For the assumption of Gaussian noise and statistically independent filters, the false alarm probability can be written as

$$p_{fa}(x > x_{thr}) = 1 - \left(\frac{1 + \mathrm{erf}(\frac{x_{thr}}{\sqrt{2}})}{2}\right)^{N_F}, \quad (8)$$

where $x$ is the correlation of a single filter with parameters $\phi(\tau)$, $f_0(\tau)$ , $\alpha(\tau)$ and $\beta(\tau)$ has with the data, $x_{thr}$ the threshold normalized to the estimated noise level of the data and $N_F$ is the total number of applied filters.

The false dismissal probability for a signal of a certain strength is defined as the probability that no correlation between filter and data passes a given threshold. The false dismissal probability for a given signal can be calculated if the correlations $\kappa_i$ of all filters with the signal with amplitude $h_0$ and $S_n$ are known. The expression is given by

$$p_{fd}(x < x_{thr}) = \Pi_{i=0}^{N_F-1}\left(\frac{1 + \mathrm{erf}\left(\frac{x_{thr} - \kappa_i X}{\sqrt{2}}\right)}{2}\right), \quad (9)$$

[3] The value for $\phi_0$ has been fixed to 0. Note, that by considering all lag parameters $\tau$ all values of $\phi(\tau)$ are sufficiently covered.

where and $\chi = h_0\sqrt{\frac{T}{S_n}}$.

The false alarm and false dismissal rates are calculated for an example scan with parameters

$$T = 1200\text{ s}, \; \Delta f = 8\text{ Hz}, \; \alpha = \pm 10^{-3}\text{ Hz/s and } \beta = \pm 10^{-5}\text{ Hz/s}^2,$$

where $T$ represents the integration time, $f_F$ the sample frequency of the filters, $\Delta f$ the bandwidth of the data which has been used for the polynomial search, $\alpha$ and $\beta$ give the range of both parameters which have been used in this search. For this particular set of parameters the amount of explicitly computed filters $n_f$ is $10^5$. Using these parameters one finds that the number of distinct applied filters per Hz of searched data $N_F = Nn_f = 1.2 \times 10^8$ filters/Hz.

The false alarm probability as a function of the threshold can be calculated directly with Eq. (8) and the number of applied filters. On the other hand the false dismissal probability depends on the value of the correlation of each filter with the signal. This means that the false dismissal probability can be estimated by using simulations.

The false dismissal probability was estimated for the parameters stated above. In the calculation of the false dismissal probability, only the filters which had $\kappa_i > 0.5$ were taken into account due to the large number of filters with smaller correlations. In this manner, the product in Eq. (9) runs over a subset of filters[4] and the calculation of the false dismissal rate as a function of signal strength can be done in a timely manner.

Fig. 2 shows the false alarm (left panel) and false dismissal (right panel) probability per Hz as a function of the threshold and of $\chi$, respectively. For a given signal strength $h = \chi\sqrt{\frac{T}{S_n}}$ and

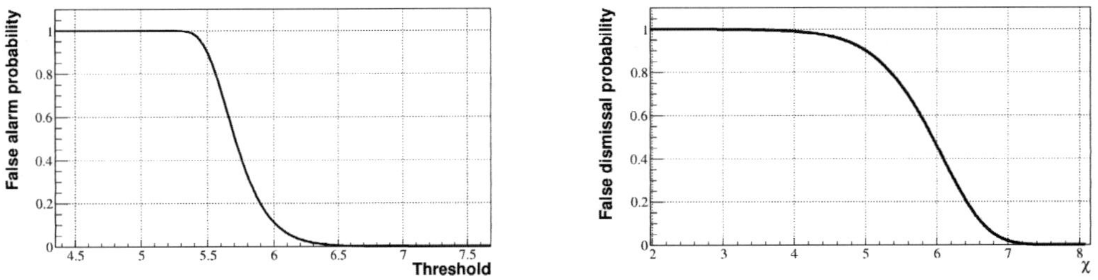


**Figure 2.** *The left panel shows the false alarm probability per Hz searched bandwidth which has been computed using the parameters given in the text. The right panel show the false dismissal probability as a function of the signal strength. A threshold of* 6.4, *corresponding to a* 1% *false alarm probability has been used.*

a given number of data chunks the probability for $k$ hits in $n$ data chunks is

$$P(k) = \frac{n!}{k!(n-k)!} p_{fd}^{n-k} (1 - p_{fd})^n, \qquad (10)$$

where $p_{fd}$ is the false dismissal probability as shown in Eq. (9) and in Fig. 2. The corresponding false alarm probability can be computed similarly.

## 5. Simulations

In order to illustrate the effectiveness of the polynomial search a binary system has been simulated with parameters

$$P = 0.1\text{ d}, \; a_p = 2\text{ ls}, \; e = 0.1, \; f_0 = 600\text{ Hz}, \; h_0 = 10^{-27},$$

[4] Of the $9.6 \times 10^8$ filters in the search band only about 500 have a $\kappa_i > 0.5$.

where $P$ represents the period of the binary in days, $a_p$ the projected semi major axis in light seconds, $e$ the eccentricity of the orbit, $f_0$ the frequency of the gravitational wave emitted by the neutron star and $h_0$ the emitted amplitude of the gravitational wave. The beam pattern function of the interferometer has not been taken into account. The total simulation contained 6 hours of data sampled at 4 kHz and 35 overlapping FFTs (Fast Fourier Transforms) were made from this data. The polynomial search was performed with the parameters shown in the previous section.

Fig. 3 shows the frequency evolution of the signal in a time-frequency diagram where the color coding shows the strain of the signal. It can be seen that per FFT the power is spread

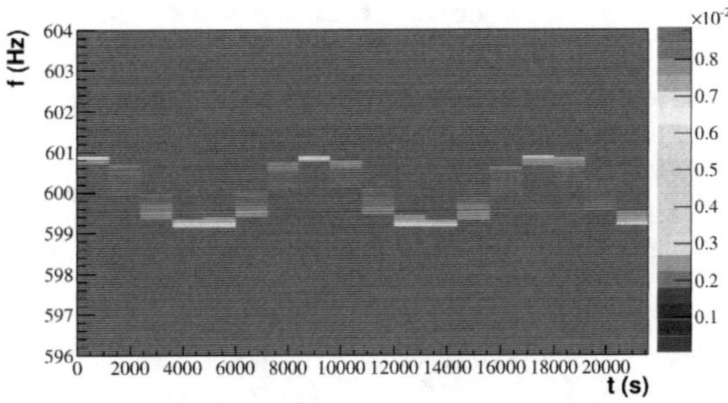


**Figure 3.** *The frequency-time diagram of the simulated signal without noise.*

out over a number frequency bins depending on the phase of the Keplerian orbit of the neutron star.

Gaussian noise has been added to the signal in the time domain with different amplitudes. The top left panel of Fig. 4 shows the frequency time diagram just like in Fig. 3 but with

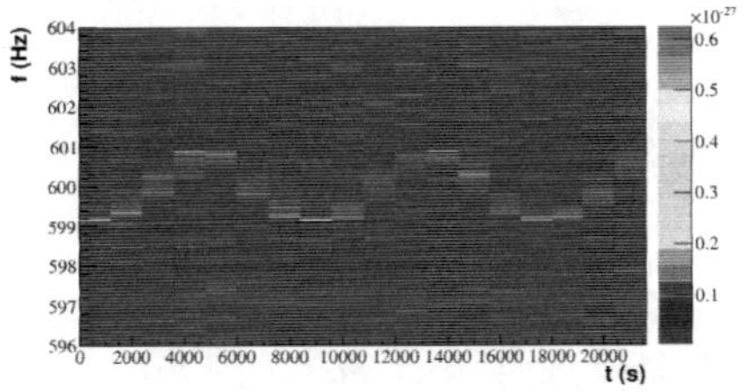


(a) *f-t diagram: noise amplitude* $10^{-25}$.

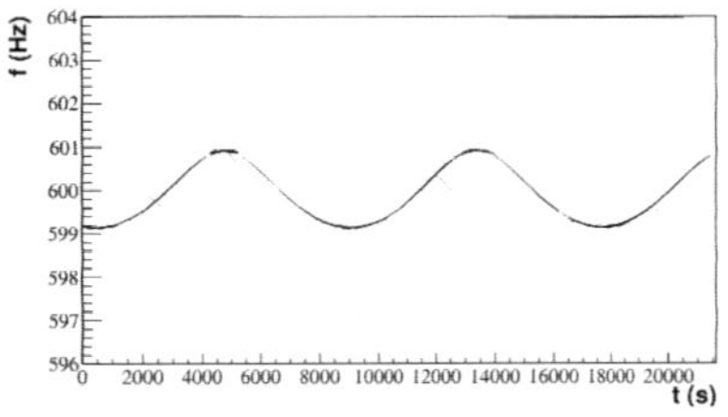


(b) *hitmap: noise amplitude* $10^{-25}$.

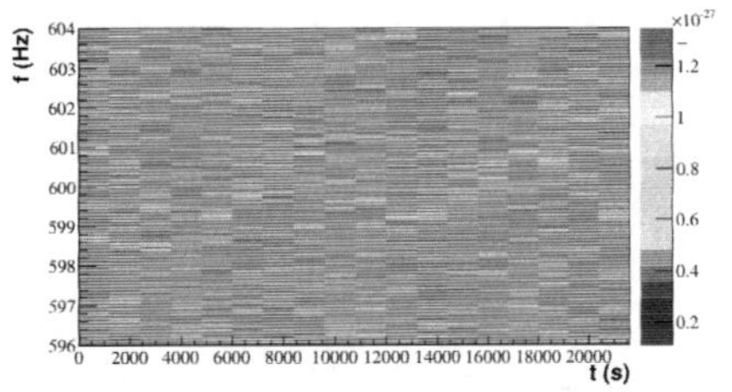


(c) *f-t diagram noise amplitude* $2 \times 10^{-25}$.

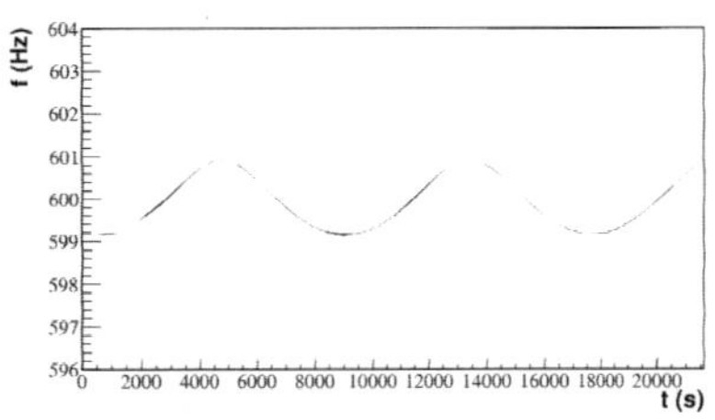


(d) *hitmap: noise amplitude* $2 \times 10^{-25}$.

**Figure 4.** *The frequency-time diagram of the simulated signal and the hit maps coming from the polynomial search which are made with a threshold of* 6.4.

Journal of Physics: Conference Series **228** (2010) 012005 doi:10.1088/1742-6596/228/1/012005

added noise with an amplitude of $10^{-25}$ in the time domain. Given the 1200 s coherent time[5], this corresponds to a SNR in amplitude of 0.35. The top right panel shows the result of the polynomial search. This plot is made using the filters which exceed the threshold of 6.4. The filters are written in the form shown in Eq. (7) and drawn as a function of time.

The bottom left plot shows the same signal, but with a noise amplitude of $2 \times 10^{-25}$ and bottom right shows the result coming from the polynomial search. Fig. 4 demonstrates the effectiveness of the polynomial search in terms of improved signal to noise ratio.

## 6. Conclusion and outlook

An algorithm for performing an all-sky search for gravitational waves from neutron stars in binary systems, called the polynomial search, has been presented and the first step in an hierarchical pipeline has been discussed. The polynomial search uses an empirical phase model which can describe the phase evolution up to third order in time, or the frequency evolution up to second order in time. Using this method, the Doppler shifts of the gravitational wave signals from neutron stars in binary systems can be dealt with and the coherent length of the individual data segments can be increased. The false alarm and false dismissal probability of the method have been presented giving a theoretical sensitivity of the search applied to a single stretch of data.

The sensitivity of any coherent continuous wave search scales as $\sqrt{T}$ where $T$ is the observation time for a single chunk of data. The coherent part of a Hough search demands that the change in frequency of the signal cannot exceed one frequency bin. The first step of the polynomial search enables the use of observation times more than a factor of 25 longer with respect to a Hough search for a single data stretch. This allows for a factor of 5 in signal to noise gain.

The polynomial search is designed for performing an all-sky search for gravitational waves coming from neutron stars in binary systems. It has been shown that the polynomial search can follow gravitational waves with high Doppler shifts and stay coherent for longer than $10^3$ *s*. The polynomial search has been shown to be able to search a large volume of the parameter space with a limited set of filters. In the simulation discussed in this paper $1.2 \times 10^8$ filters per Hertz bandwidth of data were sufficient to cover all frequency derivatives up to $\left|\frac{df}{dt}\right| \leq 2$ mHz/s and $\left|\frac{d^2f}{dt^2}\right| \leq 10\ \mu\text{Hz/s}^2$. The resulting hitmaps demonstrate that the polynomial search is able to recover the phase evolution of the signal from a neutron star in a binary system.

The computation of the hitmaps is fast; the entire search of a band of 8 Hz of data took $\approx 5$ CPUh. Since the software makes use of large computing infrastructures like the GRID, a search over all data and all frequencies up to 2 kHz can be performed in real-time with $\approx 100$ nodes.

The polynomial search as presented in this paper will be applied to the LIGO and Virgo data in the future. Using the computed false alarm rates upper limits can be set on the gravitational wave emission from neutron stars in binary systems. Furthermore, the polynomial search will be extended with a second step that links the hits in the hitmaps together and fits the physical parameters to the resulting signal.


## Acknowledgments

This work is part of the research programme of the Foundation for Fundamental Research on Matter (FOM), which is financially supported by the Netherlands Organisation for Scientific Research (NWO).


[5] The SNR would increase with $\sqrt{T}$ if the signal would stay in one bin. Since the Doppler shift prevents this, the SNR increase would be less then $\sqrt{T}$

## References

[1] Jaranowski P, Krolak A and Schutz B F 1998 *Physical Review D* **58** 063001
URL `http://link.aps.org/doi/10.1103/PhysRevD.58.063001`

[2] Brady P R, Creighton T, Cutler C and Schutz B F 1998 *Phys. Rev. D* **57** 2101–2116
URL `http://link.aps.org/doi/10.1103/PhysRevD.57.2101`

[3] Brady P R and Creighton T 2000 *Phys. Rev. D* **61** 082001
URL `http://link.aps.org/doi/10.1103/PhysRevD.61.082001`

[4] Krishnan B, Sintes A M, Papa M A, Schutz B F, Frasca S and Palomba C 2004 *Phys. Rev. D* **70** 082001
URL `http://link.aps.org/doi/10.1103/PhysRevD.70.082001`

[5] Mendell G and Wette K 2008 *Classical and Quantum Gravity* **25** 114044 (8pp)
URL `http://stacks.iop.org/0264-9381/25/114044`

[6] Abbott B *et al* 2008 *Phys. Rev. D* **77** 022001
URL `http://prd.aps.org/abstract/PRD/v77/i2/e022001`

[7] Abbott B *et al* 2009 *Phys. Rev. D* **79** 022001
URL `http://prd.aps.org/abstract/PRD/v79/i2/e022001`

[8] Manchester R N, Hobbs G B, Teoh A and Hobbs M 2005 *AJ* **129**
URL `http://www.atnf.csiro.au/research/pulsar/psrcat/`

[9] Dhurandhar S V and Vecchio A 2001 *Phys. Rev. D* **63** 122001
URL `http://link.aps.org/doi/10.1103/PhysRevD.63.12200`

# Using a cleaning technique for the search of continuous gravitational waves in LIGO data

**Paola Leaci[1], Pia Astone[2], Maria Alessandra Papa[1,3] and Sergio Frasca[2]**

[1] Max-Planck-Institut für Gravitationsphysik, Albert-Einstein-Institut, 14476 Golm, Germany
[2] University 'La Sapienza', 00185 Rome, Italy
[3] University of Wisconsin-Milwaukee, WI 53201, Milwaukee

E-mail: paola.leaci@aei.mpg.de and maria.alessandra.papa@aei.mpg.de

**Abstract.** High frequency short events, due for instance to delta-like spurious disturbances, may affect the broad band noise level and thus produce a loss in the efficiency of detection of continuous gravitational waves. We identify such events, remove them from a set of LIGO fifth science run (S5) data and characterize the resulting sensitivity improvements. We use the same parameter values as used in a previous pilot study on Virgo data, but do not observe on LIGO S5 data the same sensitivity improvements.

PACS numbers: 04.80.Nn, 07.05.Kf, 95.55.Ym

(Some figures in this article are in colour only in the electronic version)

## 1. Introduction

The detection of continuous gravitational wave (CW) signals is one of the challenges for ground based gravitational wave detectors [1, 2, 3, 4, 5]. The sensitivity of any CW search is determined by the level of noise and of spectral line contamination in the frequency band of interest and by the search technique. The noise level is entirely determined by the instrument and in fact the sensitivity of gravitational wave detectors is usually represented by graphing the power spectral density (PSD) as a function of frequency. Large short-lasting disturbances may affect the level of this noise floor. The sensitivity of CW searches may be improved by removing this type of disturbance before performing the search, irrespective of having identified their origin. Such a pre-processing procedure was first proposed in [6], which is the pilot study referred to in the abstract. In the present paper we characterize its performance on LIGO data from the 5th science run (S5). This run started on November 4 (14), 2005 at the LIGO Hanford (Livingston) Observatory and ended on October 1, 2007.

© 2010 IOP Publishing Ltd

We stress that this paper reports on the results obtained using as input only science-mode data, i.e. data segments covering a time period for which the detector is optimized for data collection and ready to observe a possible astrophysical event.

## 2. Cleaning procedure

The cleaning procedure we describe here identifies disturbances in the time domain data and removes them from the data set. Such a data set is divided in 1800 s chunks and each of these is cleaned separately. After the cleaning step, each chunk is Fourier-transformed producing data in a format that is used as input to many CW searches. This Fourier-domain data is referred to as Short time-baseline Fourier Transforms, SFTs in short. The steps carried out to accomplish the cleaning on each chunk are:

(i) A 10th-order *Butterworth filter* (at 38 Hz): it's a high-pass filter useful to minimize the contribution of the prominent low frequency component.

(ii) *Tukey window*: the 0.1% of the data are effectively lost unless the data are interlaced.

(iii) A 1st-order *high-pass filter* (at 100 Hz) and event identification: the events are identified on this data.

(iv) *Time Domain cleaning procedure*: after finding events and registering some of their characteristics, their amplitude is subtracted from the amplitude of the original time series. Note that the high-passed time series, containing all the events, is not the same as the original time series because the high-passed time series does not have the low frequency component of the original data.

In order to identify the events, we set a threshold on the critical ratio (CR) of the high-passed data (produced in the third step above). The CR is defined as

$$\mathrm{CR} = \left| \frac{y(i) - \mu_A(i)}{\sigma_A(i)} \right|, \tag{1}$$

where $y(i)$, $\mu_A(i)$ and $\sigma_A(i)$ are the data samples, the mean and the standard deviation, respectively. These last two quantities are estimated in an autoregressive way with an adaptive threshold, i.e. a threshold that changes with time as the sensitivity of the detector changes, as a consequence of the non-stationary noise. We chose a threshold value of 5. The value of the time constant for the autoregressive estimation was chosen to be 20 s, which is appropriate to track typical changes in the statistics of the data. An event starts when the amplitude of a sample goes above threshold (CR $> 5$). The last sample of an event is the last sample above threshold followed by a series of samples all lying below the threshold and covering a time longer than a given dead-time interval $\Theta$. In other words, the event identification scheme takes as belonging to the same event all samples above threshold which are closer than $\Theta$. Events begin and end with a sample over threshold. We have taken $\Theta = 0.1$ s as this is smaller than the average duration of the typical disturbances that we want to remove, which typically have a few threshold-crossings.

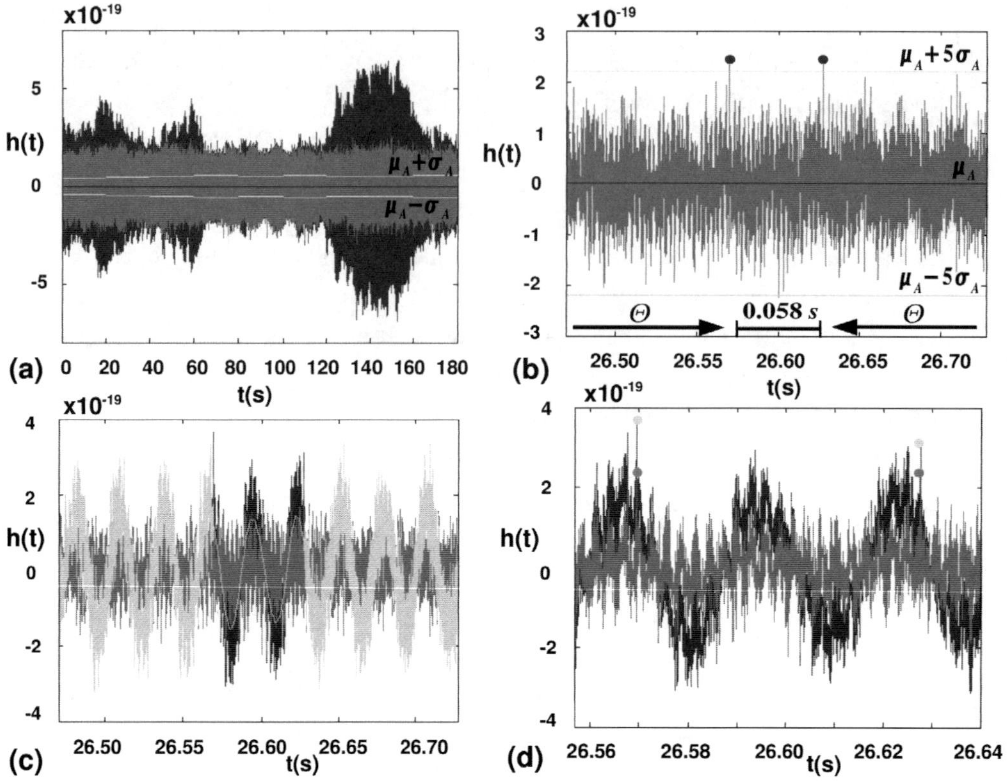


**Figure 1.** Strain amplitude h(t) curves versus the time. The blue curve represents h(t) after a Butterworth high-pass filter, with a knee frequency of 38 Hz. The red curve consists of the further application of a high-pass filter, with a cutoff frequency of 100 Hz to identify 'high frequency spikes'. The 3 horizontal lines in the panel (a) represent the autoregressive estimation, $\mu_A \pm \sigma_A$ (cyan lines) and $\mu_A$ (blue line). The green curve corresponds to the cleaned data set and all the circles in the subplots label the beginning and the end of the considered event, lasting for about 0.058 s.

After the events have been identified they are subtracted from the original series, as was explained above. To avoid discontinuities in the cleaned time series, a few samples before and after the event are used to smoothly connect the original and the cleaned data. We used ten samples, corresponding to $6 \times 10^{-4}$ s, being the sampling time of $\sim 6 \times 10^{-5}$ s.

As an example, figure 1 (a) shows data from the 4 km Hanford (H1) detector taken during S5. The original time data series h(t) is the blue curve and the time series on which the events are identified is the red curve. We further zoom in the region where an event has been identified. The event contains 954 samples, corresponding to roughly 0.058 s. As we can see in figure 1 (b), the data lies above the threshold (CR > 5) twice (blue circles). After the subtraction of the amplitudes of all the samples of the event, the cleaned time series is obtained and plotted in figure 1 (c) (green curve). A further zoom of the region where the subtraction is applied is shown in figure 1 (d). Note that the cleaned time series (green curve) is obtained as subtraction between the original

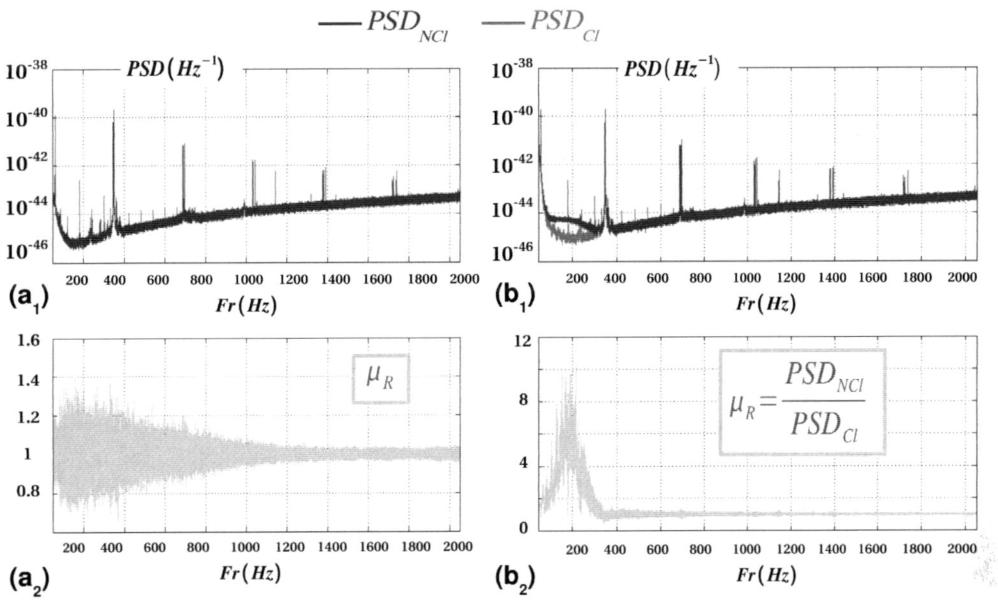


**Figure 2.** Two examples of noise floor estimation. Panels ($a_1$), ($b_1$): the PSD of uncleaned data ($PSD_{NCl}$, blue curve) is compared to the PSD of the corresponding cleaned data ($PSD_{Cl}$, red curve). It is evident that the data analyzed in the panel ($b_1$) is more disturbed. Therefore, the cleaning procedure is more effective in this case. In the bottom panels ($a_2$), ($b_2$), the ratio $\mu_R$ between the PSDs is also shown versus the frequency.

time series (blue curve) and the high-passed series (red curve) only for the amplitudes of those samples labelled as part of an event. The green and blue curves coincide for the remaining samples that do not belong to any event.

The autoregressive mean $\mu_A$ (blue line) and standard deviation $\sigma_A$ are also shown in figure 1 (a,b).

## 3. Results and conclusions

We have applied this cleaning procedure to data from the S5 run of the H1 detector taken between GPS time 846374761 (Nov 01 00:05:47 GMT 2006) and GPS time 875234357 (Mon Oct 01 00:39:03 GMT 2007). Contiguous segments of 1800 s science-mode data were selected and yielded 12590 SFTs of the data. We present here results only for data up to May 16 23:00 GMT 2007 (7007 SFTs).
The cleaning procedure, consisting of a set of C-code routines in the PSS package [7], has been integrated into some of the SFT-production software used by the LIGO Scientific Collaboration (LSC) [8].

We illustrate in figure 2 ($a_1$) and ($b_1$) the improvements that may be obtained when the data is particularly quiet and disturbed, respectively. We show the PSD of 1800 s of H1 data with and without cleaning. We can appreciate a decrease of the noise floor of a factor $\sim 5$ in the band around 200 Hz in the cleaned data with respect to the original data. As shown in figure 3, the increase in the level of the noise floor is due to a single very loud disturbance (pointed by the arrow in the top panel of figure 3) with an energy that is about 30 times greater than the mean of the other events. The gain in sensitivity (of order 2) more than counter-balances the effective loss in observation time for having effectively decreased the observation time, $\sim$ 1.1 s out of 1800 s.

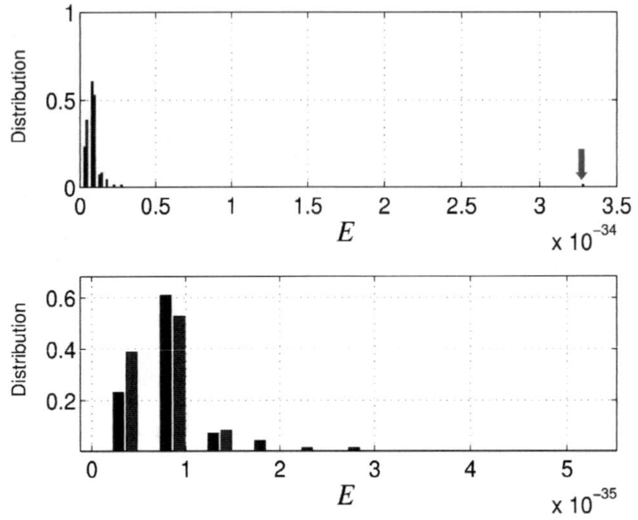


**Figure 3.** Distribution of the event energies $E = \sum_i |y(i)|^2$ of 36 (red bars) and 69 (blue bars) events found for the data whose PSDs are plotted in red in figure 2 ($a_1$) and ($b_1$), respectively. The bottom plot is a zoom of the top one. The two populations of disturbances have very similar energy distributions. What produces the increase in the noise floor of the PSD of figure 2 ($b_1$) is the very loud event at $E \sim 3.3 \times 10^{-34}$. We point out that the energy is preserved, i.e. the Parseval theorem is satisfied.

Figure 4 shows the cumulative results on 7007 SFTs. In particular, we pick four 1 Hz bands centered at 500, 857,1370 and 1630 Hz and compute the mean value $\mu_{\mathrm{R}}$ of the ratio between the PSD of 7007 uncleaned and cleaned data series. These bands do not contain spectral disturbances and have been chosen to be representative of the noise floor at nearby frequencies. The histograms of $\mu_R$ are shown in figure 4. The average value of $\mu_R$ is 1.0055, 1.0026, 1.0011 and 1.0009 for the different four bands, respectively.

The expected improvement in sensitivity for a CW search may be expressed by the lowest detectable strain amplitude $h_0$. For a coherent search this is proportional to $\sqrt{PSD/T}$, where $T$ is the coherent observation time. Hence we denote the gain in sensitivity from the cleaning procedure, $G_{h_0}$, with

$$G_{h_0} = \sqrt{\mu_R \, \frac{T - T_e}{T}}, \tag{2}$$

where $T_e$ is the sum of the duration of all the events subtracted during the time $T$. Figure 5 shows the distribution of the duration of events of 100 SFTs uniformly selected from the original data set. The mean value of $T_e$ is 0.55 s. If we use such a mean

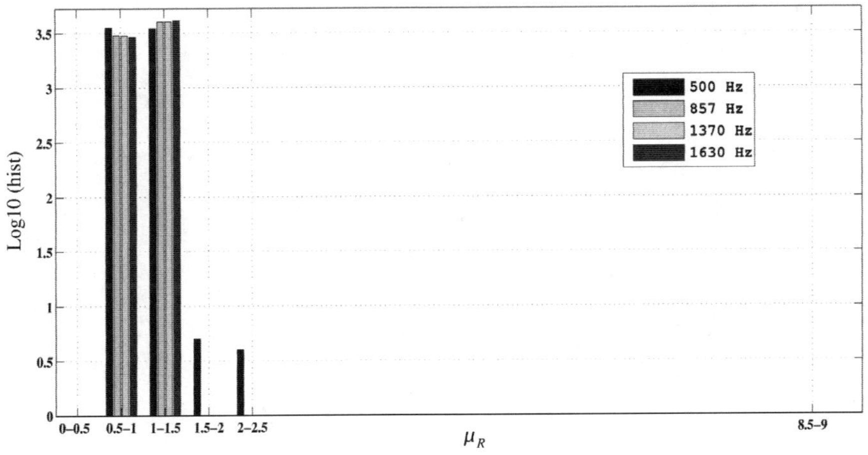


**Figure 4.** Log10 of the histogram of the mean value $\mu_R$ of the ratio between the PSD of the uncleaned and cleaned data series, computed in four 1 Hz frequency bands. For clarity the x-axis labels indicate the 0.5 wide bin intervals.

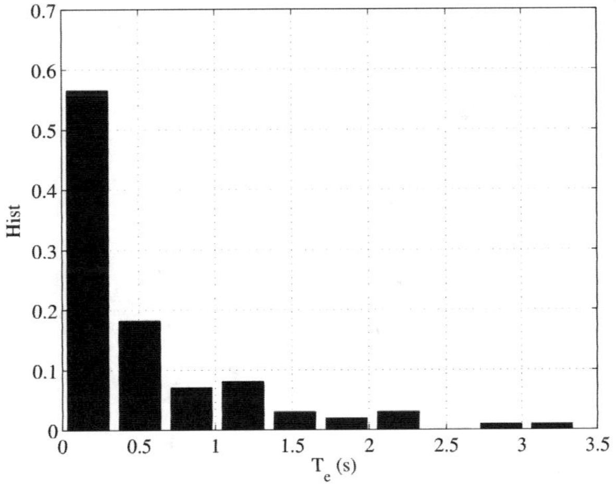


**Figure 5.** Histogram of the duration of events ($T_e$) identified by the procedure.

value in equation (2) and the 1-Hz-mean values of $\mu_R$ shown in figure 4, we obtain the distributions shown in figure 6. The effect of cleaning is to decrease the PSD but also, effectively, to decrease the observation time because also the signal is removed when one subtracts the disturbed events. Figure 6 shows that on typical LIGO science-quality S5 data the gain in sensitivity is marginal: it happens in a very small fraction of the total SFTs and not uniformly at all frequencies. The mean values of $G_{h_0}$ for the bands of figure 6 are 1.0016, 1.0010, 1.0004 and 1.0003, respectively. On data from the first Virgo science run (VSR1) improvements of the order of 1.07/1.3 at high/low frequencies were observed analyzing roughly 10 days of data due to the fact that non science-mode were used, unlike what was done with the LIGO data. A further cleaning procedure is under

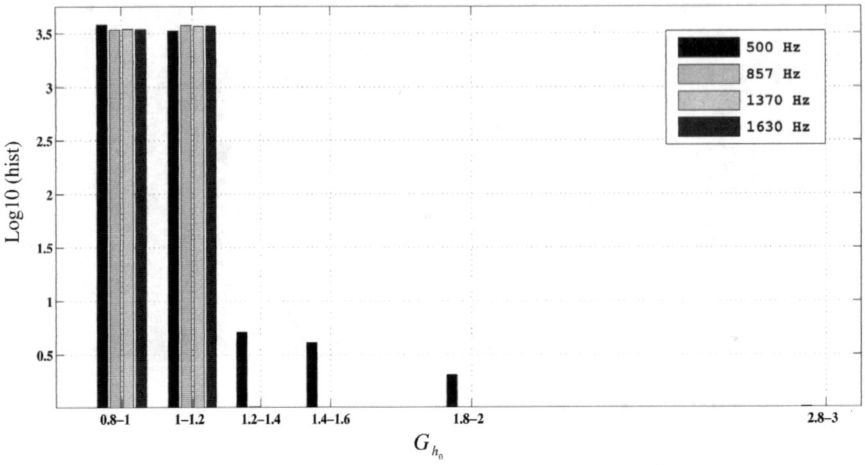


**Figure 6.** Log10 of the histogram of $G_{h_0}$ computed assuming $T_e = 0.55\ s$ and using the average $\mu_R$ values on the four 1 Hz bands centered respectively at 500, 857, 1370 and 1630 Hz.

development [9] aiming at removing disturbances in small frequency sub-bands, a few Hz each.

## Acknowledgments

The authors would like to thank B. Machenschalk for his ongoing work aimed at integrating the original PSS library cleaning routines into the LALAPPS SFT production routines. We also acknowledge the LSC-Virgo continuous waves search working group for useful discussions and input.

## References

[1] Abbott B P *et al* 2009 *Phys. Rev.* D **79** 022001
[2] Abbott B P *et al* 2009 *Phys. Rev.* D **80** 042003
[3] Abbott B P *et al* 2009 *Phys. Rev. Lett.* D **102** 111102
[4] Owen B J 2009 *arXiv:0903.2603*
[5] Owen B J 2009 *Class. Quantum Grav.* **26** 204014
[6] Astone P *et al* 2005 *Class. Quantum Grav.* **22** S1197
[7] http://wwwcascina.virgo.infn.it/cgi-bin/cvsweb/cvsweb.cgi/; http://grwavsf.roma1.infn.it/PSS/
[8] https://www.lsc-group.phys.uwm.edu/daswg/projects/lalapps.html
[9] Astone P *et al* (in preparation)

Journal of Physics: Conference Series **228** (2010) 012007 doi:10.1088/1742-6596/228/1/012007

# Searching for monochromatic signals in the ALLEGRO gravitational wave detector data

**C. Frajuca[1], R. Pires[1], F. S. Bortoli[1], R. F. Melo[1], N. S. Magalhaes[2], R. M. Marinho, Jr.[3], F. G. Oliveira[3], O. D. Aguiar[4] and W. W. Johnson[5]**

[1] Sao Paulo Federal Institute of Education, Science and Technology,
Rua Pedro Vicente 625, Sao Paulo, SP 01109-010, Brazil

[2] Universidade Federal de Sao Paulo at Diadema
Rua Prof. Artur Riedel, 275, Diadema, SP 09972-270, Brazil

[3] Technological Institute of Aeronautics,
Praça Mal. Eduardo Gomes 50, São Jose dos Campos, SP 12228-900, Brazil

[4] Astrophysics Division, National Institute for Space Research,
Av. dos Astronautas 1758, Sao Jose dos Campos, SP 12227-010, Brazil,

[5] Physics & Astronomy., Louisiana State University
271B Nicholson Hall, Baton Rouge, LA, 70803-4001 USA

E-mail: frajuca@cefetsp.br

Abstract. The detection of continuous gravitational wave (GW) signals is investigated using on a method developed for the detection of monochromatic signals in the middle of strong noise which makes power spectrum estimates using averaged modified periodograms. With this method it is possible to obtain a power spectrum for the data which preserves peaks due to monochromatic signals. In GW detection two kinds of Doppler shifts are expected, with annual and daily periods. Using the method an analysis is performed on persistent peaks, searching for frequency drifts that might present a pattern similar to the one due to Doppler shifts. For the chosen data set of the ALLEGRO gravitational wave detector it was found that none of the strongest, persistent peaks investigated presented a frequency drift with a Doppler shift pattern, thus excluding them as GW signal candidates.

## 1. Introduction

At the moment a large amount of data from gravitational wave (GW) detectors is available for analysis, including those from the ALLEGRO detector [1].

This detector was originally designed for the detection of burst signals [2] but the large amount of data taken continuously allows the search for monochromatic signals as well, like the

© 2010 IOP Publishing Ltd

Journal of Physics: Conference Series **228** (2010) 012007 doi:10.1088/1742-6596/228/1/012007

continuous emission of a non-axisymmetric neutron star. In one approach [3] the Doppler shift due to the Earth's motion was compensated and the intensity of the signals was the main parameter used for the identification of candidate signals. Using the same approach, a search for continuous signals coming from any direction in the sky was also performed [4].

In the present work a new approach will be tried. It is based on a method developed for the detection of monochromatic signals in the middle of strong noise by P. D. Welch [5, 6].

The results will be presented in Section 4. But first a theoretical introduction to the GW Doppler shift will be given in Section 2. The ALLEGRO detector data acquisition system will be briefly described in Section 3. The final Section is devoted to conclusions.

## 2. The Doppler shift of monochromatic gravitational wave signals

In GW detection one can assume that there will be a relative motion between an astrophysical source and a detector on Earth. This assumption is important when one analyzes long time series looking for periodic signals. A relative motion between source and detector changes the signal frequency by the Doppler effect. For galactic sources there are two major Doppler shifts: one due to the Earth's rotation and the other due to the Earth's translation.

Here one is interested on searching for signals that change in frequency in the ALLEGRO antenna time series. Those signals would be, in principle, gravitational wave signal candidates. From Doppler shift calculations one obtains the minimum frequency shift for an entire year.

## 3. The ALLEGRO gravitational wave detector and its data

The ALLEGRO bar detector is located at the Louisiana State University (latitude: 30.412577o N; longitude: 91.178627o W) in Baton Rouge (USA). The bar axis is directed along a line 40º24′ west of North. When in operation the system was cooled down to 4.2 K and had two modes in which the sensitivity reached maxima: ~ 895 Hz (minus mode) and ~ 919 Hz (plus mode).

ALLEGRO's data was sampled at a rate of 125 Hz. Twenty seconds worth of data (2500 samples) were assembled by software into a data block (a raw data record) and written to disk.

## 4. Applying averaged modified periodograms to ALLEGRO data

Several days of ALLEGRO's data from the year 1997 were analyzed in the search for frequency drifts.

One day with particularly low noise is day 8, Figure 1 shows the temperatures of the two modes of this day. Figure 2 presents ALLEGRO's power spectral density for day 8, 1997, obtained applying the Welch method. Some peaks have well known origin, like the minus mode of the antenna, near 895 Hz, the plus mode, near 919 Hz, and the peak near 900 Hz, which comes from a harmonic of the electric 60 Hz network. There are other peaks of unknown origin. One of them was called "Mystery Mode" by the ALLEGRO group, with frequency near 887.5 Hz, identified as X mode in that figure, the peak identified as Y has a frequency near 907.5 Hz and the peak identified as Z has a frequency of 957.6 Hz.

In order to apply Welch's method looking for an annual Doppler shift, 500 records were selected each day, among the total 4320, what gives a integration time of 2.77 hours. Their data concerning the x and the y outputs of the lock-in were stored in a MATLAB [7] file, totalizing 1,250,000 (x, y) pairs.

A MATLAB software was then developed to manipulate this data conveniently. The routine groups data as powers of 2. A 32768 (= $2^{15}$) sample group needed $\Delta t \sim 262$ s to be collected, thus yielding a frequency resolution $\Delta f \sim 3.7$ mHz. This value is much smaller than the annual GW Doppler shift. Therefore, this frequency resolution is applicable to a data analysis searching for an annual period Doppler shift.

The *pwelch* routine yielded a function in the time domain, and it was transformed into the frequency domain using the MATLAB *fftshift* routine. Figure 1 shows the result obtained for day 8. The identified peaks were found to be persistent throughout the year. However, none of them showed a modulated pattern as a Doppler shifted signal would show. In contrast, the peaks at the antenna modes (plus and minus) were monitored as well and they did not change bins at all during the observed days.

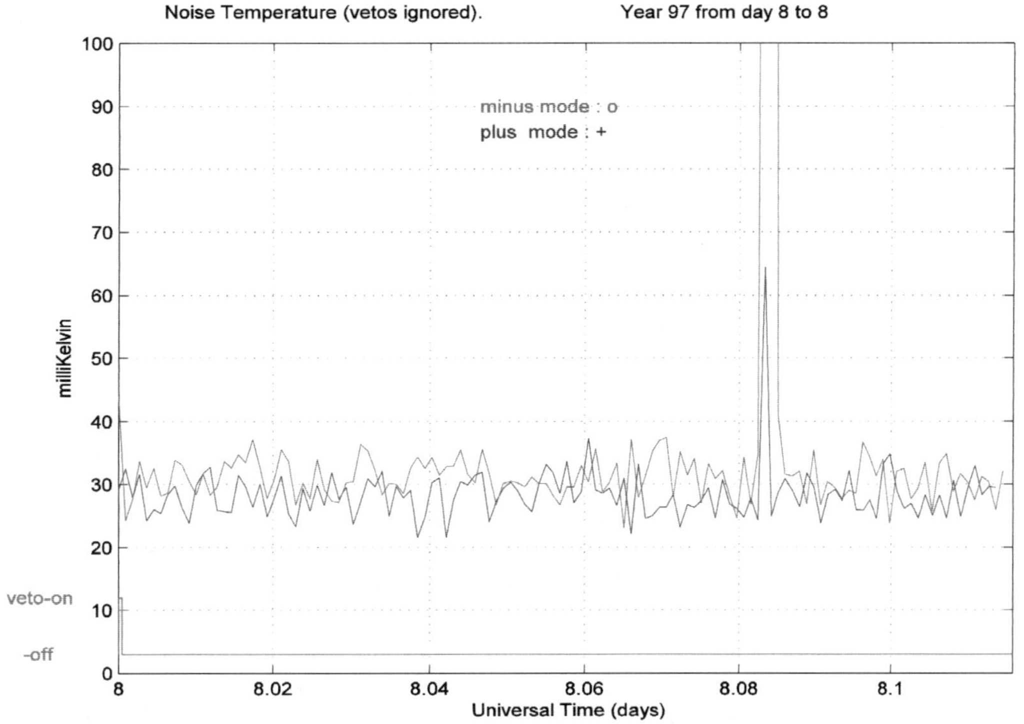


Figure. 1: Temperatures of modes plus and minus along day 8, 1997.

## 5. Conclusions

In this work a new method was applied to the search for monochromatic GW signals within data collected from the gravitational wave detector ALLEGRO. The method, due to Welch, is known in the literature but had not been applied to GW searches, as far as the authors know.

Three peaks in the data spectrum (X, Y and Z) were investigated in order to verify whether they had GW origin or not. This analysis indicates that none of the investigated peaks had GW origin, because they don´t present a behavior compatible with Doppler variations. The analysis is versatile enough to allow for future more detailed searches (in frequency resolution).

The natural continuation of this work is the investigation of frequency drifts in other peaks in the data spectrum, including those with small amplitudes.

In fact, in order to improve the investigation it is planned to analyze more days of the year 1997, as well as including days from the years 1999 and 2000, increasing the sensitivity by averaging the periodograms in a longer period of time taking in account the change in frequency due to relative motion of the source and its direction in the sky. Figure 3 shows the periods of time that can be used, where the days when the detector was very noisy have been removed. When increasing the integration new peaks could appear, and these will be a vey good candidate a gravitational wave signal.

**Acknowledgments**

The research by CF acknowledges FAPESP (grant # 09/01067-6), NSM acknowledges FAPESP (grant # 06/07316-0) and ODA acknowledges FAPESP (grant #06/56041-3) and from CNPq (grant #308759/06-0).

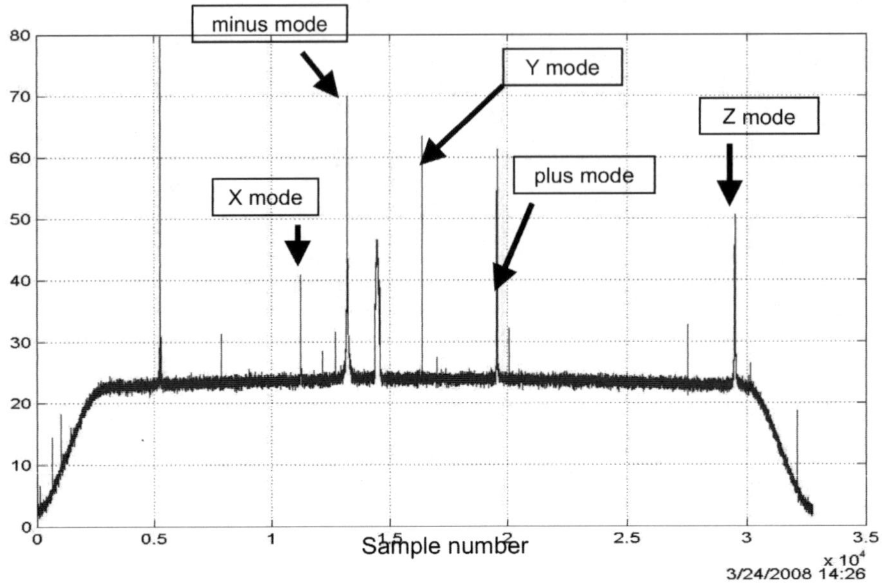


Figure 2: Averaged power spectrum for day 8.
The first sample starts at 844.810853 Hz, and the bin width is 3.824 mHz.

## References

[1] The ALLEGRO detector homepage is "sam.phys.lsu.edu".
[2] Mauceli, E. et al., *Phys. Rev.* D **54**, 1264 (1996).
[3] Mauceli, E., McHugh, M.P., Hamilton, W. O., Johnson, W. W. and Morse, ArXiv:gr-qc/0007023v1, July 11, 2000.
[4] E. S. Matos, O. D. Aguiar, R. M. Marinho Jr., W. O. Hamilton, W. W. Johnson and I. S. Heng, *Class. Quantum Grav.* **21**, S787 (2004).
[5] A. V. Oppenheim and R.W. Schafer. Discrete-Time Signal Processing. Prentice Hall International, 1989.
[6] P. D. Welch. The Use of the Fast Fourier Transform for the Estimation of Power Spectra. IEEE Transactions on Audio Electroacoustics AU-15, 70-73 (June 1970).
[7] MATLAB. Version 7.0.4.365 (R14). The MathWorks, Inc., 2005. See also at "www.mathworks.com".

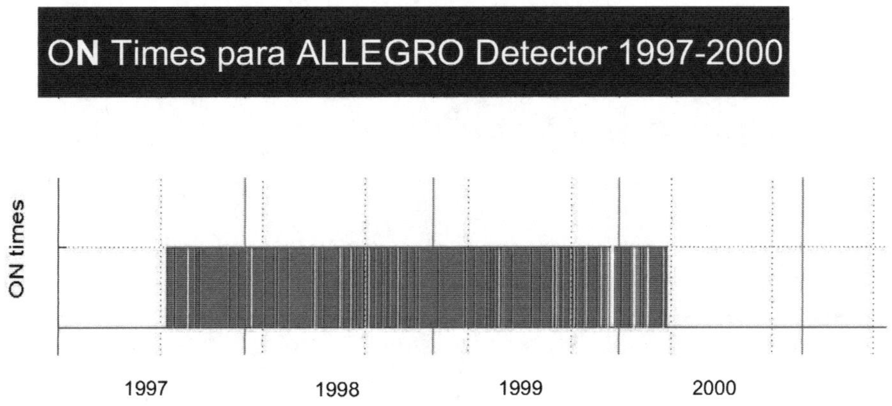


Figure 3. Periods of times when the ALLEGRO's data can be used.

# Random template placement and prior information

**Christian Röver**
Max-Plack-Institut für Gravitationsphysik (Albert-Einstein-Institut), Callinstraße 38, 30167 Hannover, Germany.

**Abstract.** In signal detection problems, one is usually faced with the task of searching a parameter space for peaks in the likelihood function which indicate the presence of a signal. Random searches have proven to be very efficient as well as easy to implement, compared e.g. to searches along regular grids in parameter space. Knowledge of the parameterised shape of the signal searched for adds structure to the parameter space, i.e., there are usually regions requiring to be densely searched while in other regions a coarser search is sufficient. On the other hand, prior information identifies the regions in which a search will actually be promising or may likely be in vain. Defining specific figures of merit allows one to combine both template metric and prior distribution and devise optimal sampling schemes over the parameter space. We show an example related to the gravitational wave signal from a binary inspiral event. Here the template metric and prior information are particularly contradictory, since signals from low-mass systems tolerate the least mismatch in parameter space while high-mass systems are far more likely, as they imply a greater signal-to-noise ratio (SNR) and hence are detectable to greater distances. The derived sampling strategy is implemented in a Markov chain Monte Carlo (MCMC) algorithm where it improves convergence.

## 1. Introduction

Signal detection, in gravitational wave detection in particular, frequently entails the problem of performing a computationally expensive numerical search over a large parameter space. The *search* here means a search for a peak in the likelihood function, or another detection statistic, based on the data at hand and varying the unknown signal parameters. A peak or a threshold excess then indicates the presence of a signal [1, 2]. Such "brute-force" searches may be implemented as *grid searches*, evaluating the detection statistic at regularly placed points in parameter space. Computing the detection statistic usually means evaluating the *match* between a *signal template* and the data; the spacing between evaluated points in parameter space is then usually based on a *template metric* which ensures that all possible signals (corresponding to points in parameter space) have at least a certain minimal match with one of the evaluated templates (corresponding to the grid points). Instead of using regularly spaced template banks (e.g. [3, 4, 5, 6]), the use of *random template banks* has recently gained popularity, as these are often very easily implemented, and have also be shown to be very efficient, especially in higher dimensions [7, 8, 9, 10]. Here the idea is to populate the parameter space randomly, but uniformly with respect to the template metric.

These template placement strategies have by now usually been based on "*minimax*" reasoning, by aiming at minimizing the maximal (worst-case) mismatch across the whole parameter space. Once one takes prior information on the unknown parameters into consideration, by accounting for a priori probabilities attached to different regions of parameter space, a decision-theoretic

© 2010 IOP Publishing Ltd

approach allows us to devise other strategies, effectively concentrating efforts on the more promising regions of parameter space in pursuit of a certain optimality criterion [11, 12]. In fact, a minimax strategy may often only exist once one imposes hard bounds on the parameter space (and by that ensuring the existence of an absolute *worst case*).

Markov chain Monte Carlo (MCMC) methods are meanwhile widely used for (Bayesian) parameter estimation in the signal processing stage for gravitational-wave signals [13, 14, 15]. MCMC algorithms are, first of all, methods for *stochastic integration* [16, 17], although by the way they work they often behave similarly to *stochastic search* algorithms as well. This is in fact a most welcome property, as part of the parameter estimation problem is usually also a search/optimization problem, as, besides integration over the parameters' posterior distribution, it requires finding the global mode or secondary modes. *Parallel tempering* [18, 19] is a variety of the Metropolis-Hastings MCMC algorithm (and a special case of Metropolis-coupled MCMC algorithm [20, 17]) aimed at enhancing these stochastic search capabilities. This is done by basically running several MCMC chains in parallel, where *tempering* at increasing temperature values is applied to subsequent chains (as in simulated annealing methods [21, 22]), and additional steps are introduced to allow for communication between chains [23]. Parallel tempering methods have been applied to gravitational-wave data analysis for binary inspiral signals in the context of ground-based [24] and space-based (LISA) measurements [23], where they have proven advantageous especially in cases of high SNR and of posterior distributions exhibiting multiple modes or degeneracies [25, 26]; they have also been adopted for the analysis of burst signals [27] and utilized for Bayesian evidence computation [28].

Among the parallel Markov chains being run at different 'temperatures' within the parallel tempering implementation, the 'cool' ones with no tempering applied produce samples from the posterior distribution for the stochastic integration part, while the high-temperature chains are producing samples for the stochastic search. The question now is how to set up the algorithm so that the search is most efficient, given our knowledge of prior and template metric, i.e., our knowledge of "where the true parameters are (un-) likely to be", and "how hard one needs to look" across the parameter space. The problem is of special interest in the context of binary inspiral signals, as prior and template metric are particularly contradictory: a priori one is most likely do detect an inspiral involving high masses, as these result in a high-SNR signal that is detectable to a greater distance. On the other hand, considering the template metric only, one might want to mostly try low-mass templates, since at low masses the template's and true signal parameters need to be in very close agreement in order for them to match, while at high masses greater discrepancies still yield a good match. What needs to be defined is the distribution to sample from in order to find the mode(s) fastest, which is very similar to setting up a random template bank, the difference being that one does not settle on some fixed number of templates, as the MCMC sampler in principle is thought to sample indefinitely.

In the following Sec. 2, we will introduce the problem for the case of binary inspiral signals, and Sec. 3 briefly introduces the parallel tempering context. In Sec. 4 the general problem is formulated in decision-theoretic terms and solved for a particular optimality criterion. Sec. 5 shows some illustrative examples, and Sec. 6 eventually closes with conclusions and perspectives.

## 2. Binary inspiral parameters

In the simplest description, a binary inspiral signal as measured by ground-based interferometers is determined by 9 parameters: sky location (declination $\delta$, right ascension $\alpha$), polaristion ($\psi$), companion masses ($m_1$, $m_2$), luminosity distance ($d_L$), time of arrival ($t_c$), phase ($\phi$) and inclination angle ($\iota$). Assuming some prior distribution for the masses (in the following simply defined to be uniform, $m_1, m_2 \in [1\,\mathrm{M}_\odot, 10\,\mathrm{M}_\odot]$), and an isotropic distribution of events across space while folding in the detectability as a function of signal-to-noise ratio (SNR), one can derive a joint prior distribution whose marginal distribution of masses is shown in Fig. 1

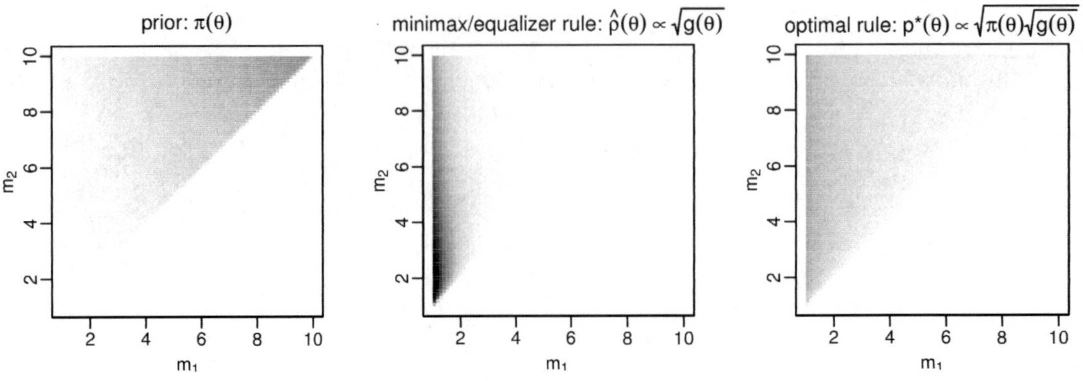


**Figure 1.** (Marginal) densities of the distributions $\pi$, $\hat{\rho}$ and $p^\star$ for the two mass parameters $(m_1, m_2)$ of a binary inspiral signal. The prior (left plot) indicates that high masses are most likely, which is because they result in stronger signals that are detectable to greater distances. The template metric on the other hand implies that low masses require a dense template spacing (middle plot).

[29, 23]. A template metric may be defined following [30, 31], assuming the metric to be constant in the space of the *Newtonian* and *1.5 PN chirp times* $\lambda_1$ and $\lambda_2$, which are functions of the mass parameters. For the remaining parameters, for now, we again assume the metric to be uniform $(t_c, \log(d_L))$ and isotropic $(\delta, \alpha, \psi, \iota, \phi)$. The implied distribution in terms of $(m_1, m_2)$ following from a uniform spacing in $(\lambda_1, \lambda_2)$ may be derived using the reparametrisation explicated in [32]. This distribution is shown in Fig. 1.

## 3. Parallel tempering

In the context of Monte Carlo integration, tempering is utilised to prevent the integration algorithm from getting stuck in local modes of the distribution from which it is sampling. A temperature parameter $T \geq 1$ is introduced, and instead of sampling from the distribution of actual interest, with density function $f(\theta)$, the modified distribution

$$f_{(T)}(\theta) \;\propto\; f(\theta)^{\frac{1}{T}} \tag{1}$$

is used. The introduced exponent is supposed to make the distribution more tractable, as it has a "flattening" effect on the density; the same effect is also taken advantage of in simulated annealing methods [22]. In the limit of $T \to \infty$, the density $f_{(T)}(\theta)$ then approaches a uniform distribution [17]. In the context of posterior inference, when the target distribution $f(\theta)$ is the product of prior $\pi(\theta)$ and likelihood $\mathcal{L}(\theta)$, it may be more sensible to use a scheme only tempering the likelihood part:

$$f_{(T)}(\theta) \;\propto\; \pi(\theta)\,\mathcal{L}(\theta)^{\frac{1}{T}}, \tag{2}$$

in which case $f_{(T)}(\theta)$ goes towards the prior $\pi(\theta)$ for $T \to \infty$ [23]. Both uniform distribution and prior distribution may in general not be the most sensible choice, as was pointed out above, since the tempering is also supposed to enhance the algorithm's *stochastic search* properties. Assume that one had a distribution $p^\star(\theta)$ available, which leads to an optimal sampling (w.r.t. to some pre-specified criterion), and which is then the desired density for $T \to \infty$. This suggests a generalized tempering parametrisation:

$$f_{(T)}(\theta) \;\propto\; p^\star(\theta)\left(\frac{f(\theta)}{p^\star(\theta)}\right)^{\frac{1}{T}} \;=\; p^\star(\theta)^{1-\frac{1}{T}}\, f(\theta)^{\frac{1}{T}} \tag{3}$$

which in the special cases of $p^\star(\theta) \propto 1$ and $p^\star(\theta) = \pi(\theta)$ again yields the tempering schemes from (1) and (2) above. The question now is how to choose such a limiting distribution $p^\star(\theta)$ based on given prior information and template metric.

## 4. The decision theoretic approach

Let $g(\theta)$ be the determinant of the template metric as a function of the signal parameters. A large value of $g$ means that that templates need to be densely spaced around $\theta$, while a smaller $g$ indicates that a coarser spacing is sufficient. The volume "covered" by a template placed at parameter $\theta$ is proportional to $g(\theta)^{-\frac{1}{2}}$, and hence the probability density to sample from for setting up a random template bank is given by $\hat{\rho}(\theta) \propto \sqrt{g(\theta)}$ [8].

Now consider the case of the true parameter value being $\theta_0 \in \Theta$. The actual value $\theta_0$ is unknown, what is known is the prior probability density $\pi(\theta)$. Whenever a template $\theta^*$ is placed in parameter space, it is considered a *match* if it was sufficiently close to the true value $\theta_0$. What exactly is "sufficiently close" is determined via mismatch considerations and is expressed through the template metric. Then the probability of a match is

$$\mathrm{P}(\mathrm{match}\,|\,\theta^*) \;=\; c\,\frac{1}{\sqrt{g(\theta^*)}}\,\pi(\theta^*), \tag{4}$$

where $c \in \mathbb{R}^+$ is a constant depending on how close a match actually is required to be. If one was to pick a *single* template $\theta^*$, the chances for success would obviously be maximal where the above product reaches its maximum. Analogously, consider the case of a *given* true value $\theta_0$ and repeated, independent "guesses" drawn from $p^\star(\theta)$. Then for each single guess the probability of success is

$$\mathrm{P}(\mathrm{match}\,|\,\theta_0) \;=\; c\,\frac{1}{\sqrt{g(\theta_0)}}\,p^\star(\theta_0). \tag{5}$$

What is desired is a distribution $p^\star$ from which to generate independent draws so that the chances of getting a match are "optimal". Whether or when one will get a match is a matter of chance, depending on both the true value $\theta_0 \in \Theta$ and the choice of $p^\star \in \mathcal{P}^\star$, where $\mathcal{P}^\star$ is the space of probability distributions over $\Theta$. Suppose we are interested in minimizing the expected number of trials $T$ (or *waiting time*) until the first match. Any choice of $p^\star$ implies a probability distribution for $T$; for a *given* true value $\theta_0$ and a sampling distribution $p^\star$, $T$ follows a geometric distribution with density and expectation:

$$\mathrm{P}(T\!=\!t\,|\,\theta_0) \;=\; \Big(1\!-\!c\,\frac{1}{\sqrt{g(\theta_0)}}\,p^\star(\theta_0)\Big)^{t-1}\Big(c\,\frac{1}{\sqrt{g(\theta_0)}}\,p^\star(\theta_0)\Big), \quad \mathrm{E}[T\,|\,\theta_0] \;=\; \frac{1}{c\,\frac{1}{\sqrt{g(\theta_0)}}\,p^\star(\theta_0)}. \tag{6}$$

In decision theoretic terms, we are given a *state-of-nature space* $\Theta$, an *action space* $\mathcal{P}^\star$, and a *loss function* $L : \Theta \times \mathcal{P}^\star \rightarrow \mathbb{R}$ with $L(\theta_0, p^\star) = \mathrm{E}_{p^\star}[T|\theta_0]$ [11, 12]. An optimal choice of $p^\star$ may now be determined by minimizing the expected loss; integrating over the possible values that $\theta_0$ could take, that (prior) expectation is

$$\mathrm{E}[T] \;=\; \frac{1}{c}\int_\Theta \frac{1}{\frac{1}{\sqrt{g(\theta_0)}}\,p^\star(\theta_0)}\,\pi(\theta)\,\mathrm{d}\theta, \tag{7}$$

which is minimized by choosing

$$p^\star(\theta) \;\propto\; \sqrt{\pi(\theta)\,\sqrt{g(\theta)}} \;=\; \sqrt{\pi(\theta)\,\hat{\rho}(\theta)}, \tag{8}$$

i.e., the optimal $p^\star$ here is proportional to the geometric mean of $\pi$ and $\hat{\rho}$, and independent of $c$.

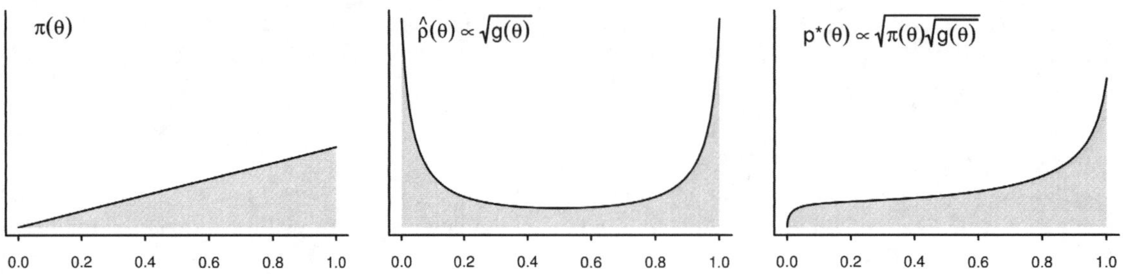


**Figure 2.** Densities of the distributions $\pi$, $\hat{\rho}$ and $p^\star$ for the toy example discussed in Sec. 5.2.

The distribution defined through the density $\hat{\rho}$ that is usually utilized for random template banks [8] plays a particular role in this context. From equation (5) one can see that by setting $p^\star := \hat{\rho}$ the probability of a match (and with that also the waiting time) becomes independent of the actual parameter value $\theta_0$, so that $\hat{\rho}$ constitutes an *equalizer rule*. From (8) it follows that $\hat{\rho}$ will be optimal in the case that the prior happens to be $\pi = \hat{\rho}$. This implies that $\pi = \hat{\rho}$ defines the "*least favourable prior distribution*" for this case, and that $p^\star = \hat{\rho}$ also constitutes the *minimax* strategy (independent from the particular prior $\pi$), as it minimizes the maximum of $\mathrm{E}[T \,|\, \theta_0]$ across all possible true values $\theta_0$ [11]. Since $p^\star = \hat{\rho}$ leads to a uniform match probability in (5), it actually constitutes the equalizer rule for the wider family of optimality criteria that are functions of $\mathrm{P}(\text{match} \,|\, \theta_0)$.

## 5. Examples

*5.1. Toy example 1: Gaussian prior*

Consider a parameter space $\Theta = \mathbb{R}$ where the prior is Gaussian with mean $\mu$ and variance $\sigma^2$: $\pi = \mathrm{N}(\mu, \sigma^2)$, and the template metric is *flat*, i.e., $g(\theta) = \gamma$ is independent of $\theta$. Then the equalizer rule $\hat{\rho}$ does not exist, and the optimal rule would be $p^\star = \mathrm{N}(\mu, 2\sigma^2)$.

*5.2. Toy example 2: Numerical simulation*

Consider a parameter space $\Theta = [0, 1]$, where the prior and template metric behave as shown in Fig. 2. For this simple case the behaviour of different sampling strategies can be simulated numerically, by drawing "true" parameter values $\theta_0$ from the prior distribution and then drawing "guesses" $\theta^*$ from either $\hat{\rho}$ or $p^\star$ in order to see how the strategies differ.

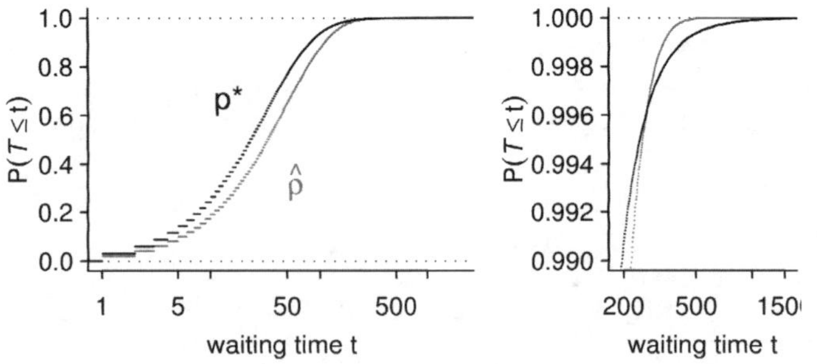


**Figure 3.** Cumulative distributions of the resulting waiting times $T$ when using sampling strategies $p^\star$ and $\hat{\rho}$ in the toy example of Sec. 5.2. The right panel shows a zoom-in on the differing tail behaviour.

Fig. 3 illustrates the distribution of the resulting times $T$, for both the minimax and optimal strategies $\hat{\rho}$ and $p^\star$. As expected, the average waiting time is lower for $p^\star$, and one can see that the minimax strategy performs better in the unlikely "worst cases".

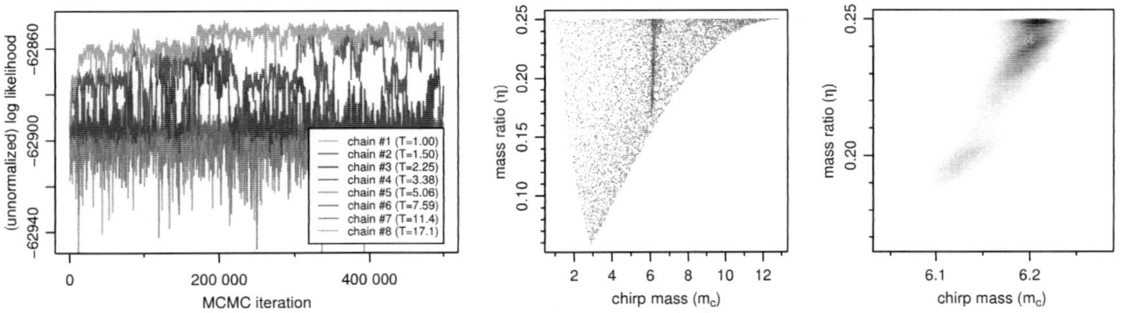


**Figure 4.** This plot illustrates the behaviour of a Parallel Tempering algorithm utilizing the distribution $p^\star$ when running on simulated data. The left panel shows how the algorithm's 'cool' chains manage to ascend to greater likelihood values while the tempered chains keep sampling at lower likelihood values. The 2nd panel is a scatter plot of mass parameter samples from all the different chains (after the algorithm's burn-in phase). The right panel eventually shows the resulting mass parameters' marginal posterior density derived from the 'cool' chain #1 alone; the cross indicates the true parameter value.

### *5.3. Binary inspiral example*

The prior $\pi$ and minimax sampling rule $\hat{\rho}$ for the mass parameters of a binary inspiral event were shown in Fig. 1. The right panel of the same figure also shows the resulting optimized sampling distribution $p^\star$. The obvious discrepancy between least favourable ($\hat{\rho}$) and actual prior ($\pi$) suggests that there actually is a gain in doing the optimization. Fig. 4 shows how a parallel tempering algorithm for parameter estimation behaves when utilizing the distribution $p^\star$ for high-temperature chains as described in Sec. 3 (3). The MCMC chains quickly converge to the true parameter values, while the higher-temperature chains keep scanning the parameter space efficiently.

## 6. Conclusions and outlook

We have applied a decision-theoretic approach in order to derive an optimized sampling distribution to be used within a parallel tempering MCMC implementation. The optimization step here provides a natural link between the parameter space metric and the prior information about the parameter values. The particular optimality criterion chosen here (the expected time until a matching template is found, $\mathrm{E}[T|\theta_0]$) turns out to be computationally convenient, as the resulting sampling distribution $p^\star$ is independent of the particular mismatch threshold $c$, and is almost trivial to implement within an MCMC application. Other criteria are conceivable though, like the probability of a missed detection within $N$ samples $\mathrm{P}(T > N|\theta_0)$ for example, which may then lead to more complicated results.

The general approach used here should also be useful in other contexts; it turns out that the distribution usually used for setting up random template banks here constitutes the special case of a *minimax* strategy, which implies that the explicit specification of particular figures-of-merit and the consideration of prior information may yield great efficiency improvements, especially in cases where the implicitly assumed *least favourable prior* greatly deviates from the actual prior information as in the binary inspiral case. In the framework discussed above, the resulting optimized sampling distribution $p^\star$ even exists for cases where the minimax rule does not (as in the example of Sec. 5.1 above). This suggests that a similar approach may also make other ad-hoc fixes like the mass parameter bounds in the binary inspiral example dispensable, as it would naturally focus in on the promising parameter range while ruling out too unlikely and too costly regions of parameter space.

## Acknowledgments
The author would like to thank Chris Messenger, Reinhard Prix and Graham Woan for helpful discussions. This work was supported by the Max-Planck-Society.

## References
[1] McDonough R N and Whalen A D 1995 *Detection of signals in noise* 2nd ed (New York: Academic Press)
[2] Wainstein L A and Zubakov V D 1962 *Extraction of signals from noise* (Englewood Cliffs, NJ: Prentice-Hall)
[3] Croce R P, Demma T, Pierro V and Pinto I M 2002 *Physical Review D* **65** 102003
[4] Babak S, Balasubramanian R, Churches D, Cokelaer T and Sathyaprakash B S 2006 *Classical and Quantum Gravity* **23** 5477–5504
[5] Prix R 2007 *Classical and Quantum Gravity* **24** S481–S490
[6] Cokelaer T 2007 *Physical Review D* **76** 102004
[7] Babak S 2008 *Classical and Quantum Gravity* **25** 195011
[8] Messenger C, Prix R and Papa M A 2009 *Physical Review D* **79** 104017
[9] Harry I, Allen B and Sathyaprakash B S 2009 *Physical Review D* **80** 104014
[10] Manca G M and Vallisneri M 2009 *Arxiv preprint 0909.0563 [gr-qc]*
[11] Berger J O 1985 *Statistical decision theory and Bayesian analysis* 2nd ed (Springer-Verlag)
[12] Ferguson T S 1967 *Mathematical statistics: A decision theoretic approach* (New York: Academic Press)
[13] Christensen N and Meyer R 1998 *Physical Review D* **58** 082001
[14] Umstätter R, Meyer R, Dupuis R, Veitch J, Woan G and Christensen N 2004 *Classical and Quantum Gravity* **21** S1655–S1665
[15] Arnaud K A *et al.* 2007 *Classical and Quantum Gravity* **24** S529–S539
[16] Metropolis N and Ulam S 1949 *Journal of the American Statistical Association* **44** 335–341
[17] Gilks W R, Richardson S and Spiegelhalter D J 1996 *Markov chain Monte Carlo in practice* (Boca Raton: Chapman & Hall / CRC)
[18] Hukushima K and Nemoto K 1996 *Journal of the Physical Society of Japan* **65** 1604–1608
[19] Hansmann U H E 1997 *Chemical Physics Letters* **281** 140–150
[20] Geyer C J 1991 *Computing Science and Statistics: Proceedings of the 23rd Symposium on the Interface* ed Keramidas E M (Fairfax Station: Interface Foundation) pp 156–163
[21] Kirkpatrick S, Gelatt C D and Vecchi M P 1983 *Science* **220** 671–680
[22] Press W H, Teukolsky S A, Vetterling W T and Flannery B P 1992 *Numerical recipes in C: The art of scientific computing* (Cambridge: Cambridge University Press)
[23] Röver C 2007 *Bayesian inference on astrophysical binary inspirals based on gravitational-wave measurements* Ph.D. thesis The University of Auckland URL `http://hdl.handle.net/2292/2356`
[24] Röver C, Meyer R and Christensen N 2007 *Physical Review D* **75** 062004
[25] van der Sluys M V, Röver C, Stroeer A, Christensen N, Kalogera V, Meyer R and Vecchio A 2008 *The Astrophysical Journal Letters* **688** L61–L64
[26] Raymond V, van der Sluys M V, Mandel I, Kalogera V, Röver C and Christensen N 2009 *Classical and Quantum Gravity* **26** 114007
[27] Key J S and Cornish N J 2009 *Physical Review D* **79** 043014
[28] Littenberg T and Cornish N J 2009 *Physical Review D* **80** 063007
[29] Röver C, Meyer R, Guidi G M, Viceré A and Christensen N 2007 *Classical and Quantum Gravity* **24** S607–S615
[30] Owen B J and Sathyaprakash B S 1999 *Physical Review D* **60** 022002
[31] Chronopoulos A E and Apostolatos T A 2001 *Physical Review D* **64** 042003
[32] Umstätter R and Tinto M 2008 *Physical Review D* **77** 082002

# Confusion background from compact binaries

**T. Regimbau**
UMR ARTEMIS, CNRS, University of Nice Sophia-Antipolis,
Observatoire de la Côte d'Azur, BP 4229, 06304, Nice Cedex 4

E-mail: regimbau@oca.eu

**Scott A. Hughes**
Department of Physics and MIT Kavli Institute,
77 Massachusetts Avenue, Cambridge, MA 02139

**Abstract.** Double neutron stars are one of the most promizing sources for terrestrial gravitational wave interferometers. For actual interferometers and their planned upgrades, the probability of having a signal present in the data is small, but as the sensitivity improves, the detection rate increases and the waveforms may start to overlap, creating a confusion background, ultimately limiting the capabilities of future detectors. The third generation Einstein Telescope, with an horizon of $z > 1$ and very low frequency "seismic wall" may be affected by such confusion noise. At a minimum, careful data analysis will be require to separate signals which will appear confused. This result should be borne in mind when designing highly advanced future instruments.

## 1. Introduction

The coalescence of two neutron stars (BNS), two black holes (BBH) or a neutron star and a black hole (NS-BH), are among the most promising sources for ground-based gravitational wave detectors due to the huge amount of energy emitted in the last phase of their inspiral trajectory. The waveform is very well modeled until the last stable orbit and a detection would provide strong constraints on the source parameters. With the third generation interferometer Einstein Telescope, the horizon of compact binaries is expected to reach cosmological distances where it may become possible to study cosmology. In particular, double neutron stars or neutron star-black holes, if associated with an electromagnetic counterpart, may be used as standard sirens to constrain dark energy [1]. At such distances however, it is likely that the sources create a confusion foreground, where the detection of a single coalescence becomes difficult and requires advanced data analysis methods [2]. In the first section, we derive the coalescence rate of BNS; in the second section we discuss the different detection regimes and the consequences for Einstein Telescope; in the third section we present our conclusions and the work in progress.

## 2. Coalescence Rate

The final merger of a compact binary occurs after two massive stars in a binary system have collapsed to form neutron stars or black holes and have inspiralled through the emission of

© 2010 IOP Publishing Ltd

gravitational waves. The cosmic coalescence rate is given by:

$$\dot{\rho}_c^o(z) \propto \int \frac{\dot{\rho}_*(z_f)}{1+z_f} P(t_d) dt_d \text{ with } \dot{\rho}_c^o(0) = \dot{\rho}_0 \quad (1)$$

In this expression, $\dot{\rho}_c$ is the star formation rate in $M_\odot Mpc^{-3}yr^{-1}$, $z_f$ the redshift at the time of formation of the binary system and $P(t_d)$ the probability distribution of the delay between the formation and the coalescence.The coalescence rate per interval of redshift $dR^o/dz(z)$ is obtained by multiplying by the element of comoving volume $dV/dz(z)$. Here, we assume a distribution of the form $P(t_d) \propto 1/t_d$ with $t_d > t_{\min}$ and take $t_{\min} = 20$ Myr as representative of BNSs. We use the star formation rates of [3, 4, 5, 6], and consider local rates in the range $0.01 - 10$ $Mpc^{-3}Myr^{-1}$, with reference models of 1 and 0.4, corresponding to the most current estimates derived from statistical studies [7] and population synthesis [9, 8].

## 3. Detection Regime

The contribution of BNS to the instrumental data falls into three statistically very different regimes, depending on the duty cycle (or the average number of sources present at the detector at the same time):

$$\Lambda(z) = \int_0^z \tau^o(z) \frac{dR^o}{dz}(z) dz \quad (2)$$

where $\frac{dR^o}{dz}(z) = \dot{\rho}^o(z)\frac{dV}{dz}$ is the coalescence rate per interval of redshift, and $\tau^o(z)$ is the typical duration of the inspiral in the detector frequency band, which depends strongly on the low frequency limit of the instrument and can last from a few minutes for advanced detectors with $f_L = 10$ Hz to a few days for the Einstein Telescope with planned low frequency bound between $1 - 5$ Hz (see Table 1 )

(i) *Shot noise ($\Lambda << 1$)*: This case describes when the number of sources is small enough that the interval between events is long compared to an individual event's duration. Measured waves are separated by long stretches of silence and can be resolved individually. This case pertains to instruments that are only sensitive to events at low redshift.

(ii) *Popcorn noise ($\Lambda \sim 1$)*: As the reach of instruments increases, the time interval between events may come closer to the duration of single bursts. Events may sometimes overlap, making it difficult to distinguish between them.

(iii) *Gaussian ($\Lambda >> 1$)*: For instruments with very large reach and excellent low frequency sensitivity, the interval between events can be small compared to the duration of an event. The signals overlap to create a confusion noise of unresolved sources.

The limit of the popcorn and the Gaussian regimes, defined by $\Lambda(z_*) = 1$ and $\Lambda(z_{**}) = 10$ are given in Table 1 [2] for different values of the low frequency bound and local coalescence rate. With actual and advanced interferometers, whose horizon is tens or hundreds of Mpc, we are able to probe only the low duty cycle regime, where sources don't overlap. With the third generation Einstein Telescope, on the other hand, the horizon is expected to extend to redshift larger than $z > 1$, and the signal may fall deep inside the confusion regime, especially between $1 - 5$ Hz, where it can last for a few days (see Table 1).

We find that if the low frequency bound is at $f_L = 1$ Hz, the transition to a Gaussian background at $z_{**}$ is well within the ET horizon; if $f_L = 5$ Hz, it most likely occurs at $z_{**} \sim 0.6 - 1.2$, but can fall beyond the detection horizon if $\dot{\rho}_c^o < 0.15$ $Myr^{-1}$ $Mpc^{-3}$. The transition to a popcorn background at $z_*$, on the other hand, always occur before the detection horizon, and more likely around $z_* \sim 0.25 - 0.4$, expect for the most pessimistic coalescence rates ($\dot{\rho}_c^o < 0.015$ $Myr^{-1}$ $Mpc^{-3}$).

**Table 1.** Threshold between resolved and unresolved BNSs for different estimates of the source rate $\dot{\rho}_c^o$ and detector lower frequency bound $f_L$. No value means that the number of sources at the detector is always $< 1$ or $< 10$. $\tau_0$ refers to the typical duration of a source located at $z = 0$.

| $f_L$ | $\tau_0$ | $\dot{\rho}_c^o$ | $z_*$ | $z_{**}$ |
|---|---|---|---|---|
| 10 | 16.7 m | 0.01 | - | - |
| | | 0.4 | 0.8-0.9 | - |
| | | 1 | 0.5-0.6 | $> 2$ |
| | | 10 | 0.2 | 0.5-0.6 |
| 5 | 1.8 d | 0.01 | - | - |
| | | 0.4 | 0.4 | 1-1.2 |
| | | 1 | 0.25 | 0.6-0.7 |
| | | 10 | 0.1 | 0.25 |
| 1 | 5.4 d | 0.01 | 0.3 | 0.8 |
| | | 0.4 | 0.08 | 0.2 |
| | | 1 | 0.06 | 0.13 |
| | | 10 | 0.03 | 0.06 |

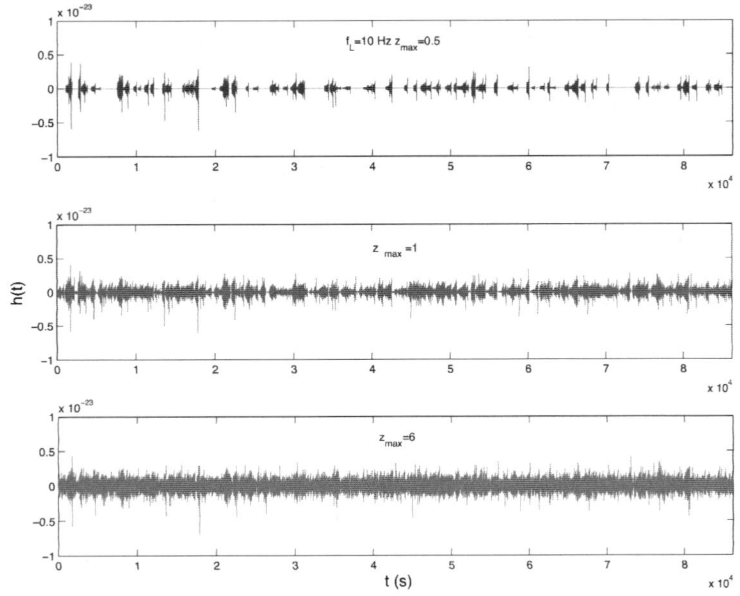


**Figure 1.** Evolution with the detector horizon (top: $z_{\max}$ =0.5 ; middle : $z_{\max}$=1 ; bottom : $z_{\max}$=6 ) of a GW time series from a simulated population of BNSs. Here we have assumed for the detector low frequency bound $f_L = 10$ Hz.

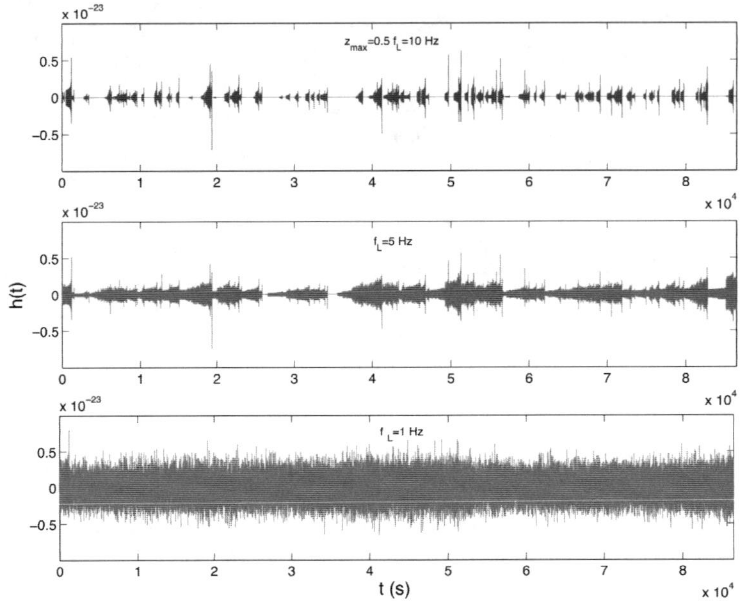


**Figure 2.** Evolution with the detector low frequency bound (top: $f_L$ =10 Hz ; middle : $f_L$=5 Hz ; bottom : $f_L$=1 Hz ), of a GW time series of the from a simulated population of BNSs. Here we have assumed for the detector horizon $z_{\rm max} = 0.5$.

## 4. Conclusions and future work

With the first and second generations of gravitational-wave detectors the chance of detection is limited by the instrumental noise, but as the sensitivity and the number of sources increase, the waveforms may start to overlap, creating a confusion background, ultimately limiting the capabilities of these instruments. Experience from the Mock LISA Data Challenges [10] and ideas developed for the Big Bang Observatory [11] prove that disentangling multiple signals in a gravitational-wave detector's data stream is certainly possible. It has been shown that Markov Chain Monte Carlo based codes, such as the Block Annealed Metropolis-Hastings algorithm, are able to extract about 20,000 out of 30 million individual white dwarf binaries from the galactic foreground, approaching the theoretical limit of 25,000 resolvable sources. In a recent paper, [12] suggested that similar algorithms could be used for ET. We have developed a Monte Carlo simulation code to generate the signal from populations of extra-galactic binaries in the output of detectors, which can be used to study the statistical properties of the background and investigate advanced data analysis strategies. The evolution of the signal with the detector horizon and with the low frequency bound is illustrated by the simulated time series in Fig. 1 and Fig. 2. This work is in progress and will be reported in a future paper.

[1] D. E. Holz and S. A. Hughes, Astrophys. J. **629**, 15 (2005).
[2] T. Regimbau and S. A. Hughes, Phys. Rev. D **79**, 062002 (2009).
[3] A. M. Hopkins and J. Beacom, Astrophys. J. **651**, 142 (2006).
[4] M. Fardal, N. Katz, D. Weinberg, and R. Davé, Mon. Not. R. Astron. Soc. **379**, 985 (2007).
[5] S. M. Wilkins, N. Trentham, and A. Hopkins, Mon. Not. R. Astron. Soc. **385**, 687 (2008).
[6] K. Nagamine, J. Ostriker, M. Fukugita, and R. Cen, Astrophys. J. **653**, 881 (2006).
[7] V. Kalogera et al., Astrophys. J. **614**, 137 (2004).
[8] R. O'Shaughnessy, K. Belczynski, and V. Kalogera, Astrophys. J. **675**, 566 (2008).

[9] J. A. de Freitas Pacheco, T. Regimbau, A. Spallici, and S. Vincent, Int. J. Mod. Phys. D **15**, 235 (2006)
[10] S. Babak et al., Class. Quantum Grav. **25**, 184026 (2008).
[11] C. Cutler and J. Harms, Phys. Rev. D **73**, 042001 (2006).
[12] L. Bosi and E. K. Porter, arXiv:0910:0380.

# Nested Sampling as a tool for LISA data analysis

**Jonathan R Gair[1], Farhan Feroz[2], Stanislav Babak[3], Philip Graff[2], Michael P Hobson[2], Antoine Petiteau[3] and Edward K Porter[4]**

[1] Institute of Astronomy, Madingley Road, CB3 0HA, Cambridge, UK
[2] Astrophysics Group, Cavendish Laboratory, JJ Thomson Avenue, Cambridge CB3 0HE, UK
[3] Max-Planck-Institut für Gravitationsphysik, Am Mühlenberg 1, 14476, Potsdam, Germany
[4] APC, UMR 7164, Université Paris 7 Denis Diderot, 10, rue Alice Domon et Léonie Duquet, 75205 Paris Cedex 13, France

E-mail: jgair@ast.cam.ac.uk

**Abstract.** Nested sampling is a technique for efficiently computing the probability of a data set under a particular hypothesis, also called the Bayesian Evidence or Marginal Likelihood, and for evaluating the posterior. MultiNest is a multi-modal nested sampling algorithm which has been designed to efficiently explore and characterize posterior probability surfaces containing multiple secondary solutions. We have applied the MultiNest algorithm to a number of problems in gravitational wave data analysis. In this article, we describe the algorithm and present results for several applications of the algorithm to analysis of mock LISA data. We summarise recently published results for a test case in which we searched for two non-spinning black hole binary merger signals in simulated LISA data. We also describe results obtained with MultiNest in the most recent round of the Mock LISA Data Challenge (MLDC), in which the algorithm was used to search for and characterise both spinning supermassive black hole binary inspirals and bursts from cosmic string cusps. In all these applications, the algorithm found the correct number of signals and efficiently recovered the posterior probability distribution. Moreover, in most cases the waveform corresponding to the best a-posteriori parameters had an overlap in excess of 99% with the true signal.

## 1. Introduction

The data stream from the proposed space-based gravitational wave detector, the Laser Interferometer Space Antenna (LISA) [1], is expected to contain thousands of overlapping gravitational wave signals from many different types of source, including galactic compact binaries, the inspirals and mergers of supermassive black hole binaries and extreme-mass-ratio inspirals. Disentangling and correctly identifying these many sources is a challenging problem for data analysis. Research in this area has been stimulated by a sequence of Mock LISA Data Challenges (MLDCs), of which there have been four to date [2, 3, 4, 5, 6]. The most recent of these, round 3, finished at the end of April 2009. Several of the LISA sources have degeneracies in their parameter spaces, which lead to the existence of multiple modes in the posterior probability surface, some of which can be nearly equal in probability to the true solution. In the early rounds of the MLDC, this caused problems for searches for extreme-mass-ratio inspirals in particular, as Markov Chain Monte Carlo algorithms tended to get stuck on secondary modes and did not find the primary [5, 6, 7]. Data sets containing multiple sources of a given type will also have highly multi-modal posterior probabilities when explored with single-source waveform models. To date, multiple-source extraction has mostly been tackled by iterative source identification

© 2010 IOP Publishing Ltd

and subtraction, but simultaneous identification of modes is likely to be computationally more efficient.

Nested sampling [8] is a technique for efficiently computing the Bayesian evidence of a data set under a particular model hypothesis, but also returns the posterior over the model parameters as a by-product. The MultiNest algorithm [9] is an efficient implementation of nested sampling, which is designed to deal with posteriors containing multiple modes of arbitrary shapes. It has proven to be effective in dealing with many varied problems in astronomy and particle physics [10, 11, 12, 13]. Moreover, the algorithm works without using any information about the type of signal space being explored, other than in the computation of the likelihood at individual points in the parameter space. As the posterior probability surfaces for LISA sources are expected to be highly multi-modal, MultiNest is well-suited to tackling data analysis for LISA as well. In this article we describe the application of MultiNest to characterization of various simulated LISA data sets, both in a controlled study [14] and for the analysis of two types of source in the last round of the MLDC. Nested sampling has previously been used in a gravitational wave context for model selection for ground based interferometers [15], and was a key element in a hybrid evolutionary algorithm that also proved effective for exploring the multi-modal posteriors of LISA data sets [16]. However, the work described here represents the first use of MultiNest in a gravitational wave data analysis context.

The article is organised as follows. In section 2 we describe the MultiNest algorithm and in section 3 we describe results for our controlled study and for the analysis of two of the MLDC round 3 data sets. We finish with a discussion of future applications of the algorithm in section 4.

## 2. The MultiNest Algorithm

The Bayesian evidence associated with a hypothesis, $\mathcal{H}$, for data, $\mathbf{D}$, depending on parameters, $\mathbf{\Theta}$, is the factor that appears in the denominator of Bayes theorem

$$\begin{aligned} P(\mathbf{\Theta}|\mathbf{D},\mathcal{H}) &= \frac{P(\mathbf{D}|\mathbf{\Theta},\mathcal{H})P(\mathbf{\Theta}|\mathcal{H})}{P(\mathbf{D}|\mathcal{H})} \\ P(\mathbf{D}|\mathcal{H}) \equiv \mathcal{Z} &\equiv \int P(\mathbf{D}|\mathbf{\Theta},\mathcal{H})P(\mathbf{\Theta},\mathcal{H})\,\mathrm{d}^N\Theta \end{aligned} \tag{1}$$

The evidence, $\mathcal{Z}$, can be used to choose between alternative hypotheses, but it is time-consuming to evaluate as it is a multi-dimensional integral. Nested sampling is a Monte Carlo method for evaluating evidences efficiently [8], which works by replacing the multi-dimensional integral (1) with a one dimensional integral over the prior volume, $X$,

$$X(\lambda) = \int_{\mathcal{L}(\mathbf{\Theta})>\lambda} P(\mathbf{\Theta}|\mathcal{H})\mathrm{d}^N\Theta \qquad \Rightarrow \qquad \mathcal{Z} = \int_0^1 \mathcal{L}(X)\mathrm{d}X \tag{2}$$

Here, $\mathcal{L}(\mathbf{\Theta}) \equiv P(\mathbf{D}|\mathbf{\Theta},\mathcal{H})$ is the likelihood. If the likelihood is evaluated on a nested sequence of contours of decreasing prior volume, the integral (2) can be computed using, e.g., the trapezium rule. This is illustrated in Figure 1.

This algorithm is implemented in practice using a set of N live points initially drawn from the prior. At each iteration $i$ the point of lowest likelihood, with likelihood $\mathcal{L}_i$, is removed from the set and replaced by a point of higher likelihood, $\mathcal{L} > \mathcal{L}_i$, sampled from the prior. At each iteration the prior volume is thus reduced by a factor $t$ drawn from $P(t) = Nt^{N-1}$. The algorithm travels through nested shells of likelihood as the prior volume reduces.

The challenge is to efficiently sample new points from the prior with likelihood $\mathcal{L} > \mathcal{L}_i$. The MultiNest algorithm [9] achieves this by partitioning the current set of live points into a set of (possibly overlapping) ellipsoids. An ellipsoid is then chosen at random and a new point sampled uniformly from within it. This is repeated until a point with likelihood $\mathcal{L} > \mathcal{L}_i$ is found. This

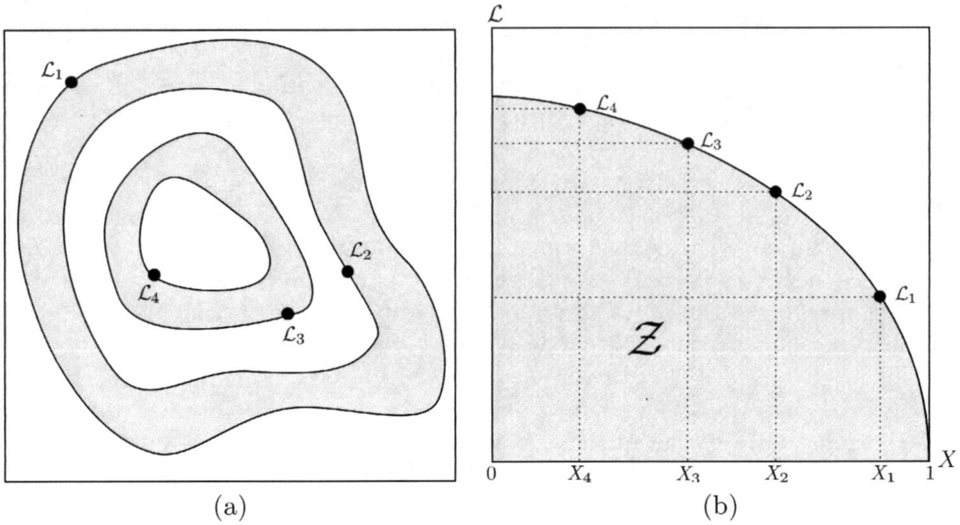


**Figure 1.** Cartoon illustrating (a) the posterior of a two dimensional problem; and (b) the transformed $\mathcal{L}(X)$ where the prior volumes, $X_i$, are associated with each likelihood $\mathcal{L}_i$.

ellipsoidal decomposition allows MultiNest to identify and separate different modes in the posterior, making it an ideal algorithm for exploring multi-modal posterior probability surfaces. Although designed for evidence computation, the algorithm also returns the posterior and it is for characterising multi-modal posteriors that we have used the algorithm so far. We note that MultiNest is not problem-specific in the sense that the particular model being used only enters in the likelihood evaluation and not in how the algorithm updates the live point set.

## 3. Results

### *3.1. Non-spinning black holes*

As a first application of MultiNest to gravitational wave data analysis, we generated a simulated LISA data set containing two signals from non-spinning supermassive black hole mergers [14]. One of the sources coalesced during the observation, while the second coalesced shortly after the observation ended. Due to the nature of the LISA response, there is a degeneracy between positions on the sky that are antipodal to one another. We therefore expected the posterior to contain at least four modes. We searched for these signals in two stages — the first stage involved searching only for the five intrinsic parameters of the source (two masses, time of coalesence and sky position) and used the generalised $\mathcal{F}$-statistic [17, 18] to maximize automatically over the four extrinsic parameters (inclination, distance, polarization and phase at coalesence); the second stage involved searching the full nine dimensional parameter space, but with tight priors on the intrinsic parameters based on the results of the first stage.

The search took $\sim 2.5$ hours on a single 3GHz processor and successfully identified 11 posterior modes, of which four were the two true and two antipodal modes that we expected to see. The parameters for both sources were recovered to very high precision, well within the expected error as estimated from the Fisher Matrix. Full results and further details can be found in [14].

### *3.2. Mock LISA Data Challenge Round 3*

The third round of the MLDC consisted of five data sets. The MultiNest algorithm was used successfully to search two of these — challenge 3.2 which was a 2 year data stream containing $4-6$ signals from inspiralling spinning supermassive black hole binaries; and challenge 3.4 which

| Source | 3.2.1 | 3.2.2 | 3.2.3 | 3.2.4 | 3.2.5 |
|---|---|---|---|---|---|
| Overlap A | 0.992 | 0.999 | 0.999 | 0.961 | 0.449 |
| Overlap E | 0.992 | 0.999 | 0.999 | 0.967 | 0.410 |

**Table 1.** Overlaps between the true waveforms and the maximum a-posteriori solutions identified by MultiNest for each of the five spinning black hole binary sources in the MLDC challenge 3.2 data set. Overlaps are given for both the $A$ and $E$ TDI data channels.

was a 1 month data set containing an unknown number (drawn from a Poisson distribution with mean 5) of bursts from cosmic string cusps.

*3.2.1. Challenge 3.2: spinning supermassive black hole binary inspirals* For this search, we divided the data set into sections according to estimates of the plunge times for the sources as computed from a time-frequency analysis of the data. These sections were then searched with MultiNest to obtain parameter estimates. We identified and recovered parameters for 5 sources, which was the correct number. The fifth source, the weakest, was only found after subtracting the best estimate for the fourth source from the data stream. The overlaps of the recovered sources with the true signals are tabulated in Table 1. Sources 3.2.2–3.2.4 were correctly recovered, although the dominant mode identified for 3.2.2 and 3.2.4 was the mode antipodal on the sky to the true solution. Source 3.2.1 was recovered with high overlap, but given this source was the brightest in the data set, with SNR$\sim$ 2000, a 0.8% mismatch leaves a significant residual. This arose in part because of the approximations to the detector response model that we used, and subsequent runs after the MLDC deadline using the full detector response have been much closer to the true parameters. Source 3.2.5, which was a low-SNR source that did not coalesce within the observation window, was not correctly identified at the time of the MLDC deadline, but this was due to a shortage of time. This source has since been recovered correctly. More details on the search and results will be presented elsewhere.

*3.2.2. Challenge 3.4: cosmic string bursts* For this search, we divided the 1 month release data into segments of 8192s in length and these were searched with MultiNest. We also used segments of length 16384s and 32768s to cross-check the results. We included the LISA response in our waveform model of the cosmic string burst using the static LISA approximation [19]. We identified three sources and this was again the true number that were present in the data set. The overlaps of the solutions identified by MultiNest with the true waveforms were in excess of 99.7% for both the $A$ and $E$ channels in all three cases. This is sufficient to ensure that the residual after subtraction of the sources would be smaller than the instrumental noise in the detector. The recovered maximum a-posteriori parameters did not match the true values very closely, however. This was a consequence of the large degeneracies in the cosmic string parameter space, which arise because for burst sources the direction to the source cannot be well determined. This is illustrated in Figure 2, which shows the 2D and 1D marginalised posteriors for each of the six parameters that describe the cosmic string burst. We show these posteriors, as recovered by MultiNest, for the second of the three burst signals present in the MLDC challenge data release. Degeneracies can be clearly seen in the posteriors for sky position and time of coalescence, which arise from symmetries in the response of the detector at low frequency. These features were also seen and discussed in [20]. Further details of our methods and results can be found in a separate paper [21].

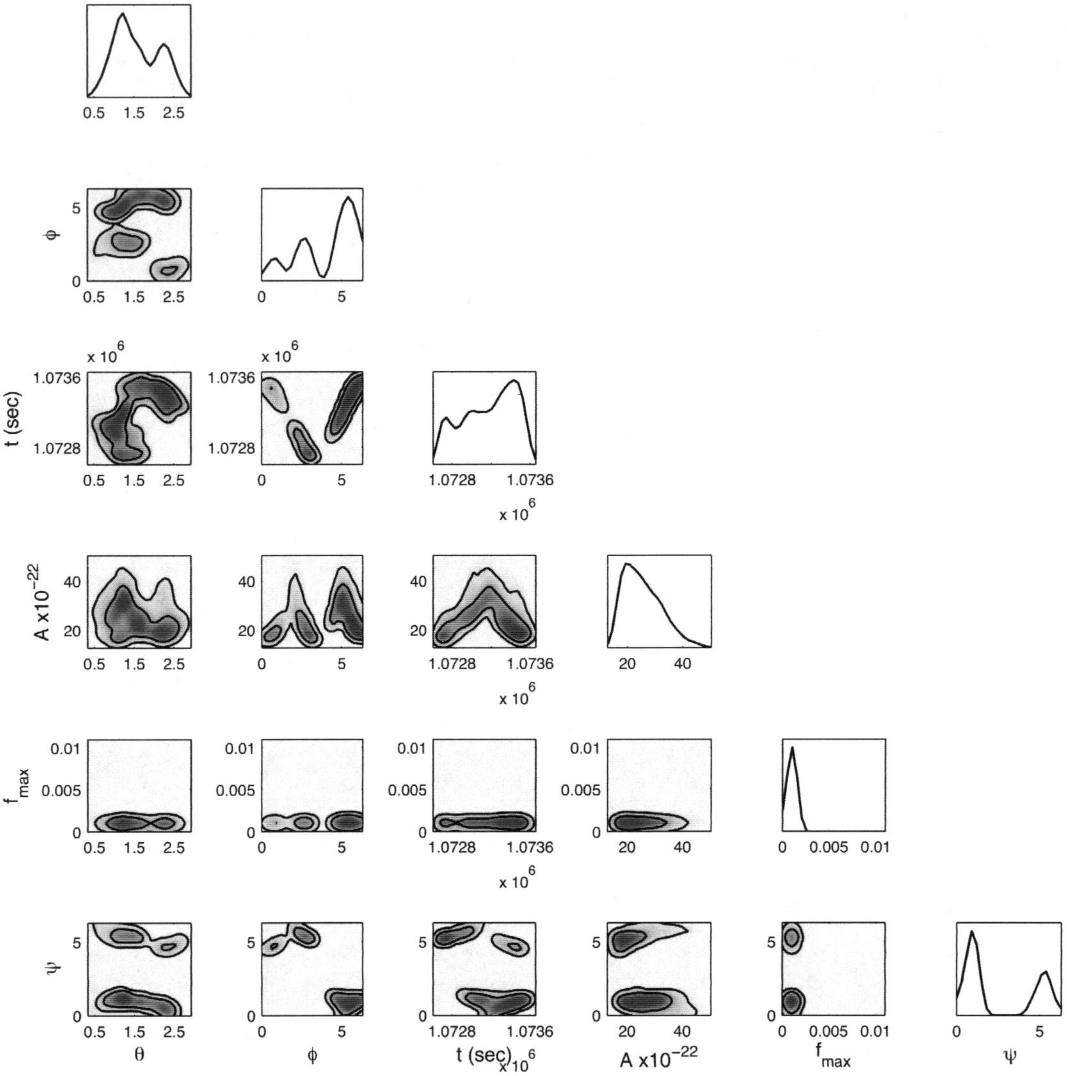


**Figure 2.** Two-dimensional marginalised posteriors as recovered by MultiNest in the search for the second of the cosmic string bursts in the MLDC challenge 3.4 data set. The parameters, from top-to-bottom and left-to-right, are colatitude, longitude, burst time, burst amplitude, burst break frequency and waveform polarization. At the top of each column we also show the one-dimensional posterior for the column parameter.

## 4. Discussion

We have described the application of MultiNest, a multi-modal nested sampling algorithm, to various problems in gravitational wave data analysis for LISA. Nested sampling is a method for computing evidences and posteriors efficiently, and MultiNest is an implementation of nested sampling which is optimized for analysing posterior probability surfaces containing multiple modes. While MultiNest is designed for evidence evaluation, it also returns the posterior distribution over the model parameters and it is for that purpose that we have used it to date. We have applied the algorithm to the characterisation of several different gravitational wave data sets — a data set containing two non-spinning supermassive black hole inspiral signals; a data set containing spinning supermassive black hole inspirals and a data set containing bursts from cosmic string cusps. In all three cases, we found that MultiNest was able to quickly and accurately characterise the main modes of the posterior. Moreover, the algorithm is model-independent in the sense that the specific problem being tackled enters only in the likelihood computation, and does not affect how the live point set is updated.

These promising initial results suggest that it is worth exploring the application of MultiNest to further problems in gravitational wave data analysis. These include

- Using the evidence computed by the algorithm for model selection, e.g., to identify the number of sources of a particular type in the data stream, or for testing general relativity against alternative theories of gravity.
- Using MultiNest to search for and characterise extreme-mass-ratio inspiral signals, which are known to have very complex posteriors containing many secondary maxima.
- Applying MultiNest to the analysis of data from ground-based interferometers (LIGO) — as a search tool, to recover posteriors and for model selection.

The next round of the MLDC [22] will consist of a single data set containing, simultaneously, all of the sources considered separately in round 3. We will apply MultiNest to the analysis of that data in order to further explore the utility of the algorithm for LISA data analysis.

## Acknowledgments

This work used the Darwin Supercomputer of the University of Cambridge High Performance Computing Service (`http://www.hpc.cam.ac.uk/`), provided by Dell Inc. using Strategic Research Infrastructure Funding from the Higher Education Funding Council for England and we would like to thank Dr. Stuart Rankin for computational assistance. JG's work is supported by the Royal Society. FF is supported by Trinity Hall. The work of SB and AP was supported in parts by DFG grant SFB/TR7 "Gravitational Wave Astronomy" and by DLR (Deutsches Zentrum fur Luft-und Raumfahrt). PG is supported by the Gates Cambridge Trust.

## References

[1] Danzman K *et al.* 1998 *LISA Laser Interferometer Space Antenna, Pre-Phase A Report*, Report MPQ 233
[2] Arnaud K A *et al.* American Institute of Physics Conference Series, 873, 619 (2006).
[3] Arnaud K A *et al.*, Class. Quantum Grav. **24** S529 (2007).
[4] Arnaud K A *et al.*, Class. Quantum Grav. **24** S551 (2007).
[5] Babak S *et al.*, Class. Quantum Grav. **25** 114037 (2008).
[6] Babak S *et al.*, Class. Quantum Grav. **25** 184026 (2008).
[7] Babak S, Gair J R & Porter E K, Class. Quantum Grav. **26** 135004 (2009).
[8] Skilling J, American Institute of Physics Conference Series, 735, 395 (2004).
[9] Feroz F, Hobson M P & Bridges M, MNRAS **398** 1601 (2009).
[10] Feroz F, Marshall P J & Hobson M P, preprint arxiv:0810.0781 (2008).
[11] Feroz F, Hobson M P, Zwart J T L, Saunders R D E & Grainge K J B, MNRAS **398** 2049 (2009).
[12] Feroz F, Allanach B C, Hobson M P, Abdus Salam S S, Trotta R & Weber A M, JHEP, 10, 64 (2008).
[13] Trotta R, Feroz F, Hobson M P, Roszkowski L & Ruiz de Austri R, JHEP, 12, 24 (2008).
[14] Feroz F, Gair J R, Hobson M P & Porter E K, Class. Quantum Grav. **26** 215003 (2009).
[15] Veitch J & Vecchio A, Class. Quantum Grav. **25** 184010 (2008).
[16] Gair J R & Porter E K, Class. Quantum Grav. **26** 225004 (2009).
[17] Jaranowski P, Królak A & Schutz B F, Phys. Rev. D **58** 063001 (1998).
[18] Cornish N J & Porter, E K, Class. Quantum Grav. **24** 5729 (2007).
[19] Cornish N J & Rubbo L J, Phys. Rev. D **67** 022001 (2003); Erratum-ibid. D**67** 029905 (2003).
[20] Key J S & Cornish N J, Phys. Rev. D **79** 043014 (2009).
[21] Feroz F, Gair J R, Graff P, Hobson M P & Lasenby A, preprint arxiv:0911.0288 (2009).
[22] Babak S *et al.*, preprint arxiv:0912.0548 (2009).

# A physical criterion for validating the method used to design mechanical impedance matchers for Mario Schenberg's transducers

**F S Bortoli[1] , C Frajuca[1] , N S Magalhaes[2] and E N Duarte[1]**

[1] Sao Paulo Federal Institute of Education, Science and Technology, Rua Pedro Vicente 625, Sao Paulo, SP 01109-010, Brazil

[2] Department of Exact and Earth Sciences, Federal University of Sao Paulo - Diadema Campus, Rua Prof. Artur Riedel, 275, Diadema, SP 09972-270, Brazil

E-mail: bortoli@cefetsp.br

**Abstract**. This work reports improvements made in the modelling of mechanical impedance matchers with mushroom shape using the finite elements method when shell elements type were used instead of tetrahedron elements type. Also, it is presented here an original methodology which makes use of the symmetry of the system and its influence on the mechanical vibrational modes to validate the modelling that was the base for the simulations performed.

## 1. Introduction

"Mario SCHENBERG"[1, 2, 3, 4, 5] is a spherical resonant-mass gravitational wave (GW) detector weighting 1.15ton, being built in the Department of Materials Physics at the University of Sao Paulo (Brazil). The sphere, with 65cm in diameter, is made of an alloy with 94% copper and 6% aluminum (94%Cu-6%Al). In a spherical resonant-mass detector the signal is generated when the GW passes through it and causes vibrations in the spherical mass.

The distribution of six sensors on the surface of the sphere is chosen purposefully [6, 7]: they are arranged on the sphere surface in a half-dodecahedron distribution. These transducers are located as if in the center of the 6 connected pentagons in a dodecahedron surface.

The resonant frequencies of this array will be around 3.2 kHz with a bandwidth of about 200Hz. Each transducer amplifies the motion of the region of the sphere where it is attached to. By analyzing the signal of such transducers the intensity and the direction of the incoming gravitation wave can be obtained.

The impedance mechanical matchers, which are part of such transducers, filter and amplify part of the movement of the region of the sphere to which they are connected. Thus the transducer project is of great importance in order to guarantee that the transducer is tuned to the frequency of the sphere (quadrupole mode), that resonates with the GW.

© 2010 IOP Publishing Ltd

The improvements obtained in the modeling of mechanical impedance matchers were confirmed by the validation method used. The symmetry of the antenna system relative to its vertical axis "z" made possible the validation procedure used here, which would be impossible to be made experimentally. There would be no other way to do this validation, since an experimental procedure is not viable. Far as the authors of this paper know such a procedure to validate the simulation method used is new. We believe that this method can be used to validate other simulations of systems that have symmetries, using the finite element method or other method of numerical simulation.

## 2. Mechanical Impedance Matchers

In Schenberg each transducer has a 2-mode mechanical impedance matcher [9]. This study used 1-mode mechanical impedance matchers [9, 10] with mushroom shape, arranged in the same way as on the Schenberg detector (figure 1). The design of the mechanical impedance matchers with a mushroom shape was made using the finite elements method [11].

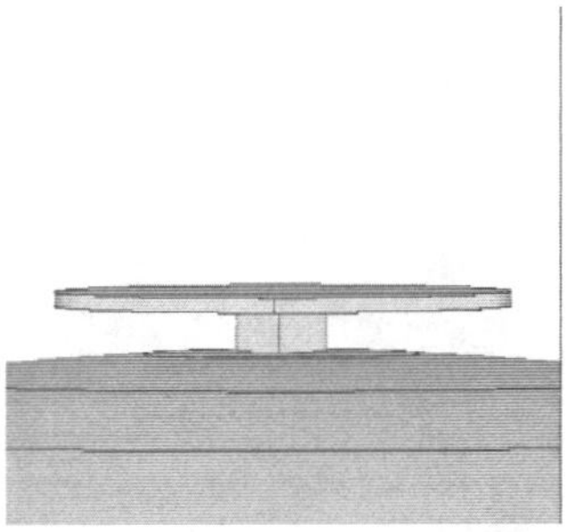

**Figure 1.** The illustration shows a mushroom impedance matcher attached on the sphere surface. The diaphragm has diameter of 50mm, thickness from 0.5 to 2.5mm and height of 7 mm.

## 3. Initial phase of this study

In the first phase of this study only tetrahedral elements type were used in the preparation of models for the simulations with the finite element method (FEM). The reasons for this choice were the good results obtained in the simulations with the bare resonant-mass sphere (the sphere without impedance matchers) [12]. The values found were close to the experimental values obtained modeling the sphere with holes that will be used to house the transducers[13]. Furthermore, the standard deviation of the values found was 0.03%, which suggests that the modes are degenerate.

The best outcome of this phase of the study can be seen in figure 2, which shows, through 66 snapshots, the maximum amplitudes of each of the six mushrooms coupled to the sphere, in the 11 normal modes of vibration of the sphere-mushrooms array or sphere-mechanical impedance matchers.

In figure 3 is shown the graph of the frequencies obtained for the normal modes of the sphere-mushrooms array.

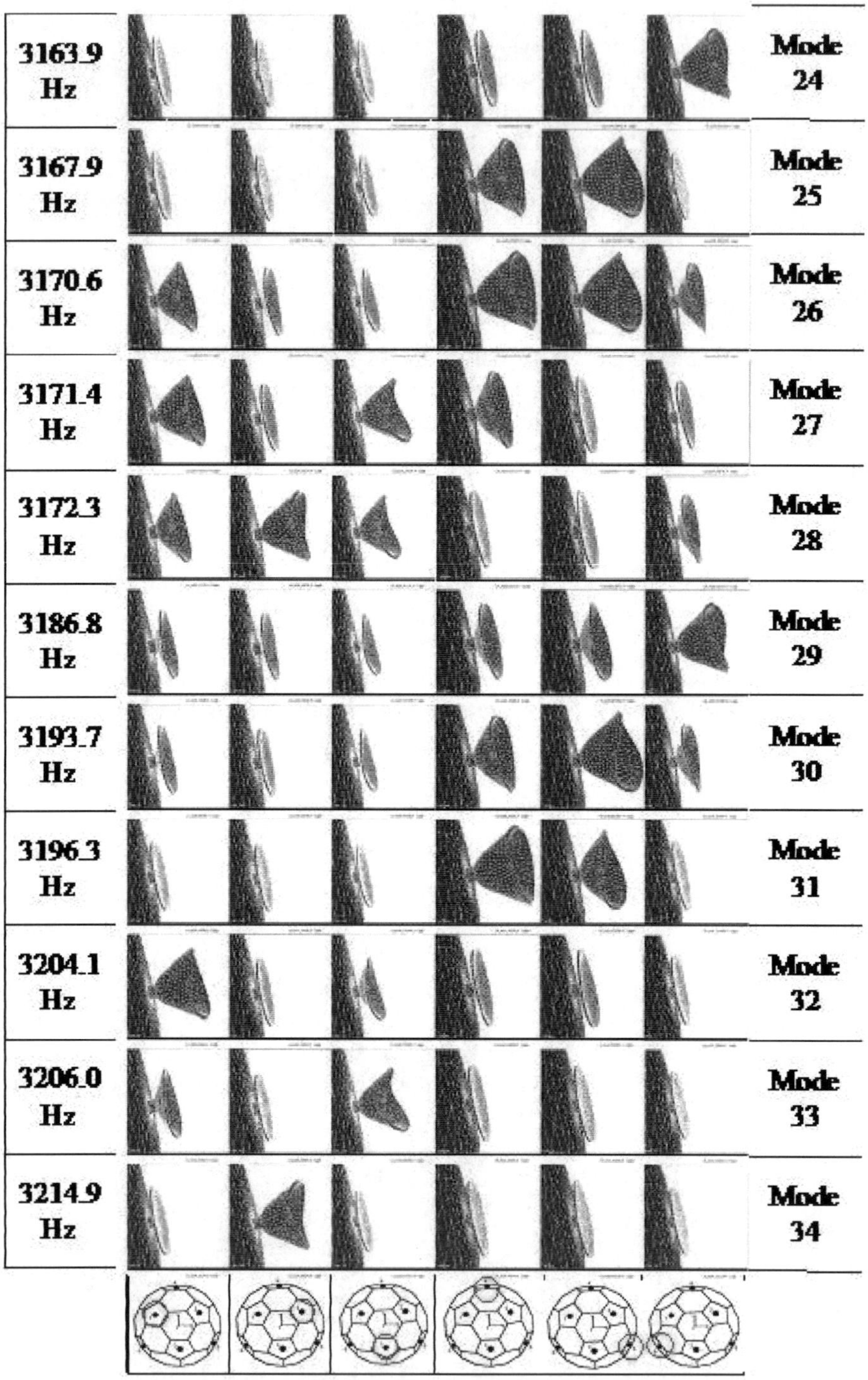


**Figure 2.** From the first phase of this study: snapshots obtained in models for which only tetrahedral elements type were used. Shown is the maximum amplitude of each of the six mushrooms coupled to the sphere, in the 11 normal modes of vibration of the sphere-mushrooms array. The 11 frequencies of the modes are shown at the left side of this figure. The location of the mushroom is shown in the lower part of the figure.

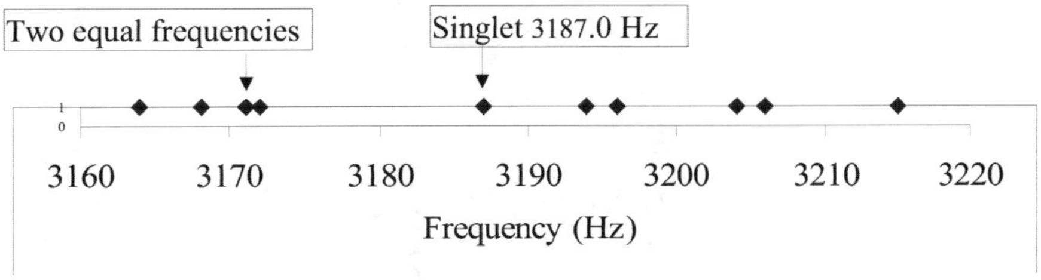


**Figure 3.** From the first phase of this study: the frequencies of the 11 modes of the sphere-mushrooms array.

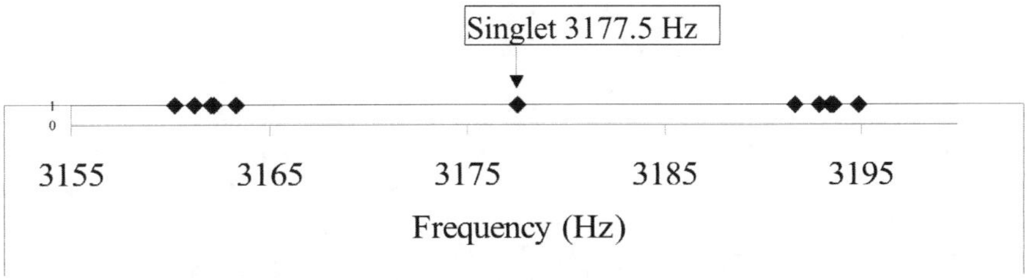


**Figure 4.** From the current phase of this study: the frequencies of the 11 modes of the sphere-mushrooms array.

## 4. Current phase of this study

In the current phase of this study the models for the sphere used in the simulations using the finite elements method have also been developed using elements of the tetrahedral type. However, in developing the models for the diaphragms of the mushrooms, shell-type elements were used.

The best outcome of this phase of the study can be seen in figures 4 and 5.

In figure 4 is shown the graph of the frequencies obtained for the normal modes of the sphere-mushrooms array

Figure 5 shows, through 66 snapshots, the maximum amplitude of each of the mushrooms – 6 coupled to the sphere, in the 11 normal modes of vibration of the sphere-mushroom array.

## 5. Results

Unlike the results obtained in the initial phase of the study (figures 2 and 3), the results obtained in the simulations made with the models developed with shell-type elements (figures 4 and 5) were quite symmetrical.

As the system has several vertical planes of symmetry, all parallel to the vertical axis z of the reference system, an asymmetrical result relative to the frequencies of normal vibrational modes of the system were not expected, as the one obtained in the first phase of this study.]

Another aspect arising from this symmetry can be seen in snapshots in Figure 5. Mushrooms on both sides of the sphere tend to have their maximum amplitudes in the same mode. Notice the amplitudes of mushrooms 1 and 5, or 2 and 6, or 3 and 4 in the same mode (same line) and compare with other mushrooms of the same mode in Figure 2.]

We believe that these symmetries were obtained because the model was appropriate for the current system and the principles of mechanical vibration [14] involved.]

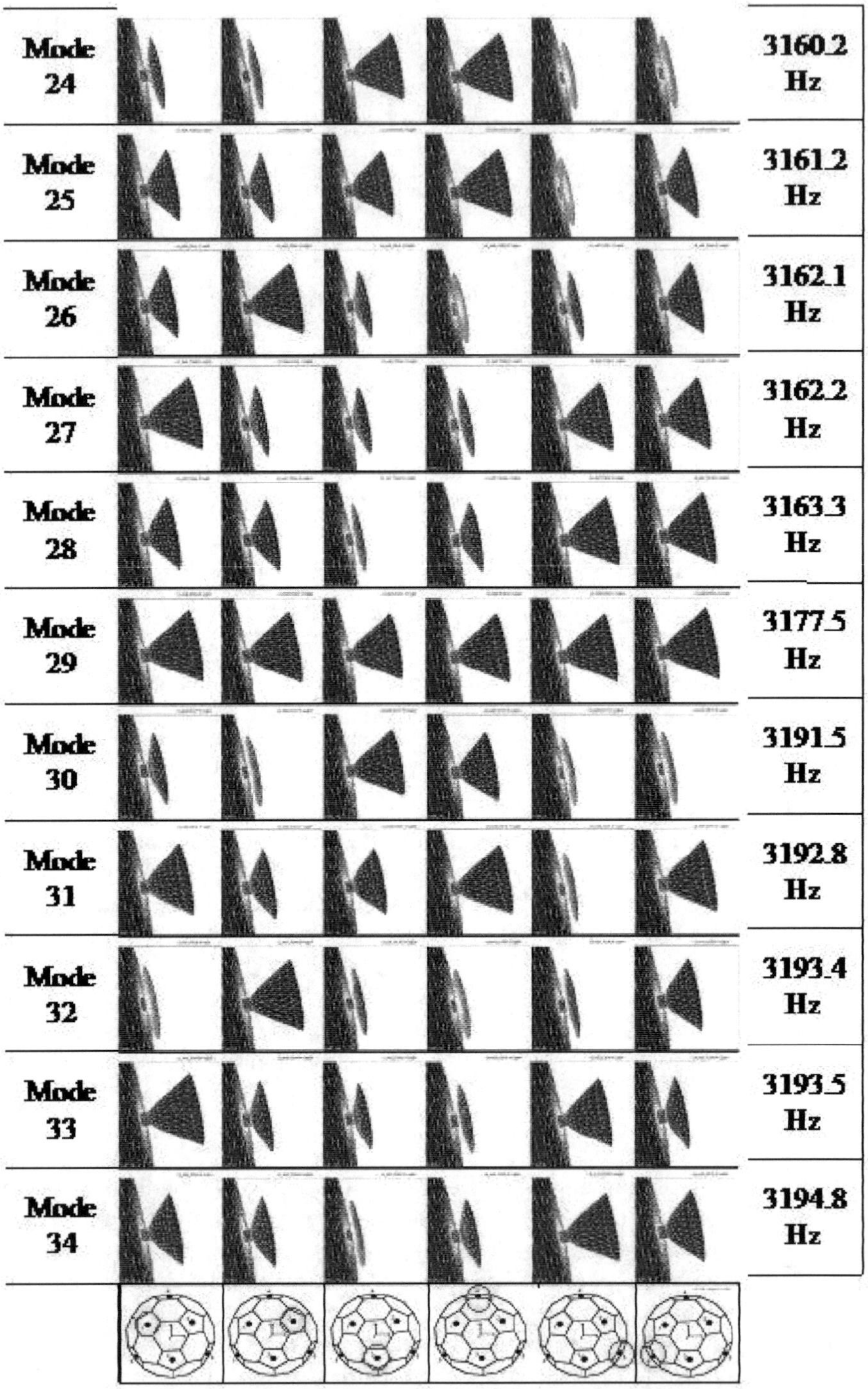


**Figure 5.** From the current phase of this study: snapshots obtained with models that used shell-type elements in the diaphragms of the mushroom. Shown is the maximum amplitude of each of the six mushrooms coupled to the sphere, in the 11 normal modes of vibration of the sphere-mushrooms array. The 11 frequencies of the modes are shown at the right side of this figure. The location of the mushroom is shown in the lower part of the figure.

**References**

[1]Aguiar O D et al. 2008 The Schenberg spherical gravitational wave detector: the first commissioning runs *Classical and Quantum Gravity* **25** 114042
[2]Frajuca C, Magalhaes N S, Bortoli F S and Horiguti A M 2008 Study of six mechanical impedance matchers on a spherical gravitational wave detector *Journal of Physics. Conference Series (Online)* **122** 012029
[3]Aguiar O D et all 2005 The Brazilian gravitational wave detector Mario Schenberg: progress and plans *Classical and Quantum Gravity* **22** S209-14
[4]Aguiar O D et al. 2004 The Brazilian spherical detector: progress and plans *Classical and Quantum Gravity* **21** n. 5 S457-63
[5]Frajuca C 2001 (in Portuguese) Mario Schenberg Gravitational Wave Detector *Sinergia - Revista do Centro Federal de Educação Tecnológica de São Paulo* **2** n. 1 14-8
[6]Merkowitz S M and Johnson W W 1993 *Phys. Rev. Lett.* **70** 2367
[7]Magalhaes N S, Aguiar O D, Johnson W W and Frajuca C 1997 *Gen. Relat. Grav.* **29** 1511
[8]Frajuca C 1996 (in Portuguese) *Optimization of two Mechanical Modes Transducers for Gravitational Wave Detectors* Ph.D. thesis (Sao Paulo, SP, Brazil: Sao Paulo University)
[9]Frajuca C and Bortoli F S 2006 Planning to improve the mechanical quality factor in the transducer impedance matchers for Mario Schenberg detector *Journal of Physics. Conference Series* **32** 315-8
[10]Frajuca C, Bortoli F S and Magalhaes N S 2005 Resonant transducers for spherical gravitational wave detectors *Brazilian Journal of Physics* (São Paulo, SP: Brasilian Society of Physic) **35** n. 4B 1201-3
[11]Frajuca C, Bortoli F S and Magalhaes N S 2006 Studying a new shape for mechanical impedance matchers in Mario Schenberg transducers *Journal of Physics. Conference Series* **32** 319-22
[12]Bortoli F S 2006 (in Portuguese) *Study of Mechanical Impedance Matchers for Microwave Parametric Transducers on a Spherical Gravitational Waves Detector* M.Sc. dissertation, (Sao Paulo, SP, Brazil: Sao Paulo University)
[13]Aguiar O D et al 2004 *Classical Quantum Gravity* **21** 457
[14]Marion J B 1975 (in Spanish) *Dinámica Clasica de las Particulas y Sistemas* (Barcelona: Reverte)

# The upgrade of GEO 600

**H. Lück[1], C. Affeldt[1], J. Degallaix[1], A. Freise[2], H. Grote[1], M. Hewitson[1], S. Hild[3], J. Leong[1], M. Prijatelj[1], K.A. Strain[1,3], B. Willke[1], H. Wittel[1] and K. Danzmann[1]**

[1] Max-Planck-Institut für Gravitationsphysik (Albert-Einstein-Institut) and Leibniz Universität Hannover, Callinstr. 38, 30167 Hannover, Germany
[2] School of Physics and Astronomy, The University of Birmingham, Edgbaston, Birmingham, B15 2TT, United Kingdom
[3] Institute for Gravitational Research, University of Glasgow, Glasgow, G12 8QQ, United Kingdom, United Kingdom
Institut für Gravitationsphysik, Leibniz Universität Hannover, Callinstr. 38, 30167 Hannover, Germany

E-mail: harald.lueck@aei.mpg.de

**Abstract.** The German / British gravitational wave detector GEO 600 is in the process of being upgraded. The upgrading process of GEO 600, called GEO-HF, will concentrate on the improvement of the sensitivity for high frequency signals and the demonstration of advanced technologies. In the years 2009 to 2011 the detector will undergo a series of upgrade steps, which are described in this paper.

## 1. Introduction

GEO 600, the British / German gravitational wave (GW) detector[1], is one of the worldwide network of GW detectors. The main initial goal of GW detectors is the first detection of GW signals. In order to improve the chances for detecting a signal, the world-wide collaboration of GW scientists typically operates more than one detector in coincidence. GEO 600 finished its latest data run in July 2009 after taking data for about 3.5 years, partly together with LIGO[2] and Virgo[3]. To further improve the chances of detecting a signal the detectors have been - and will further be - upgraded to increase their sensitivity. The first upgrades for the LIGO and Virgo detectors to the 'enhanced' versions happened from late 2007 to July 2009. In this period, called Astrowatch, GEO 600 took data to cover the downtime of the two 4 km LIGO detectors and the Virgo detector. In the Astrowatch period GEO 600 was partly joined in data taking by the 2 km LIGO detector located near Hanford, Washington, forming a small detector network. The performance of GEO 600, and GEO 600 related activities during Astrowatch are described elsewhere[4]. Rather than just detecting that there was a signal at all, the goal of future generations of GW detectors is to do GW astronomy, i.e. to focus on extracting astro-physically interesting science from the GW signals. The upgrades of GEO 600, which are described in this paper, will firstly enable GEO 600 to take data during a part of the upgrading time of LIGO and Virgo to the 'advanced' state with scientifically interesting sensitivity and secondly to test and prove advanced technologies that can be used in future generations of GW detectors, e.g. Squeezing. This paper outlines the possibilities, and plans to enhance the sensitivity of the GEO detector. Section 2 gives a short overview of the GEO 600 configuration before the

© 2010 IOP Publishing Ltd

upgrades. The main steps in the upgrade will be the injection of squeezed states into the output port, tuned Signal Recycling with DC readout, installation of an output mode cleaner, the change of the Signal Recycling bandwidth and a power increase inside the interferometer in combination with the installation of a thermal compensation system. The individual steps of the upgrading process are described in section 3. The sensitivity improvement mainly concerns the **h**igh **f**requency part, which is why the upgrading process is called GEO-**HF**. At lower frequencies thermal noises become a limit which cannot significantly be lowered with currently available techniques.

## 2. Status Quo

GEO 600 is a dual recycled (i.e. Power- and Signal-Recycled) Michelson type interferometer with an arm length of 600 m. The arms are folded once giving an effective arm length of 1200 m.

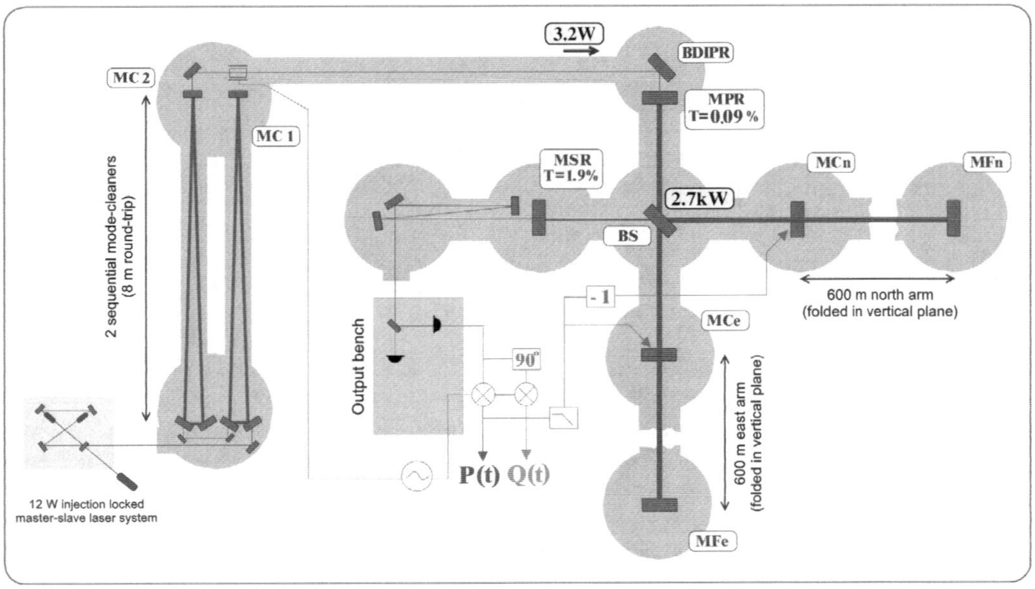


**Figure 1.** Schematic layout of GEO 600

In the past few years GEO 600 used Schnupp modulation[8], see also section3.1, for a heterodyne read–out, Signal Recycling detuned to a frequency of 530 Hz, a light power at the input to the Power Recycling cavity of about 3 W with a Power Recycling factor of about 1000. A schematic layout with the relevant parameters is given in figure 1. The light from a 12 W Nd-YAG laser with a wavelength of 1064 nm is filtered by two optical resonators, the Mode Cleaners, and then sent through modulators and Faraday isolators to the Power Recycling cavity. The peak sensitivity of GEO 600, $2.2 \cdot 10^{-22}[1/\sqrt{Hz}]$, is reached at the tuning frequency of 530 Hz (see figure 2). Above this frequency the sensitivity of GEO 600 is mostly shot noise limited. Between roughly 100 Hz and 500 Hz a yet unknown noise source limits the sensitivity in 'detuned RF readout', whereas below 100 Hz technical alignment related noise sources dominate the spectrum. The latest status of recent GEO commissioning is described in [4]. In recent 'tuned DC readout' experiments all of the observed noise could be explained by simulated (e.g. thermal noise, quantum noise) and measured (e.g. alignment feedback noise) contributions from known sources.

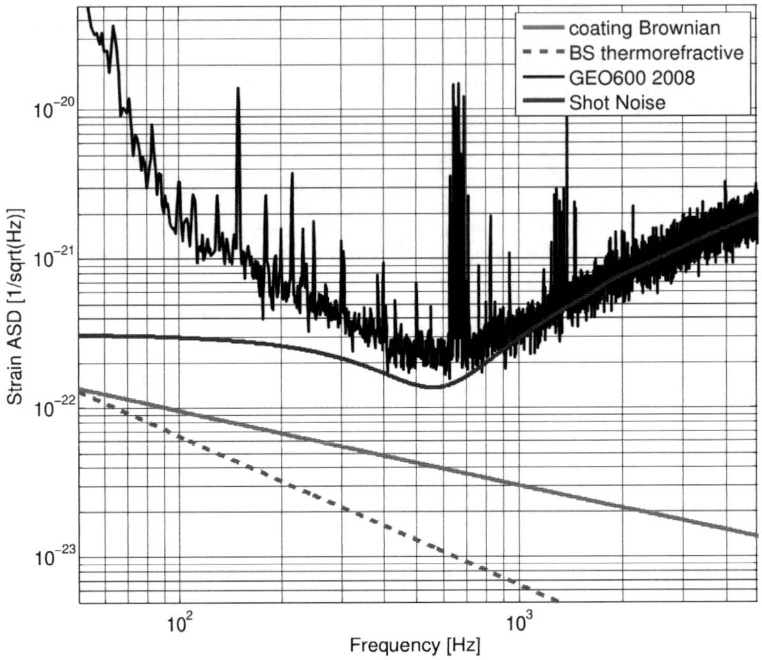


**Figure 2.** Sensitivity of GEO 600

## 3. Upgrading plan

The upgrade of GEO 600 will be done in a staged way, checking the performance after each individual step. The upgrading is aimed at frequencies above 500 Hz where roughly an order of magnitude can be gained in sensitivity by lowering shot noise. Below 500 Hz GEO 600 will be limited by thermal noises, mainly thermo-refractive noise at the beam splitter and coating Brownian noise. Techniques to reduce these noise sources are available[6],[7],[5] but are beyond the scope of the envisioned GEO upgrading plans for the next few years. The upgrade will therefore concentrate on lowering the shot noise.

Shot noise in an interferometric gravitational wave detector originates mainly from the vacuum fluctuations entering the output port. The contributions that come from the fluctuations on the input laser beam are small compared to those entering the output port because the input laser beam is well stabilized, filtered by the Power Recycling cavity and the Schnupp asymmetry and the imbalance in the losses of the arms of the interferometer are small. Reducing the vacuum fluctuations entering the output port in the phase quadrature can therefore reduce the observed shot noise, see section 3.3.

### *3.1. From RF readout to DC readout*

The GEO 600 interferometer is held at a 'dark fringe' operating point. At this operation point the arm length difference is arranged such that the light fields returning from the interferometer arms interfere destructively at the output port; hence it is **dark**. A control system maintains this condition. Right on the 'dark fringe' there is no dependence of the light power at the output port on the arm length difference, hence the light power can neither be used as an error signal for the control system nor as a signal sensing GWs. Differentially modulating the arm length around the interference minimum and demodulating the resulting power detected at the output port solves this problem. A physical modulation of the mirror position would introduce too much noise. So called 'Schnupp modulation' opens a way around this problem. If the laser light is frequency modulated and if the arm lengths are slightly different (13.5 cm in the GEO 600 case) the same effect as differentially modulating the arms lengths is achieved. In GEO 600 the modulation frequency is about 14.9 MHz and therefore the readout technique is called **R**(adio)-**F**(requency) readout or heterodyne readout. Unfortunately this method collects additional vacuum noise from twice the modulation frequency, which, beating with the modulation sidebands at the modulation frequency, produces noise at the modulation frequency which gets demodulated into the signal frequency band[13]. This RF-readout technique was chosen for the first generation of GW detectors because the lasers used were shot noise limited only in the MHz frequency range. The excellent amplitude and frequency stability of (stabilized) solid state ND-YAG lasers in the detection frequency band in addition to the filtering properties of the Power Recycling cavity now allows to switch to a self–homodyne readout technique, also called DC readout. In this case the operation point is slightly (in the range of about 5 pm to 50 pm) detuned from the dark fringe, which yields a dependence of the light power at the output port on the arm length difference. The shot noise for DC readout is lower than for RF readout. All large interferometric gravitational-wave detectors will or have switched to DC readout with the upgrade from the first generation to the next upgrade step. The first experiences with DC readout in the GEO 600 detector are reported in [4],[9] and [10]. DC readout is also beneficial for the injection of squeezed states as only the correct phasing at the laser carrier frequency and the respective signal sidebands has to be taken into account; see also section 3.3.

In GEO 600 Schnupp modulation will still be used for generating the error signals of the differential wave–front sensing for the Auto–Alignment system. As a status of January 2010 GEO 600 is routinely operating with DC readout.

### *3.2. Output Mode Cleaner*

The light leaving the output port of GEO 600 will be sent through an Output Mode Cleaner (OMC)[10] before being detected on the photo-detector. The OMC is a four mirror cavity with a round trip length of roughly 66 cm as shown in figure 3. The mirrors are glued onto a fused silica base plate to reduced thermal expansion to a level which allows the use of a small range ($\approx 3\,\mu m$) piezo-electric actuator. The Finesse of about 150, giving a FWHM of about 3 MHz, assures that the modulation sidebands (14.9 MHz) are reflected by the OMC when it is tuned to carrier resonance. With proper mode matching the light in the TEM–00 mode will get transmitted and the unwanted light in higher TEM modes will be reflected. One of the mirrors is mounted on a piezo-electric actuator. Modulating the OMC length with this Piezo-actuator allows the generation of error signals for the length control and for the Auto–alignment system aligning the beam onto the OMC. More details on this technique can be found in [11]. At the end of January 2010 the OMC is installed and stably operating.

### *3.3. Squeezing*

With the shot noise contribution being the main noise source limiting the GEO600 sensitivity, the upgrade plans are aiming at lowering the relative shot noise level. The interferometer is kept close to a 'dark fringe' operating point (DC readout; see section 3.1) where the output port is almost dark. For energy conservation reasons all the light must be reflected towards the input port. For symmetry reasons the same must be valid for fields entering the output port. The vacuum fluctuations entering the output port therefore get reflected from the interferometer and interfere with the carrier light getting to the output port due to the offset from the dark fringe for DC readout. As it is the fluctuations in the phase quadrature that cause the shot noise, the injection of squeezed light, where the noise in the phase quadrature is decreased, can lower the

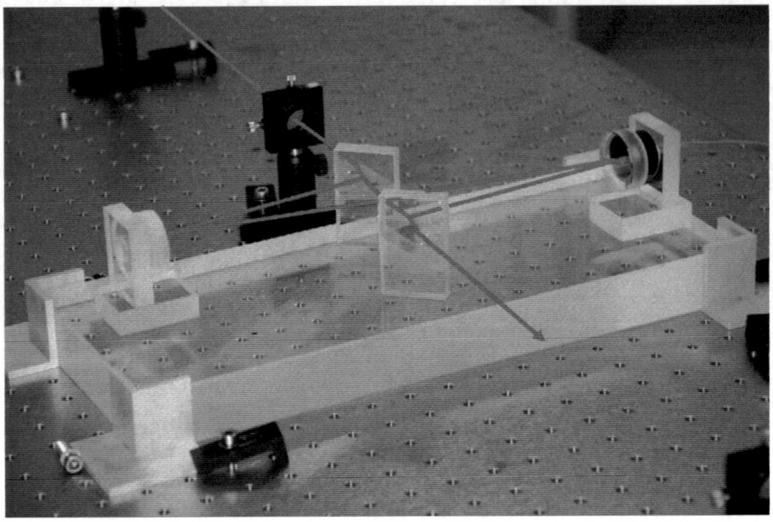

**Figure 3.** The output mode cleaner for GEO 600

shot noise detected by the photo detector. GEO 600 will use a squeezed vacuum field with a squeezing level of about 10dB to be injected into the output port. The injection will be done via a Faraday isolator in an additional vacuum vessel added to the GEO vacuum system. This vessel contains some mode matching optics for the Output Mode Cleaner, the Output Mode Cleaner itself, a Faraday isolator and a photo diode. The vacuum level is about $5 \cdot 10^{-3}$ mbar. All the optical elements are placed on a platform seismically isolated with a Minus-K® isolation system.

To make optimal use of the squeezing over the full detection frequency range, the squeezing ellipse, i.e. the phase of the injected squeezed light, must be optimally oriented with respect to the light leaving the output port for all frequencies within the detection bandwidth. In case of a detuned (with respect to the carrier) Signal Recycling cavity this can be achieved by sending the squeezed light through additional filter cavities[15] or using Twin Signal Recycling[16]. For tuned (zero frequency detuning) Signal Recycling the orientation of the squeezing ellipse is frequency independent and no filtering of the squeezed light is required. The improvement in sensitivity by using squeezing (together with the series of other changes) is shown in figure 5. In January 2010 the squeezing breadboard, which provides the squeezed input for GEO 600, is providing strong, broadband squeezing, is undergoing some long term stability tests and will be brought to the site in late February. The additional vacuum vessel with all its components is in place and operative. A schematic layout of the output section of GEO 600 is shown in figure4

### *3.4. Tuned Signal Recycling*

In the future GEO 600 will use tuned Signal Recycling together with the squeezing to improve the sensitivity over the full frequency range[17]. In this case the Signal Recycling mirror is shifted such that the Signal Recycling cavity is resonant for the laser carrier light instead of the 530 Hz signal sidebands as in the detuned mode GEO 600 operated in before. The achievable sensitivity is shown in figure 5. Tuned Signal Recycling resonantly enhances both signal sidebands, providing higher optical gain on resonance, which results in lower shot noise at the resonance frequency than detuned Signal Recycling. In the case of GEO 600 this does not influence the total sensitivity as at low frequencies the total noise is dominated by thermal noise and technical noises anyway.

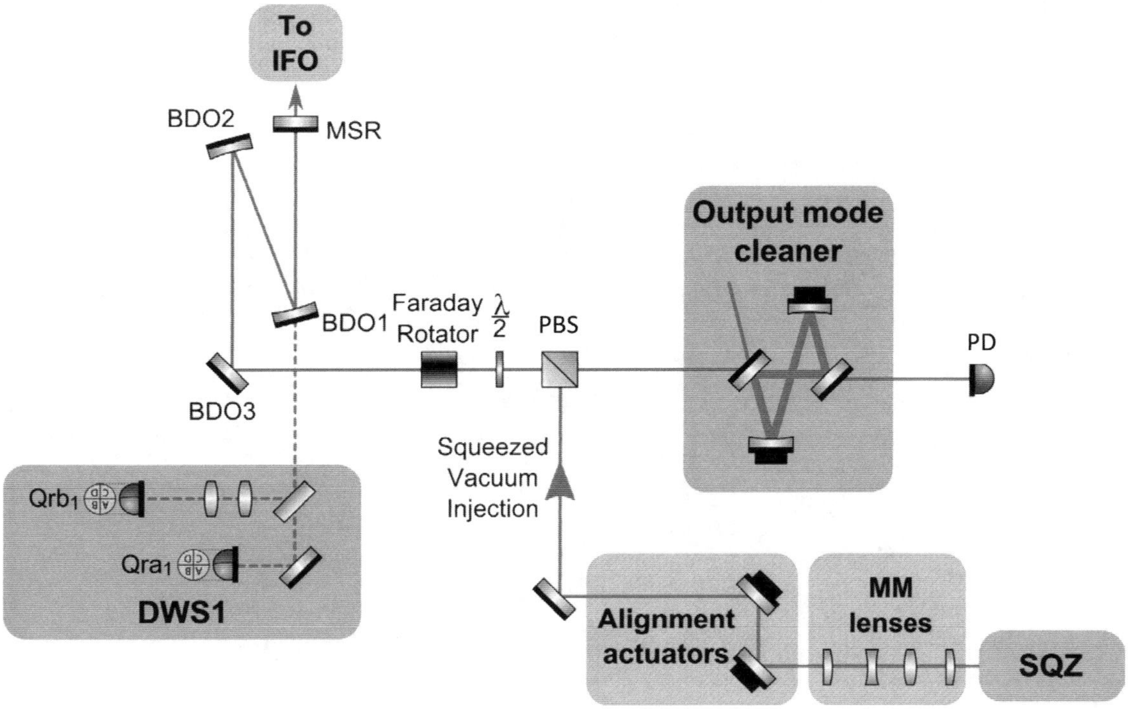


**Figure 4.** Schematic layout of the output optics of GEO 600, showing the squeezing input and the output mode cleaner. The output signal of the interferometer is transmitted through the Signal Recycling mirror (MSR) and de-magnified by the curved mirror BDO1. BDO2 and BDO3 serve as steering mirrors aligning the output mode to the Eigenmode of the OMC. The half-wave plate, the Faraday rotator and the polarizing beam splitter (PBS) serve as an optical diode to inject the squeezing from the one direction and let the output signal pass through the OMC to the main photo diode (PD) in the other direction. The mode matching lenses and the beam steerers between the PBS and the squeezing system align and match the squeezing mode to the interferometer mode. The differential wave-front sensing system (DWS1) gives the error signals for aligning the two beams of the interferometer arms to each other.

### *3.5. Increasing the Signal Recycling Transmission*

Widening the bandwidth of GEO 600 by increasing the transmission of the Signal Recycling mirror from 2% to 10% improves the high frequency shot noise at the expense of the low frequency one. Being limited by thermal noises at frequencies below 700 Hz anyway this influences the overall noise for low frquencies only marginally. The crossover point between coating thermal noise and shot noise only moves from 700 Hz (curve ④ in figure 5) to 800 Hz (curve ⑤ in figure 5), while the high frequency total noise drops by a factor of about two. A further increase in bandwidth would raise the low frequency shot noise to an undesirable level while giving marginal improvement at high frequencies.

### *3.6. Power Increase*

The final step in the series of GEO 600 upgrades, foreseen to be done in late 2010 and in 2011, is an increase in light power inside the interferometer by a factor of about ten. To achieve this several changes become necessary. In the following two sections the changes that will lead to a power increase in the interferometer by a factor of about 8 will be described. The modifications

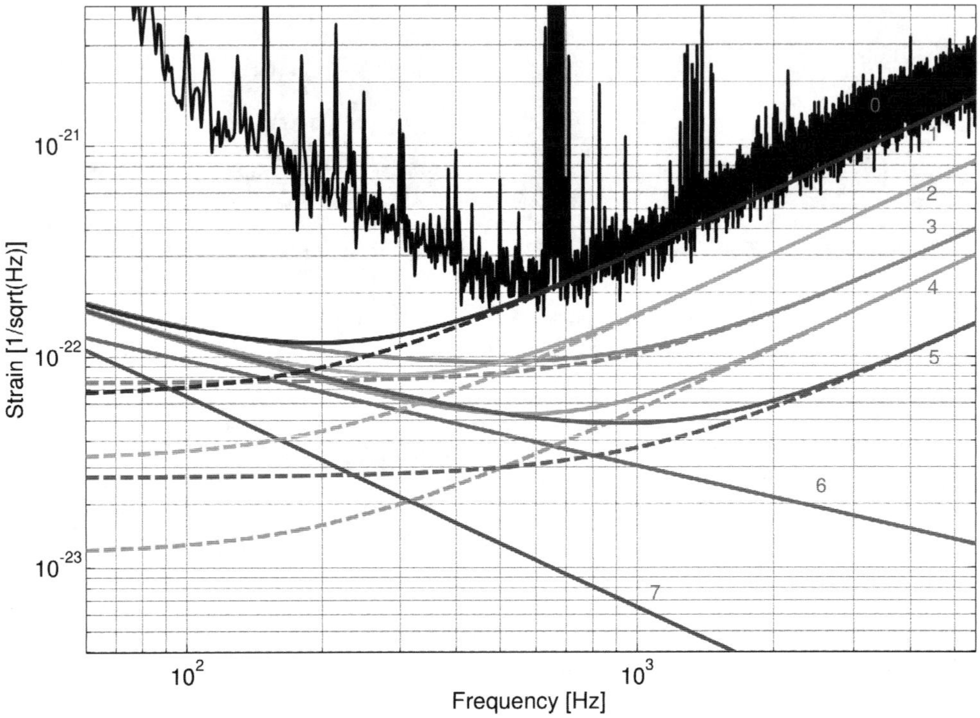


**Figure 5.** Sensitivity of GEO 600 for different configurations.
Solid lines indicate total noise, dashed lines represent shot noise only. Technical noise sources are neglected here.
⓪ GEO 600 sensitivity 2009
① DC, tuned SR, $T_{SR}$=2%, P=3.2 W
② DC, tuned SR, $T_{SR}$=2%, P=3.2 W, 6 dB Squeezing
③ DC, tuned SR, $T_{SR}$=10%, P=3.2 W, 6 dB Squeezing
④ DC, tuned SR, $T_{SR}$=2%, P=20 W, 6 dB Squeezing
⑤ DC, tuned SR, $T_{SR}$=10%, P=20 W, 6 dB Squeezing
⑥ Coating thermal noise
⑦ Thermo-refractive noise of the beam splitter

shown in sections 3.6.3 and 3.6.4 are needed to stably operate the detector at a power level of about 20 kW in the Power Recycling cavity.

*3.6.1. Laser Power increase* Currently GEO 600 is operated using a 10 W master-slave Nd-YAG laser with a wavelength of 1064 nm. In the current operation mode the laser is attenuated to about 6 W before being sent into the vacuum system. Due to losses in the Mode Cleaners (see below) only about 3.2W arrive at the Power Recycling mirror as indicated in figure1. The laser will be exchanged for a master-slave-amplifier system delivering about 30 W, which has been developed by the Laser Zentrum Hannover. It is the same kind of laser that is being used in the 'enhanced' versions of the other large gravitational wave detectors Enhanced LIGO[18] and Virgo+[19]. The increase in laser power will yield a power increase of about 5 in the Power Recycling cavity.

*3.6.2. Mode Cleaner Modifications* Due to scattering losses of the mirrors of the GEO Mode Cleaners (i.e. two optical resonators, which remove higher order TEM modes from the laser beam and reduce angular beam fluctuations) about 50% of the light is lost in the Mode Cleaners. At the current level of laser power, radiation pressure already needs to be compensated when bringing the GEO Mode Cleaners into resonance. Hence increasing the laser power would lead to undesirable problems with radiation pressure in the Mode Cleaners when keeping the same cavity finesse. Reducing the reflectivity of the coupling mirrors and with it the power build-up inside the Mode Cleaners, improves the throughput and reduces radiation pressure problems. The filtering properties of the initital Mode Cleaners were very conservative[12] and will still yield a sufficiently high suppression level after the decrease in finesse[12]. This mirror exchange will give a power increase in the Power Recycling cavity of a little less than two. The high-power compatibility of the auxiliary optics still needs to be confirmed.

*3.6.3. Local Control Changes* The mechanical resonances of the triple pendula for seismic isolation of the main interferometer mirrors are actively damped by a local control system using shadow sensors to monitor the movement of the uppermost mass[20]. These shadow sensors are operated with a DC light source and unfortunately detect the light scattered by the interferometer mirrors. Hence slight fluctuations in the circulating light power, which can originate from a slight misalignment of one of the interferometer mirrors, results in a signal at the shadow sensor which is regarded as a movement of the upper mass and a 'compensation' signal is fed back to the actuator. This misaligns the mirror controlled by this sensor and can lead to an instability. The coupling increases with the light power inside the interferometer. In GEO 600 this coupling has been observed and would make the foreseen increase in laser power impossible. Operating the light source of the shadow sensor (an infra-red LED) with modulated current with frequencies in the kHz range and demodulating the signal of the photo-detector can make the system insensitive to low frequency light power fluctuations. If the system works well all local control systems of the main GEO optics will be converted to AC operation. First experiments have already shown a satisfactory noise performance.

*3.6.4. Thermal Compensation System* In contrast to the other gravitational wave detectors, which use resonators in the interferometer arms for power buildup, GEO 600 achieves high circulating light power with a high Power Recycling factor of about 1000. As a consequence the full light power in the interferometer arms has to pass through the beam splitter. Although GEO 600 uses fused silica glass with very low absorption of 0.5 ppm/cm (Suprasil® 311SV), about 30 mW are absorbed inside the beam splitter. The local temperature increase causes a thermal lens which limits the maximum power that can be used. A system shining infra-red light, which is absorbed by the beam splitter surface, will be used to reduce the thermal lensing. Instead of using a $CO_2$ laser like in the other gravitational wave detectors[21],[22] an incandescent light source will be used in GEO 600 minimizing the fluctuations in heating power to the required level. In GEO 600 the required power stability for the heating source is beyond the current possibilities of a $CO_2$ laser.

## 4. Conclusions

The upgrades for the GEO 600 detector foresee a sensitivity enhancement for frequencies above 1 kHz of about one order of magnitude within the next two years, i.e. until the middle of 2011. In late January 2010 part of the upgrades have already been done: DC readout, Tuned Signal

Recycling operation,and the Output Mode Cleaner are routinely operating. Locking stretches on the order of 10 hours have been shown. Squeezing is about to be installed in GEO 600 in February 2010.

As far as it can be foreseen now, in the period from 2011 to 2015 GEO600 will be the only interferometer in the world that can be on-line for astrophysical observations for a substantial amount of time. The three LIGO interferometers as well as Virgo will be upgrading to the advanced versions of their instruments. Therefore, current planning is, that GEO 600 will spend more and more time in observation mode, as all the upgrades foreseen for GEO 600 at this time are implemented, and the sensitivity would approach the planned goal.

## 5. Acknowledgments

The authors are grateful for support from the Science and Technology Facilities Council (STFC) in the UK, the BMBF, Max Planck Society (MPG) and the state of Lower Saxony in Germany. This work was partly supported by DFG grant SFB/Transregio 7 Gravitational Wave Astronomy. This document has been assigned LIGO Laboratory document number LIGO-P0900122.

## References

[1] H. Grote et al 2008 The status of GEO 600, CQG 25 114043
[2] R. Adhikari et al. 2009 LIGO Interferometers' Status the Amaldi conference, submitted to CQG
[3] B. Swinkels et al. 2009 Commissioning status of the Virgo interferometer the Amaldi conference, submitted to CQG
[4] H. Grote 2009 The GEO 600 status, submitted to CQG
[5] G. M. Harry 2007 Titania-doped tantala/silica coatings for gravitational-wave detection, CQG 24 405415
[6] S. Chelkowski et al. 2009 Prospects of higher-order Laguerre-Gauss modes in future gravitational wave detectors, Phys. Rev. D 79, 122002
[7] J. Agresti et al. Proc. 2006 Optimized multilayer dielectric mirror coatings for gravitational wave interferometers SPIE 6286 628608, LIGO-P060027-00-Z.
[8] L. Schnupp 1988 Presentation at European Collaboration Meeting on Interferometric Detection of Gravitational Waves, (Sorrent, Italy, Oct 1988) (unpublished)
[9] S. Hild et al. 2009 DC-readout of a signal-recycled gravitational wave detector, CQG 26 055012
[10] J. Degallaix 2009 Commissioning of the tuned DC readout at GEO600, JPCS, published in this issue
[11] M. Prijatel et al. Control and automatic alignment of the output mode cleaner of GEO600, published in this issue
[12] W. Winkler et al. 2007, The core optics of GEO600, Optics Communications 280, 491-499
[13] A. Buonanno et al 2003 Quantum noise in laser-interferometer gravitational-wave detectors with a heterodyne readout scheme Phys. Rev. D 67 122005
[14] S. Chelkowski et al. 2005 Experimental characterization of frequency-dependent squeezed light, Phys. Rev. A **71** 013806
[15] J. Harms et al. 2003 Squeezed-input, optical-spring, signal-recycled gravitational-wave detectors, Phys. Rev. D **68**, 042001
[16] A. Thüring, et. al, broadband queezing of quantum noise in a Michelson interferometer with Twin-Signal Recycling, Optics Letters, 34, (6), (2009)
[17] S. Hild et al. 2007 Demonstration and comparison of tuned and detuned signal recycling in a large-scale gravitational wave detector, CQG 24 1513
[18] J. Smith 2009 The path to the enhanced and advanced LIGO gravitational-wave detectors, CQG 26 114013
[19] F Acernese et al. 2008 The Virgo 3 km interferometer for gravitational wave detection, J. Opt. A: Pure Appl. Opt. 10 064009
[20] B. Willke et al. 2002 The GEO 600 gravitational wave detector, CQG 19 (7) 1377-1387
[21] S. J. Waldman, 2006 Status of the LIGO at the start of the fifth science run, CQG 23 S653S660
[22] F. Acernese et al. 2008 Status of Virgo, CQG 25 114045

Journal of Physics: Conference Series **228** (2010) 012013 doi:10.1088/1742-6596/228/1/012013

# Commissioning of the tuned DC readout at GEO 600

**J. Degallaix[1], H. Grote[1], M. Prijatelj[1], M. Hewitson[1], S. Hild[2,3], C. Affeldt[1], A. Freise[2], J. Leong[1], H. Lück[1], K. A. Strain[1,3], H. Wittel[1], B. Willke[1] and K. Danzmann[1]**

[1] Max-Planck-Institut für Gravitationsphysik (Albert-Einstein-Institut) and Leibniz Universität Hannover, Callinstr. 38, 30167 Hannover, Germany

[2] School of Physics and Astronomy, The University of Birmingham, Edgbaston, Birmingham, B15 2TT, United Kingdom

[3] Institute for Gravitational Research, University of Glasgow, Glasgow, G12 8QQ, United Kingdom, United Kingdom

E-mail: `jerome.degallaix@aei.mpg.de`

**Abstract.**
Recent experimental results from GEO600 operating with a DC readout and a tuned signal recycling cavity are reported. Compared to the S5/Astrowatch setup, two major changes in the configuration have been implemented: the control readout to keep the interferometer on the dark fringe is changed from heterodyne to homodyne readout and the signal recycling cavity is shifted from a 550 Hz detuning to a 0 Hz detuning (also called tuned). As preliminary experiments showed, the tuned DC readout sensitivity is similar to the heterodyne one. To take advantage of the new DC readout detection scheme, an Output Mode Cleaner (OMC) has to be installed. The design, building and testing of the GEO OMC, which consists of a 4 mirrors monolithic ring cavity, will also be presented in this article.

## 1. Introduction

The transition between GEO [1] and GEO-HF [2] will be achieved after the completion of several major upgrades. Among those upgrades, the error signal used to keep the interferometer on the dark fringe will no longer be derived from a heterodyne readout but from a homodyne detection scheme called DC readout. The advantages of DC readout over heterodyne readout are numerous [3] but the full benefits of this technique can only be achieved with the presence of an Output Mode Cleaner (OMC). At the same time as changing the readout technique, the GEO signal recycling cavity resonance will be shifted from 550 Hz (detuned case) to 0 Hz (tuned case). This shift of the signal recycling mirror lowers the contribution of most technical noises to the sensitivity [4] and it allows a frequency independent injection of broadband squeezed light in order to lower the shot noise limited sensitivity [5].

Experimentally, the advantage of DC readout has first been reported by the 40m prototype interferometer at Caltech [6]. Regarding GEO, the motivation and technicality of tuned DC readout have also already been published [3] and will not be reminded here.

This paper is part of a series of articles about the recent development of GEO 600. Additional details regarding the interferometer can be found in the following papers. The meaning of Astrowatch as well as a comparison of the GEO noise budget for heterodyne and homodyne

© 2010 IOP Publishing Ltd

readout is reported by Grote [7]. The longitudinal control and the autoalignment of the output mode cleaner is described in Prijatelj *et al.* [8]. Finally for the long term vision and to understand the role of tuned DC readout in GEO-HF, the reader can refer to Lück *et al.* [9].

This article, about the preparation of tuned DC readout for GEO-HF, is divided into two parts. In the first part, the experiments and the preliminary sensitivity of GEO in a tuned DC readout configuration are reported. The second part is about the design and the building of the output mode cleaner which has been tested in the Albert Einstein Institute in Hannover before being installed at the GEO site.

## 2. Experimental results of tuned DC readout

### *2.1. Background*

From November 2007 to July 2009, GEO participated in Astrowatch while the LIGO (4k) and Virgo interferometers were being upgraded. GEO achieved its Astrowatch mission with a duty cycle of more than 85%. This high duty cycle was possible due to restrictions placed on commissioning. Only reversible and limited hardware changes were permitted and were carried out during the day in such a way as to allow the detector to operate in low noise condition over the nights and weekends. Even with these constraints, we managed to stably lock on tuned DC readout to futher investigate this configuration.

Due to Astrowatch, we were not able to test the complete DC readout setup planned for GEO-HF. In particular, we did not have an OMC installed, so the higher order optical modes and the modulation sideband fields were still present at the dark port with large amplitudes (contributing a few tens of milliwatts to the detected power). As a result, for the carrier field from the dark fringe offset to dominate the dark port, two actions had to be taken. First, the modulation index of the Michelson sidebands was reduced, reducing the power of the sidebands by a factor 50. Secondly, a large dark fringe offset of 20 pm was implemented.

### *2.2. Lock acquisition*

It would have taken too long to develop a new locking procedure to directly lock to tuned signal recycling and DC readout, so it was naturally decided to first lock the interferometer using detuned heterodyne readout. Thus, the same robust locking script implemented for the previous S5/Astrowatch configuration is also used for DC readout.

After the interferometer is locked with a detuned signal recycling and a low laser input power, the metamorphosis to DC readout can begin. First, control of the interferometer's longitudinal degree of freedom is shifted from low noise photodiodes to photodiodes with large dynamical which are detecting attenuated beams. Second, the signal recycling mirror is (abruptly) shifted from the detuned position to the tuned one [4], the typical jump involved moving the signal recycling mirror by 1.4 nm over 8 millisecond. Third, the RF modulation index is reduced and the gains of the different control loops are adjusted to compensate. Fourth, a dark fringe offset is introduced that, together with the reduced RF sideband power, ensures the fundamental mode dominates the dark port power. Fifth, the dark fringe control is shifted from the heterodyne to the DC readout error signal. And, finally, the input laser power is increased to the nominal power."

Presently, during normal weather conditions, the success rate of going from detuned heterodyne to tuned DC readout is around 70%. The most critical steps of the locking procedure are the controlled jump of the signal recycling mirror from detuned to tuned position and the switching on of the DC control with the high power photodiode.

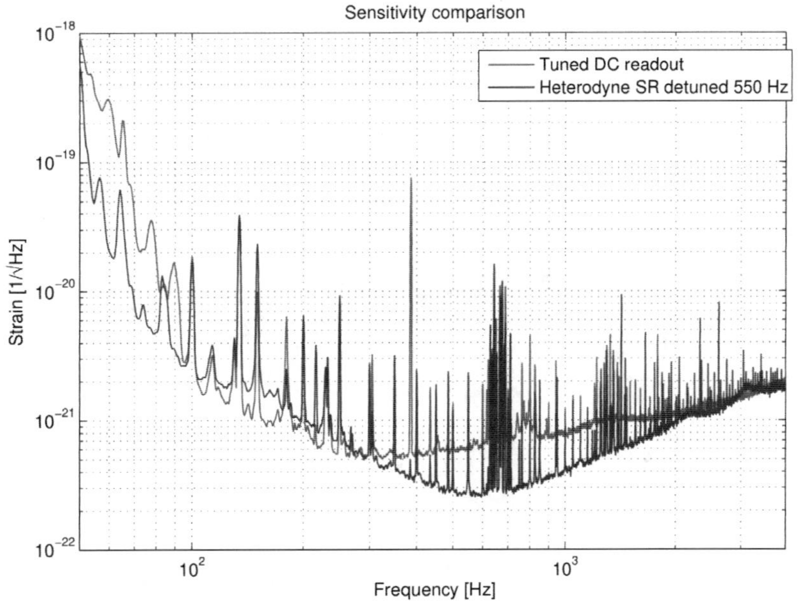


**Figure 1.** Sensitivity comparison between the preliminary tuned DC readout configuration and the heterodyne readout used during S5 and Astrowatch.

### *2.3. The DC readout sensitivity*

The sensitivity achieved for tuned DC readout is presented in figure 1. Compared to the heterodyne sensitivity which has been used since the beginning of GEO and so has been heavily optimised and tested, the sensitivity of tuned DC readout is promising. We obtained a better sensitivity in tuned DC readout compared to heterodyne readout in the frequency range from 100 Hz to 300 Hz and similar sensitivity at high frequencies above 2 kHz, where the sensitivity is limited by the shot noise.

For frequencies below 100 Hz, where we are limited by the Michelson autoalignment feedback noise, DC readout yields a slightly worse sensitivity. That is a direct consequence of the reduction of the sideband amplitude since these sidebands are also used to derive the alignment error signal from the differential wavefront sensing scheme. The decrease in the sideband power is compensated by an increase in the autoalignment gain resulting in higher noise at low frequency, due to the amplification of the electronic noise.

In the frequency range between 300 Hz and 2 kHz, the preliminary tuned DC sensitivity is worse compared to the heterodyne readout. That is not fully understood but recently this noise has been greatly decreased (not shown here but presented in [7]) with better amplitude stabilization of the Schnupp modulation sidebands compatible with the reduced modulation index used and by placing the detection photodiode inside a closed vacuum tank (but still in air).

### *2.4. Noise budget*

The noise budget of tuned DC readout is presented in figure 2. As mentioned earlier, at low frequencies (below 100 Hz) we are limited by the autoalignment feedback noise. In the middle frequency range, the dominant known noise source is the laser amplitude noise. As in the case of heterodyne readout, the noise budget does not explain fully the measured sensitivity since a gap is remaining between the sum of the projected noise and the sensitivity. The large coupling of

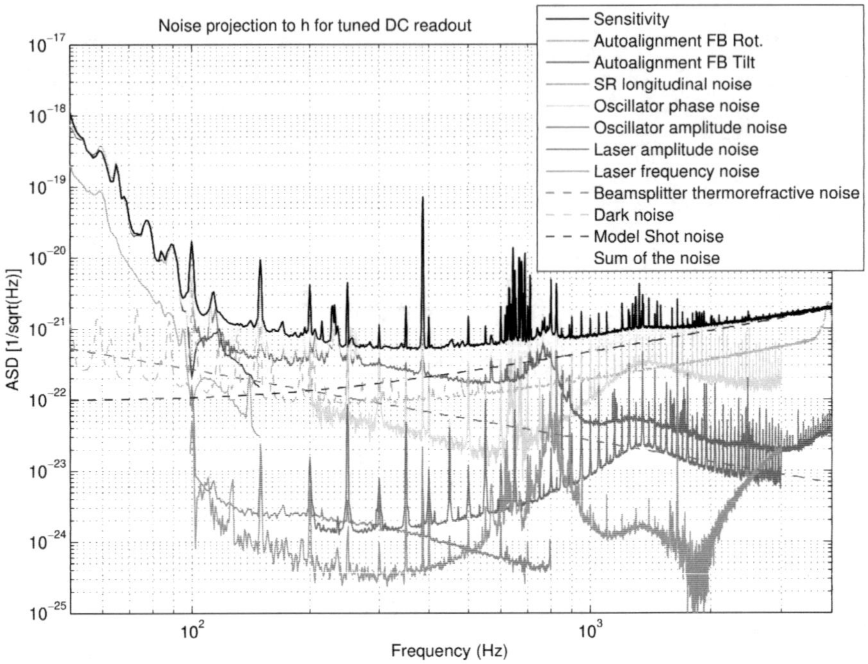


**Figure 2.** Noise budget of the tuned DC readout configuration.

laser amplitude noise is a direct consequence of the dark fringe offset. And at high frequencies, the sensitivity is shot noise limited.

Compared to detuned heterodyne readout, the tuned DC readout scheme lowers the coupling of several technical noises to the sensitivity as shown in figure 3. The transfer functions presented here can simply simply be understood as the sensitivity curve divided by the noise spectra. They are measured by injecting a dominating white noise into the noise source of interest. The absolute magnitude of the transfer functions is arbitrary since it depends how the noise is detected, whiten and then recorded. However the noise detection process is common to heterodyne and DC readout, so a relative comparison of the 2 configurations is still meaningful.

The noise transfer functions of the signal recycling longitudinal actuator, of the laser frequency noise, and of the sidebands oscillator amplitude noise (not shown) are reduced by a factor up to 10. The sidebands oscillator phase noise transfer function has roughly the same level but with different features. Only the coupling of the laser amplitude noise appears higher in the new configuration.

In DC readout, the laser laser amplitude noise couples stronger to the sensitivity than in heterodyne readout. To be explained, it must first be reminded that the calibrated strain sensitivity curve is derived from the error signal used to keep the Michelson interferometer on the dark fringe. For the DC readout configuration, this error signal is simply the power spectra of the light detected at the dark port and so is also a direct measurement of the input laser amplitude [10] (minus the low pass filtering effect of the power recycling cavity). Moreover, the use of a dark fringe offset also reduce the common mode rejection from the Michelson interferometer, increasing slightly the level of the laser amplitude noise at the dark port.

For GEO-HF, we expect an improvement in the sensitivity compared to the one presented here since after the installation of the OMC, a smaller dark fringe offset will be implemented and the sidebands will be restored to their optimal amplitudes while not reaching the detection photodiode.

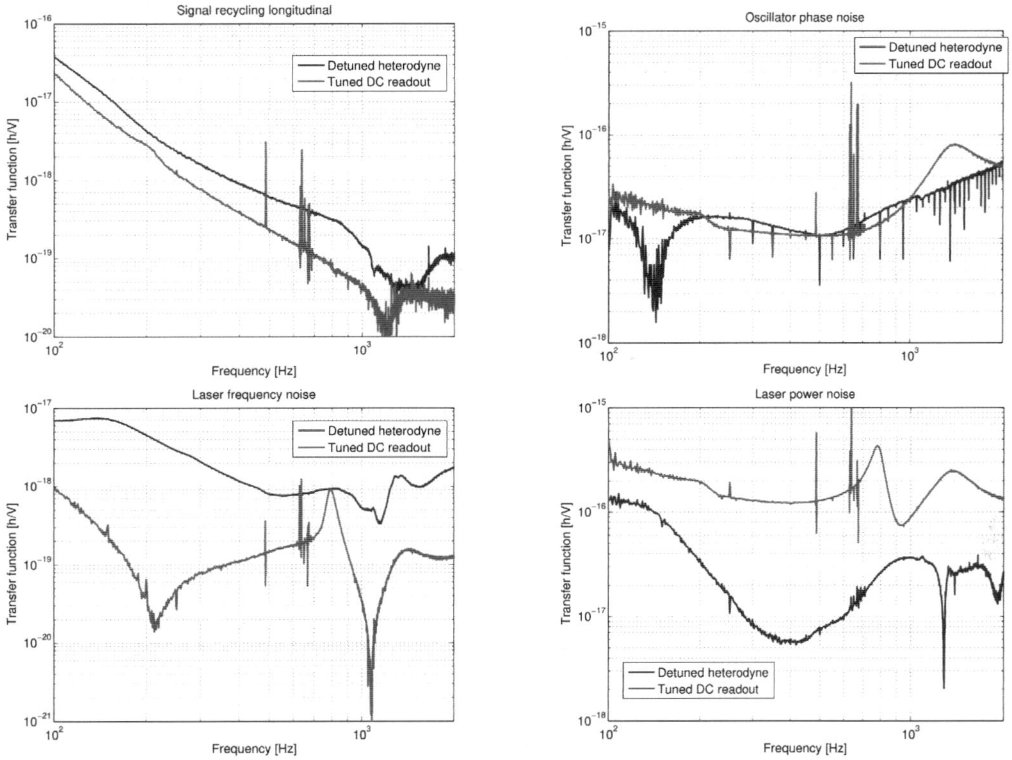


**Figure 3.** Comparison of the noise transfer functions to h between detuned heterodyne (blue curves) and tuned DC readout (red curves). The peaks in the noise transfer function for tuned DC readout at 785 Hz and 1365 Hz are due to the resonance in the signal recycling cavity of respectively, the signal recycling control sidebands and the Michelson control sidebands. Such peaks will be eliminated after the addition of the output mode cleaner which prevents the control sidebands to reach the detection photodiode.

## 3. The GEO 600 output mode cleaner

During the installation of GEO-HF, the input laser will be upgraded and up to 10 times more light will circulate in the interferometer compared to the present GEO setup. Due to the presence of higher order modes and of the sidebands fields kept for the autoalignement system, more than 400 mW of total power will be available for detection at the dark port. To handle that amount of power, multiple photodiodes will have to be used and a large dark fringe offset will be required. To avoid this solution, an Output Mode Cleaner (OMC) will be installed, reducing the incident power on the detection photodiode by a factor 20.

The Output Mode Cleaner (OMC) is a filter cavity transmitting the DC field in the fundamental mode to the photodiode and attenuating the higher order modes and the modulation sidebands. The details of the OMC can be found in the following sections.

### *3.1. Requirements*

The OMC must satisfy two main requirements:

(i) The power of the RF sidebands at 15 MHz must be reduced by at least a factor 100 in transmission. This requirement tells us how selective the mode cleaner must be, so it gives us an upper limit on the mode cleaner bandwidth. Practically, for a light round trip length

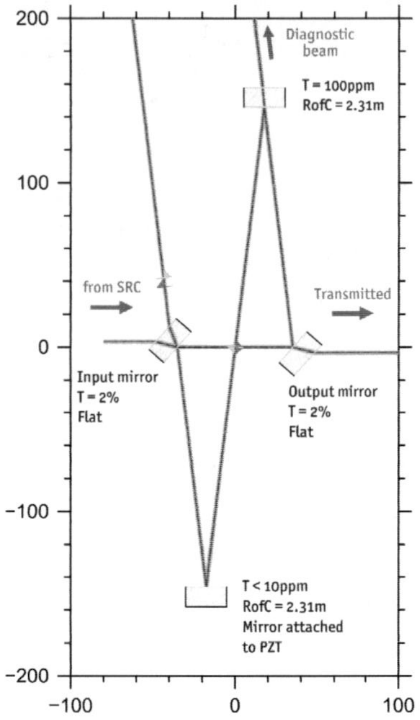


**Figure 4.** Optical setup of the OMC with the dimension in millimeter. The beam leaving the signal recycling cavity is coming from the left.

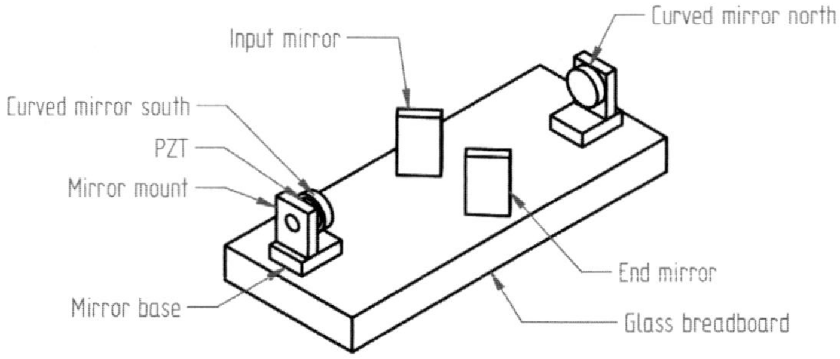


**Figure 5.** Isometric view of the OMC. All parts are mode of fused silica expect for the PZT.

of 66 cm, this requirement is satisfied for a finesse above 150.

(ii) The power of the higher order optical modes must be reduced by at least a factor 100 in transmission. This requirement sets a constraint on the finesse of the cavity and on the g-factor. With a finesse of 150, the requirement is met for a g-factor of 0.3 or 0.7, for suppression of the higher order modes up to the $6^{\text{th}}$ order and for an even number of mirrors in the cavity.

### *3.2. Optical and mechanical design*

For the optical parameters of the OMC, we chose the minimal acceptable finesse i.e. 150. The preferred g factor is 0.735 which allows a larger beam size on the mirror than a g factor of 0.3.

The design of the OMC must also take into account that the optics will be installed on an isolated platform where the weight of the components must be carefully distributed. For this reason, the OMC is composed by 4 mirrors equally distributed on either side of the incoming beam as shown in the left part of figure 4. A 4-mirror configuration presents also two additional advantages: first the vertical and horizontal resonances are degenerate [11], so the number of optical mode resonances is halved compared to a mode cleaner with an odd number of mirrors. Second, the mirror which is not used for the input or output port and not glued to the opaque PZT can be partially transmissive, so a low power beam can exit the OMC and be used for diagnostic without the introduction of an additional beam splitter.

A piezo actuator (PZT) is glued to one of the curved mirror to control the length of the OMC. The PZT is composed of low voltage multi layer stack and will be used to longitudinally lock the OMC to the incoming fundamental mode. The control and the autoalignment as well as the length noise requirements of the OMC are detailed in the paper by M. Prijatelj et al. [8].

For thermal stability and low thermal noise, it was decided that all the parts of the OMC will be made of fused silica, except for the PZT. Instead of using a spacer, the optics are mounted on a rectangular cuboid glass breadboard. The input and output mirrors which are flat, have a rectangular shape, whereas the two curve mirrors are cylindrical.

The low frequency thermal drift of the optical path is further minimized since more than 70% of the thermal expansion of the glass breadboard is compensated by the thermal expansion of the PZT for homogenous temperature change.

The glass breadboard is relatively thick, 38 mm. This dimension was optimized to minimize the nuisance of the mechanical modes which can modulate the optical path length of the OMC. In particular, the first mechanical mode of the OMC breadboard is chosen to be at 1.3 kHz where the sensitivity of the interferometer is already slightly degraded by the second harmonic of the violin modes of the suspension fibers.

The substrate for the optics is made of highly homogenous, low inclusion synthetic fused silica (Schott 7980-0-A), the optical mounts and mirror bases are made with a slightly lower quality fused silica (Schott 7980-3-C) and finally the glass breadboard is made of industry-grade fused silica (Schott Lithosil QT). The optics are superpolished and IBS coated. The other parts are simply polished. In the specification, special attention was drawn to the perpendicularity of the different surfaces to ensure a tilt free assembly.

### *3.3. Building and testing*

The different parts have been glued with the epoxy Optocast 3553-UTF-LV-HM. This UV-cured epoxy was particularly suitable since a very thin layer can be applied and it has very low residual outgassing. The mirrors were first positioned to their mounts using mechanical templates. The use of templates was inspired from the experience developed with the bonding of LISA pathfinder glass breadboard [12]. Since the epoxy is only cured after UV light is applied, all the parts can be accurately positioned without hurry. To position the mirrors on the glass breadboard, two steps were required. First a template was used to glue 3 out of the 4 OMC mirrors. Then to glue the last mirror, a carefully aligned laser was sent through the cavity and the last mirror was positioned manually to close the cavity.

The OMC is controlled via the LIGO Control and Data System (CDS) designed for Advanced LIGO. The error signal to keep the OMC on resonance is derived by dithering the PZT at 20 kHz and then demodulating the transmitted light in phase. Stable locks were achieved, in the AEI clean room with the feedback applied to the input laser. This setup is different from what is planned for GEO-HF, since for GEO-HF the feedback will be applied to the PZT itself [8]. Nevertheless, the test gave us some valuable insight and experience about the OMC and the LIGO CDS.

By scanning the cavity over one free spectral range, two essential parameters of the OMC cavity can be derived. First, the finesse of the cavity, derived from the measurement of the cavity linewidth was found to be 147, and second, the g-factor, derived from the frequency spacing of the higher order modes was measured to be 0.731. Both the finesse and the g factor are within 2% of the design value.

## 4. Conclusion and outlook

A DC readout control scheme has been implemented for the GEO600 interferometer. The preliminary sensitivity achieved without an output mode cleaner is comparable to heterodyne readout in the middle frequency range (100 Hz - 300 Hz) and at high frequency ($\geq$ 2 kHz). In parallel to this work, an output mode cleaner has been build and tested off site. The output mode cleaner actuator is controlled by the LIGO CDS system, showing the path for further digital implementation in GEO.

As of September 2009, the plan is to first have a short commissioning of DC readout without the OMC, then inject squeezed vacuum from the dark port and finally around mid-autumn, install the OMC in vacuum in front of the DC readout photodiode. Only then, intensive commissioning and optimization of tuned DC readout for GEO-HF will be endeavored.

## Acknowledgments

We would like to thank R. Schilling for his precious help with Optocad and the design of the OMC. We also thank Sam Waldman for useful comments on this manuscript. The authors are grateful for support from the Science and Technology Facilities Council (STFC) in the UK, the BMBF, Max Planck Society (MPG) and the state of Lower Saxony in Germany and the European Gravitational Observatory (EGO). This work was partly supported by DFG grant SFB/Transregio 7 "Gravitational Wave Astronomy". This document has been assigned LIGO Laboratory document number LIGO-P0900119.

## References

[1] Grote H (for the LIGO Scientific Collaboration) 2008 *Classical and Quantum Gravity* **25** 114043
[2] Willke B, Ajith P, Allen B, Aufmuth P, Aulbert C, Babak S, Balasubramanian R, Barr B W, Berukoff S, Bunkowski A, Cagnoli G, Cantley C A, Casey M M, Chelkowski S, Chen Y, Churches D, Cokelaer T, Colacino C N, Crooks D R M, Cutler C, Danzmann K, Dupuis R J, Elliffe E, Fallnich C, Franzen A, Freise A, Gholami I, Gossler S, Grant A, Grote H, Grunewald S, Harms J, Hage B, Heinzel G, Heng I S, Hepstonstall A, Heurs M, Hewitson M, Hild S, Hough J, Itoh Y, Jones G, Jones R, Huttner S H, Kotter K, Krishnan B, Kwee P, Lück H, Luna M, Machenschalk B, Malec M, Mercer R A, Meier T, Messenger C, Mohanty S, Mossavi K, Mukherjee S, Murray P, Newton G P, Papa M A, Perreur-Lloyd M, Pitkin M, Plissi M V, Prix R, Quetschke V, Re V, Regimbau T, Rehbein H, Reid S, Ribichini L, Robertson D I, Robertson N A, Robinson C, Romano J D, Rowan S, Rudiger A, Sathyaprakash B S, Schilling R, Schnabel R, Schutz B F, Seifert F, Sintes A M, Smith J R, Sneddon P H, Strain K A, Taylor I, Taylor R, Thuring A, Ungarelli C, Vahlbruch H, Vecchio A, Veitch J, Ward H, Weiland U, Welling H, Wen L, Williams P, Winkler W, Woan G and Zhu R 2006 *Classical and Quantum Gravity* **23** S207–S214
[3] Hild S, Grote H, Degallaix J, Chelkowski S, Danzmann K, Freise A, Hewitson M, Hough J, Lück H, Prijatelj M, Strain K A, Smith J R and Willke B 2009 *Classical and Quantum Gravity* **26** 055012
[4] Hild S, Grote H, Hewitson M, Lück H, Smith J R, Strain K A, Willke B and Danzmann K 2007 *Classical and Quantum Gravity* **24** 1513–1523
[5] Chelkowski S, Vahlbruch H, Hage B, Franzen A, Lastzka N, Danzmann K and Schnabel R 2005 *Physical Review A* **71** 013806
[6] Ward R L, Adhikari R, Abbott B, Abbott R, Barron D, Bork R, Fricke T, Frolov V, Heefner J, Ivanov A, Miyakawa O, McKenzie K, Slagmolen B, Smith M, Taylor R, Vass S, Waldman S and Weinstein A 2008 *Classical and Quantum Gravity* **25** 114030
[7] Grote H 2009 Status of GEO600 LIGO Technical Document No. P0900199, submited to CQG
[8] Prijatelj M, Grote H, Degallaix J, Hewitson M, Hild S, Affeldt C, Freise A, Leong J, Lück H, Strain K A, Wittel H, Willke B and Danzmann K 2009 Control and automatic alignment of the output mode cleaner of GEO 600 LIGO Technical Document No. P0900156, submited to JPCS
[9] Lück H, Affeldt C, Degallaix J, Freise A, Grote H, Hewitson M, Hild S, Leong J, Prijatelj M, Strain K, Willke B, Wittel H and Danzmann K 2009 The transition to GEO-HF LIGO Technical Document No. P0900122, submited to JPCS
[10] Smith J R, Degallaix J, Freise A, Grote H, Hewitson M, Hild S, Lück H, Strain K A and Willke B 2008 *Classical and Quantum Gravity* **25** 035003
[11] Barriga P, Zhao C and Blair D G 2005 *General Relativity and Gravitation* **37** 1609–1619
[12] Middleton K F, Caldwell M E, Edeson R, Neeson C, Munro G, Eley C, Sandford M, Braxmaier C, Johann U, Killow C, Robertson D, Ward H, Hough J, Heinzel G and Wand V 2006 *Optical Engineering* **45** 125601

Journal of Physics: Conference Series **228** (2010) 012014 doi:10.1088/1742-6596/228/1/012014

# Control and automatic alignment of the output mode cleaner of GEO 600

**M. Prijatelj[1], H. Grote[1], J. Degallaix[1], M. Hewitson[1], S. Hild[2,3], C. Affeldt[1], A. Freise[2], J. Leong[1], H. Lück [1], K. A. Strain[1,3], H. Wittel[1], B. Willke[1] and K. Danzmann[1]**

[1] Max-Planck-Institut für Gravitationsphysik (Albert-Einstein-Institut) and Leibniz Universität Hannover, Callinstr. 38, 30167 Hannover, Germany

[2] School of Physics and Astronomy, The University of Birmingham, Edgbaston, Birmingham, B15 2TT, United Kingdom

[3] Institute for Gravitational Research, University of Glasgow, Glasgow, G12 8QQ, United Kingdom, United Kingdom

E-mail: mirko.prijatelj@aei.mpg.de

**Abstract.** The implementation of a mode cleaner at the output port of the GEO 600 gravitational wave detector will be part of the upcoming transition from GEO 600 to GEO-HF. Part of the transition will be the move from a heterodyne readout to a DC readout scheme.

DC readout performance will be limited by higher order optical modes and control sidebands present at the output port. For optimum performance of DC readout an output mode cleaner (OMC) will clean the output beam of these contributions.

Inclusion of an OMC will introduce new noise sources whose magnitudes needed to be estimated and for which new control systems will be needed.

In this article we set requirements on the performance of these control systems and investigate the simulated performance of different designs.

## 1. Introduction

The upcoming GEO-HF [1] upgrade to GEO 600 [2] consists among other things in the change from the current heterodyne readout scheme to a DC readout scheme [3]. DC readout performance will be limited by higher order optical modes (HOMs) and modulation sidebands present in the output beam of GEO 600.

HOMs and sidebands contribute to shot noise but no gravitation wave signal will be extracted from them. In combination with the planned upgrade of the laser power by a factor of 10 the presence of HOMs and sidebands would require multiple photo diodes to accommodate for the increased light power. To reject sidebands and HOMs from the output beam an output mode cleaner (OMC) will be installed. The OMC will be located on a suspended bench. For more details see [3]. Transmissivity of the OMC for the $TEM_{00}$ carrier will depend on the frequency mismatch $\Delta f$ between the GEO 600 output beam and the OMC resonance frequency. It will also depend on the alignment of the output beam onto the OMC.

Fluctuations in $\Delta f$ and the alignment will create fluctuations in the power transmitted through the OMC. In DC readout these fluctuations couple directly to the GEO 600 strain measurement. The impact of these two new noise sources are investigated and compared to other

© 2010 IOP Publishing Ltd

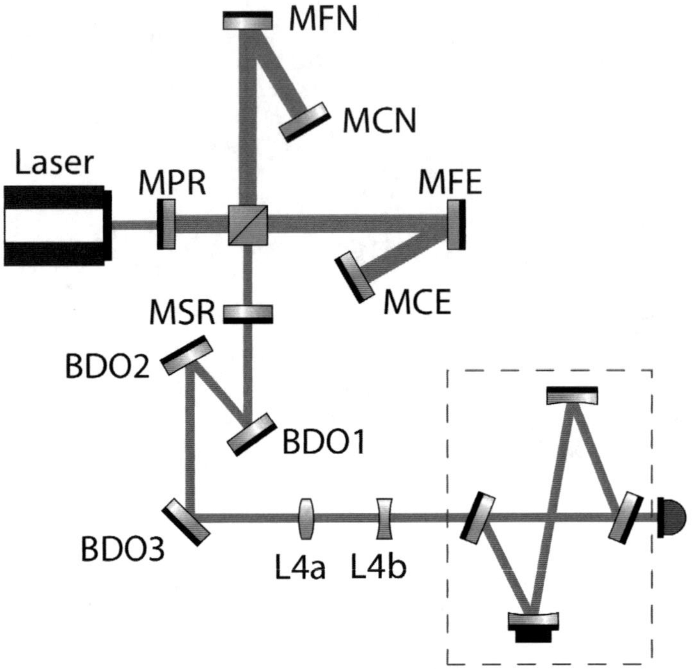


**Figure 1.** Simplified optical layout of the GEO-HF gravitational wave detector. The interferometer is formed by the mirrors MPR, MCN, MCE, MFN, MFE and MSR. BDO1 to BDO3 steer the output beam. L4a and L4b are used to mode match the output beam to the OMC. The output mode cleaner is shown in the lower right corner inside the dashed box.

noise sources in section 2. These investigations lead to requirements onto the control systems that minimize $\Delta f$ and misalignment. In section 3 different designs for control systems are presented and expected performances are compared to the requirements derived in the previous section. In section 4 the current status of the preliminary experiments with the OMC under air are layed out and a outlook is given.

## 2. Investigations regarding noise sources introduced by the OMC

The magnitude of the noise sources introduced by the OMC are investigated. Requirements are formulated to ensure that they are below shot noise for the frequencies of interest.

The frequency of the GEO 600 laser is locked to the GEO 600 power recycling cavity (PRC) [4]. The PRC is formed by the mirrors MPR, MCE and MCN (see fig. 1). These mirrors are seismically isolated by multiple pendulum stages with a resonance frequency around 1 Hz [5]. If we omit frequencies below 0.3 Hz their residual motion is dominated by frequencies around 1 Hz. This is well below the GEO-HF measurement band. The amplitude of these low frequency fluctuations can be assumed to be $\pm 2\,\mu$m. Given the length of the PRC of 1200 m this results in a frequency change of the laser of $\pm 470$ kHz. Since fluctuations in the OMC length around 1 Hz are negligible due to its monolithic design this introduces a frequency mismatch $\Delta f_{\mathrm{LF}}$ between the output beam of GEO 600 and the OMC resonance frequency of the same amount. At higher frequencies in the GEO-HF measurement band length fluctuations of the PRC can be neglected due to the elaborate seismic isolation system of the mirrors forming the PRC.

In addition to $\Delta f_{\mathrm{LF}}$ there are higher frequency fluctuations $\Delta f_{\mathrm{audio}}$ in the frequency mismatch dominated by length changes of the OMC. The actual value of $\Delta f_{\mathrm{LF}}$ defines the coupling coefficient for $\Delta f_{\mathrm{audio}}$ into the power transmitted through the OMC. Since fluctuations in the length of the OMC at $\Delta f_{\text{audio}}$ happen in the measurement band of GEO-HF they contribute to noise in the strain signal. It is therefore important to minimize the low frequency frequency mismatch to minimize the coupling coefficient. Coupling coefficients were calculated using FINESSE software [6]. We want the fluctuations in the power transmitted through the OMC due to $\Delta f_{\text{audio}}$ to be less than 1/10 shot noise. The resulting requirement for $\Delta f_{\text{audio}}$ was calculated:

$$\Delta \mathrm{f}_{\text{audio}} \leq 10.79 \frac{\mathrm{Hz}}{\sqrt{\mathrm{Hz}}} \frac{470\,\mathrm{kHz}}{\Delta f_{\mathrm{LF}}}. \tag{1}$$

The OMC length will be 65.8 cm. The required stability in the measurement band was calculated as

$$\text{OMC length stability} \leq 7.69 \cdot 10^{-16} \frac{\mathrm{m}}{\sqrt{\mathrm{Hz}}} \frac{470\,\mathrm{kHz}}{\Delta f_{\mathrm{LF}}}. \tag{2}$$

The OMC will be located on a suspended bench. We assume that the OMC will achieve a length stability of $1 \cdot 10^{-15}\mathrm{m}/\sqrt{\mathrm{Hz}}$. This requires $\Delta f_{\mathrm{LF}}$ to be less than 100 kHz. This requires the closed loop suppression at the dominant frequencies around 1 Hz to be at least 5. This should be achievable without great difficulty.

### *2.1. Alignment requirements*

Determination of alignment requirements follows the same structure as the determination of the frequency mismatch requirements. Again the margin of allowed fluctuations in the measurement band is determined by lower frequency fluctuations. Alignment fluctuations in and below the measurement band are both caused by fluctuations in the rotation and tilt of the GEO 600 mirrors MPR, MCE, MCN, MSR, BDO1, BDO3 and translation of the OMC itself. Due to the GEO 600 configuration it is useful to investigate motions of the mirrors MCE and MCN as the degrees of freedom (DOFs) MCE-MCN and MCE+MCN. MCE-MCN is the differential misalignment. Low frequency rotations of the mirrors are assumed to be ±250 nrad. Mirror fluctuations in the measurement band were estimated by multiplying seismometer data from the GEO site with the respective pendulum transfer functions. Pendulum transfer functions were assumed to be of magnitude 1 up to the pendulum resonance frequency at 1 Hz and $f^{-2n}$ with n the number of pendulum stages for higher frequencies. The number of pendulum stages for the different mirrors can be seen in table 1. The suspension points of the pendulums are isolated by stack isolators consisting of layers of metal and rubber. These stack isolators provide additional isolation above approximately 10 Hz.

Coupling of alignment fluctuations in the measurement band in any DOF into the power transmitted through the OMC is determined by the current misalignments of all other DOFs. To simulate the coupling coefficients of a DOF the other DOFs were assumed to be misaligned by discrete values in steps of 50 nrad between -250 nrad and 250 nrad. This leads to 23716 possible configurations: (7 DOFs times 11 misalignment values times 2 axes)$^2$. To reduce the computational needs only 2 DOFs at a time were evaluated yielding less than 1000 configurations. For each DOF the biggest coupling coefficient was recorded and the misalignment fluctuations in the measurement band were multiplied by it. The resulting fluctuations in the power transmitted through the OMC were summed coherently. The resulting noise is shown in figure 2. It can be seen that noise due to misalignments falls below shot noise at 300 Hz (see 'all DOFs' trace). This is for a power transmitted through the OMC of 40 mW. It should be noted that this noise estimation is very conservative. The DOFs of the MPR mirror and MCE-MCN are controlled by automatic alignment (AA) systems that reduce their misalignments. If perfect alignment for these DOFs is assumed this results in a alignment noise as shown in trace 'w/o AA-controlled DOFs' in figure 2. Noise introduced by the OMC will fall below shot noise above frequencies of 100 Hz to 300 Hz. Therefore only a drift control is needed.

Journal of Physics: Conference Series **228** (2010) 012014 doi:10.1088/1742-6596/228/1/012014

**Table 1.** Number of pendulum seismic isolation stages of different GEO 600 mirrors.

| | MPR | MCE | MCN | MSR | BDO1 | BDO3 |
|---|---|---|---|---|---|---|
| # of pendulum stages | 2 | 3 | 3 | 3 | 1 | 1 |

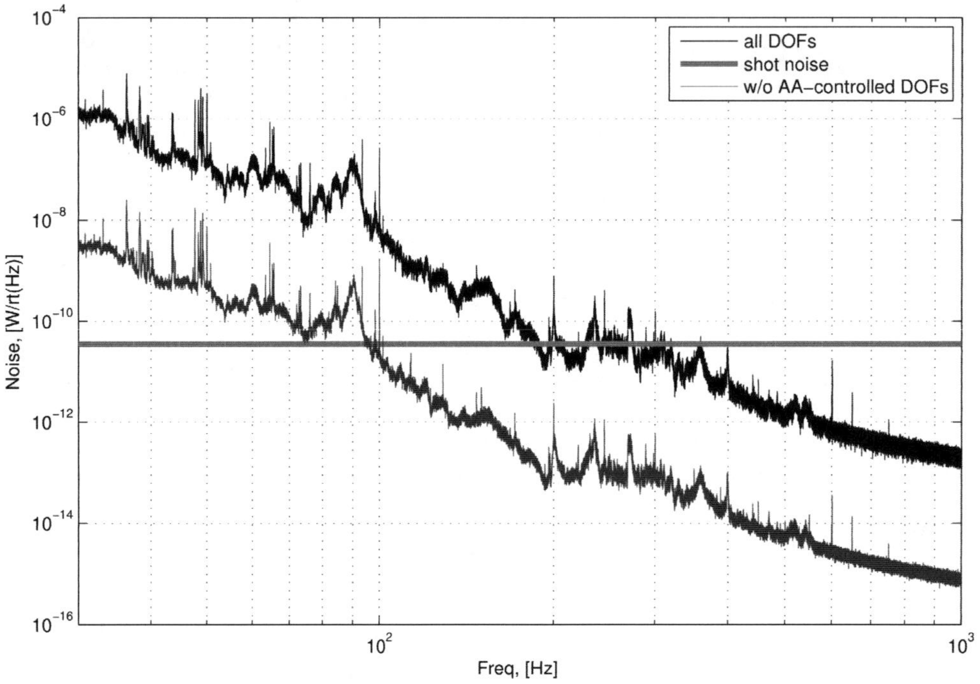


**Figure 2.** Noise in the power transmitted through the OMC induced by misalignments. Uncalibrated spectrum at the detector behind the OMC. Transmitted power 40 mW.

## 3. Potential output mode cleaner control systems

### *3.1. Longitudinal control*

As has been stated in the previous section the frequency mismatch between the OMC and the PRC resonance frequencies needs to be controlled and minimized. The error signal for this control system is generated via dither locking. The OMC length is dithered (modulated) at $f_{\mathrm{dither}}$. At zero frequency mismatch this leads to a modulation of the power transmitted through the OMC at $2 \cdot f_{\mathrm{dither}}$. With increasing frequency mismatch, a modulation at $f_{\mathrm{dither}}$ appears. It is possible to either make the OMC follow the PRC or to make the PRC follow the OMC. We will first make the OMC follow the PRC. This has the benefit of avoiding adding additional feedback into the complex control loops controlling the suspended GEO 600 mirrors. If it should prove necessary to extend the actuation range actuation on the PRC length will however be possible. Tuning the length of the OMC is done with a piezo actuator (Physik Instrumente S-303) attached to one of the OMC mirrors (see fig. 3). Actuation range of the PZT is $3\,\mu$m

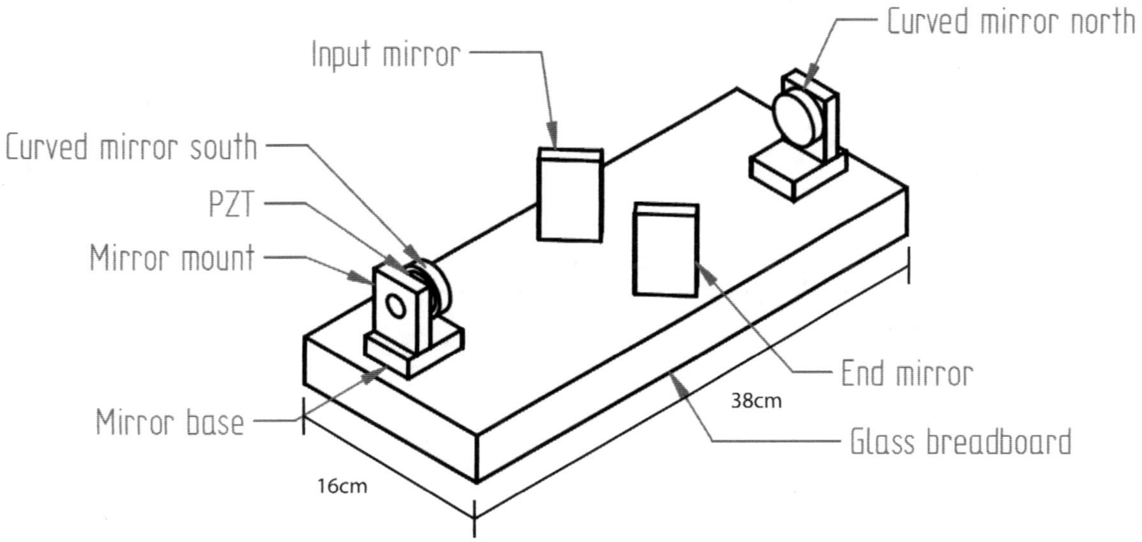


**Figure 3.** Schematic view of the output mode cleaner. Modified from [3].

corresponding to a change in resonance frequency of the OMC of 1.3 MHz. We do not expect lengh-to-angle coupling to be a problem.

*3.2. Alignment control*

All possible alignment configurations can be represented as points in a 4D parameter space. Misalignments in one plane can be represented as a transversal shift and a rotation. For two planes this results in a 4D parameter space. A requirement for the automatic alignment system is to have actuators available that can cover the whole 4D alignment parameter space. Promising actuators in our case are BDO1 and BDO3. Actuators can be represented by vectors in the alignment parameter space. These vectors are the rows of the control matrix. The angle between the two actuator vectors in the alignment space is a measure the controllability of the system. As a rule of thumb a bigger angle leads to a more robust control system. This separation angle is influenced by the optical configuration of the output optics. Several configurations for the output optics are possible: keep the existing BDO mirrors and add two lenses or change the BDOs ROCs and add either one or no lens. For simplicity we have decided to keep the existing BDOs and accept the non-optimal separation angle. Separation angles for several configurations are compared in table 2. The benefit of keeping the existing BDO mirrors convinced us to choose the two lens configuration.

Several designs for the automatic alignment system were investigated:

1. Dither lock with four frequencies. Dithering is applied to BDO1 and BDO3:

Both tilt (pitch) and rotation (yaw) degrees of freedom for BDO1 and BDO3 respectively are dithered at separate frequencies $f_1$ to $f_4$. At perfect alignment the power transmitted through the OMC will be modulated at frequencies $(2 \cdot f_1)$ to $(2 \cdot f_4)$. Proportionally to misalignment modulations at $f_1$ to $f_4$ appear. In the current coil driver design, the maximum possible dither amplitude is limited to

$$\text{Dither amplitude} \leq 1.14 \cdot 10^{-7}\,\text{rad}\frac{100\,\text{Hz}^2}{{f_\text{dither}}^2} \tag{3}$$

Journal of Physics: Conference Series **228** (2010) 012014 doi:10.1088/1742-6596/228/1/012014

**Table 2.** Comparison of optical configurations of the output optics and resulting separation angles for the actuators BDO1 and BDO3. $R_{BDO1}$ and $R_{BDO2}$ are the radii of curvature of BDO1 and BDO2. $f_{L4a}$ and $f_{L4b}$ are the focal lengths of L4a and L4b.

| Configuration | $R_{BDO1}$ [m] | $R_{BDO2}$ [m] | $f_{L4a}$ [m] | $f_{L4b}$ [m] | Sep. angle [°] |
|---|---|---|---|---|---|
| Two lens | 6.27 | $\infty$ | 0.25 | -0.052 | 17.44 |
| One lens | 6.27 | -6.86 | -0.50 | $\infty$ | 9.47 |
| Zero lens | 4.12 | -1.14 | $\infty$ | $\infty$ | 64.69 |

**Table 3.** Shot noise limited sensitivity of different designs for the output mode cleaner automatic alignment system in terms of BDO1 and BDO3 misalignments.

| | Design 1 | Design 2 | Design 3 |
|---|---|---|---|
| BDO1 rotation [rad] | $4.88 \cdot 10^{-12}$ | $9.00 \cdot 10^{-3}$ | $2.04 \cdot 10^{-9}$ |
| BDO1 tilt [rad] | $1.95 \cdot 10^{-10}$ | $9.30 \cdot 10^{-4}$ | $1.48 \cdot 10^{-9}$ |
| BDO3 rotation [rad] | $1.05 \cdot 10^{-9}$ | $7.26 \cdot 10^{-2}$ | $2.70 \cdot 10^{-8}$ |
| BDO3 tilt [rad] | $9.78 \cdot 10^{-8}$ | $1.19 \cdot 10^{-1}$ | $3.15 \cdot 10^{-8}$ |

In the simulation we assumed a dither amplitude of $1 \cdot 10^{-7}$ rad which corresponds to a dither frequency of about 105 Hz. A lower dither frequency can also be implemented.

2. Dither lock with a common frequency. Dithering is applied to BDO1 and BDO3:
Again the BDO mirrors are dithered. This time with a common frequency. The light reflected off of the OMC is split 50:50 and detected by two quadrant photo detectors. The Gouy phase shift between them is $\pi/2$. Essentially this is differential wavefront sensing [7] with sidebands created by dithering the BDO mirrors instead of by a electro-optic modulator. HOM-content of the GEO 600 output beam was assumed to be 1.4 mW. Ideal mode matching was assumed.

3. Dither lock with a common frequency. Dithering is applied to the OMC length:
In this configuration the OMC length is dithered. The same OMC length dither can be used for longitudinal and alignment control. The full width at half maximum of the OMC resonance is about 7 nm. A realistic dither amplitude can be assumed to be 1 nm. Just as in design 2 the light reflected off of the OMC is split 50:50 and detected by two quadrant photo detectors.
The shot noise limited sensitivities of designs 1 to 3 were estimated using the FINESSE simulation software for a power of 40 mW transmitted through the OMC. Their sensitivities are presented in table 3. Since the sensitivities of design 1 and 3 are considerably better than the requirements we derived on mirror movements in section 2, both designs seem promising. Sensitivities and requirements should scale proportionally for different noise levels. We intend to test designs 1 and 3 in GEO-HF.

## 4. Preliminary tests of the output mode cleaner

The output mode cleaner has been assembled in a clean room. The frequency locking scheme has been successfully tested by locking the OMC to a laser (Innolight Mephisto 500 NE). The frequency mismatch has been minimized in this test setup by actuating on the laser frequency. The dithering, demodulation and the control loop filter were implemented with a real-time computer control system identical to the one developed for the Advanced LIGO detector. The unity gain frequency of this control loop was 145 Hz and closed loop suppression at 1 Hz was above 30 dB.

It has become obvious that the OMC can easily lock onto HOMs instead of the $TEM_{00}$ mode. Since HOMs do not contain an easily usable gravitational wave signal the OMC must not lock onto HOMs when integrated into GEO 600. Future investigations will show if this will pose a problem in the GEO 600 implementation.

## Acknowledgments

The authors are grateful for support from the Science and Technology Facilities Council (STFC) in the UK, the BMBF, Max Planck Society (MPG) and the state of Lower Saxony in Germany and the European Gravitational Observatory (EGO). This work was partly supported by DFG grant SFB/Transregio 7 "Gravitational Wave Astronomy". This document has been assigned LIGO Laboratory document number LIGO-P0900156-v2.

## References

[1] Lück H, Affeldt C, Degallaix J, Freise A, Grote H, Hewitson M, Hild S, Leong J, Prijatelj M, Strain K A, Willke B, Wittel H, Danzmann K 2009 Commissioning of the tuned DC readout at GEO 600 *J. Phys.: Conf. Series* published in this issue
[2] Grote H (for the LIGO Scientic Collaboration) 2008 *Class. and Quantum Grav.* B **25** 114043
[3] Degallaix J, Grote H, Prijatelj M, Hewitson M, Hild S, Affeldt C, Freise A, Leong J, Lück H, Strain K A, Wittel H, Willke B and Danzmann K 2009 *J. Phys.: Conf. Series* published in this issue
[4] Wilke B et al 2002 *Class. and Quantum Grav.* **19** 1377
[5] Goßler S 2004 Phd thesis *http://edok01.tib.uni-hannover.de/edoks/e01dh04/389559768.pdf*
[6] Freise A *http://www.rzg.mpg.de/adf/*
[7] Morrison E et al 1994 *Appl. Opt.* **33** 5041-5049
[8] Risk W P, Gosnell T R, Nurmikko A V 2003 *Compact blue-green lasers* (Cambridge: Cambridge University Press)

Journal of Physics: Conference Series **228** (2010) 012015 doi:10.1088/1742-6596/228/1/012015

# Virgo calibration and reconstruction of the gravitationnal wave strain during VSR1

**T. Accadia[12], F. Acernese[6ac], F. Antonucci[9a], S. Aoudia[15a], K. G. Arun[11], P. Astone[9a], G. Ballardin[2], F. Barone[6ac], M. Barsuglia[1], Th. S. Bauer[14a], M.G. Beker[14a], A. Belletoile[12], S. Bigotta[8ab], S. Birindelli[15a], M. Bitossi[8a], M. A. Bizouard[11], M. Blom[14a], C. Boccara[3], F. Bondu[15b], L. Bonelli[8ab], R. Bonnand[13], L. Bosi[7a], S. Braccini[8a], C. Bradaschia[8a], A. Brillet[15a], V. Brisson[11], R. Budzyński[17b], T. Bulik[17cd], H. J. Bulten[14ab], D. Buskulic[12], C. Buy[1], G. Cagnoli[4a], E. Calloni[6ab], E. Campagna[4ab], B. Canuel[2], F. Carbognani[2], F. Cavalier[11], R. Cavalieri[2], G. Cella[8a], E. Cesarini[4b], E. Chassande-Mottin[1], A. Chincarini[5], F. Cleva[15a], E. Coccia[10ab], C. N. Colacino[8a], J. Colas[2], A. Colla[9ab], M. Colombini[9b], A. Corsi[9a], J.-P. Coulon[15a], E. Cuoco[2], S. D'Antonio[10a], A. Dari[7ab], V. Dattilo[2], M. Davier[11], R. Day[2], R. De Rosa[6ab], M. del Prete[8ac], L. Di Fiore[6a], A. Di Lieto[8ab], M. Di Paolo Emilio[10ac], A. Di Virgilio[8a], A. Dietz[12], M. Drago[16cd], V. Fafone[10ab], I. Ferrante[8ab], F. Fidecaro[8ab], I. Fiori[2], R. Flaminio[13], J.-D. Fournier[15a], J. Franc[13], S. Frasca[9ab], F. Frasconi[8a], A. Freise*, M. Galimberti[13], L. Gammaitoni[7ab], F. Garufi[6ab], G. Gemme[5], E. Genin[2], A. Gennai[8a], A. Giazotto[8a], R. Gouaty[12], M. Granata[1], C. Greverie[15a], G. M. Guidi[4ab], H. Heitmann[15], P. Hello[11], S. Hild**, D. Huet[2], P. Jaranowski[17e], I. Kowalska[17c], A. Królak[17af], N. Leroy[11], N. Letendre[12], T. G. F. Li[14a], M. Lorenzini[4a], V. Loriette[3], G. Losurdo[4a], J. M. Mackowski[13], E. Majorana[9a], I. Maksimovic[3], N. Man[15a], M. Mantovani[8ac], F. Marchesoni[7a], F. Marion[12], J. Marque[2], F. Martelli[4ab], A. Masserot[12], C. Michel[13], L. Milano[6ab], Y. Minenkov[10a], M. Mohan[2], J. Moreau[3], N. Morgado[13], A. Morgia[10ab], S. Mosca[6ab], V. Moscatelli[9a], B. Mours[12], I. Neri[7ab], F. Nocera[2], G. Pagliaroli[10ac], L. Palladino[10ac], C. Palomba[9a], F. Paoletti[8a,2], S. Pardi[6ab], M. Parisi[6b], A. Pasqualetti[2], R. Passaquieti[8ab], D. Passuello[8a], G. Persichetti[6ab], M. Pichot[15a], F. Piergiovanni[4ab], M. Pietka[17e], L. Pinard[13], R. Poggiani[8ab], M. Prato[5], G. A. Prodi[16ab], M. Punturo[7a], P. Puppo[9a], O. Rabaste[1], D. S. Rabeling[14ab], P. Rapagnani[9ab], V. Re[16ab], T. Regimbau[15a], F. Ricci[9ab], F. Robinet[11], A. Rocchi[10a], L. Rolland[12], R. Romano[6ac], D. Rosińska[17g], P. Ruggi[2], B. Sassolas[13], D. Sentenac[2], R. Sturani[4ab], B. Swinkels[2], A. Toncelli[8ab], M. Tonelli[8ab], O. Torre[8ac], E. Tournefier[12], F. Travasso[7ab], J. Trummer[12], G. Vajente[8ab], J. F. J. van den Brand[14ab], S. van der Putten[14a], M. Vavoulidis[11], G. Vedovato[16c], D. Verkindt[12], F. Vetrano[4ab], A. Viceré[4ab],**

© 2010 IOP Publishing Ltd

**J.-Y. Vinet[15a], H. Vocca[7a], M. Was[11], M. Yvert[12]**

[1]AstroParticule et Cosmologie (APC), CNRS: UMR7164-IN2P3-Observatoire de Paris-Université Denis Diderot-Paris 7 - CEA : DSM/IRFU
[2]European Gravitational Observatory (EGO), I-56021 Cascina (Pi), Italy
[3]ESPCI, CNRS, F-75005 Paris, France
[4]INFN, Sezione di Firenze, I-50019 Sesto Fiorentino[a]; Università degli Studi di Urbino 'Carlo Bo', I-61029 Urbino[b], Italy
[5]INFN, Sezione di Genova; I-16146 Genova, Italy
[6]INFN, sezione di Napoli [a]; Università di Napoli 'Federico II'[b] Complesso Universitario di Monte S.Angelo, I-80126 Napoli; Università di Salerno, Fisciano, I-84084 Salerno[c], Italy
[7]INFN, Sezione di Perugia[a]; Università di Perugia[b], I-6123 Perugia,Italy
[8]INFN, Sezione di Pisa[a]; Università di Pisa[b]; I-56127 Pisa; Università di Siena, I-53100 Siena[c], Italy
[9]INFN, Sezione di Roma[a]; Università 'La Sapienza'[b], I-00185 Roma, Italy
[10]INFN, Sezione di Roma Tor Vergata[a]; Università di Roma Tor Vergata[b]; Università dell'Aquila, I-67100 L'Aquila[c], Italy
[11]LAL, Université Paris-Sud, IN2P3/CNRS, F-91898 Orsay, France
[12]Laboratoire d'Annecy-le-Vieux de Physique des Particules (LAPP), IN2P3/CNRS, Université de Savoie, F-74941 Annecy-le-Vieux, France
[13]Laboratoire des Matériaux Avancés (LMA), IN2P3/CNRS, F-69622 Villeurbanne, Lyon, France
[14]Nikhef, National Institute for Subatomic Physics, P.O. Box 41882, 1009 DB Amsterdam[a]; VU University Amsterdam, De Boelelaan 1081, 1081 HV Amsterdam[b], The Netherlands
[15]Université Nice-Sophia-Antipolis, CNRS, Observatoire de la Côte d'Azur, F-06304 Nice[a]; Institut de Physique de Rennes, CNRS, Université de Rennes 1, 35042 Rennes[b]; France
[16]INFN, Gruppo Collegato di Trento[a] and Università di Trento[b], I-38050 Povo, Trento, Italy; INFN, Sezione di Padova[c] and Università di Padova[d], I-35131 Padova, Italy
[17]IM-PAN 00-956 Warsaw[a]; Warsaw Univ. 00-681 Warsaw[b]; Astro. Obs. Warsaw Univ. 00-478 Warsaw[c]; CAMK-PAN 00-716 Warsaw[d]; Białystok Univ. 15-424 Bial ystok[e]; IPJ 05-400 Świerk-Otwock[f]; Inst. of Astronomy 65-265 Zielona Góra[g], Poland
[*]University of Birmingham, Birmingham, B15 2TT, United Kingdom
[**]University of Glasgow, Glasgow, G12 8QQ, United Kingdom

E-mail: `rollandl@in2p3.fr`

**Abstract.** Virgo is a kilometer-length interferometer for gravitationnal waves detection located near Pisa. Its first science run, VSR1, occured from May to October 2007. The aims of the calibration are to measure the detector sensitivity and to reconstruct the time series of the gravitationnal wave strain $h(t)$.

The absolute length calibration is based on an original non-linear reconstruction of the differential arm length variations in free swinging Michelson configurations. It uses the laser wavelength as length standard. This method is used to calibrate the frequency dependent response of the Virgo mirror actuators and derive the detector in-loop response and sensitivity within $\sim 5\%$.

The principle of the strain reconstruction is highlighted and the $h(t)$ systematic errors are estimated. A photon calibrator is used to check the sign of $h(t)$. The reconstructed $h(t)$ during VSR1 is valid from 10 Hz up to 10 kHz with systematic errors estimated to 6% in amplitude. The phase error is estimated to be 70 mrad below 1.9 kHz and 6 $\mu$s above.

## 1. Introduction

The Virgo detector [1] is a kilometer-length interferometer (ITF) located near Pisa (Italy). It is designed to search for gravitational waves (GW) in the frequency range from 10 Hz to a few kHz. Expected astrophysical sources of detectable GWs are compact objects such as neutrons stars or black holes.

The Virgo first science run (VSR1) was performed from May 18th to October 1st 2007 with a sensitivity close to its nominal one, and in coincidence to the end of the fifth science run (S5) of the three LIGO detectors [2]. In order to reduce the false detection rate and improve the estimation of the parameters of potential sources, the data of all the detectors are used together to search for a GW signal.

The purpose of the Virgo calibration is to measure the parameters needed (i) to estimate the ITF sensitivity to GW strain as a function of frequency, $\tilde{h}(f)$, and (ii) to reconstruct from the ITF data the amplitude $h(t)$ of the GW strain signal. To achieve optimum sensitivity, the positions of the different mirrors are controlled [3] to have, in particular, beam resonance in the cavities, destructive interference at the ITF output port and to compensate for environmental noise. The controls would also attenuate the effect of a passing GW below a few hundreds hertz. A synopsis of the longitudinal control loop and its components is given in figure 1. The effects of the longitudinal controls have thus to be precisely calibrated in the frequency domain. Above a few hundreds hertz, the mirrors behave as free falling masses in the longitudinal direction. The main effect of a passing GW would then be a frequency-dependent variation of the power at the ITF output. characterized by the ITF optical response. Finally, the readout electronics of the output power and its timing precision have also to be calibrated.

In this paper, the main method used to get an absolute length calibration is first described and the mirror actuator calibration results are highlighted. The way the Virgo sensitivity is measured is then shown. In the last section, the principle of the h(t) reconstruction is described as well as some measurements performed to estimate the errors on the h(t) signal.

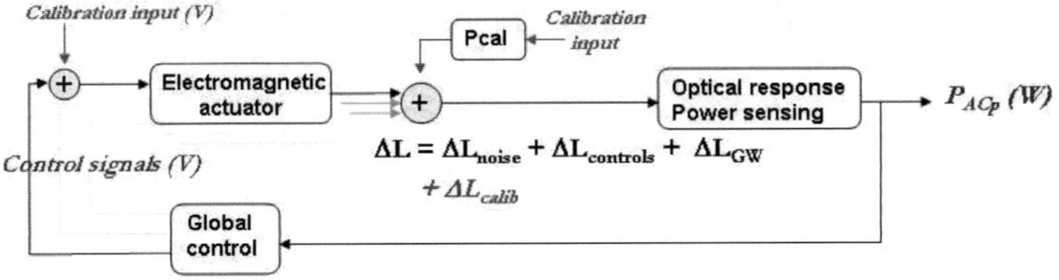


**Figure 1.** Synopsis of the Virgo longitudinal control loop. A variation of the differential length of the arms of the ITF, $\Delta L$, is sensed through the optical response which converts the length variation into laser power variation at the ITF output port. The power impinging on photodiodes is read and sampled at 20 kHz in the channel $P_{ACp}$ (W). This channel is used as error signal to compute the control signals (V) sent to the different mirror actuators.
Calibration signals can be injected into the loop to force differential length variations $\Delta L_{calib}$. The standard path is through the electromagnetic actuators. An auxiliary setup, the photon calibrator (pcal), can also be used (see section 4.3).

## 2. Absolute length measurement for mirror actuation calibration

The ITF calibration is based on absolute length measurements. The displacement induced by the mirror actuators is reconstructed using the ITF as a simple Michelson. The non-linear method used to reconstruct the arm differential motion in the freely swinging Michelson with passing fringes is described. The laser wavelength is used as length standard.

Mirrors of the Virgo ITF are mis-aligned to get a simple Michelson configuration. Different configurations are used to calibrate the beam-splitter and arm mirrors actuations (comprising the beam-splitter, one end mirror and the input mirror in the other arm). The typical results of the mirror actuation calibration are given.

*2.1. Absolute length measurement*

In a simple Michelson ITF, the phase difference $\Delta\Phi$ between the two interfering beams is function of the differential arm length $\Delta L$, using the laser wavelength $\lambda = 1064\,\text{nm}$ as standard: $\Delta\Phi(t) = \frac{4\pi}{\lambda}\Delta L(t)$.

In a simple Michelson ITF with frontal phase-modulation, the power of the output beam ($\mathcal{P}_{DC}$) as well as the demodulated power ($\mathcal{P}_{AC}$) signals are functions of the phase difference $\Delta\Phi$ between the two interfering beams: $\mathcal{P}_{DC} = \beta(1 - \gamma\cos(\Delta\Phi))$ and $\mathcal{P}_{AC} = \alpha\sin(\Delta\Phi)$, where $\alpha$ and $\beta$ are proportional to the laser power and $\gamma$ is proportional to the ITF contrast. Therefore, in the $(\mathcal{P}_{DC}, \mathcal{P}_{AC})$ plane, the signal follows an ellipse as shown in figure 2.

The measurement of the differential arm length $\Delta L$ requires a non-linear reconstruction. The ellipse is fitted using the method described in [4]. The fit gives the ellipse center position (theoretically ($\beta$,0)) and the axis widths (theoretically ($\beta\gamma, \alpha$)). The variations of the parameters $\alpha$, $\beta$ and $\gamma$ are monitored and found to be of the order of 0.5% during a few-minute dataset. $\Delta\Phi'$ can then be estimated directly from the ITF signals as the angle between the ellipse axis along DC and the line from the ellipse center to the current point position $(\mathcal{P}_{DC}, \mathcal{P}_{AC})$. Using a suitable ellipse tour counting, the right number of $2\pi$ is added to $\Delta\Phi'$ to recover completely the angle $\Delta\Phi$. The differential arm length $\Delta L_{rec}$ is then computed from equation 2.1.

Figure 3 is an illustration of the method. In the window $\Delta t$, 6 interfringes passed on the DC signal: It indicates a differential arm elongation of $6 \times \frac{\lambda}{2} = 3.19\,\mu\text{m}$. In the same window, the reconstructed $\Delta L$ varies by $\sim 3.18\,\mu\text{m}$.

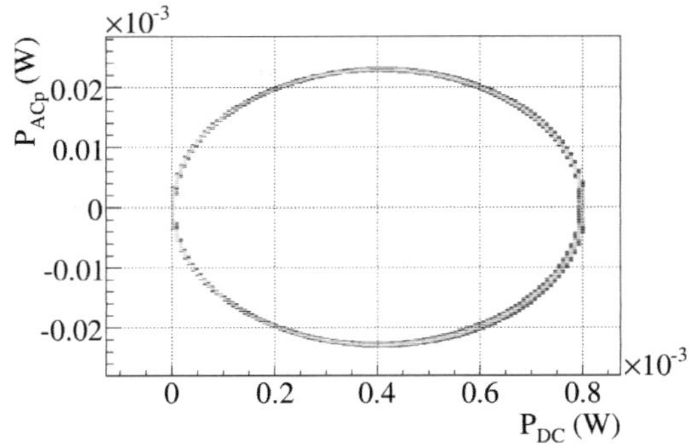


**Figure 2.** $\mathcal{P}_{AC}$ vs $\mathcal{P}_{DC}$ measured in free swinging Michelson data (configuration comprising the beam-splitter and the two arm input mirrors).

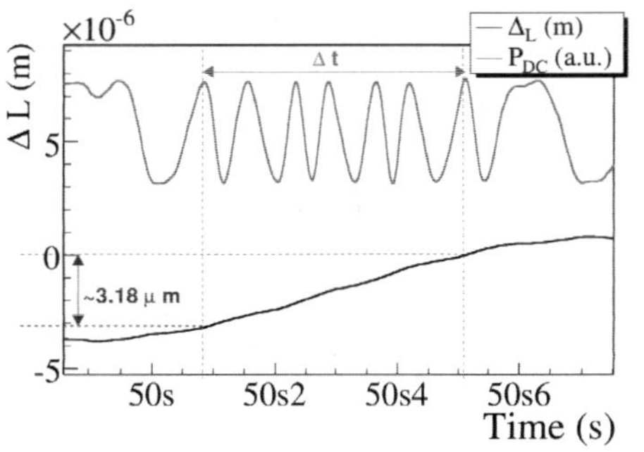


**Figure 3.** Red curve (top): measured time series of $\mathcal{P}_{DC}$ in free swinging Michelson data, in arbitrary units. Black curve (bottom): reconstructed time series of $\Delta L$, in meters.

*2.2. Mirror actuator calibration*

The Virgo electromagnetic actuators are used to induce a longitudinal motion to the suspended mirrors from a voltage signal. As shown in figure 4, the signals pass through some electronics and pairs of coil-magnet ($A$) and the pendulum mechanical response ($P$). Their calibration consists in measuring the transfer function between the input voltage $V_{inj}$ and the induced motion $\Delta L_{inj}$. This is done setting the ITF in free swinging Michelson configurations: sine signals $V_{inj}$ are injected to the actuators and the motion $\Delta L_{rec}$ is resconstructed using the method described in section 2.1. The complete response of the ITF is shown in figure 4: the reconstructed $\Delta L_{rec}$ has to be corrected for some effects to get the true induced mirror motion $\Delta L_{inj}$, mainly some delays due to the light propagation time in the arms ($O$) and due to the output power readout electronics ($S$).

Journal of Physics: Conference Series **228** (2010) 012015 doi:10.1088/1742-6596/228/1/012015

Thirteen sine signals have been injected from 5 Hz to 1. kHz. Such injections were done at different amplitudes in order to check the linearity of the measurements. They were also performed every two weeks during VSR1 in order to monitor their time stability. The mirror actuation responses $\Delta L_{inj}/V_{inj}$ are measured with a statistical error better than 1%. A regular monitoring of the actuation has shown $\sim$ 2% variations and different measurements had up to $\sim$ 2% offsets from the standard measurements. Systematic errors of the order of 4% on the modulus and 50 mrad on the phase have been derived.

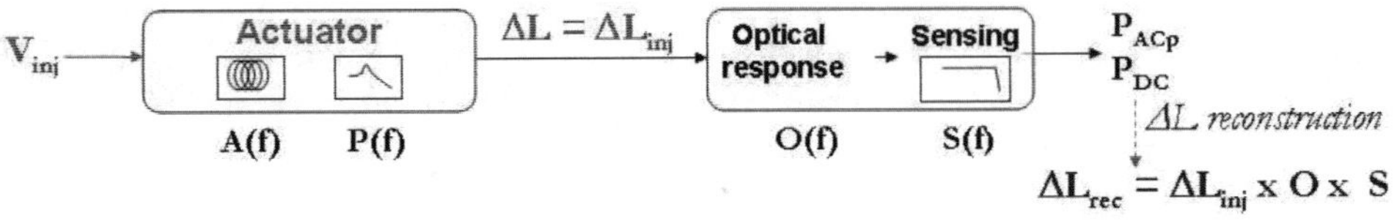


**Figure 4.** Signal processing in the free swinging Michelson data used for the mirror actuation calibration. See text for details.

## 3. Virgo ITF response and sensitivity measurements

The first goal of the calibration is to measure the Virgo sensitivity. It is based on the knowledge of a mirror actuator response calibrated as shown in section 2. The ITF response in closed-loop is first measured and then used to estimated the ITF sensitivity.

### *3.1. Virgo ITF response*

The ITF response, in W/m, is the the transfer function from the differential arm length variation $\Delta L$ to the ITF output power $\mathcal{P}_{ACp}$. When the Virgo ITF is in closed-loop (see figure 1), some white noise $V_{inj}$ is injected up to 1.5 kHz through an end mirror actuator. Using the actuator calibration, the induced $\Delta L_{inj}$ is derived. The transfer function $\Delta L_{inj}/\mathcal{P}_{ACp}$ (inverse of the ITF response) is then measured up to $\sim$ 1.5 kHz.

The longitudinal controls do not act above a few hundred Hz: the ITF response is only the optical response and the output power sensing. A simple model is thus fit to the measured transfer function from 800 Hz to 1 kHz. The optical response is modeled by a simple pole at 500 Hz due to the arm cavities with finesse 50. The finesse of the arm cavities changes over time by as much as 3%. These variations are ignored at this stage. The gain of the optical response is let free in the fit. The measurements of the output power sensing are not described in this paper: the fitted model of the sensing lies within 2% in modulus and 20 mrad in phase from the measured response up to 6 kHz.

An example of a measured inverse ITF response during VSR1 is given in figure 5.

### *3.2. Virgo sensitivity*

The Virgo sensitivity is measured from the ITF response in closed-loop, in m/W, and from the spectrum of the output signal $\mathcal{P}_{ACp}$, in W: they are multiplied to get the sensitivity in meters. It is then divided by the Virgo arm length, 3 km, to get the sensitivity in strain as shown in figure 6. The sensitivity is measured a few minutes after the ITF response in order to avoid possible variations of the optical gain.

Note that the sensitivity is measured with systematic errors of the order of 4% coming from the mirror actuator calibration below 1 kHz. The errors above 1 kHz are higher (5–10%) due to (i) the use of a fixed cavity finesse, (ii) the power sensing model errors and (iii) the fact that the controls were not totally negligible in the fitted frequency range during VSR1.

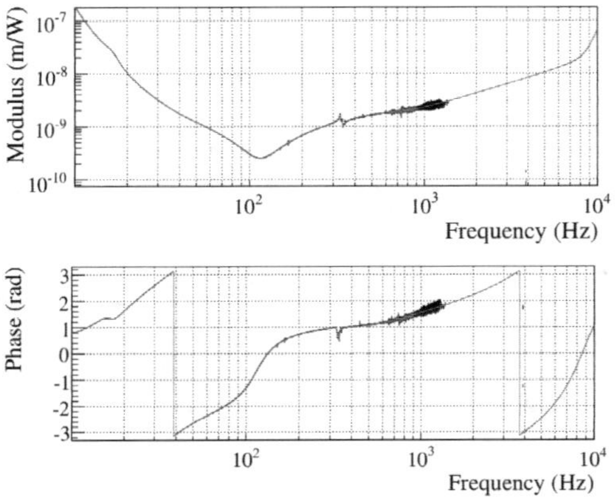


**Figure 5.** Virgo ITF response measured in September 2007. Black: measured transfer function up to 1.5 kHz (with increasing statistical uncertainties above a few 100's Hz). Red: extrapolated transfer function.

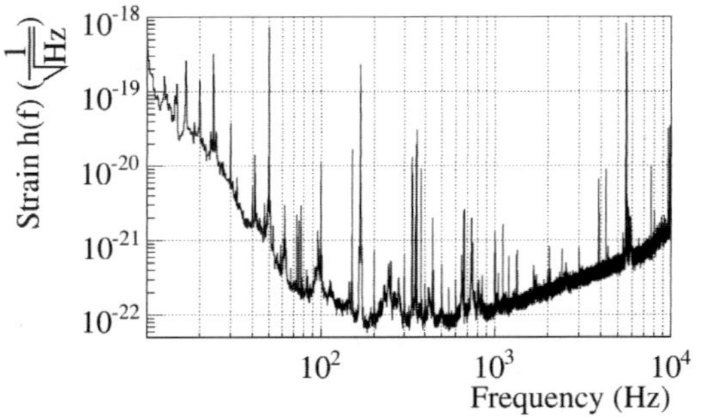


**Figure 6.** Virgo sensitivity measured in September 2007.

## 4. h(t) strain reconstruction

The second aim of the calibration process is the reconstruction of the GW strain time series $h(t)$. The principle of the reconstruction used for the VSR1 data is described. Some estimations of the errors and the sign of $h(t)$ are then given.

### 4.1. Reconstruction principle

The principle of the reconstruction of $h(t)$ from the ITF output power and the mirror longitudinal controls is given in figure 7. The effective arm differential length variation $\Delta L_{effective}$ is reconstructed in the frequency domain from the output power $\mathcal{P}_{ACp}$ and the calibrated optical response and power sensing. It contains the contribution from the noise, from the controls and possibly from the GWs. The contributions from the controls are reconstructed in the frequency domain from the correction signals sent to the mirror actuators knowing the actuator responses. They are then subtracted from $\Delta L_{effective}$ to get the final $\Delta L$ with the noise and GW contributions only. The result is then divided by 3 km and put back in the time domain to get the strain $h(t)$. The contribution from the power line (50 Hz and harmonics) is finally subtracted.

### 4.2. Error estimation using mirror actuators

In order to estimate the errors on the reconstructed strain $h(t)$, some datasets with sine signals $V_{inj}$ injected to an out-of-loop mirror (an input mirror) were taken every week during VSR1. It corresponds to an injected strain $h_{inj} = \Delta L_{inj}/(3000\,\mathrm{m})$. The transfer function of the reconstructed over injected strain $h/h_{inj}$ is then measured. This transfer function is expected to be flat: at 0 in phase and at 0.97 in modulus (since the motion of the input mirror changes both the arm length and the Michelson length, the optical response of Virgo is 3% lower for an input mirror than for an end mirror).

The differential arm length variations $\Delta L_{inj}$ induced by the injected signal is derived from the mirror actuation calibration, with errors of the order of 4% in modulus and 50 mrad in phase.

The transfer function measured every week during VSR1 is shown in figure 8. The observed

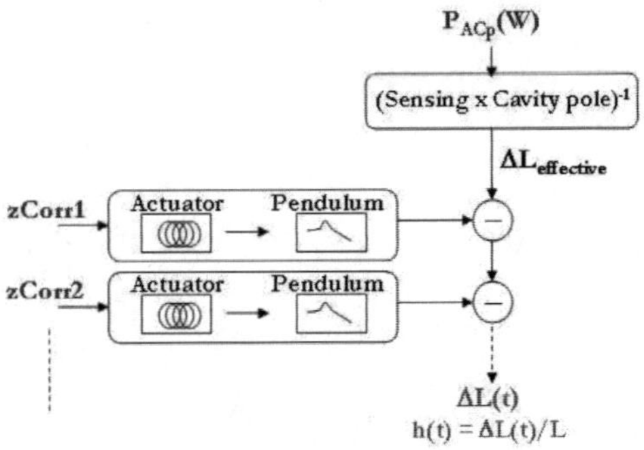


**Figure 7.** Synopsis of the reconstruction of $h(t)$ from the ITF output power $\mathcal{P}_{ACp}$ and the longitudinal correction signals sent to the mirror actuators ($zCorr$ channels). See text for details.

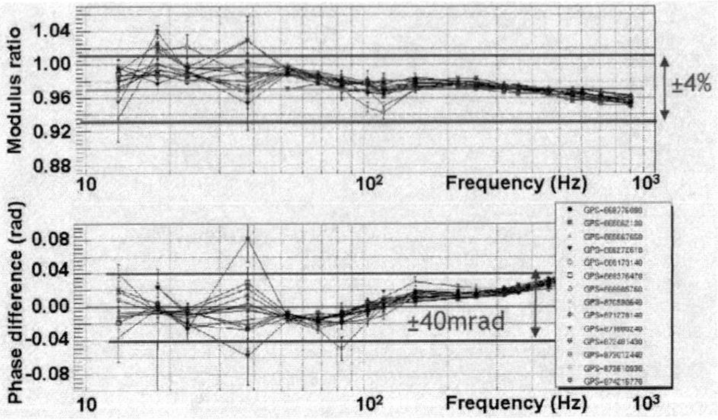


**Figure 8.** Top curve: ratio between the reconstructed amplitude of the lines in $h(t)$ and the expected injected amplitude. Bottom curve: phase difference for the two signals. The difference symbols correspond to different measurements performed once per week during VSR1.

time variations at a given frequency are compatible with statistical errors. The difference between the measurements and the expected values gives an estimation of the systematic errors on the reconstructed strain $h(t)$. They are of the order of 4% in modulus and 50 mrad in phase below 1 kHz. This is compatible with the estimated systematic errors of the mirror actuation calibration used in the $h(t)$ reconstruction.

### *4.3. "Photon calibrator" and sign of h(t)*

As in GEO [5] and LIGO [6, 7], a photon calibrator (pcal) was setup in Virgo during VSR1. It is used as an independent mirror actuator: the radiation pressure of an auxiliary laser is used to displace one input mirror. Its main use during VSR1 has been to check the sign of $h(t)$, defined in common with LIGO as $h = \frac{L_x - L_y}{L}$ where, in Virgo, $x$ and $y$ are the north and west arms respectively.

Sine signals injections were performed with the Virgo pcal, corresponding to an expected signal $h_{inj}$. The phase between the reconstructed $h$ and the expected one $h_{inj}$ has been measured below 1 kHz. It is expected to be 0. The phase at low frequency converges to 0 within 15 mrad: this proves that the sign of the reconstructed $h(t)$ is correct.

Note that the pcal has been used also to cross-check the mirror actuation standard calibration. It agrees within the large pcal systematic errors of 20%.

## 5. Conclusion

The Virgo standard calibration is based on a non-linear reconstruction of the error signal in free swinging Michelson configurations. It relies on the laser wavelength as length standard. It has been used to calibrate the mirror electromagnetic actuators within 4% in modulus and 50 mrad in phase.

The Virgo photon calibrator has been mainly used to check the sign of $h(t)$.

The principle of the reconstruction of the strain $h(t)$ applied on the VSR1 data has been given along with some estimations of the errors. Some other sources of errors must be taken into account, especially from the ITF output power sensing and from some timing issues. Taking them into account, the reconstructed $h(t)$ during VSR1 is valid from 10 Hz up to the Nyquist frequency of the channel (2048 Hz, 8192 Hz or 10000 Hz) with systematic errors estimated to 6% in amplitude. The phase error is estimated to be 70 mrad below 1.9 kHz and 6 $\mu$s above.

## References

[1] The Virgo collaboration 1997 *Final Design Report* Virgo Technical Note VIR-TRE-DIR-1000-13
[2] Abbott B et al 2004 *Nuclear Instruments and Methods for Physics Research* **517** pp 154–179
[3] Acernese et al (Virgo collaboration) 2008 *Astroparticle Physics* **30** pp 29–38
[4] Halíř R and Flusser J 1998 *Proc. of the 6th International Conference in Central Europe on Computer Graphics, Visualisation and Interactive Digital Media. WSCG 98* (Bory, Czech Republic) pp 125–132
[5] Hild S et al 2007 *Class. Quantum Grav.* **24** pp 5681–5688
[6] Goetz E et al submitted to CQG
[7] Savage R *these proceedings*

Journal of Physics: Conference Series **228** (2010) 012016 doi:10.1088/1742-6596/228/1/012016

# Design of the Advanced Virgo non-degenerate recycling cavities

**M Granata[1], M Barsuglia[1], R Flaminio[2], A Freise[3], S Hild[4] and J Marque[5]**

[1] Laboratoire Astroparticule et Cosmologie (APC)
10 rue Alice Domon et Léonie Duquet, 75013 Paris, France
[2] Laboratoire des Matériaux Avancés (LMA), IN2P3/CNRS
F-69622 Villeurbanne, Lyon, France
[3] School of Physics and Astronomy, University of Birmingham
Birmingham, B15 2TT, United Kingdom
[4] Institute for Gravitational Research, Department of Physics and Astronomy
University of Glasgow, Glasgow, G12 8QQ, United Kingdom
[5] European Gravitational Observatory (EGO)
I-56021 Cascina (Pi), Italy

E-mail: granata@apc.univ-paris7.fr

**Abstract.** Advanced Virgo is the project to upgrade the interferometric gravitational wave detector Virgo, and it foresees the implementation of power and signal non-degenerate recycling cavities. Such cavities suppress the build-up of high order modes of the resonating sidebands, with some advantage for the commissioning of the detector and the build-up of the gravitational signal. Here we present the baseline design of the Advanced Virgo non-degenerate recycling cavities, giving some preliminay results of simulations about the tolerances of this design to astigmatism, mirror figure errors and thermal lensing.

## 1. Introduction

Interferometric gravitational wave detectors of the first generation like Virgo [1] and LIGO [2] have successfully completed their first long-duration data taking runs, and a new joint run is ongoing presently. Upgrades of the first generation detectors are currently under study, in order to improve the chance of detecting gravitational waves from astronomical sources and open a new observational window on the universe.

Advanced Virgo (adVirgo) [3] is the project to upgrade the present Virgo detector to a second generation instrument. Compared to the optical layout of Virgo, the design of adVirgo features two main changes: the inclusion of the signal recycling technique, and the move from marginally stable recycling cavities to non-degenerate power and signal recycling cavities (NDRC).

In this paper we present the baseline design of the adVirgo NDRC: we first describe the main motivations for having NDRC in adVirgo in section n.2; then, in section n.3, we list the main issues of the NDRC, and the features of the baseline design accompliant with these needs. In section n.4 we report on some preliminary simulation results about the tolerances of the NDRC to errors on mirrors' radii of curvature (RoC) and thermal effects.

© 2010 IOP Publishing Ltd

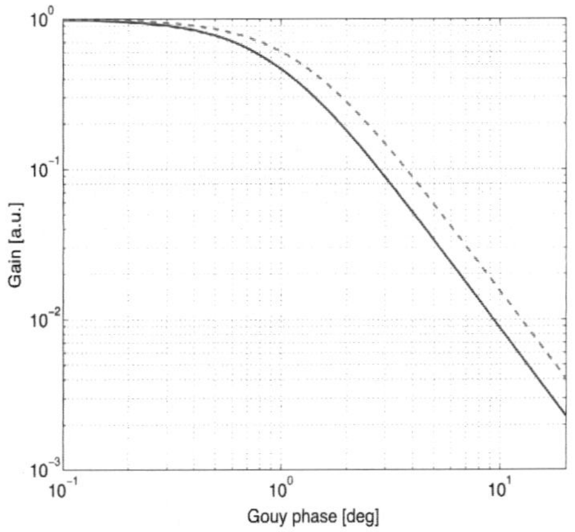


Figure 1: $TEM_{01}$ build-up suppression as a function of the cumulated Gouy phase $\Delta\Phi_G$: the solid curve refers to a resonator with finesse F = 70, the dashed curve refers to F = 40. For increasing values of $\Delta\Phi_G$ the build-up is suppressed of some orders of magnitude.

## 2. Motivations

In a marginally stable optical resonator, like the present power recycling cavity of Virgo, the high order modes (HOMs; please refer to Appendix A for the related notation) can resonate as a consequence of misalignment of the mirrors, mirror figure errors or thermal aberrations. If the contirbution of HOMs is not negligible, the optical gain of the recycling resonator for the fundamental mode is reduced and the control of the detector becomes more difficult, as it has already been experienced during the commissioning of Virgo [4].

The build-up (optical gain) of the resonator is a function of the Gouy phase, $\Phi_G$, and also of the complex transmissivity $t_i$ and reflectivity $r_i$ of the input ($i = 1$) and the end mirror ($i = 2$), which define the finesse $F$ of the resonator. The optical gain $G_{mn}$ for the $TEM_{mn}$ mode can then be written as follows:

$$G_{mn}(t_1, r_1, r_2, \Delta\Phi_G) = \left| \frac{t_1}{1 - r_1 r_2 \cdot exp\Big[-i2(m+n+1)\Delta\Phi_G\Big]} \right|^2 \qquad (1)$$

The normalized optical gain for the $TEM_{01}$ mode ($m = 0$, $n = 1$) is shown in fig.1; this figure clearly illustrates that, given the finesse $F = 70$ and $F = 40$ for power and signal recycling respectively (as stated in ref.[3]), the optical gain of the first high order mode in the NDRC can be reduced of more than two order of magnitude for $\Delta\Phi_G = 20$ degrees.

The NDRC would allow a lower HOMs content in the radio frequency sidebands used for the interferometer signal readout and control, and they would also lower HOMs content in the signal sidebands, reducing gravitational wave signal loss [5]. The HOMs build-up in the resonator would be suppressed with a proper choice of the cumulated Gouy phase, $\Delta\Phi_G$, for the beam resonating in the NDRC.

## 3. Baseline design

The design concept of the adVirgo NDRC has been inspired by the advanced LIGO NDRC design [6]: both power and signal recycling cavities have a folded design, and a mode matching telescope is placed inside the cavities. PRM1 (SRM1) is the power (signal) recycling mirror, and PRM3 (SRM3) and PRM2 (SRM2) form the fast telescope for the focusing of the large beam ($w$ = 5.5 cm, see ref.[3]) coming from the Fabry-Perot cavities in the long arms of the interferometer: PRM3 is a strongly converging mirror, and PRM2 is a diverging mirror for the tuning of the beam parameters (see fig.2).

In this design most of the Gouy phase $\Delta\Phi_G$ is cumulated between PRM2 and PRM1, and the size of the beam on PRM1 is greater than 1 mm, in order to avoid a too large power density on the power recycling mirror.

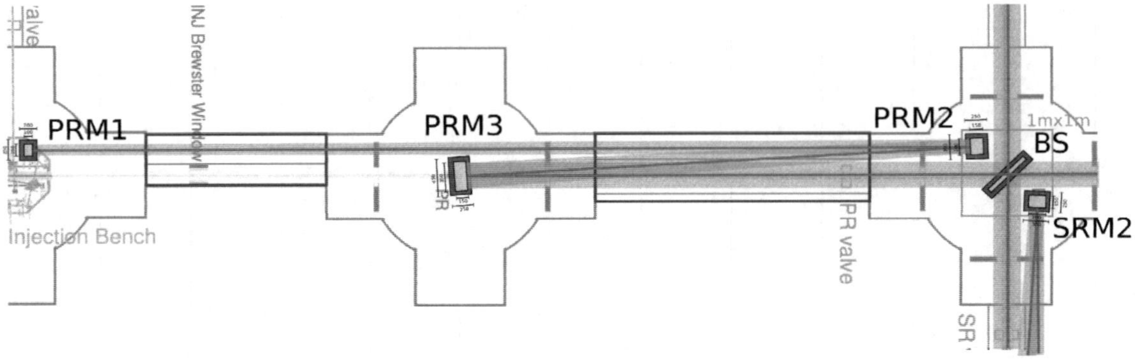


Figure 2: Baseline design of the Advanced Virgo NDRC. Only the power recycling cavity is shown, the signal recycling cavity has the same configuration. The power recycling cavity mirrors are shown as PRM$i$, while signal recycling cavity mirrors are indicated as SRM$i$ (i = 1, 2, 3); BS is the beamsplitter.

Here we present the main features of the baseline design for the NDRC of adVirgo (shown in fig.2), drawn by the adVirgo Optical Simulation and Design group [7]. In the design process, the main constraints came from the already existing Virgo infrastructures, in particular from the vacuum system, and from interferometer sensing and control (ISC) requirements on sideband resonance conditions.

The total length of the NDRC baseline design will fit the present Virgo vacuum tubes, and three mirrors will be suspended in the beamsplitter tower: the beam tuning mirrors PRM2 and SRM2, and the beamsplitter itself. For this purpose, the present Virgo suspension system, the SuperAttenuator, will be adapted for off-axis and multiple mirror suspension. The power recycling mirror PRM1 will be placed on the injection bench, which will be suspended; the suspension of PRM1 itself from the bench is currently under study, and the same configuration will be adopted for SRM1.

The main parameters of the design can be adapted to achieve any $\Delta\Phi_G$ desired value; for $\Delta\Phi_G = 20$ degrees, these parameters are listed in table 1.

## 4. Stability, matching and tolerances

For the following analysis, the parameters of the beam resonating in the NDRC and in the arm cavities of adVirgo have been computed using a Matlab based code, developed by two of the authors (MG and MB); a set of ABCD ray matrices (for both $x$ and $y$ directions in the case of the NDRC) [8] have been used to calculate the stable resonating mode of each cavity. The results given in this section are valid for power and signal recycling, and in the following we will

Table 1: List of parameters of the adVirgo NDRC baseline design, for $\Delta\Phi_G = 20$; the total length of both power and signal recycling cavities is 27.5 m. PRM$i$ (SRM$i$) are the mirrors inside the power (signal) recycling cavity, ITM is the input test mass of the arm cavity.

| RoC (m) | Lengths (m) | Angle of incidence (deg) |
|---|---|---|
| PRM3 = 12.80 | ITM to PRM3 = 11.5 | 1.7 |
| PRM2 = -2.04 | PRM3 to PRM2 = 5.5 | 1.7 |
| PRM1 = -2.04 | PRM2 to PRM1 = 10.5 | - |

refer mainly to the power recycling cavity for simplicity.

### *4.1. Coupling mismatch analysis*

The non-degenerate power recycling cavity is coupled to the beam coming from the injection bench and to the beam of the Fabry-Perot (FP) cavities in the arms; the non-degenerate cavity for signal recycling is coupled to the beam going towards the detection bench and to the beam of the FP cavities. Any coupling mismatch can decrease the power/signal build-up of the fundamental mode in the NDRC themselves or in the FP cavities.

Since in the baseline design the incidence of the laser beam on telescope mirrors is not perpendicular, the beam resonating in the NDRC will be astigmatic. The waist size and the distance from the cavity waist of the NDRC beam will have different values on each of the two transverse directions, $x$ and $y$, and the associated complex parameters $q_x$, $q_y$ will be different in general from those of the beams from the injection (detection) and from the FP.

We can evaluate the coupling between two beams, $U(q_1(z))$ and $U(q_2(z))$, by means of the overlap integral $\gamma$, defined as the scalar product of the beams over the transversal plane $x-y$:

$$\gamma(q_1, q_2) = \langle U(q_2)|U(q_1)\rangle \tag{2}$$

As a first approximation we will suppose that all the cavities (NDRC and FP) of the detector are decoupled, so that we can neglect the mutual effects of the mismatch on the build-up of the fundamental mode in the coupled cavities. If we do so, then we can consider the overlap integral as a coupling coefficient; if we indicate with $\gamma_{NDRC}$ the coupling between the NDRC and the input (output) beam, and with $\gamma_{FP}$ the NDRC coupling with the arm beam, we can write the coupling losses $L$ as follows:

$$L = 1 - \gamma_{NDRC}^2 \cdot \gamma_{FP}^2 \tag{3}$$

With this hypothesis, the coupling losses of the baseline design due to astigmatism are 6.8 %; in any case we would like to stress the fact that this is just a preliminary estimation, and further investigation is presently ongoing to correctly evaluate the impact of the coupling losses on the sensitivity of adVirgo.

### *4.2. Stability and tolerances*

The mode-matching telescope of the NDRC is very sensitive to errors on the RoC of the mirrors: this can be seen in fig.3a, where a colourscale plot shows the stability of the NDRC as a function of the percentage error on the RoC of the telescope mirrors.

For any point in the couloured region of fig.3a, the NDRC is stable on both $x$ and $y$ transverse directions, and the colourscale indicates the percentage coupling losses due to astigmatism and RoC errors; in the white regions of the plot the NDRC is unstable at least on one transverse direction. The working point of the baseline design is the $(0, 0)$ point at the centre of the plot, indicated by the red arrow.

The RoC of PRM3 is a critical parameter: with the present technology the manufacturers can realize mirrors of $\sim$ 10 m RoC like PRM3 with an accuracy of $\pm$ 0.1 %, while in this case an accuracy of $\pm$ 0.02 % is needed. Manufacturer's errors on RoC could be overcome by tuning the distance L2 that separates the two mirrors of the telescope, as shown in fig.3b: a percentage variation of L2 of the same order of the error on the mirror RoC is needed. Given the parameters listed in table 1, this means that in presence of an error of +1 cm on the RoC of PRM3 (corresponding to an error of about +0.1 %), L2 should be elongated of about 5.5 mm (corresponding to a variation of about +0.1 %).

The tolerances for the NDRC stability of the other parameters of the design are listed in table 2, together with the tolerances for the Gouy phase to be within the following boundaries: 10 deg $< \Delta\Phi_G <$ 30 deg. The RoC of the other mirrors are not so critical as for PRM3, since manufacturers can realize optics with an accuracy of 0.04 % for mirrors whose RoC is $\sim$ 1m.

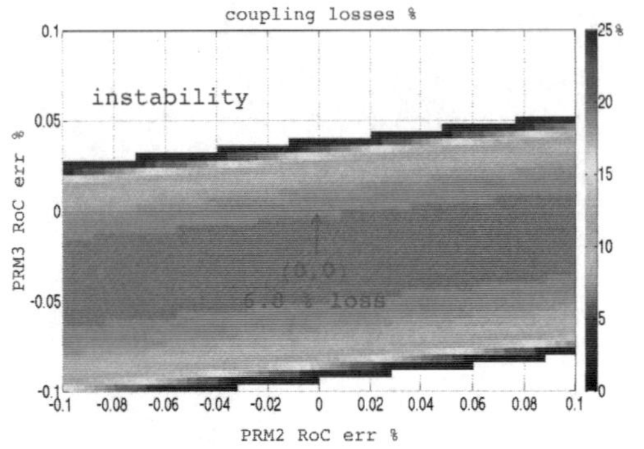


(a) NDRC stability as a function of the RoC of the telescope mirrors.

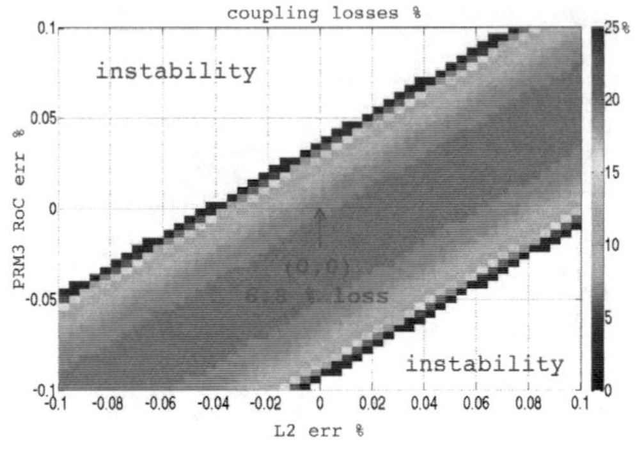


(b) NDRC stability as a function of the distance between the telescope mirrors.

Figure 3: Stability plots of the NDRC baseline design: the cavity is stable only in the coloured region. The colourscale indicates the coupling losses, in percentage; the working point of the detector is (0,0), and the corresponding coupling losses are shown.

### *4.3. Thermal effects*

According to the baseline design of adVirgo [3], the power resonating in the non degenerate power recycling cavity will be of the order of 3 kW, while in the arms the power will be 760 kW. The absorption in the input test masses (ITM) of a small fraction of the circulating power will induce thermal lensing effects, changing the optical path of the light beam; we can then include in our simulations an equivalent lens, giving the same optical path variation, in order to study the 'hot' state of the interferometer for increasing circulating power inside its cavities.

Our simulations indicate that the NDRC baseline design will become unstable for an equivalent lens with focal length f smaller than 7 km, corresponding to an absorption in the ITM of more than 0.45 W. Another preliminary simulation, run by the adVirgo thermal compensation system (TCS) group, shows that in the full-power 'hot' state 0.50 W will be absorbed in the ITM, giving an equivalent thermal lens of $f = 4.5$ km.

Fig.4 shows the results of our simulation for this last value of absorption: it clearly illustrates that NDRC design (conceived for a cold configuration where only a fraction of the nominal power is circulating and thermal lensing is negligible) will be no more stable at full power.

Table 2: List of stability and Gouy phase (10 deg $< \Delta\Phi_G <$ 30 deg) tolerances of the advanced Virgo non-degenerate recycling cavity baseline design.

| | Stability tol. | Gouy phase tol. |
|---|---|---|
| RM3 | ± 0.03 % | ± 0.02 % |
| RM2 | ± 0.3 % | ± 0.1 % |
| RM1 | ± 20 % | ± 11 % |
| L3 | ± 100 % | ± 100 % |
| L2 | ± 0.04 % | ± 0.02 % |
| L1 | ± 4.0 % | ± 1.0 % |

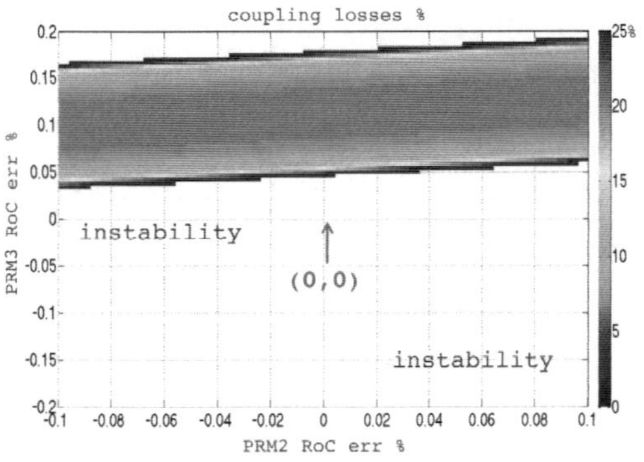


Figure 4: Stability plot with the thermal lensing, f = 4.5 km. The working point (0,0) of the 'hot' configuration is out of the stability region.

This result indicates that a TCS will be needed, once adVirgo will be operating at full power. For example, the TCS may include ring heaters placed around the lateral surface of the ITM, and compensation plates placed in front of the ITM [3].

Some thermal effects will be potentially present also on PRM1, and they have not been considered so far; this issue will require further investigation.

## 5. Alternative design

In parallel with this preliminary baseline design analysis, an alternative design has also been considered in order to decrease astigmatism coupling losses. This alternative design features longer cavities and smaller angles of incidence on telescope mirrors. The telescopes would be placed between the beamsplitter and the Fabry-Perot input mirrors, i. e. in the small Michelson arms (fig.5). For this design, the simulated coupling losses due to astigmatism are $L = 3.3$ %.

This alternative configuration would also allow to remove the compensation plates of the TCS, which would be only composed of ring heaters and/or $CO_2$ lasers for the actuation on the RoC of the telescope mirrors.

The main drawback of this alternative design is the potential arise of differential noise, eventually introduced by the suspension of two additional mirrors in each arm of the small Michelson; this issue has not been considered so far, and an exaustive treatment of this topic will be addressed in the near future.

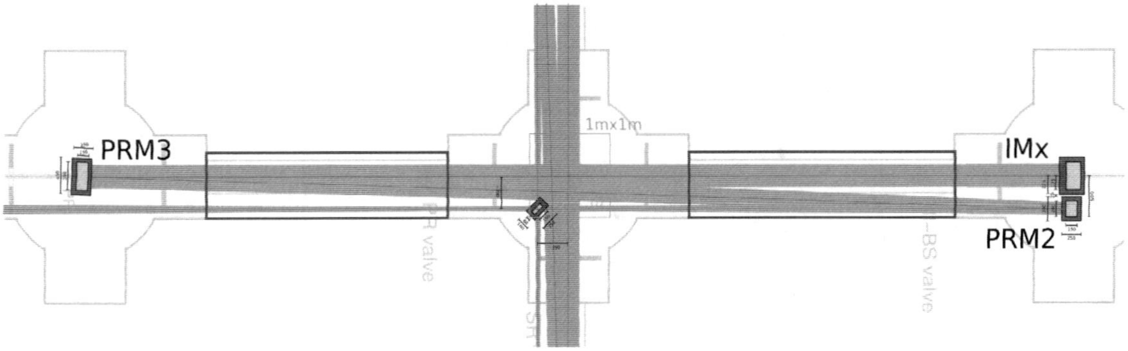


Figure 5: Alternative design for the adVirgo NDRC. The mode-matching telescope is placed in the differential part of the detector; IMx is the input test mass of the Fabry-Perot cavity of the $x$ arm.

## 6. Conclusions

AdVirgo will have non-degenerate power and signal recycling cavities, and the baseline design of the NDRC has been selected. This baseline design will reduce the HOMs content in the sidebands used for detector control and in the signal sidebands, being also compliant with the requirements of the sideband resonance condition and the infrastructure constraints.

According to the simulation results presented in this paper, the coupling losses of the NDRC baseline design will be 6.8 %, and the impact of this losses on the detector sensitivity is currently under investigation. Our tolerance analysis shows that the stability of the NDRC basline design can be settled by tuning the distance that separates the telescope mirrors, and that a TCS will be required for compensating the thermal lensing given by the full circulating power.

An alternative NDRC design has been considered, for which the astigmatism coupling losses would be lowered to 3.3 %, but it may introduce some additional source of noise; this issue will require further investigations.

## Acknowledgments

The authors would like to thank all the members of the adVirgo OSD and ISC groups, and Viviana Fafone and Alessio Rocchi of the TCS group for their suggestions and comments during the development of this preliminary analysis.

## Appendix A. Laser high order modes

On the plane perpendicular to the direction of propagation $z$, any intensity distribution of a laser beam can be described by a complete set of orthogonal fields [9], the Hermite-Gauss modes $U_{mn}(x, y, z)$. Their complex amplitude is given by [9, 10]

$$\begin{aligned} U_{mn}(x,y,z) \quad = \quad & \Big(\frac{2}{\pi\ m!\ n!\ 2^{m+n}}\Big)^{1/2} \cdot \tfrac{1}{w(z)} \cdot exp\Big[-\tfrac{x^2+y^2}{w^2(z)}\Big]\cdot \\ & \cdot exp\Big[-i\Big(k\tfrac{x^2+y^2}{2R(z)} - (m+n+1)\Phi_G\Big)\Big]\cdot \\ & \cdot H_m\Big(\tfrac{x\sqrt{2}}{w(z)}\Big)H_n\Big(\tfrac{y\sqrt{2}}{w(z)}\Big) \end{aligned} \tag{A.1}$$

In equation (A.1) $w(z)$ is the beam radius, $R(z)$ the beam front curvature, $\Phi_G = arctan(z/z_0)$ is the Gouy phase, $z_0 = \pi w_0^2/\lambda$ is the Rayleigh range, $w_0$ is the beam waist; the $H_{m,n}$ are the Hermite polynomials, and $q(z) = z + iz_0$ is the beam complex parameter. These modes are also referred to as $\text{TEM}_{mn}$; the $\text{TEM}_{00}$ is the fundamental gaussian mode, while for $m \neq 0$, $n \neq 0$ the $\text{TEM}_{mn}$ are high order modes.

## References

[1] Acernese F et al 2008 Virgo Status *Class. Quantum Grav.* **25** 184001
[2] Kawabe K et al 2008 Status of LIGO *J. Phys.: Conf. Ser.* **120** 032003
[3] Virgo Collaboration 2009 Advanced Virgo baseline design *Virgo internal note* VIR-007A-09
[4] Acernese F et al 2008 Lock acquisition of the Virgo gravitational wave detector *Astropart. Phys.* **30** 29
[5] Pan Y 2006 Optimal degeneracy for the signal-recycling cavity in advanced LIGO *Preprint* gr-qc/0608128v1
[6] Arain M A and Mueller G 2008 Design of the advanced LIGO recycling cavities *Optics Expr.* **16** 10018
[7] Freise A, Hild S and Marque J 2009 Advanced Virgo design: layout options for the non-degenerate recycling cavities *Virgo internal note* VIR-0025A-09
[8] Massey G A and Siegman A E 1969 Reflection and refraction of gaussian light beams at tilted ellispoidal surfaces *Appl. Opt.* **8** 975
[9] Kogelnik H and Li T 1966 Laser beams and resonators *Appl. Opt.* **5** 1550
[10] Siegman A E 1986 *Lasers* (Sausalito: University Science Books) p 642

Journal of Physics: Conference Series **228** (2010) 012017 doi:10.1088/1742-6596/228/1/012017

# The dynamics of monolithic suspensions for advanced detectors: a 3-segment model

**F Piergiovanni[1,2], M Lorenzini[2], G Cagnoli[2,3], E Campagna[1,2], E Cesarini[1,2], G Losurdo[2], F Martelli[1,2], F Vetrano[1,2], A Viceré[1,2]**

[1] Università di Urbino, Via S.Chiara 27, 61029 Urbino (PU), Italy
[2] INFN, Istituto Nazionale di Fisica Nucleare, Sez. di Firenze, via G. Sansone 1, 50019 Sesto Fiorentino (FI), Italy
[3] SUPA, University of Glasgow, Department of Physics and Astronomy, Kelvin Building G12 8QQ Glasgow, Scotland

E-mail: piergiovanni@fi.infn.it

**Abstract.**
In order to reduce the suspension thermal noise, the second generation GW interferometric detectors will employ monolithic suspensions in fused silica to hold the mirrors. The fibres are produced by melting and pulling apart a fused silica rod, obtaining a long thin wire with two thicker heads. The dynamics of such a fibre is in principle different from that of a cylindrical, regular fibre, because most of the deformation energy is stored in the neck region where the diameter is variable. This is an advantage, since adjusting the neck tapering, a thermoelastic noise cancellation effect can be obtained. Therefore, a careful study of the suspensions behavior is necessary to estimate the overall noise and to optimize the control strategy. To simplify the control design, a simple three segment model for the silica fibres has been developed, fully equivalent to the beam equation at low frequencies. The model, analytically proved for a regular cylindrical fibre, can be extended to a fibre with tapered necks, provided that the equivalent bending length is suitably measured. We developed a tool to measure the position of the bending point for each fibre, thus allowing to experimentally check the validity of the model. A numerical code has been written to solve the beam equation for wires with varying diameter. This code confirms the validity of the three segment model. Moreover, it is possible to extend the solution to higher frequencies thus computing the transfer function and the energy distribution of the suspension system and estimating the thermal noise contribution.

## 1. Introduction

The modeling of the suspension system of GW interferometric detectors is necessary to verify the static equilibrium conditions and to design the best control strategies. Different approaches can be used to model the behavior of suspension fibres: elastic equations can be solved analytically or numerically, a finite element analysis can be performed or just simple pendulum can be considered, neglecting the elasticity. The elastic equations solution and the FEM analysis are computationally expensive and the simple pendulum approach is inaccurate for many purposes. The simplest model that describes with sufficient accuracy the low frequency elastic behavior of suspension fibers, is the so-called 3-segment model: each fiber is replaced by 3 rigid segments connected at the bending points[1]. The model provides a great simplification of the equations for the elastic forces and torques at the ends of the fibre. The simplicity of this model allows

© 2010 IOP Publishing Ltd

Journal of Physics: Conference Series **228** (2010) 012017 doi:10.1088/1742-6596/228/1/012017

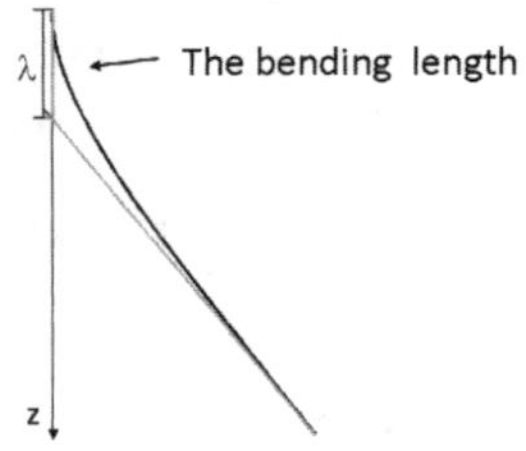


**Figure 1.** The bending point is the intersection of two lines that are tangent to the first and the middle part of the fibre. The distance of this point from the fibre end is the bending length ($\lambda$). Each fibre end has an associated bending point and a bending length.

a straightforward study of the changes in the suspension behavior with respect to the fibers parameters.

## 2. The 3-segment model

Considering the shape of a slightly bent fibre, the bending point is the intersection of two lines that are tangent to the first and the middle part of the fibre[2]. The distance of this point from the fibre end is the bending length ($\lambda$) (figure 1). The two bending lengths ($\lambda_1$ and $\lambda_2$) and the fibre total length ($L_{tot}$) are the only three parameters used by the model. A triple pendulum composed by the 3 segments of length $\lambda_1$, $L_{tot} - (\lambda_1 + \lambda_2)$, $\lambda_2$ is considered (figure 2). The deformation is described by the transverse coordinates X , Y and by the tilt angles $\theta_x$, $\theta_y$, while the excitations are due to the force F and the torque $\tau$. Since each fibre is linked to the rest of the suspension through its ends, the equivalence of the model with respect to the elastic fibre, means that the relationships between the coordinates and the excitations in the model at the fibre ends are the same obtained from the elastic equations at first order of approximation. The model is valid under the following conditions: the oscillation wavelength must be greater than the bending length (low frequencies), the transverse displacement is small compared with the total length (small deflections) and the bending length has to be smaller than the total length. These conditions are all largely satisfied in GW detector suspensions in a range of frequency below some tens of Hz.

## 3. Analytical proof of the model equivalence for cylindrical fibres

For a cylindrical fibre with a constant cross section, it is possible to verify analytically the validity of the model solving the elastic equations for a slightly deflected cylindrical rod stretched by a tension T[3]. For harmonic excitation a Fourier transform can be performed:

$$EIX^{IV}(z) - TX^{II}(z) - \rho S\omega^2 X(z) = 0 \qquad (1)$$

where $E$ and $\rho$ are the Young modulus and the density of the material, $I$ the momentum of inertia, $S$ the cross section, $T$ the fibre tension, $\omega$ the angular frequency and $X$ the Fourier transform of the deflection x in x-z plane. The Roman numerals in suffix mean the order of the derivative with respect to the z coordinate. Obviously the equation is the same for the deflection y in plane y-z. A general solution of the equation (1) is:

$$X(z) = A\exp^{(-z/\lambda)} + B\exp^{-(L-z)/\lambda} + C\cos(p\lambda) + D\sin(p\lambda) \qquad (2)$$

where:

$$\lambda \equiv \left(\frac{T + \sqrt{T^2 + 4EI\rho S\omega^2}}{2EI}\right)^{-1/2} \quad \text{and} \quad p \equiv \left(\frac{-T + \sqrt{T^2 + 4EI\rho S\omega^2}}{2EI}\right)^{-1/2} \qquad (3)$$

Boundary conditions at the ends ($X(0), X^I(0), X(L), X^I(L)$) fix the values of the constants $A, B, C, D$. In the low frequency regime, equations (3) show that $p$ goes to zero and $\lambda \approx \sqrt{EI/T}$, where $\lambda$ is the above mentioned bending length. Forces and torques can be obtained by boundary

Journal of Physics: Conference Series **228** (2010) 012017 doi:10.1088/1742-6596/228/1/012017

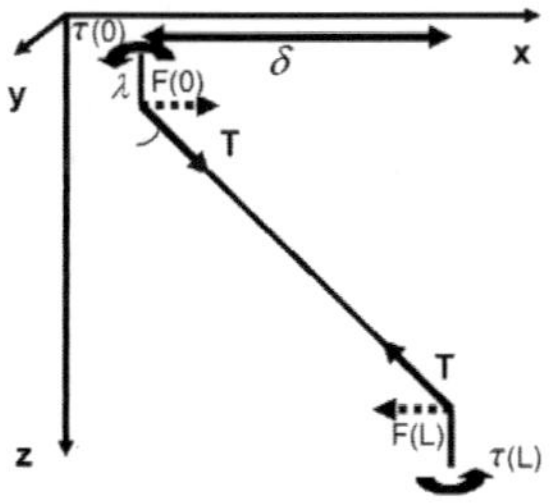


**Figure 2.** A 3-segment model scheme for a deflection X(L)=$\delta$ in x-z plane. The relationships obtained by the model, between displacement and force and torque are the same as those in equation (4)

equations: $F([0,L]) = -EIX^{III}([0,L]) + TX([0,L])$ and $\tau([0,L]) = -EIX^{II}([0,L])$. Within the above approximation, it is possible to compute the transfer matrix between linear and angular displacements on one side and forces and torque on the other. For deflections in x-z plane the transfer matrix is the following:

$$\begin{pmatrix} F_x(0) \\ F_x(L) \\ \tau_y(0) \\ \tau_y(L) \end{pmatrix} = \begin{pmatrix} -\frac{T}{L-2\lambda} & \frac{T}{L-2\lambda} & -\frac{T\lambda}{L-2\lambda} & \frac{T\lambda}{L-2\lambda} \\ \frac{T}{L-2\lambda} & -\frac{T}{L-2\lambda} & \frac{T\lambda}{L-2\lambda} & -\frac{T\lambda}{L-2\lambda} \\ -\frac{T\lambda}{L-2\lambda} & \frac{T\lambda}{L-2\lambda} & -\frac{T\lambda(L-\lambda)}{L-2\lambda} & \frac{T\lambda^2}{L-2\lambda} \\ -\frac{T\lambda}{L-2\lambda} & \frac{T\lambda}{L-2\lambda} & -\frac{T\lambda^2}{L-2\lambda} & -\frac{T\lambda(L-\lambda)}{L-2\lambda} \end{pmatrix} \begin{pmatrix} X(0) \\ X(L) \\ \theta_y(0) \\ \theta_y(L) \end{pmatrix} \quad (4)$$

Equation (4) contains the relationships between displacements and excitations. Identical relationships can be much more easily obtained from the 3-segment model. A scheme of computation for a X(L)=$\delta$ displacement in x-z plane is shown in figure 2. Restoring forces are the x-components of the tension $T$, while torques are evaluated using $\lambda$ as lever arm. The same results are obtained by putting $X(0) = 0$, $X(L) = \delta$, $\theta_y(0) = 0$, $\theta_y(L) = 0$ in equation (4).

## 4. The neck in real suspension fused silica fibres

Fused silica shows very low bulk loss angle and thermoelastic losses at room temperature[4][5][6]. To minimize the thermal noise contribution in GW interferometric detectors, the fused silica mirrors are suspended by fused silica fibres in the so called 'monolithic suspension'. At present only GEO600 uses that solution[7], but monolithic suspensions will be implemented in Virgo+ in the near future[8][9] and they will be used in second generation GW detectors[10][11][12]. Fibres are produced by an oxy-hydrogen flame or a $CO_2$ laser starting from an initial fused silica rod few millimeters thick[13]. The transition regions where the diameter changes from the rod value to the fibre one (usually few hundred $\mu$m) are called *necks*[14][8]. The length of the necks ranges from few millimeters to tens of millimeters. Size and shape of the neck can be optimized to further depress the thermoelastic contribution to thermal noise[15].

### *4.1. A numerical validation of the model for fibres with necks: a finite element model*

The bending energy for a real suspension fibre is concentrated in a region around the bending points, close to the fibre's necks[3]. For the presence of a variable cross section the dynamical behavior of these fibres is different from that of the perfectly cylindrical ones, even if cylindrical symmetry is preserved. In this case no analytical solution of the elastic equations can be found. To test the validity of the 3-segment scheme, a finite element model of the fibre was realized and the results were compared. Meshing is obtained by dividing the fibre into N 1-dimensional elements of length $h = L/N$ separated by nodes (figure 3)[16][17]. For each element $j$ a local non-dimensional coordinate $\zeta \equiv j - z/h$ is defined and four cubic interpolating functions $\phi_{1..4}(\zeta)$ are introduced. The displacement inside the element $j$ (x($\zeta$)), can then be written in terms of slopes and displacement at the nodes ($x_{j-1}$, $\theta_{j-1}$, $x_j$ and $\theta_j$):

$$x(\zeta) = \phi_1(\zeta)x_{j-1} + \phi_2(\zeta)h\theta_{j-1} + \phi_3(\zeta)x_j + \phi_4(\zeta)h\theta_j \quad (5)$$

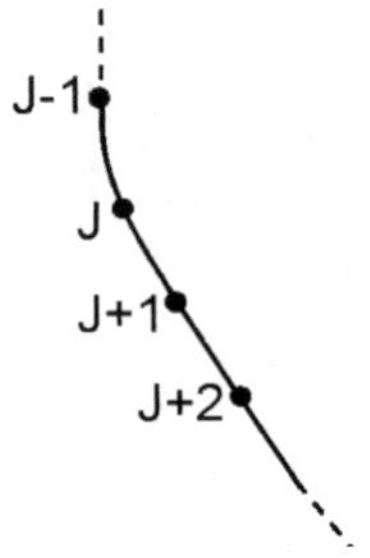


**Figure 3.** The mesh of the fibre. The element $j$ is limited by the nodes $j$ and $j-1$.

Matching conditions at the nodes determine the four constants of each cubic polynomial $\phi_{1..4}$.

The kinetic and potential energies of the fibre transverse motion are:

$$E^{kin} = \frac{1}{2}\int_0^L \rho(z)\,[\dot{x}(z)]^2\,dz; \qquad E^{pot} = \frac{1}{2}\int_0^L \left(EI\left[x^{II}(z)\right]^2 + T\left[x^{I}(z)\right]^2\right)dz \tag{6}$$

Using the defined interpolating functions the integrals can be decomposed in a sum of terms:

$$E_j^{kin} = \frac{1}{2}\dot{X}_j^{\,T} M_j \dot{X}_j; \qquad E_j^{pot} = \frac{1}{2}X_j^T K_j X_j \tag{7}$$

where $X_j \equiv (x_{j-1},\ h\theta_{j-1},\ x_j,\ h\theta_j)^T$. The mass and the stiffness matrix, respectively $\mathrm{M}_j$ and $\mathrm{K}_j$, are fully defined by the integrals in equation(6), i.e.

$$M_j^{mn} = h\rho_j \int_0^1 \phi_m(\zeta)\phi_n(\zeta)d\zeta; \quad K_j^{mn} = \frac{EI_j}{h^3}\int_0^1 \phi_m^{II}(\zeta)\phi_n^{II}(\zeta)d\zeta + \frac{T}{h}\int_0^1 \phi_m^{I}(\zeta)\phi_n^{I}(\zeta)d\zeta \tag{8}$$

Variable profile of the fibre is obtained specifying different $\rho_j$ and $I_j$ for each segment. For each element the equation of motion can be written as $M_j\ddot{X}_j = -K_jX_j + F_j^{ext}$ where $F_j^{ext} \equiv (F_{j-1}, \tau_{j-1}, F_j, \tau_j)$. Introducing vectors $Q$ and $F^{ext}$ whose elements are displacements, slopes, forces and torques at each node, the equations of motion for the whole fibre are $M\ddot{Q} + KQ = F^{ext}$. $M$ and $K$ are built from $M_j$ and $K_j$ keeping into account the coincidence of the lower and the upper end of two contiguous elements.

Some typical fibre diameter profiles are shown in figure 4 where a gaussian neck shape is considered. When a shift $X(L) = \delta$ is imposed, the bending length can be fitted from the bent fibre shape obtained by the FEM (figure 5). Forces and the torques predicted by the simulation for various neck profiles are in good agreement with the results of the 3-segment model. The relative error is of the order of $10^{-5}$ for forces and $10^{-4}$ for torques for any typical value of $\sigma$ (figure 6).

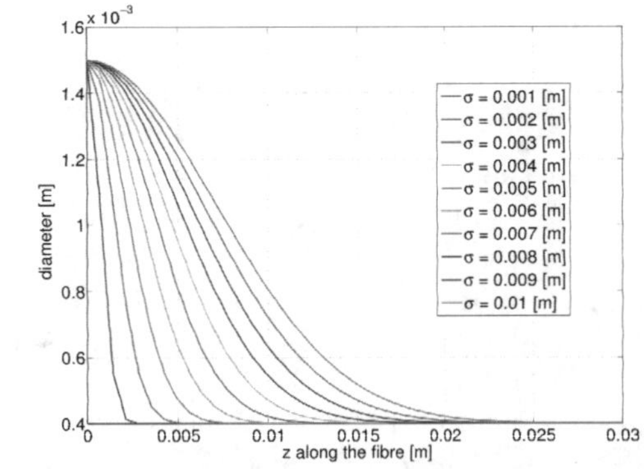


**Figure 4.** Diameter profiles for gaussian necks with different width $\sigma$.

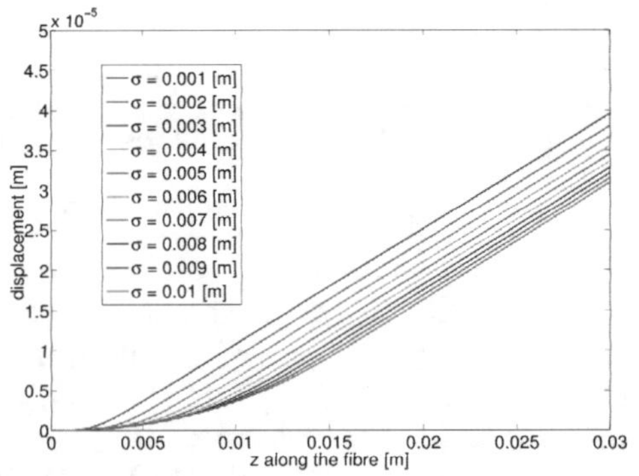


**Figure 5.** Bending of the firsts mms of the fibre. The bending length increases with $\sigma$.

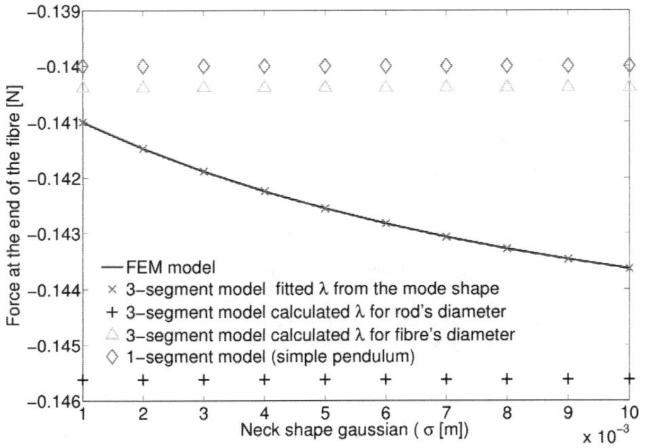


**Figure 6.** The line shows the prediction of the FEM for the force, for a displacement $X(L) = \delta$. Crosses are the results of a 3-segment model with $\lambda$ length that is fitted from the bent shape or calculated from equation (3) for rod's diameter (plus) or fibre diameter (triangle). Diamond dots are the results for a simple pendulum.

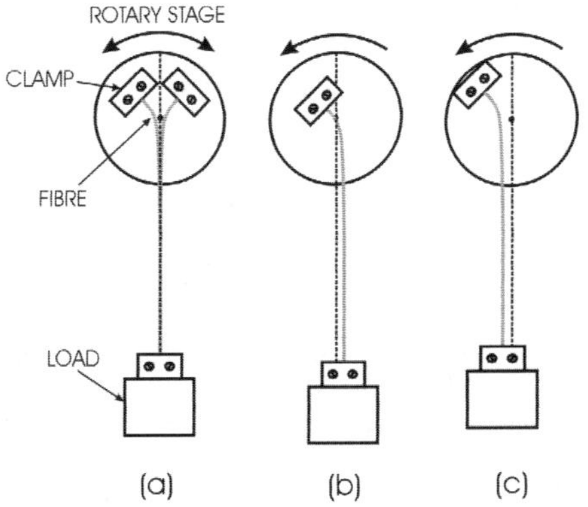


**Figure 7.** If the bending point is exactly on the rotation axis the lower part of the fibre does not move from the vertical line.

## 5. Experimental check of the model for real fibres

A good agreement between the 3-segment model and the FEM is shown in figure 7. Such comparison is made after the bending length of the fibre has been worked out through a fitting analysis of the bent fibre as it comes from the FEM model. If $\lambda$ value is computed from equation (3) with rod or fibre's diameter, the comparison is worse. For a real fibre, it is not possible to calculate analytically the bending length, hence a direct measurement is needed.

The so called $\lambda$ machine is used to determine the positions of the $\lambda$ points along the fibre[1]. The working principle is to rotate slightly and slowly the upper part of a loaded fibre and to detect the horizontal displacement of the lower part. When the rotation axis is adjusted to the bending point, the load does not shift horizontally. In figure 7, the working principle of the lambda machine is shown. The machine was realized at INFN Firenze and it is operational in EGO (European Gravitational Observatory). A typical fused silica suspension fibre was clamped and loaded as in figure 8. A computation of the proper angular frequencies with the 3-segment model gives:

$$\omega_{\pm}^2 = \frac{g}{2LI_c}\left(Ml^2 + I_c + MlL \pm \sqrt{(Ml^2 + I_c + MlL)^2 - 4MlLI_c}\right) \tag{9}$$

where $L$ is the distance between the bending points, $d$ the distance between the end fibre and the load's center of mass, $M$ and $I_c$ are the mass and the momentunm of inertia of the load, $\lambda_t$ and $\lambda_b$ the upper and lower bending lengths and $l = \lambda_b + d$. The load mass is designed to allow the variation of the position of the center of mass ($d$) with a threaded rod and the momentum of inertia ($I_c$) adding extra-loads (figure 9). Measurements of resonant frequencies (figure 10) are consistent with the expected values for each oscillation mode, even though a more complete characterization would require heavier extra-loads, to appreciate the effects of $I_c$ variations.

Journal of Physics: Conference Series **228** (2010) 012017 doi:10.1088/1742-6596/228/1/012017

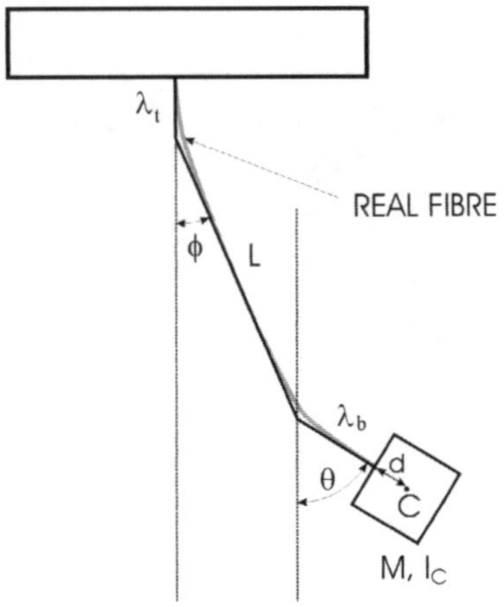


**Figure 8.** The scheme of a 3-segment modelized fibre, clamped at one end and loaded at the other with a mass $M$ and a momentum of inertia $I_c$

**Figure 9.** The load mass. A threaded rod gives the possibility to shift the center of mass position and $I_c$ can be modified adding some extra loads.

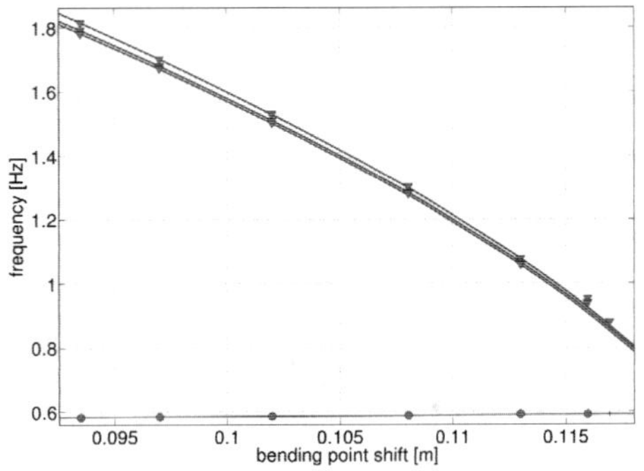


**Figure 10.** Experimental results (triangles and circles) are consistent with the expected lines computed by equation (9) for $\omega_+$ (blue) and $\omega_-$ (red). Three $\omega_+$ lines correspond to different values of $I_c$. The bending point shift in x axis is the decrement, from a reference value, of the parameter $d$ cited in the text.

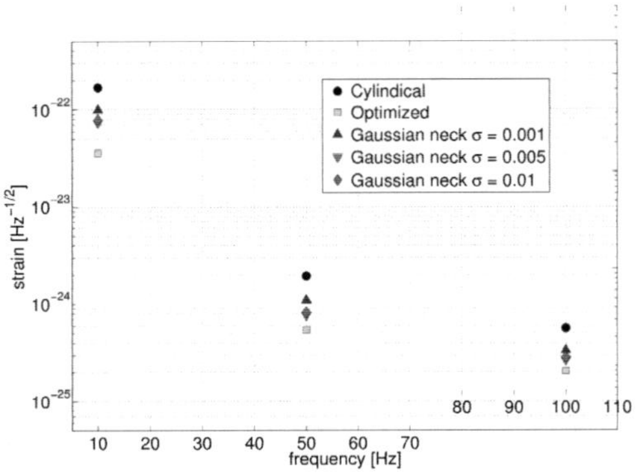


**Figure 11.** Fibre thermal noise predictions. Circles and squares refers to cylindrical and optimized fibres, respectively. Results for gaussian neck fibres are shown as triangles and diamonds. Each fibre is loaded with a quarter of the AdVirgo mirror mass (40 kg) and it has distance of 700 mm between the bending points.

## 6. An estimation of the thermal noise of fibres with necks.

Because most of the elastic energy is stored in the neck regions, the necks determine the dynamical behavior of a suspension fibre[3][9]. Hence the mechanical impedance strongly depends on the neck profile rather than on the middle part of the fibre. This means that the thermal noise contribution of a fibre with neck is different from that of a perfectly cylindrical fibre with stepped transition to larger diameter. To estimate thermal noise up to some hundreds of Hz, the FEM described in section 4.1 can be used. To compute the mechanical impedance, an oscillating force is applied to the mirror and thereby to one fibre's end. The mirror is suspended by four fibres hence the angular displacement is suppressed, this means that the tilts $X^I(0)$ and $X^I(\mathrm{L})$ are zero and the mirror only shifts horizontally. For a harmonic excitation, the amplitude of the shift results proportional to the amplitude of the force. The ratio between the time derivative of the shift (the speed) and the force, multiplied by the four suspension fibres, is the admittance of the system. As usual, the dissipation is evaluated adding the loss angle as an imaginary part to the Young modulus. The total loss angle is the sum of bulk, surface and thermoelastic contributions. The bulk loss angle is a property of the material, while surface and thermoelastic losses depend on the geometry, in this case on the diameter at each position along the fiber. Thus, the losses are computed separately for each element of the model mesh, then the total sum is weighted with the bending energy derived from the FEM analysis. Finally, thermal noise can be calculated from the resulting complex mechanical admittance, using the fluctuation-dissipation theorem[18]. Thermal noise contribution is plotted in figure 11, for a fibre with a diameter of 400 $\mu$m and gaussian neck's profile, loaded with the AdVirgo mirrors[12]. The figure also shows thermal noise analytically computed for a cylindrical fibre (with a diameter of 400 $\mu$m) and for an optimized fibre (named tapered or dumbbell-shaped[19]) designed to minimize thermal noise (two end sections, of diameter 800 $\mu$m, and a middle part of 400 $\mu$m). All fibres present a distance between the bending points of 700 mm. For neck profiles similar to the real ones, thermal noise turns out to be smaller than for cylindrical fibre and not so far from the values given by optimized fibre, less than a factor two above.

## 7. Conclusions

In future detectors, test masses will be suspended with fused silica fibres. The statics and the low frequencies dynamical behavior of the fibres can be simulated with a simple 3-segment model. The model validity was analytically proved for cylindrical fibre's profile. A finite element modellization and experimental results lead to extend the validity of the 3-segment model for real fibre that presents necks at the ends. Moreover, for real fibres with typical diameter profiles, the FEM predicts a lower thermal noise with respect to perfectly cylindrical ones.

## References

[1] Lorenzini M 2007 *Suspension thermal noise issues for advanced GW interferometric detectors* PhD thesis, Università di Firenze, available at http://www.infn.it/thesis/PDF/getfile.php?filename=2460-Lorenzini-dottorato.pdf
[2] Cagnoli G, Hough J, DeBra D, Fejer M M, Gustafson E, Rowan S and Mitrofanov V 2000 *Phys. Lett.* A **272** 39–45
[3] Landau L D and Lifshitz E M 1970 *Theory of Elasticity* 2nd Edition (Oxford: Pergamon Press)
[4] Braginsky V B, Mitrofanov V P and Tokmakov K V 1995 *Phys. Dokl.* **40** 564
[5] Rowan S, Hutchins R, McLaren A, Robertson N A, Twyford S M and Hough J 1997 *Phys. Lett.* A **227** 153
[6] Gretarsson A M and Harry G M 1999 *Rev. Sci. Instrum.* **70** 4081
[7] Willke B *et al* 2004 *Class. Quantum Grav.* **21** S417
[8] Lorenzini M *et al* 2009 *The monolithic suspension for the interferometer Virgo. Proc. 8th Amaldi Conf. on Gravitational Waves (New York)* in preparation
[9] Piergiovanni F, Punturo M and Puppo P 2009 *The thermal noise of the Virgo+ and Virgo Advanced Last Stage Suspension (The PPP effect)* Virgo Document VIR-015A-09
[10] Robertson N *et al* 2006 *Advanced LIGO Suspension System Conceptual Design* LIGO Document T010103-05
[11] Barton M *et al* 2008 *Proposal for baseline change from ribbons to fibres in AdvLIGO test mass suspension monolithic stage* LIGO Document T080091-00
[12] The Virgo Collaboration 2009 *Advanced Virgo Baseline Design* Virgo Document VIR-027A-09
[13] Cagnoli G *et al* 2006 *J. Phys.: Conf. Series* **32** 386–392
[14] Heptonstall A, Martin I, Cumming A, Cantley C A, Cagnoli G, Jones R and Crooks D R M 2005 *Production and characterization of synthetic fused silica ribbons for Advanced LIGO suspensions* LIGO Document T050206-00
[15] Cagnoli G and Willems P A 2002 *Phys. Rev.*B **65** 174111
[16] Meirovitch L 1997 *Principles and techniques of vibrations* (Upper Saddle River, New Jersey: Prentice Hall)
[17] Viceré A 1999 *Proc. Int. Summer School on Experimental physics of Gravitational waves (Urbino)* (Singapore: World Scientific) p 349
[18] Callen H B and T A Welton 1951 *Phys. Rev.* **83** 35
[19] Willems P 2002 *Phys. Lett.* A **300** 162–168

Journal of Physics: Conference Series **228** (2010) 012018 doi:10.1088/1742-6596/228/1/012018

# Characterization of electrostatic actuators for suspended mirror control with modulated bias

**R De Rosa[1,2], F Garufi[1,2], L Milano[1,2], S Mosca[1,2], G Persichetti[1,2]**

[1]Università di Napoli and [2]INFN - Sezione di Napoli, Complesso Universitario di Monte S.Angelo, via Cintia, 80126, Napoli - Italy

E-mail: simona.mosca@na.infn.it

**Abstract.** Electrostatic actuators are one of the most promising devices for mirror control in advanced gravitational waves detectors. An accurate characterization of such actuators is required for a correct design, able to satisfy the requirement of the control system, both in term of low noise content as well as to fit the required dynamic range. To this aim a simple and effective experimental set-up was developed, consisting in a suspended mirror which displacement, induced by an electrostatic actuator, is measured by using an optical lever. The effect of stray patch charge on the mirror was minimized by using an alternate voltage as bias reference for the actuator. Different working conditions were investigated, in particular by varying the mirror-actuator distance and the bias amplitude. The experimental results were compared to the prediction of a numerical model taking into account the actuator geometry and the working conditions.

## 1. Introduction

The first generation of interferometric detectors for gravitational waves detection have currently reached their target performances. In particular the VIRGO [1], LIGO [2] and GEO 600 [3] detectors have been involved, in the last years, in cooperative and long term scientific data taking, while the TAMA [4] detector is upgrading to enhance its sensitivity in low frequency range.

In about one year from now, most of the detectors are expected to start their upgrade to the *Advanced* configuration, with the aim to enhance their sensitivity by about one order of magnitude. This second generation of detectors will take advantage from some technical solution that allow to reduce the effect of some fundamental noises on the interferometer sensitivity. Nevertheless, at the same time, it is necessary to reduce the effect of other contributions, like control and environmental noises.

To this aim, a non-negligible contribution could come from the use of electrostatic actuators (EA) for the control of the last suspended mass, a solution already applied for the GEO 600 detector [5] [6] that is under study also for other detectors, currently adopting coil-magnets pair actuators. In particular, the use of EA allows to keep the mirrors under control without the need of gluing the magnets on the mirror bulk, saving, in this way, the mechanical quality of the test masses and, as a consequence, reducing the final thermal noise [7]. Another advantage is the strong reduction of the coupling with external magnetic fields, that is an important issue for magnetic actuators, since no direct coupling is anymore possible in the case of EA.

© 2010 IOP Publishing Ltd

In this paper a simple electrostatic actuator, working on one degree of freedom, is described. The characterization is performed by using a suspended mass with a simple optical lever as position sensor. The experimental set-up is placed under vacuum to reduce the charging effects due to the air. The characterization was performed for different actuator-mass distances and in two different bias conditions; this in order to fulfill a double target: to verify the effectiveness of a modulated bias in reducing the effect of the residual charges present on the test mass, and to check the validity of a theoretical model available for the computation of the actuator capacitance.

## 2. Operation theory

The working principle of an electrostatic actuator is simply described by standard electrostatic, giving, for a device of capacitance $C$, polarized at fixed voltage $V$, a resulting force along the $x$ axis equal to:

$$F_x = -\frac{1}{2}\left|\frac{\mathrm{d}C}{\mathrm{d}x}\right| V^2 \tag{1}$$

where the capacitance is supposed to vary by changing the system characteristics along the main axis while the minus sign is due to the characteristic of such forces, that are always attractive. In the case of actuator for suspended dielectric mirrors, such devices mainly consist in a set of close conductive strips, arranged in a suitable geometry, alternately polarized at two different voltages. The strips, together with the dielectric suspended test mass, placed at distance a $x$ with respect to the actuator plane, constitute a capacitor, with a capacitance variable by changing the distance of the test mass with respect to the actuator.

The theoretical expression of the capacitance is described in [8], here we want just to recall the most important characteristics. For the simplest geometry, i.e. a set of $N$ parallel conductive strips with period $b$, rectangular in shape, of length $L$ and width $a$, laying on a substrate with relative dielectric constant $\epsilon_s$, placed at distance $x$ from the test mass having a relative dielectric constant $\epsilon_m$, the capacitance can be written as:

$$C(x) = C_\infty \alpha_m\left(\tilde{a}, \tilde{x}, \epsilon_m\right) \tag{2}$$

where $\tilde{a} = a/b$ is the normalized strip width, $\tilde{x} = x/b$ is the normalized distance and $\alpha_m$ is a function of the listed parameters describing the effect of the mirror at distance $x$, while $C_\infty$ is:

$$C_\infty = N\pi^2 L\frac{\epsilon_s + 1}{4}\epsilon_0 \alpha_s\left(\tilde{a}\right) \tag{3}$$

is the capacitance of the isolated actuator, being $\alpha_s$ a function of the normalized strip width that is proportional to the capacitance linear density of the electrodes in free space, while the first term, depending on $\epsilon_s$, takes into account the effect of the actuator substrate. It is important to underline that the expression (2) is calculated in the approximation of infinitely long strips and taking into account only the contribution of the first image charges, both for the substrate and for the mirror. As a consequence the capacitance of real devices become different from the this value for small values of $x$ with respect to $b$ due to the increasing weight of border effects and image charges as the distance decreases [8].

By substituting the capacitance (2) in the expression of the force (1), assuming that the displacement of the test mass is small with respect to the static distance $d$ from the actuator, one obtains:

$$F_x = -N\pi^2 L\frac{\epsilon_s + 1}{8}\epsilon_0 \alpha_s \left|\frac{\mathrm{d}\alpha_m}{\mathrm{d}x}\right|_{x=d} V^2 = -\alpha V^2 \tag{4}$$

being $\alpha$ the coupling constant of the electrostatic actuator in ($\mathrm{N/V^2}$) where all characteristics, but the polarization voltage, are included.

The last expression has to be modified to consider also the presence of a stray electric charge $q$ on the dielectric mass. In this case, by making the simple approximation that the electric field is proportional to the polarization voltage applied to the actuator, it is possible to write:

$$F_x = -\alpha V^2 + \beta V \tag{5}$$

where the factor $\beta$ is, in general, a function of the charge $q$, the distance $x$ and the geometry of the actuator. The effects of this term were already observed on a similar set-up [9], and some techniques for its mitigation were already developed [10].

The main purpose of the EA, like any actuation system for the test masses, is the mirror driving to keep the interferometer locked on the right working point, but the expression (5) clearly shows a non linear behavior of the force with respect to the applied voltage. Since in the framework of standard control systems it is more effective to work with linear actuation systems, a simple approach consists in applying the square root of the driving signal, resulting in the desired linear driving force because of the square voltage in equation (5). Moreover, to reduce the effect of the stray charges, it is possible to modulate the driving signal, to obtain a zero averaged contribution of the linear term of the actuation force even in presence of charges on the test masses. This driving technique was already successfully experimented in the control of a bench top Michelson interferometer with a suspended mirror controlled by a such EA [11].

To clarify this approach, let $A(t)$ be the driving signal we want to apply on the test mass, $A_{DC}$ the voltage bias and $2\pi\omega_M$ the modulation frequency of the full driving signal. The square root is computed and sent, with the modulation, to the actuator driver. In this way the voltage applied to the actuator is:

$$V = G\sqrt{A_{DC} + A(t)}\cos\omega_M t \tag{6}$$

where $G$ is the gain of the EA driver. With this voltage, the force exerted on the test mass becomes:

$$F = -\frac{1}{2}\alpha G^2 \left(A_{DC} + A(t)\right)\left(1 + \cos 2\omega_M t\right) + \beta G\sqrt{A_{DC} + A(t)}\cos\omega_M t \tag{7}$$

If the modulation frequency is chosen at enough high frequency to have negligible effects on the test mass motion and the frequency content of the driving signal is much smaller with respect to $\omega_M$, the force (7) only consists of a DC bias term and of a term proportional to the original driving signal $A(t)$.

## 3. Experiment

The experimental set-up used to characterize the actuator consists of a suspended dielectric mass with a mirror attached on the front face. The mass of the suspended element is $m = 1.312$ Kg, while the length of the suspension wires is $l = 0.18$ m. On the rear face a simple electrostatic actuator, composed by $N = 20$ strips, with length $L = 8$ cm, width $a = 3.2$ mm, and period $b = 4$ mm, was placed. The EA substrate has a relative dielectric constant $\epsilon_s = 4.47$, while the suspended dielectric mass has $\epsilon_m = 3.7$. The actuator was mounted on a translation stage in order to easily change the relative distance from the suspended mass. The mass position was sensed with an optical lever, using as source a 1 mW superluminescent LED emitting at 850 nm. The reflected beam impinges on a lens and then on a 2D position sensing photodiode (PSD) placed in the image plane of the beam spot on the mirror. In this way the displacement signal detected by the PSD is dominated by the mass translation along the optical axis, while the contribution of the rotation around the vertical axis is negligible.

Both the driving and position signals are generated or processed by a digital DAQ system, running at a sampling frequency $F_s = 1$ kHz. The ADC and DAC resolutions are 16 bit, while their range is $\pm 10$ V. The driving signal (6), is digitally generated by the CPU managing the

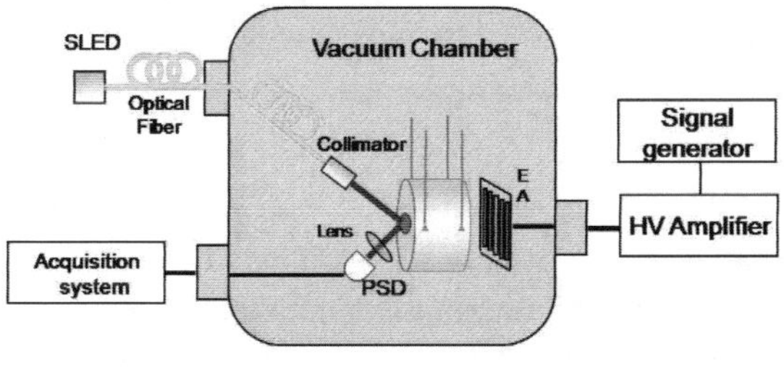


**Figure 1.** Schematic of the experimental set-up.

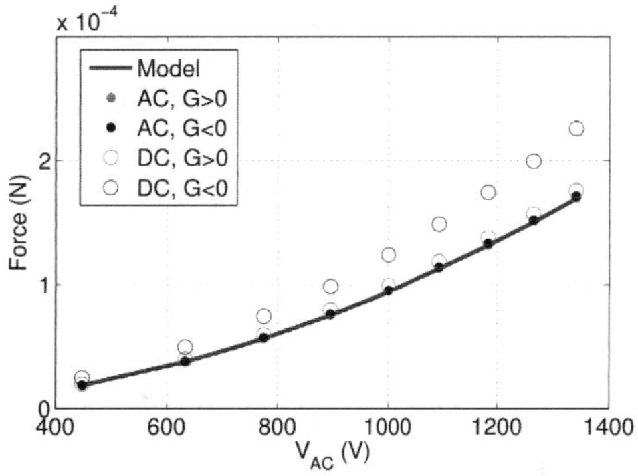


**Figure 2.** Comparison between the model and the force measured, in different bias conditions, for a excitation with $f = 0.1$ Hz.

DAQ system, and is sent, through the DAC, to a commercial HV amplifier, with a gain $G = 200$, a maximum output voltage of 2 kV and a bandwidth of 1 kHz. Because of the limited sampling rate, the modulation frequency for the driving signals was fixed at $f_M = 100$ Hz, that is not a very high value with respect to the suspended mass resonance frequency $f_0$ (around 1 Hz), but it is high enough for our purposes. The signals coming from the PSD were processed by using an amplifier that also allowed to minimize the effect of the quantization noise due to the analog to digital converter [12].

All the system, including the PSD, was placed in a vacuum chamber. The probe beam of the optical lever was sent in the chamber by using an optical fiber and guided to the mirror attached on the test mass by means of a suitable collimator. A schematic view of the experimental set-up is shown in figure 1. The characterization of the actuator was performed by applying, to the HV amplifier, a voltage of the form indicated in equation (6), with a simple line at frequency $f = 2\pi\omega$ as driving signal:

$$V = \sqrt{A_{DC} + A_{AC}\cos\omega t}\cos\omega_M t \tag{8}$$

In this way the component at angular frequency $\omega$ of the force applied on the suspended mass is:

$$F_\omega = \frac{1}{2}\alpha G^2 A_{AC} = \frac{1}{2}\alpha V_{AC}^2 \tag{9}$$

where

$$V_{AC} = G\sqrt{A_{AC}} \tag{10}$$

The effect of this force was measured by calculating the displacement spectrum of the mass, using the data coming from the PSD. Finally the displacement was converted in force by taking into account the theoretical transfer function of the suspended mass.

The measurements were performed at fixed distance between the actuator and test mass of $d = 3.5$ mm, using a fixed bias $A_{DC} = 50$ V$^2$, that corresponds to a force equal to the half of the maximum force for the given output range of the HV amplifier. Two excitation frequencies were used, namely 0.1 Hz and 20 Hz, that are both far enough from $f_0$ to allow us to neglect the transfer function details around the resonance of the suspended mass. For each frequency a set of excitation amplitudes $A_{AC}$, in the range $5 \div 45$ V$^2$ were used. All the measurements were repeated in DC bias ($f_M = 0$) and alternate bias ($f_M = 100$ Hz) conditions. In particular, for each set of measurements in DC bias, also the sign of the voltage applied to the HV amplifier was changed to enhance the effects due to the stray charges on the mass, following the suggestions given in [10]. This change can be formally taken into account by changing the sign of $G$ in equation (7). A change in the electrostatic force should arise only in the case $\beta \neq 0$ and $\omega_M = 0$.

## 4. Results and Discussion

The most interesting results are related to some observed deviation from the theoretical model developed in section 2. The measurements performed at $f = 0.1$ Hz are shown in figure 2. This set shows the best signal to noise ratio, since that frequency is well below the resonance frequency $f_0$. The filled dots represent the force measured in AC bias, both with positive or negative $G$, while the open circles are the force measured in DC bias with different $G$ sign. The deviation from the foreseen behavior, clearly visible for all the points in DC bias, in particular in the case of negative $G$, is due to the presence of spurious charges on the dielectric suspended mass. In fact in this case, equation (7) becomes:

$$F = -\alpha G^2 (A_{DC} + A_{AC}\cos\omega t) + \beta G\sqrt{A_{DC} + A_{AC}\cos\omega t} \tag{11}$$

and a not negligible contribution could arise from the last term. Moreover this contribution depends on $G$ as confirmed by the results. The two opposite polarizations, for the AC bias, give instead the same results, as the experimental measurements are practically overlapped. This confirms the effectiveness of the alternate bias technique that is insensitive to any static stray charge present on the test mass. It is important to underline that the points reported in figure 2, representing the measurements in AC bias, were multiplied by a factor 2 to take into account the presence of the modulation at $\omega_M$ that introduce a factor 1/2 in the force expression (eq. 9). In this way it is possible to compare the two different bias condition on a single graph. Of course all the measurements were performed in the same run, without opening the vacuum chamber to not intentionally change the charge distribution on the test mass.

Following the (11) a larger disagreement would be expected also for the case of DC bias with positive $G$, but one should consider that the description of the electrical field between the EA and the test mass is very roughly approximated in the model; moreover a slight dependence of $\beta$ from the sing of $G$ is expected. More investigations are need in this direction which also require some upgrade in the experimental set-up, as the possibility to change the distance between the EA and the test mass without opening the chamber and changing, in this way, the amount of charges on the mass. Similar result were obtained also for the $f = 20$ Hz excitation, but in this case the measurements were affected by a larger error, since the suspension filter out most of the force and the resulting displacement, at 20 Hz, is not very large respect to the noise floor and the finale error on the force estimation is of the order of 20% making very difficult to decide if the points follow the model or not. The displacement spectra, in case of measurements in DC bias conditions, also show additional lines placed at multiple frequency respect to the one injected by the signal. These lines disappear if the measurement is performed in AC bias as shown in figure 3. This behavior is easily explained by the form of the driving signal (8) that corresponds to the force in equation (11). By calculating the Fourier coefficient of such force, it is possible to prove that the last term of this equation can be written as:

$$\beta G\sqrt{A_{DC} + A_{AC}\cos\omega t} = G\beta\gamma\sqrt{1 + a\cos\omega t} = G\beta\gamma\left(c_0 + 2\sum_{n=1}^{+\infty} c_n \cos n\omega t\right) \tag{12}$$

where $\gamma = \sqrt{A_{DC}}$, $a = A_{AC}/A_{DC}$ and the generic Fourier coefficient is:

$$c_n = -\frac{\Gamma(n-\frac{1}{2})}{2\sqrt{\pi}\Gamma(n+1)}\left(-\frac{a}{2}\right)^n {}_2F_1\left(\frac{n}{2}-\frac{1}{4}, \frac{n}{2}+\frac{1}{4}, n+1, a^2\right) \tag{13}$$

where $\Gamma$ is the gamma function and ${}_2F_1$ is the Gauss Hypergeometric function. Since for real cases $a < 1$, this relation foresees decreasing contribution for all the harmonics of the injected

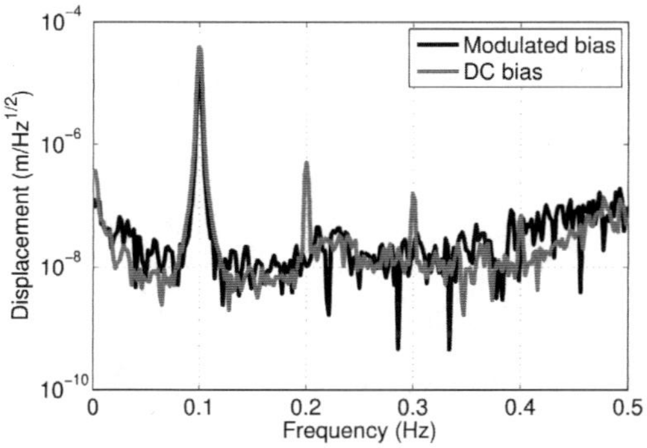


**Figure 3.** Spectra of the test mass displacement with the same actuation force, at $f = 0.1$ Hz in two different bias conditions.

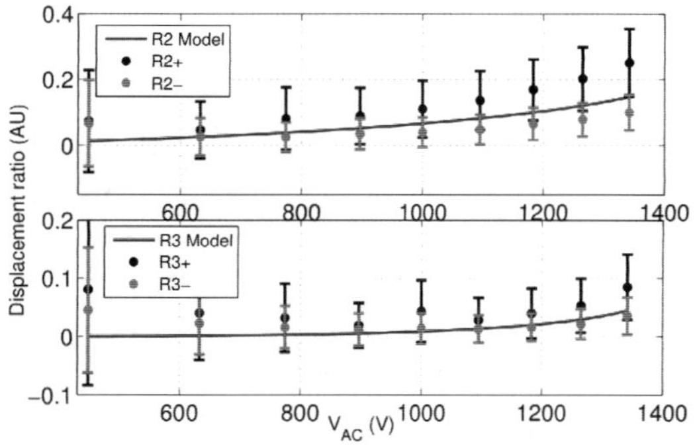


**Figure 4.** Ratios of the displacement harmonics compared with the model.

signal. In fact those lines are clearly visible in the spectrum at least until $n = 3$. Equation (12) also foresees a contribution on the fundamental frequency, that corresponds to the disagreement between the DC and AC measurements reported in figure 2.

Thanks to this behavior, the chosen form of the driving, allows to easily detect the presence of charges on the test mass, simply by looking for the harmonics of the injected line.

In figure 4 the measured displacements ratio between the 2$^{nd}$ and 1$^{st}$ harmonics (upper graph) and the 3$^{rd}$ and 1$^{st}$ harmonics (lower graph) are reported as a function of the voltage $V_{AC}$. The measurements were done both for positive and negative $G$. The comparison with the model, based on the expression (13) is also given: $R_i = c_i/c_1$. The large error do not allow to state any definite conclusion, but the measurements largely follow the model.

## 5. Conclusion

A driving technique for electrostatic actuators, making use of alternate bias to minimize the effect of stray charges present on the dielectric test mass, was developed and characterized. The result confirms the effectiveness of this technique compared with the standard DC bias. A device driven in this way show a very good agreement with the theoretical model of electrostatic actuation. Moreover this technique also allows to easily detect the presence of spurious electric charges on the test mass and gives quantitative indications on the level of charging.

## Acknowledgments

This work was supported by the Italian *Ministero dell'Università e della Ricerca* under the *PRIN 2007* Research Program Framework and by the the European Commission under the Framework Programme 7, *Einstein Telescope* (ET) design study (Grant Agreement 211743).

## References

[1] F. Acernese et al, Class. Quantum Grav. **25** (2008) 184001.
[2] B.P. Abbott et al, Rep. Prog. Phys. **72** (2009) 076901.
[3] B Willke et al, Class. Quantum Grav. **24** (2007) S389.
[4] Ando M et al, Class. Quantum Grav. **22** (2005) S881.
[5] H. Grote, Making it Work: Second Generation Interferometry in GEO600 !, PhD thesis, Hannover 2003
[6] M. Hewitson, H. Grote, G. Heinzel, K. A. Strain, H. Ward and U. Weiland, Class. Quant. Grav. **20** (2003) S885.
[7] P. Amico, L. Bosi, L. Carbone, L. Gammaitoni, M. Punturo, F. Travasso, and H. Vocca, Class. Quant. Grav. **19** (2002) 1663.

[8] S. Grasso, C. Altucci, F. Barone, V. Ragozzino, S. Solimeno, T. Pham, J. Y. Vinet, R. Abbate, Phys. Lett. A **244** (1998) 360.
[9] M. J. Mortonson, C. C. Vassiliou, D. J. Ottaway, D. H. Shoemaker, and G. M. Harrya, Rev. Scientific Instruments **74** (2004) 4840.
[10] M. Hewitson, K. Danzmann, H. Grote, S. Hild, J. Hough, H. Luck, S. Rowan, J.R. Smith, K.A. Strain and B. Willke, Class. Quantum Grav. **24** (2007) 6379.
[11] F. Acernese, F. Barone, A. Boiano, R. De Rosa, F. Garufi, L. Milano, S. Mosca, A. Perreca, G. Persichetti, R. Romano, Jou. of Physics: Conf. Series **122** (2008) 012011.
[12] F. Acernese, E. Calloni, R. De Rosa, L. Di Fiore, L. Milano, Class. Quant. Grav. **22** (2005) S279.

# Silicate bonding properties: investigation through thermal conductivity measurements

**M Lorenzini[1], E Cesarini[1,2], G Cagnoli[1,3], E Campagna[1,2], K Haughian[3], J Hough[3], G Losurdo[1], F Martelli[1,2], I Martin[3], F Piergiovanni[1,2], S Reid[3], S Rowan[3], A A van Veggel[3] and F Vetrano[1,2]**

[1] INFN, Istituto Nazionale di Fisica Nucleare, Sez. di Firenze, via G. Sansone 1, 50019 Sesto Fiorentino (FI), Italy
[2] Università di Urbino, Via S.Chiara 27, 61029 Urbino (PU), Italy
[3] SUPA, University of Glasgow, Department of Physics and Astronomy, Kelvin Building G12 8QQ Glasgow, Scotland

E-mail: `lorenzini@fi.infn.it`

**Abstract.** A direct approach to reduce the thermal noise contribution to the sensitivity limit of a GW interferometric detector is the cryogenic cooling of the mirrors and mirrors suspensions. Future generations of detectors are foreseen to implement this solution. Silicon has been proposed as a candidate material, thanks to its very low intrinsic loss angle at low temperatures and due to its very high thermal conductivity, allowing the heat deposited in the mirrors by high power lasers to be efficiently extracted. To accomplish such a scheme, both mirror masses and suspension elements must be made of silicon, then bonded together forming a quasi-monolithic stage. Elements can be assembled using hydroxide-catalysis silicate bonding, as for silica monolithic joints. The effect of Si to Si bonding on suspension thermal conductance has therefore to be experimentally studied. A measurement of the effect of silicate bonding on thermal conductance carried out on 1 inch thick silicon bonded samples, from room temperature down to 77 K, is reported. In the explored temperature range, the silicate bonding does not seem to affect in a relevant way the sample conductance.

## 1. Introduction

In designing a cryogenic, high power GW interferometric detector, the selection of materials for the realization of the mirrors and of their suspension stages is a critical item; fused silica masses and monolithic suspensions are no longer suitable, due to a broad dissipation peak in silica around 40 K which spoils its performances [1]. Among the suggested materials, silicon [2] has a very low intrinsic level of dissipation especially at low temperatures ($\phi(300\,\mathrm{K}) = 2.8 \times 10^{8}$, $\phi(77\,\mathrm{K}) = 5 \times 10^{9}$ and $\phi(4.2\,\mathrm{K}) = 6 \times 10^{10}$ [3] [4]), large bulk tensile strength (about 7 GPa, surface effects can redduce it to about 200 MPa [5]) and its thermal expansion coefficient $\alpha(\mathrm{T})$ vanishes at about 123 K and 18 K (see figure 1), being almost zero at lower temperatures. Thermal effects coupled with $\alpha$, like the thermoelastic noise in mirrors and suspension elements, are also zeroed at these temperatures.

Moreover, silicon thermal conductivity is very high. Efficient thermal conduction is needed for the suspension elements to extract the heat deposited within the optics by high power lasers

© 2010 IOP Publishing Ltd

used to reduce shot noise. In an interferometer operating at cryogenic temperatures, with about 1 MW of circulating power and mirrors with absorption of few ppm, silicon ribbons with a cross section on the order of 100 $mm^2$ in a monolithic arrangement are suitable for extracting a sufficient amount of heat with just 2 K of temperature difference along the 0.7 m length of the ribbon.

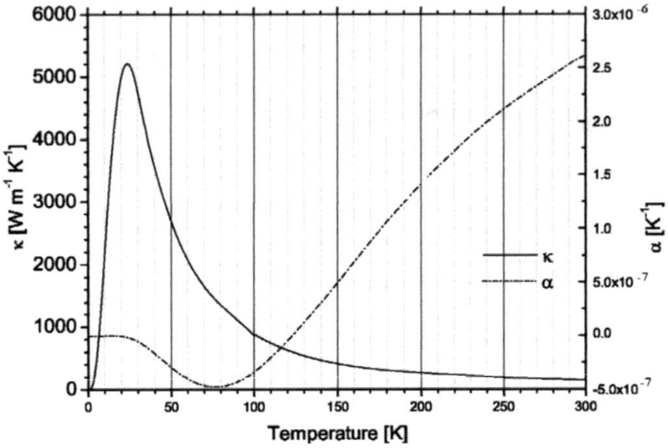


**Figure 1.** Typical curves of the thermal conductivity $\kappa(T)$ and thermal expansion $\alpha(T)$ for silicon [6] [7].

A typical thermal conductivity curve versus temperature for pure silicon is shown in figure 1. Below 100 K the thermal conductivity $\kappa$ strongly depends on doping, specimen geometry and crystalline regularity. In particular, in a perfect silicon crystal the peak position is related to the specimen dimensions which limit the mean free path of the phonons, that is dominated by Umklapp scattering at room temperatures. The free path is also sensitive to isotopic mixing: very high thermal conductance has been observed in isotopically pure silicon-28 specimens [8], corresponding to a peak value of $\kappa \simeq 3 \times 10^4$ at 20 K.

In order to realize a monolithic silicon suspension stage, mirror and suspending ribbons have to be joined together. To avoid spoiling the silicon advantages, joints must have high breaking strength and must not introduce intolerably high mechanical dissipation. A highly promising chemistry-based technique for the construction of cryogenic and ultra-low loss monolithic suspensions is silicate bonding. The bonding is realized with a very thin layer of thermal oxide grown on the two silicon surfaces with high flatness (within $\lambda/10$). A drop of aqueous hydroxide solution such as KOH or NaOH is deposited among the surfaces; a silicate gel is formed, that solidifies into a rigid and robust siloxane network. To bond Si samples, surfaces must have $SiO_2$ coatings about 100 nm thick. The thickness of the obtained bond is on the order of 50 nm, with breaking stresses at room and cryogenic temperatures (77 K) reported to be $39 \pm 14$ MPa and $39 \pm 8$ MPa respectively [9].

A complete thermomechanical characterization of the bonding requires the evaluation of its effect on the thermal conductivity across the bonded parts. This feature has to be studied in cryogenic conditions, to check that the heat extraction capability of the suspension is not compromised by it.

## 2. Measurement technique

In the Firenze INFN Virgo laboratory, we implemented a facility for the measurement of thermal conductivity of materials from 300K down to 4K. The measurement is performed by recording at each temperature $T$ the regime $\Delta T$ set along a sample portion of length $L$ and cross section $\Sigma$ by a constant heat flux $P$. The sample is placed in a liquid nitrogen cryostat under vacuum at $10^{-8}$ mbar, so that heat conduction in air is negligible. Then the conductivity $\kappa$ is computed as:

$$\kappa(T) = \frac{P}{\Delta T}\frac{L}{\Sigma}$$

For this expression to be valid, the heat flux along the sample must be homogeneous, so that cross sections $\Sigma$ are isothermal surfaces. To study the effect of the bonding layer, the sample under investigation is built by silicate bonding the flat faces of two silicon cylinders. The bonding of the cylinders was performed in the laboratory of the University of Glasgow using Prolog Semicor, P/Boron doped silicon. To have a situation close to the expected geometry of silicon monolithic suspensions, a cross section of diameter 1" was chosen. We measure the temperature difference both along the entire sample, thus including the bonding contribution, and along one of the two silicon cylinders, to compare with the known silicon curve and to check the goodness of the measurement. The two differences are conveniently obtained by placing three temperature sensors, two on the same cylinder, the last one on the other, beyond the bonding layer.

The presence of damages or inhomogeneities in the bonding layer could perturb the heat flux. To allow the measurement of the overall conductance of the specimen even in such case, the temperature difference must be sensed in places where the flux is unperturbed. Therefore, we performed FEA simulations to draw the isothermal surfaces in case of a small detachment (few mm$^2$) of the bonding. We found (see figure 2) that at 1 cm from the detachment the flux can be regarded as homogeneous. To place the three sensors at least that distance from contact and bonding surfaces, the bonded cylinders are cut 5 cm and 3 cm long.

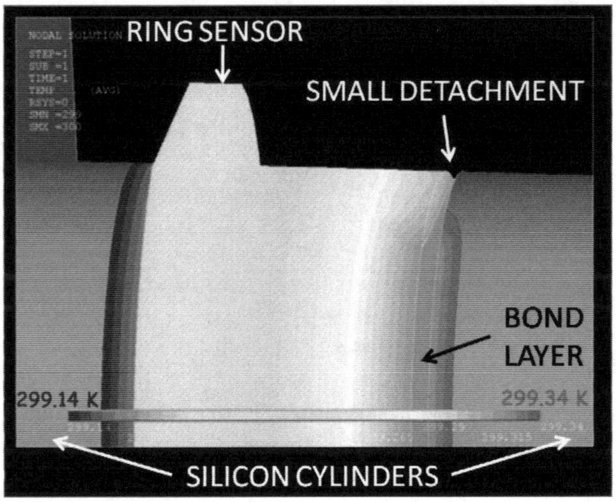


**Figure 2.** FEA simulation showing in colours the isothermal surfaces when a small detachment in the bond is produced. The ring sensor (see figure 3) is 1 cm from the bonding layer. The bonding layer is indicated by the black arrow. The simulation was performed at room temperature with 1 K temperature difference across the sample.

The experimental setup is shown in figure 3. The constant heat flux is provided by Joule effect in a resistive coil. The flux is made homogeneous by passing through an aluminium heater before reaching the sample. The specimen is placed in contact with the heater and with the

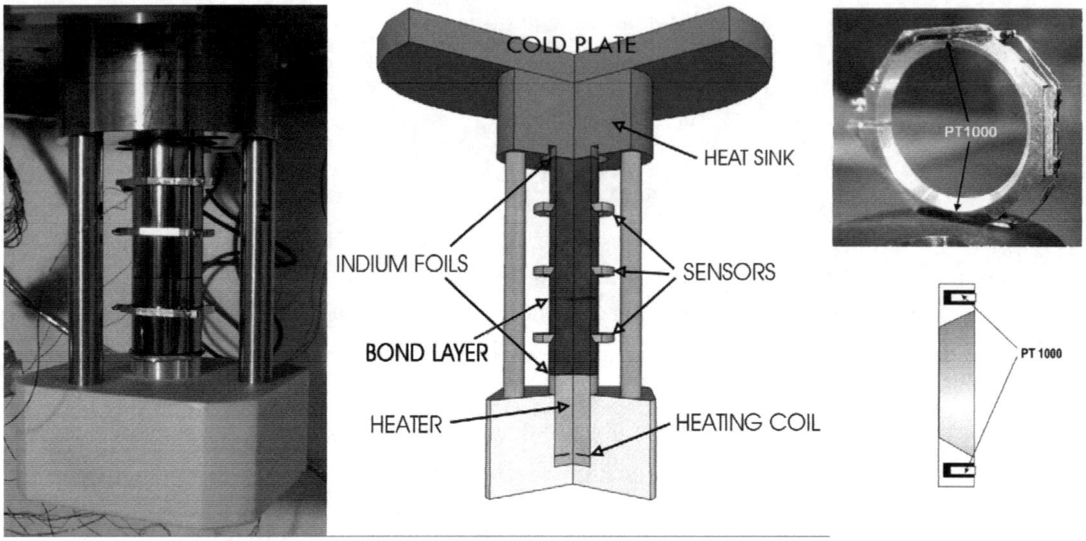


**Figure 3.** Picture and scheme of the experimental setup described in the text. On the right, a view of an aluminum ring sensor (inner diameter 1") is shown.

heat sink, connected to the cold plate of the cryostat. The quality of the contact is guaranteed by inserting thin indium foils.

Three aluminium rings are placed along the sample, with the inner sharp edge placed in contact with the silicon surface. The temperature of the rings is acquired by PT1000 resistive sensors glued to them. The rings allow a good accuracy in the determination of $L$.

After the sample was cooled down to 77 K, the measurements were performed during the free rise of the temperature to room value. For each measurement, the collected data enable calculation of the conduction of both a bond-free section of the silicon cylinder and a section that contains a bond.

## 3. Results and discussion

### 3.1. Measurement uncertainties

As from the latter equation, the estimated relative uncertainty on the measurement is given by several contributions:

$$\frac{\Delta\kappa}{\kappa} = \frac{\Delta P}{P} + \frac{\Delta L}{L} + \frac{\Delta(\Delta T)}{\Delta T} + \frac{\Delta\Sigma}{\Sigma}$$

The power $P$ is estimated by measuring the voltage and the current in the heating resistive coil. In the typical measurement conditions ($P$=20 mW), the relative error on $P$ is less than 1%. Due to the very high conductance of the sample, power losses through the coil wires or by radiation are negligible.

The distances $L$ (17.8 mm along pure silicon and 44.8 mm across bonding) have been measured within $\pm 0.3$ mm, and the radius of the cross section within $\pm 0.05$ mm. Therefore, the contribution to the relative error due to both these measurements is at most 3%.

The temperature difference $\Delta T$ is evaluated by measuring the resistance of PT1000 sensors. The resistance is measured in an AC bridge, then converted to a temperature. Therefore, we have:

$$\Delta T = \alpha G \Delta V$$

where $\alpha$ is a conversion factor given by the PT1000 calibration curve, $G = dR/dV$ is the AC bridge gain and $\Delta V$ is the bridge voltage. The relative error on $\alpha$ is less than 1%. The bridge gain was determined by unbalancing the bridge with a known resistance, and its reproducibility was within 1%. The fluctuations of $\Delta V$ are very small, so that $\sigma_{\Delta V}/\Delta V \simeq 1\%$ in the worse conditions, that is, at low $T$ where the $\Delta T$ is very small (about 4 mK). The setup is sensitive to temperature differences as low as a few tens of $\mu$K.

The estimated relative error of the measurement is therefore about 6%.

### *3.2. Results*

The measured values for $\kappa(T)$ are presented in figure 4. The obtained silicon conductivity is compared with pure silicon literature values [10] [7] and with measured values for a P/Boron doped specimen with cross section 6x6 mm$^2$ found in [11]. The agreement at temperatures not far from room temperature, where the effects of doping and geometry of the specimen are less important, proves that there is no evidence of perturbations due to non-homogeneity of the thermal flux. Also, a curve, computed modeling the bonding layer (plus the oxide layers) as a cylindrical section of glass 300 nm thick, is drawn for comparison.

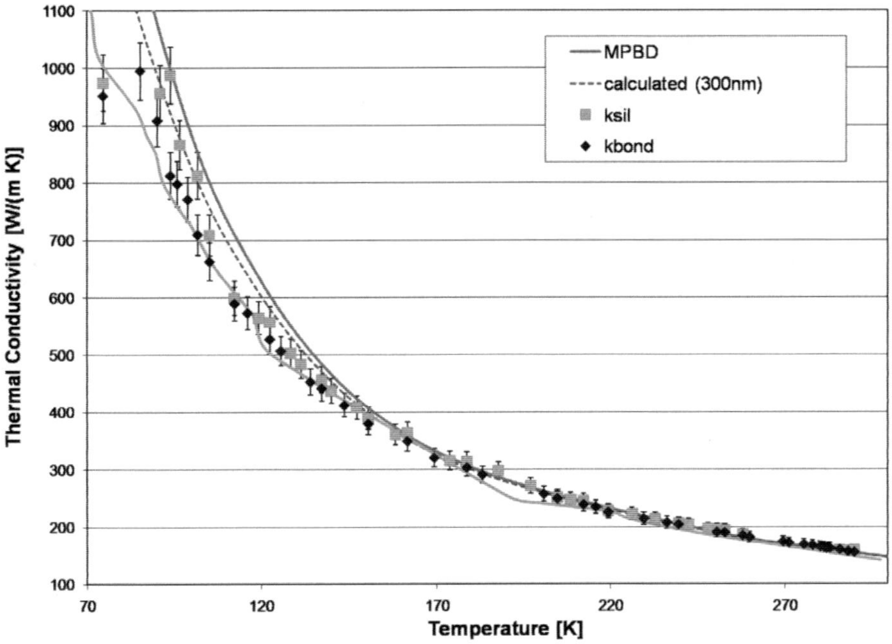


**Figure 4.** Conductivity measurement results. MPDB is the pure silicon curve from Material Properties Database. "Calculated" is the curve obtained by modeling the bonding as a cylindrical section of glass 300 nm thick. "Ksil" and "Kbond" are the measured values for silicon and across the bonding. For comparison, the green curve is an interpolation of the measured conductivity values for a P/Boron doped specimen with cross section 6x6 mm$^2$.

In the explored temperature range, the silicate bonding does not seem to affect in a relevant way the sample conductance. In figure 5, the conductivity values measured including the bonding layer, normalized to the measured silicon $\kappa$ are presented. The effect of the bonding is lower than the measurement error, in agreement with the estimation obtained from the 300 nm glass section model, except for temperatures in the range 77÷100 K.

However, measurements from 77 K to 100 K were hard to be obtained in this configuration: due to the rapid rise of the sample temperature in this range, the temperature varied significantly within a measurement, thus making impossible to reach a stationary $\Delta T$. As this error source cannot be simply estimated, it is not included in the error bars of figure 4. Therefore, in this range of temperatures, our measurements cannot be conclusive.

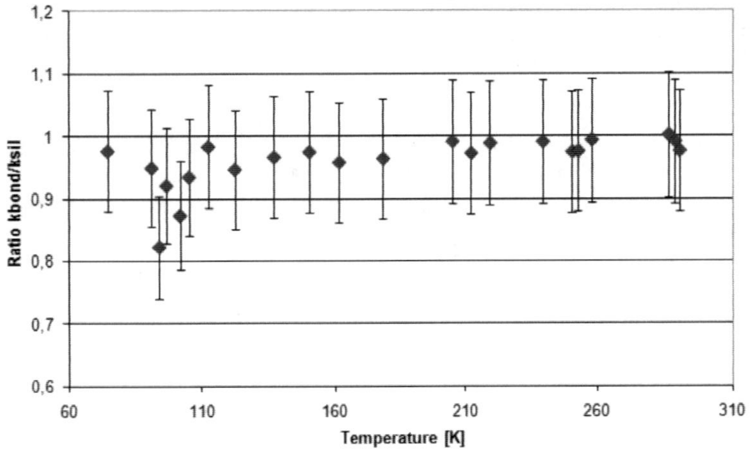


**Figure 5.** Conductivity values measured including the bonding layer, normalized to the measured silicon conductivity.

The study needs to be extended to lower temperatures in order to fully understand the effect of bonding in cryogenic conditions. Therefore, we plan to carry out the measurements on this sample down to 20 K with liquid Helium and the present setup. After that, new sensing elements and liquid Helium will be used for reaching 4 K. To avoid the problems due to the free rise of the temperature up to 100 K, a temperature control system is going to be implemented in the setup.

## References

[1] Rowan S, Hough J and Crooks D R M 2005 *Phys. Lett.* A **347** 25-32
[2] Reid S, Cagnoli G, Crooks D R M, Hough J, Murray P, Rowan S, Fejer M M, Route R and Zappe S 2006 *Phys. Lett.* A **351** 205-211
[3] Lam C C and Douglas D H 1981 *Phys. Lett.* **85** 41
[4] McGuigan D F, Lam C C, Gram R Q, Hoffman A W, Douglas D H and Gutche H W 1978 *J. Low Temp. Phys.* **30** 621
[5] Hu S M 1982 *J. Appl. Phys.* **53** 3576
[6] Swenson C A 1983 *J. Phys. Chem. Ref. Data* **12** 179
[7] Ho C Y, Powell R W and Liley P E 1972 *J. Phys. Chem. Ref. Data* **1** 279
[8] Ruf T, Henna R W, Asen-Palmera M, Gmelina E, Cardonaa M, Pohlb H J, Devyatychc G G and Sennikovc P G 2000 *Solid State Communications* **115** 243
[9] Beveridge N, van Veggel M, Hough J, Rowan S, Nawrodt R, Reid S, Besentzek B, Davidson J and Nicholson D 2009 *2nd ET Annual Workshop presentation* available at
`http://agenda.infn.it/contributionDisplay.py?contribId=41&sessionId=3&confId=1564`
[10] Material properties database (MPDB). `www.jahm.com`.
[11] Touloukian Y S and Buyco E H 1970 *Thermophysical Properties of Materials* (New York: Plenum)

Journal of Physics: Conference Series **228** (2010) 012020 doi:10.1088/1742-6596/228/1/012020

# Multitechnique investigation of $Ta_2O_5$ films on $SiO_2$ substrates: comparison of optical, chemical and morphological properties.

**M Prato[1], E Cesarini[3,4], M Lorenzini[4], A Chincarini[1], G Cagnoli[3,4], M Canepa[2,1], F Vetrano[3,4] and G Gemme[1]**

[1] Istituto Nazionale di Fisica Nucleare, Sezione di Genova, Via Dodecaneso 33, 16146 Genova, Italy

[2] Dipartimento di Fisica, Università di Genova, Via Dodecaneso 33, 16146 Genova, Italy

[3] Università di Urbino, Via S. Chiara 27, 61029 Urbino, Italy

[4] Istituto Nazionale di Fisica Nucleare, Sezione di Firenze, Via G. Sansone 1, 50019 Sesto Fiorentino (FI), Italy

E-mail: mirko.prato@ge.infn.it

**Abstract.** $Ta_2O_5$ mechanical losses seem to be the main cause of mirror thermal noise, limiting current interferometric gravitational wave detectors sensitivity in the 50-300 Hz frequency range. Work is in progress for the identification of these relaxation processes probably related with lattice defects and impurities that are distributed both in the mirror bulk and at the surface, in order to introduce step by step the suitable modifications in the samples until a stable "optimum performance" is obtained both from the optical and the thermo-mechanical point of view. Here we present our first results of a multitechnique characterization of $Ta_2O_5$ films deposited on $SiO_2$ substrates. Optical, chemical and morphological properties have been investigated by means of Spectroscopic Ellipsometry, X-ray Photoelectron Spectroscopy and Atomic Force Microscopy. Measurements carried out on pure bulk $Ta_2O_5$ samples will be also reported for comparison.

## 1. Introduction

The sensitivity of current interferometric gravitational wave detectors is limited in the 50-300 Hz frequency range by the mirror thermal noise. More precisely, the dielectric coating deposited on the mirror surface was found to induce a significant degradation of the high quality factor, Q ($10^6 \div 10^9$ at room temperature) of the silica substrates [1–3]. A stack of $Ta_2O_5$ and $SiO_2$ layers is presently used to obtain the required optical performances, and there is evidence that losses in the $Ta_2O_5$ are the major source of noise [4–8]. Several attempts have been made for improving the global (substrate + coating) quality factor of the mirrors. A good approach is to optimize the dimensions of each layer [9, 10]; this approach may be considered not alternative but complementary to other strategies aiming to improve the performances of the layers by modifying their structure and/or composition: in this sense experiments have been performed introducing a dopant in the $Ta_2O_5$ layers, leading to interesting results [8, 11].

Any strategy aimed at modifying the structure of the layers, either internal or at the interface surface between layers, requires a preliminary and accurate knowledge of the physical-chemical properties of the $Ta_2O_5/SiO_2$ repeat unit.

© 2010 IOP Publishing Ltd

In this paper, we will present our first results obtained on as-received $Ta_2O_5/SiO_2$ films. Optical, chemical and morphological properties have been investigated by means of spectroscopic ellipsometry (SE), X-ray photoelectron spectroscopy (XPS) and atomic force microscopy (AFM). XPS measurements carried out on pure bulk $Ta_2O_5$ samples will be also reported for comparison. Measurements of quality factor obtained using a new kind of nodal suspension (GeNS, see ref. [12]) will be also reported.
Our work in progress is aimed to characterize defects/non-idealities both at the surface (using AFM, XPS and SE) and in the bulk (mainly through the application of SE) of the $Ta_2O_5$ layers, and to relate them, if possible, to the mechanical response of the samples.

## 2. Experimental

### *2.1. Materials*

$Ta_2O_5$ films (200 nm nominal thickness) were deposited on Corning 7980® substrates by Double Ion Beam sputtering (ATFilms) and used without any further cleaning procedure. $Ta_2O_5$ powder (Aldrich, 99.99% trace metal basis, particle size $< 5$ $\mu$m) was used as-received.

### *2.2. Methods*

Standard SE measurements were performed on a recently upgraded rotating compensator ellipsometer (M-2000, J.A. Woollam Co. Inc.). The instrument allows simultaneous measurements at 674 different wavelengths in the range 245-1700 nm. Principles of SE are described at length in several books [13–15]. In brief, the output of ellipsometry is the complex reflection coefficient $\rho = \frac{r_p}{r_s} = \tan\Psi \exp(i\Delta)$, where $r_p$ and $r_s$ are the Fresnel reflection coefficients for p- and s-polarization, respectively. Analysis of ellipsometric spectra allows to gather information on the optical and morphological properties of films with thickness ranging from some micrometers down to monolayer level [16].
XPS analysis was carried out with a PHI ESCA 5600 MultiTechnique apparatus. The system consists of an X-ray Al-monocromatised source ($h\nu = 1486.6$ eV) and a spherical capacitor electron energy analyser, used at a constant pass energy of 5.85 eV. In the standard configuration the analyser axis formed an angle (take-off angle) of 68° with the sample surface. The binding energy (BE) scale was calibrated setting the adventitious C 1s level at 284.8 eV.
Tapping mode AFM measurements were performed using a Multimode/Nanoscope IV system (Digital Instruments) and Si cantilevers (OMCL-AC160 TS, Olympus).
Mechanical quality factor measurements were performed exciting a resonant mode of the sample and measuring the ring down. The damping envelope is an exponential decay whose rate is

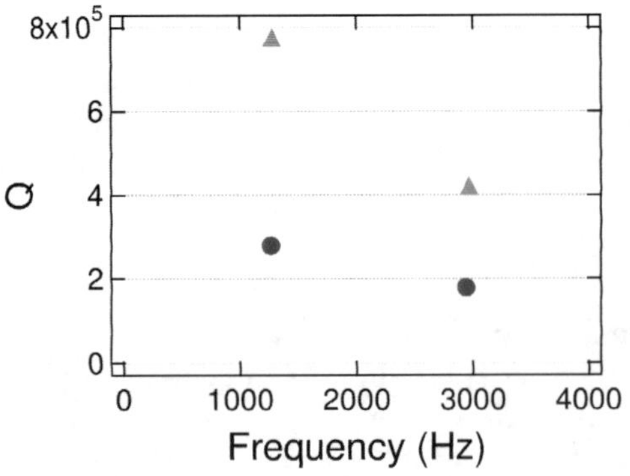


**Figure 1.** Maximum Q values obtained using GeNS [12] on 50 mm diameter/0.5 mm thickness Corning disks: bare substrates (triangles) and 200 nm thick $Ta_2O_5$ film covered ones (full circles).

inversely proportional to the quality factor at the frequency of the excitation. All the mechanical characterization were performed using GeNS [12], a new kind of nodal suspension designed to minimize the effect of the external losses due to the experimental apparatus on the measurement of Q. Figure 1 shows the maximum quality factor of the first and second butterfly mode measured on 200 nm thick $Ta_2O_5$ film on Corning glass compared with the value of bare Corning substrate (disks diameter = 50 mm, disks thickness = 0.5 mm). It is possible to observe how the presence of the film reduces the mechanical quality of a factor $\approx$2.5.

## 3. Results and discussion

### *3.1. Spectroscopic Ellipsometry*

Information on optical properties and thickness of thin films can be derived by comparing the experimental data with simulations based on realistic optical models of the system under investigation [14, 15, 17]. A good knowledge of the optical response of the substrate is hence of utmost importance. Figure 2 (panels (a) and (b)) shows typical $\cos\Delta$ and $\tan\Psi$ experimental

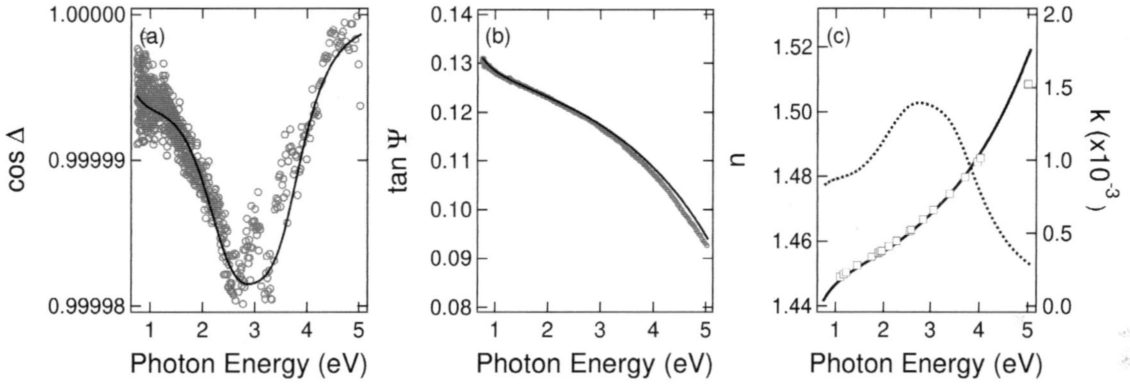


**Figure 2.** Panels (a) and (b): typical $\cos\Delta$ and $\tan\Psi$ experimental (grey markers) and generated (black curves) patterns collected on Corning 7980® bare substrates at 60° of incidence. Data range of panel (a) has been properly scaled to emphasize the very small dip in the $\cos\Delta$ pattern. Panel (c): refractive index (n, full line) and extinction coefficient (k, dotted line) obtained for the Corning 7980® substrate. A very low (k of the order of $10^{-3}$) absorption over the whole spectral range was assumed (see text). Refractive index provided by Corning Inc. company [19] (grey open squares) is reported for comparison.

patterns (grey markers) collected on Corning 7980® bare substrates at 60° of incidence. Measurements were performed following the *index matching* method described in ref. [18] to suppress backside reflections occurring when dealing with transparent substrates. The use of adhesive tape (Scotch Magic® or similar) applied to the back surface of soda-lime glass was proved being effective over the full measured spectral range [18].

The best fit (black solid curves in panels (a) and (b) of figure 2) of the experimental data required the simulation of a very weak absorption to reproduce the very small dip visible in panel (a). A model was adopted which takes effectively into account the effect of inclusions (i.e. trapped air bubbles, traces of metal coming from the melting procedure, solid impurities from the raw material), the very low surface roughness of the samples (order of 0.1 ÷ 0.2 nm, evaluated by AFM) and even some possible non-idealities of the transfer function of the instrument. The resulting optical properties are reported in panel (c) of figure 2, compared to the refractive index provided by Corning Inc. company [19].

Figure 3 shows typical $\cos\Delta$ and $\tan\Psi$ experimental patterns (collected at 60° of incidence) measured on $Ta_2O_5$ films together with simulated data obtained using a four-phases-model (surface roughness/dense $Ta_2O_5$ film/lower density $Ta_2O_5$/Corning substrate). The best fit was

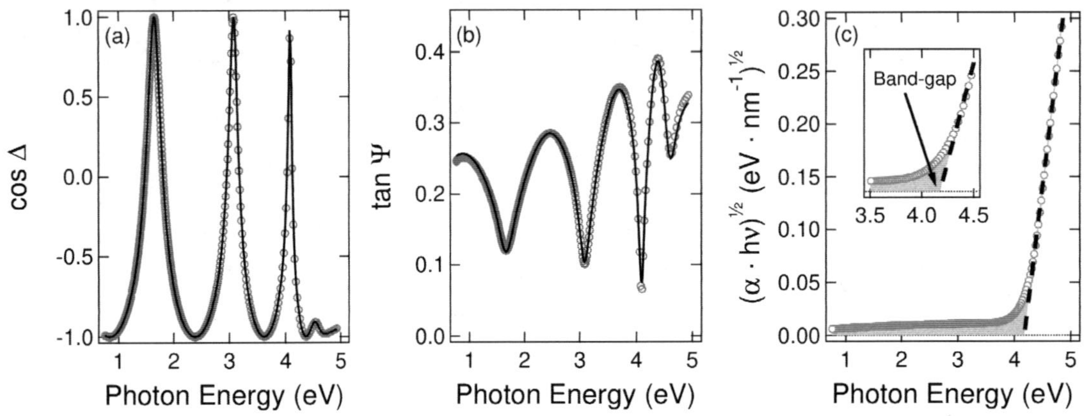


**Figure 3.** Panels (a) and (b): typical $\cos\Delta$ and $\tan\Psi$ experimental (grey markers) and generated (black curves) patterns, collected at 60° of incidence on $Ta_2O_5$ films. The best fit was obtained through the four-phases-model described in the text. Panel (c): Tauc plot (see text). $Ta_2O_5$ layers show a low but significant absorption (grey area) in the energy region below the band-gap, a behavior symptomatic of non-stoichiometric $Ta_2O_5$ [21].

obtained with a surface roughness layer (50% voids and 50% material) thickness of $(1.6 \pm 0.2)$ nm and a total thickness of the $Ta_2O_5$ layer of $(195.6 \pm 0.2)$ nm, in good agreement with the nominal thickness. Ellipsometric surface roughness was confirmed by AFM measurements on several zones of the samples. To properly describe the optical response of the $Ta_2O_5$ layer, it was necessary to take into account a small fraction of voids (approximately 1.5%) in a 20 nm-thick region next to the Corning substrate; this "low density" layer could be related to faults in the growth process.
The optical properties of the $Ta_2O_5$ layer were obtained through a Cody-Lorentz model with Urbach tail absorption in the region below the band-gap energy [20]. This model was shown to well describe the optical behavior of amorphous materials in the region around the band-gap energy. A low intensity Drude absorption was also added, in order to take into account the absorptions in the low energy range. Since stoichiometric $Ta_2O_5$ should be non-absorbing and non-dispersive [21] at energies below the band-gap, our ellipsometric characterization seems to indicate the presence in the film of species of low stoichiometry. This unexpected behavior could be emphasized by looking at the so called Tauc plot [21, 22] (panel (c) of figure 3), in which the quantity $(\alpha \cdot h\nu)^{1/2}$ is plotted against the photon energy $h\nu$. $\alpha$ is the absorption coefficient, $\alpha = 4\pi\, k/\lambda$, where $k$ is the extinction coefficient of $Ta_2O_5$ obtained from the best fit of the SE data. The energy gap of the material (approximately 4.2 eV) can be obtained through a linear fit of the high energy part of the plot, as sketched in the figure (dashed line). A significant absorption tail is present in the energy region below the band-gap (grey area).

### *3.2. X-ray Photoelectron Spectroscopy*

In order to evaluate the stoichiometry/chemical properties of our samples, XPS measurements were performed. Figure 4 (panels (a) and (b)) shows typical spectra obtained on the Ta 4f and O 1s regions. For each Ta 4f doublet, a 1.91 eV energy splitting and a 4:3 intensity ratio between

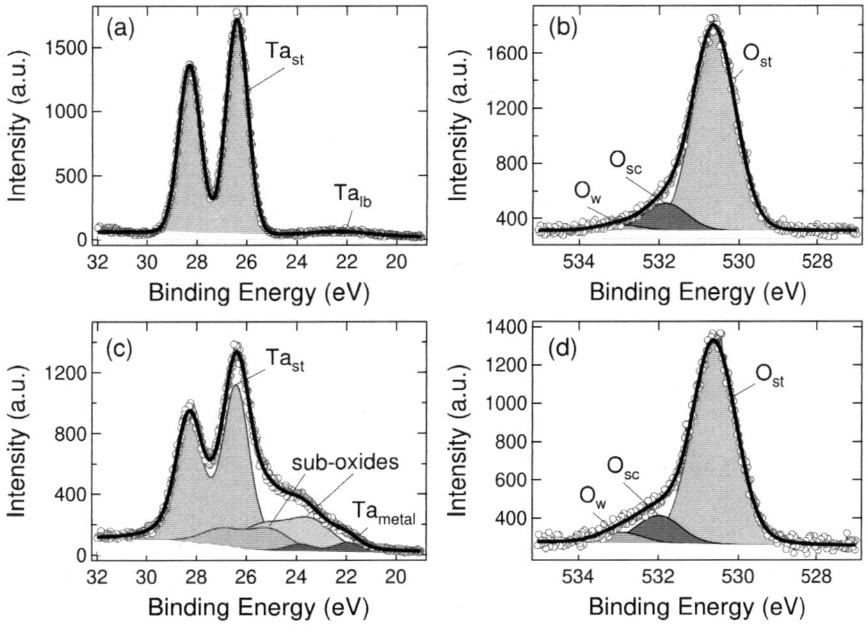


**Figure 4.** Panels (a) and (b): Ta 4f and O 1s experimental (grey markers) and generated (black lines) data collected on $Ta_2O_5$ films. For comparison, data collected in the Ta 4f region of a sputtered $Ta_2O_5$ film and O 1s region of ultra-pure $Ta_2O_5$ powder (panel (c)and (d) respectively) are reported.

the 7/2 and 5/2 components were applied.
Analysis of the Ta 4f region points out the presence of an intense doublet with the 7/2 component at 26.4 eV ($Ta_{st}$), typical of the stoichiometric form $Ta_2O_5$ [23, 24]. A hardly visible peak centered at around 22 eV is also present ($Ta_{lb}$). Metallic Ta and sub-oxides can be identified in this low binding energy region (see panel (c) of figure 4), looking at the Ta 4f spectra collected on a sputtered sample (12 seconds, 3 KeV $Ar^+$), confirming the indication obtained by the ellipsometric characterization.
Concerning the O 1s region, at least three different species should be considered in order to obtain a good fit of the experimental data: the first, $O_{st}$, intense, centered at around 530.6 eV, is ascribable to oxygen of the stoichiometric $Ta_2O_5$ [23]; the second one, $O_{sc}$, at higher binding energy (around 532 eV), is usually attributed to surface contaminations [24]; the third, $O_w$, centered at 533 eV is likely related to water contaminations [23]. However, a similar interpretation scheme does not take into account oxygen species of the sub-oxides. Moreover, the same species, with the same $O_{st}/O_{sc}$ intensities ratio, can be observed in the data collected on ultra-pure $Ta_2O_5$ powder, a sample with an higher surface-to-volume ratio (panel (d) of figure 4). In addition, on sputtered samples the $O_{sc}$ species is still present (although decreased in intensity), despite all the carbon contaminations were removed. Further investigations are needed to clearly identify this $O_{sc}$ oxygen; nevertheless, the results of these first XPS measurements seem to indicate that both $O_{st}$ and $O_{sc}$ are bound to Ta atoms and they are peculiar of $Ta_2O_5$ material. If confirmed, the possibility of thermally activated transitions between these two different species could be considered as a candidate source of mechanical losses in this material.

## 4. Conclusions and Perspectives

In this paper, we have presented the first results of the characterization of $Ta_2O_5/SiO_2$ films through SE, XPS and AFM. We have implemented an experimental protocol capable to detect and characterize defects/non-idealities in these films; in order to confirm the results here reported, our protocol has to be tested on a statistically relevant number of samples. Our studies will be focused on understanding the origin of these defects and on finding their possible implication on the degradation of the quality factor of gravitational wave detectors mirrors.

## Acknowledgments

This research has been supported by Istituto Nazionale di Fisica Nucleare - Italy, MIUR PRIN 2007T7AC3L and Fondazione Carige. MP acknowledges the experimental support of M. Neri and G. Pizzigoni in this first stage of the work.

## References

[1] Gretarsson A M, Harry G M, Penn S D, Saulson P R, Schiller J J and Startin W J 2000 *Gravitational Waves* **523** 306
[2] Crooks D R M, Sneddon P, Cagnoli G, Hough J, Rowan S, Fejer M M, Gustafson E, Route R, Nakagawa N, Coyne D, Harry G M and Gretarsson A M 2002 *Class. Q. Grav.* **19** 883
[3] Harry G M, Gretarsson A M, Saulson P R, Kittelberger S E, Penn S D, Startin W J, Rowan S, Fejer M M, Crooks D R M, Cagnoli G, Hough J and Nakagawa N 2002 *Class. Q. Grav.* **19** 897
[4] Crooks D R M, Cagnoli G, Fejer M M, Gretarsson A, Harry G, Hough J, Nakagawa N, Penn S, Route R, Rowan S and Sneddon P H 2004 *Class. Q. Grav.* **21** S1059
[5] Numata K, Ando M, Yamamoto K, Otsuka S and Tsubono K 2003 *Phys. Rev. Lett.* **91** 260602
[6] Penn S D, Sneddon P H, Armandula H, Betzwieser J C, Cagnoli G, Camp J, Crooks D R M, Fejer M M, Gretarsson A M, Harry G M, Hough J, Kittelberger S E, Mortonson M J, Route R, Rowan S and Vassiliou C C 2003 *Class. Q. Grav.* **20** 2917
[7] Crooks D R M, Cagnoli G, Fejer M M, Harry G, Hough J, Khuri-Yakub B T, Penn S, Route R, Rowan S, Sneddon P H, Wygant I O and Yaralioglu G G 2006 *Class. Q. Grav.* **23** 4953
[8] Harry G M, Abernathy M R, Becerra-Toledo A E, Armandula H, Black E, Dooley K, Eichenfield M, Nwabugwu C, Villar A, Crooks D R M, Cagnoli G, Hough J, How C R, MacLaren I, Murray P, Reid S, Rowan S, Sneddon P H, Fejer M M, Route R, Penn S D, Ganau P, Mackowski J-M, Michel C, Pinard L and Remillieux A 2007 *Class. Q. Grav.* **24** 405
[9] Agresti J, Castaldi G, De Salvo R, Galdi V, Pinto I M and Pierro V 2008 LIGO G080082-00-R
[10] Pierro V, Pinto I M, Principe M, De Salvo R and Dannenberg R 2009 LIGO-G0900205
[11] Martin I, Armandula H, Comtet C, Fejer M M, Gretarsson A, Harry G, Hough J, Mackowski J-M M, MacLaren I, Michel C, Montorio J-L, Morgado N, Nawrodt R, Penn S, Reid S, Remillieux A, Route R, Rowan S, Schwarz C, Seidel P, Vodel W and Zimmer A 2008 *Class. Q. Grav.* **25** 055005
[12] Cesarini E, Lorenzini M, Campagna E, Martelli F, Piergiovanni F, Vetrano F, Losurdo G and Cagnoli G 2009 *Rev. Sci. Instrum.* **80** 053904
[13] Azzam R M A and Bashara N M 1987 *Ellipsometry and Polarized Light* (Amsterdam: North Holland)
[14] Tompkins H G and Irene E A *Handbook of Ellipsometry* 2005 (Heidelberg: Springer-Verlag)
[15] Fujiwara H 2007 *Spectroscopic Ellipsometry: Principles and Applications* (Chichester: Wiley & Sons Ltd)
[16] Prato M, Moroni R, Bisio F, Rolandi R, Mattera L, Cavalleri O and Canepa M 2008 *J. Phys. Chem.* C **112** 3899
[17] Gonella G, Cavalleri O, Emilianov I, Mattera L, Canepa M and Rolandi R 2002 *Mater. Sci. Eng. C* **22** 359
[18] Synowicki R A 2008 *phys. stat. sol. (c)* **5** 1085
[19] Corning Incorporated, http://www.corning.com
[20] Ferlauto A S, Ferreira G M, Pearce J M, Wronski C R, Collins R W, Deng X and Ganguly G 2002 *Journal of Applied Physics* **92** 2424
[21] Demiryont H, Sites J R and Geib K 1985 *Applied Optics* **24** 490
[22] Tauc J 1974 *Amorphous and Liquid Semiconductors* (New York: Plenum)
[23] NIST X-ray Photoelectron Spectroscopy Database, http://srdata.nist.gov/xps/
[24] Atanassova E and Spassov D 1998 *Applied Surface Science* **135** 71

# OSCAR a Matlab based optical FFT code

**Jérôme Degallaix**

Max-Planck-Institut für Gravitationsphysik (Albert-Einstein-Institut) and Leibniz Universität Hannover, Callinstr. 38, 30167 Hannover, Germany

E-mail: `jerome.degallaix@aei.mpg.de`

**Abstract.**
Optical simulation softwares are essential tools for designing and commissioning laser interferometers. This article aims to introduce OSCAR, a Matlab based FFT code, to the experimentalist community. OSCAR (Optical Simulation Containing Ansys Results) is used to simulate the steady state electric fields in optical cavities with realistic mirrors. The main advantage of OSCAR over other similar packages is the simplicity of its code requiring only a short time to master. As a result, even for a beginner, it is relatively easy to modify OSCAR to suit other specific purposes. OSCAR includes an extensive manual and numerous detailed examples such as simulating thermal aberration, calculating cavity eigen modes and diffraction loss, simulating flat beam cavities and three mirror ring cavities. An example is also provided about how to run OSCAR on the GPU of modern graphic cards instead of the CPU, making the simulation up to 20 times faster.

## 1. Overview of OSCAR

OSCAR is a FFT code which is able to simulate Fabry Perot cavities with arbitrary mirror profiles. One of the key features of OSCAR is the possibility to easily modify the code to tailor specific simulation purposes. OSCAR is written with the Matlab scripting language, one can import/export files (mirror maps or cavity eigen modes profile for example), create a ring cavity, create batch file or plot 2D optical field with little programming skill. The core of the code is only 400 lines long (including comments) and the manual provides five detailed examples.

The first version of OSCAR was written with the software IGOR[1] in 2005. This code was then translated to Matlab and used to calculate diffraction losses by Pablo Barriga[2]. Finally, the Matlab code has been rewritten to decrease the computational time and to add new functionality.

OSCAR is mainly intended for people who want to quickly simulate only one cavity with non Gaussian fundamental eigenmodes or input beam. The code and the manual can also be used as an educational tool to understand how internally FFT code works.

### *1.1. Possible simulations*

OSCAR is a versatile tool to simulate Fabry Perot cavities. The following is a (non-exhaustive) list of the results which can be obtained with OSCAR:

- calculate the Gouy phase shift between higher order optical modes. It may be useful for flat beams for example, where no analytical calculations of the Gouy phase shift has been derived yet

© 2010 IOP Publishing Ltd

- calculate the coupling loss between the input beam and the cavity eigen modes in the case of mode mismatching
- calculate the circulating beam (intensity and profile) for stable and also unstable cavities
- calculate diffraction loss and eigen modes of a cavity with arbitrary mirror profiles and imperfect optics.

### *1.2. Restrictions*
OSCAR is designed to simulate anything which can be derived from the steady state, classical, optical field circulating inside a Fabry Perot cavity. It means OSCAR does not take into account radiation pressure or quantum effects. OSCAR (in the present version) can not simulate coupled cavities. For more complex simulation other FFT codes such as SIS[3] or darkF[4] exist.

## 2. Principle
In this section, we briefly introduce the concept of optical simulations using the Fourier transform. Unfortunately, due to the limited length of this article, no demonstration is included but the justification can be found in the references or in the OSCAR manual.

### *2.1. Propagation of an arbitrary optical field*
It is possible to propagate any arbitrary coherent optical field under the paraxial approximation by the use of a Fourier transform. Typically such an operation requires 3 steps [5]:

(i) Decomposition of the complex amplitude of the electric field into a sum of elementary plane waves. Mathematically, this step is achieved by a 2D Fourier transformation.
(ii) Propagation of each plane wave, which is equivalent to adding a phase shift in the frequency domain. The phase shift depends of the distance of propagation and the spatial frequency of the plane wave.
(iii) Recomposition of the electric field from the propagated plane waves. This step is in fact a 2D inverse Fourier transformation.

The above 3 steps make up the basis of optical FFT codes. The pseudo code shown here allows the propagation in free space of any arbitrary optical field, independently of any optical basis (Hermite or Laguerre Gauss) or assumption on the beam shape. In optical FFT codes, only the propagation requires a transformation into the spatial frequency domain, all the other operations (e.g. reflection by a mirror or transmission through an aperture) are performed directly on the complex electric field.

### *2.2. Adding realistic optics*
The reflection by a mirror or the transmission through a lens can be described as a change in the optical field wavefront. For example, we can consider an input laser field $E_i$ passing through an element inducing a wavefront distortion characterized by the optical path $\Delta OP(x, y)$. In this case, the transmitted field $E_t$ can be written as[6]:

$$E_t(x, y) = E_i(x, y) \times \exp\left(-jk\Delta OPL(x, y)\right) \tag{1}$$

with $k$ the constant of propagation. An aperture used to represent finite size mirrors can also be easily implemented by a 2D transmission matrix $A$. Practically, an aperture $A(x, y)$ is represented by a matrix of zeros and ones. A 0 at the position $(x, y)$ indicates that the light is blocked (falls outside the mirror) and a 1 indicates that the light is fully reflected or transmitted.

So once again, numerically, the reflection or transmission through a finite size mirror can be described as:

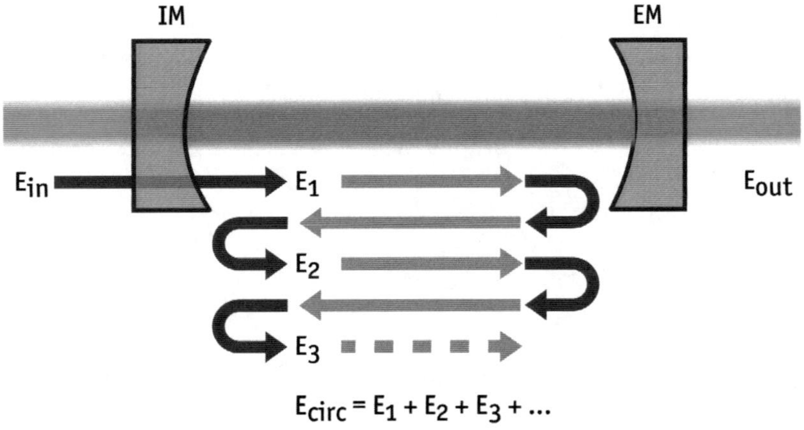


**Figure 1.** Description of the algorithm used in OSCAR to calculate the circulating field in a Fabry Perot cavity. The violet arrows represent a change in phase for the light field which is described in section 2.1 and the green arrows represent the propagation of the light field using a FFT code. From $E_i$ to $E_{i+1}$, the light field has undergone one round trip in the cavity. IM and EM stand respectively for Input Mirror and End Mirror. $E_{circ}$ is the circulating power inside the cavity.

$$E_o(x, y) = E_i(x, y) \times \exp\left(-jk\Delta OPL(x, y)\right) \times A(x, y) \tag{2}$$

*2.3. Simulating a Fabry-Perot cavity*

A Fabry-Perot cavity is made up of two mirrors facing each other. Between these two mirrors, a light field is circulating, bouncing back and forth between the two reflective coatings. One of the main interest of the Fabry-Perot cavity is that the optical power of the circulating field can be much higher than the power of the input field. With OSCAR it is possible, for a given input field, to calculate the total circulating power, reflected power and transmitted power as well as the spatial profile of all the light fields.

OSCAR calculates the circulating field by propagating back and forth the laser beam between the two mirrors and then summing all the transient fields at one reference plane [7] as shown in figure 1. This method can also be used to analytically calculate the circulating power in a Fabry Perot cavity [8].

In more detail, we can write the OSCAR algorithm used to compute the circulating power. Using the notation from the figure 1 the different consecutive steps can be described as:

(i) Define the cavity parameters as well as the mirror profiles and the input beam.

(ii) Propagate the input beam $E_{in}$ through the input mirror . For this purpose the input mirror can be exchanged with a lens, so the equation (2) can be used. As a result, we obtain the field $E_1$.

(iii) After one round trip in the cavity, the field $E_1$ becomes $E_2$. One round trip in the cavity consists specifically of one propagation through the cavity length using the FFT code, one reflection on the end mirror, another propagation back to the input mirror and then finally a reflection on the input mirror.

(iv) Repeat the last operation to create the set of electric field $\{E_i\}$.

(v) Then sum all the field $E_i$ to have the cavity circulating power $E_{circ}$. The number of light field $E_i$ to be considered to have an accurate result depends on the finesse of the cavity.
(vi) The transmitted field $E_{out}$ is simply the circulating field transmitted through the ETM.

In this above pseudo-code, we did not mention any resonance condition to maximize the circulating power in the cavity. Practically, we should always define the round trip phase shift for the field in the cavity (or a microscopic position shift for the cavity length) before calculating the cavity circulating power. The round trip phase shift allows us to set the cavity to be resonant for the fundamental mode or any other optical modes if necessary.

## 3. Applications

The OSCAR package is provided with several detailed examples to show the different possible simulations that can be carried out using optical FFT codes. In this section, we will focus on two examples: how to simulate flat beam cavities and how to calculate diffraction loss.

### *3.1. Coupling loss for flat beams cavity*

In this example, we are interested in calculating the coupling loss (also called mode matching loss) between an input Gaussian beam and a Fabry-Perot cavity supporting flat (or mesa) beams[9]. Since flat beams exhibit lower power density compared to fundamental Gaussian beams, they have often been presented as a possible way to reduce the mirror thermal noise level[10] as well as the amplitude of thermal lensing.

It is possible to directly import the mirror profiles to support flat beam into OSCAR, profiles which are usually derived using the relatively complicated analytical formula. A simpler approach can be implemented: the mirror profiles are given by the wavefront curvature of the flat beam at the mirror position. Since the flat beam is an eigen mode of the cavity, the curvature of the mirrors must match the wavefront of the incoming beam.

For this example, the length $L$ of the cavity simulated is 2 km. The power transmission of the input mirror is 0.5 %, 50 ppm for the end mirror, and the loss per reflection on each mirror is also 50 ppm. The input beam is a Gaussian fundamental mode with a beam radius of 4 cm and a wavefront curvature of 1 km. The input power is 1 W and the wavelength $\lambda$ is 1064 nm. The mirror profiles are symmetric and created to support a concentric flat beam of width $b = 2w_0$ with $w_0 = \sqrt{L\lambda/2\pi}$ following the notation from[11].

For the above setup, the steady state circulating power can be calculated with OSCAR and is found to be 546 W. Now, to calculate the coupling loss, the circulating power has to be compared to the circulating power in the case of perfect mode matching. That can be done analytically in this simple case or it can also be done in OSCAR by replacing the input beam by the beam used to create the mirror profile. The circulating power in case of perfect mode matching is 752 W, so for the simulated setup the coupling loss is around 27 %[1].

### *3.2. Diffraction loss inside a mode cleaner*

One of the most promising application of FFT optical codes is to calculate the diffraction loss of the circulating cavity field (or cavity eigen mode). As a reminder, the diffraction loss is the loss due to the finite size of the mirrors. This loss is also sometime referred to as clipping loss. Since in theory, Gaussian beams have an infinite spatial extent, diffraction losses are always present. However the loss can become negligible for large diameter mirrors and small laser beam radius.

The optical setup under consideration is a 3 mirror ring cavity with the input and output mirrors flat and the third mirror curved. The distance between the 2 flat mirrors is 6 cm and

[1] It is just a numerical example, the coupling loss between input Gaussian beam and flat beam cavity can be much lower

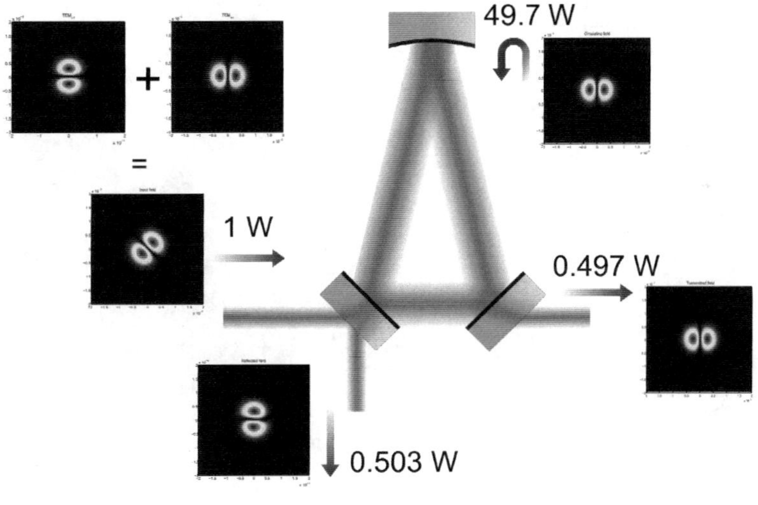


**Figure 2.** Optical setup of the mode cleaner with the input beam coming from the left side. The optical power in the different fields is also mentioned.

the distance between the curved mirror and the optical segment between the 2 flat mirrors is 20 cm. The radius of curvature of the curved mirror is 1 m, giving a g-factor of 0.77 and a waist size of 0.38 mm located between the flat mirrors. The finesse of the cavity is 310. All the mirrors have a diameter of 3 mm (this value is not realistic but interesting for the simulations).

To make the experiment more interesting, the input beam could be the sum of a $TEM_{10}$ and a $TEM_{01}$ as shown if figure 2. Both beams of power 0.5 W are well matched to the cavity (but not perfectly matched).

We select the mode cleaner to be resonant for the mode $TEM_{10}$ and as a result the mode $TEM_{01}$ is reflected. The diffraction loss is derived by calculating the round trip loss of the circulating mode when the reflectivities of all the cavity mirrors are set to 1. In this manner, the diffraction loss is found to be 15 ppm per round trip.

## 4. Calculation on the GPU

Graphical Process Unit (GPU) have recently gained special interest in the scientific community thanks to their parallel computing ability in conjunction with the availability of high level programming languages. OSCAR has also been adapted to run its most intensive calculations on the GPU using the Matlab library called GPUmat[12]. GPUmat is freeware and it requires the CUDA[13] toolkit to be installed[2]. The GPU library allows the transfer of variables from the CPU to the GPU and overloads common Matlab operators and functions which can then be executed directly on the GPU without any need of another programming language.

To test the GPU capability, we ran OSCAR on a dedicated machine with an Intel Quad core CPU (Q8200) with 8 Gb of RAM. The GPU is housed within a Telsa C1060 graphic card with 4 Gb of memory. On the software end, the operating system is Linux Ubuntu 9.04 and we use Matlab2009a, with GPUmat version 0.1 and the CUDA drivers version 2.3.

The comparison in execution time between the CPU and GPU for 100 light round trips in a cavity is shown in figure 3. For large size matrices, the GPU is more than 20 times faster than the CPU. However, to be fair, it should be mentioned that the calculation on the GPU is done with single precision numbers whereas on the CPU it is done with double precision (even

[2] so it only works with modern Nvidia graphic cards

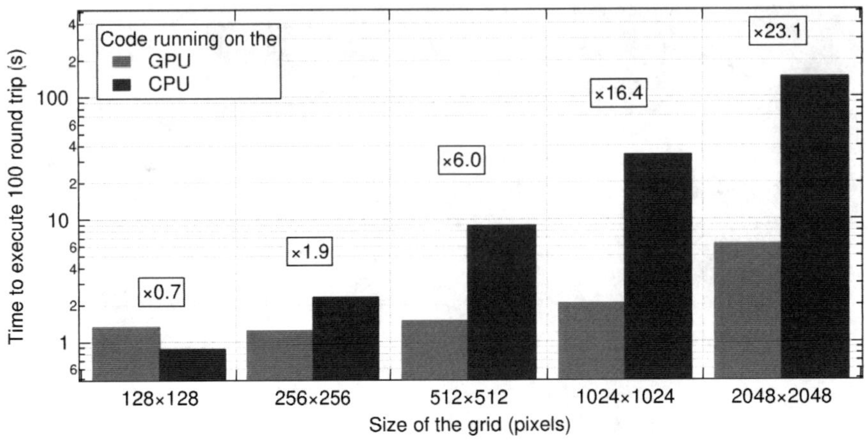


**Figure 3.** Speed comparison for the same code running on the CPU (blue bars) and on the GPU (red bars). The number in the box, indicates the gain factor of running OSCAR on the GPU. For small grid size, the GPU is slower since a large amount of time is taken by the transfer of variables between the CPU and the GPU memory.

if the variable is declared as a single in Matlab). The use of single precision or double precision number does not bring any difference in the results from OSCAR.

Regarding the future of FFT codes on the GPU, we can already take note that recent GPU cards support double precision numbers (at the price of a slower execution time) and using the Matlab Parallel Computing Toolbox, it is possible to run a single program across several GPU cards[14].

## 5. Conclusion and outlook

In this article we have presented OSCAR a versatile tool to simulate steady state Fabry-Perot cavity with imperfect mirrors. OSCAR is written for Matlab and special attention was drawn to make the code clear and easily modifiable. Moreover a fully documented manual is provided with examples.

In the next versions of OSCAR coupling cavities will be implemented and ultimately a dual recycling interferometerswith Fabry-Perot arm cavities could be simulated. Interested users may download OSCAR from two websites quoted in references [15, 16].

## References

[1] IGOR software http://www.wavemetrics.com/ 1
[2] Barriga P, Bhawal B, Ju L and Blair D G 2007 *J. Opt. Soc.Am. A* **24** 1731–1741 1
[3] Yamamoto H SIS talk at LMA https://dcc.ligo.org/cgi-bin/private/DocDB/ShowDocument?docid=5871 2
[4] Laval M Presentation of DarkF http://wwwcascina.virgo.infn.it/commissioning/OPC/Talks/DarkFstatus_08_03_2006.ppt 2
[5] Siegman A E 1986 *Lasers* (University Science Books) 2
[6] Bochner B 1991 *Modelling the Performance of Interferometric Gravitational-Wave Detectors with Realistically Imperfect Optics* Ph.D. thesis MIT 2
[7] Saha P 1997 *J. Opt. Soc.Am. A* **14** 2195–2202 3
[8] Silfvast W T 2004 *Laser Fundamentals* (Cambridge University Press) see section 11.2 3
[9] Bondarescul M and Thorne K S 2006 *Phys. Rev. D* **74** 082003 4
[10] Vinet J Y 2005 *Classical and Quantum Gravity* **22** 1395–1404 4

[11] Agresti J 2007 *Researches on Non-standard Optics for Advanced Gravitational Waves Interferometers.* Ph.D. thesis University of Pisa 4
[12] GPU mat library `http://gp-you.org/` 5
[13] CUDA website `http://www.nvidia.com/object/cuda_home.html` 5
[14] Jacket v12 for Matlab `http://www.accelereyes.com/release_notes.php` 6
[15] Matlab central website `http://www.mathworks.com/matlabcentral/fileexchange/20607` 6
[16] GEO simulation group wiki `http://www.sr.bham.ac.uk/dokuwiki/doku.php?id=geosim:oscar` 6

Journal of Physics: Conference Series **228** (2010) 012022 doi:10.1088/1742-6596/228/1/012022

# Lateral input-optic displacement in a diffractive Fabry-Perot cavity

**J. Hallam[1], S. Chelkowski[1], A. Freise[1], B.W. Barr[2], S. Hild[2], K.A. Strain[2], O. Burmeister[3] and R. Schnabel[3]**

[1] School of Physics and Astronomy, University of Birmingham, Edgbaston, Birmingham B15 2TT, United Kingdom

[2] Institute for Gravitational Research, Department of Physics and Astronomy, University of Glasgow, Glasgow, G12 8QQ, United Kingdom

[3] Max-Planck-Institut für Gravitationsphysik (Albert-Einstein-Institut) and Leibniz Universität Hannover, Callinstr. 38, D–30167 Hannover, Germany.

E-mail: jmh@star.sr.bham.ac.uk

**Abstract.** All-reflective configurations have been suggested for future laser-interferometric gravitational wave detectors. Diffraction gratings would be required as core elements of any such design. In this paper, coupling of lateral grating displacement to the output ports of a diffractive Fabry-Perot cavity was derived using a steady-state technique. The signal to noise ratio versus a potential gravitational wave signal was compared for each cavity output port. For a cavity featuring parameters similar to the planned Advanced Virgo instrument, we found the forward-reflected port offers the highest SNR at low frequencies. Furthermore, the lateral isolation suspension requirements were relaxed by a factor of twenty at a frequency of 10 Hz versus the transmitted port.

## 1. Introduction

The sensitivity of proposed gravitational wave detectors utilising the laser-interferometric technique might be improved by reducing the number of transmissive optical components and at the extreme adopting an all-reflective optical configuration [1]. Particular focus falls on the beam-splitter and cavity input mirrors of a Michelson interferometer with Fabry-Perot arm cavities as are used in LIGO [2] and Virgo [3] and which will be used for the upgrades to those detectors.

It is expected that an all-reflective configuration will eliminate transmissive thermal noises and free optical substrates to be opaque. This would widen the range of materials that could be chosen allowing the selection of other desirable properties. Further, proposals to reduce the shot noise contribution by increasing laser power are likely to introduce technical problems such as thermal lensing which are only significant with transmissive optics.

Diffraction gratings are the only possible replacements for the cavity input mirrors and beam-splitters that would still allow an all-reflective interferometer to function in a Michelson-esque way [4]. Herein we consider the case where a diffraction grating is used in place of a Fabry-Perot cavity input mirror [5].

Prior work has found that state of the art gratings cannot achieve diffraction efficiencies approaching the reflection efficiencies available from multilayer dielectric coatings [6, 7]. Thus

© 2010 IOP Publishing Ltd

Journal of Physics: Conference Series **228** (2010) 012022 doi:10.1088/1742-6596/228/1/012022

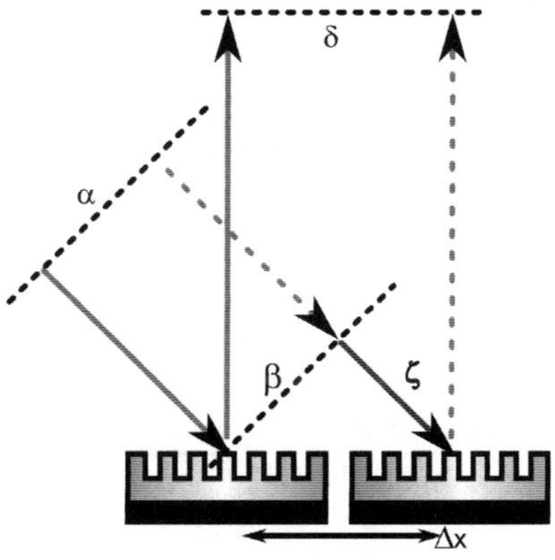


**Figure 1.** Wavefronts coupling through a diffraction grating. The wavefront (propagation vector shown in solid red) originates at $\alpha$, continues to $\beta$ and is diffracted to $\delta$. Displacing the grating by a lateral distance of $\Delta x$, causes the wavefront (propagation vector shown red-dashed) to pass through the additional path length $\zeta$ (propagation vector shown in solid blue).

the high-finesse Fabry-Perot cavities required for gravitational wave detection cannot be formed by recoupling the diffracted light back into the cavity; rather, a diffraction grating with low diffraction efficiency must be used with the diffractive order serving as the incoupler (equivalent to a conventional cavity's input mirror transmissivity). In the case of a three-port, second-order-Littrow configured cavity (as shown in Figure 2) the finesse is determined by the zeroth-order diffraction efficiency (sometimes called reflection efficiency) of the grating, which can be made equivalent to a conventional mirror reflectivity using the same multilayer coating techniques.

## 2. Lateral grating displacement phase noise

By contrast to a standard Fabry-Perot cavity using a transmissive input mirror the geometry enforced by the use of a diffraction grating causes additional coupling of grating displacement into phase noise [8].

This phase noise stems purely from the geometry of the optics and hence the grating equation alone suffices to calculate it, so long as the geometry of the set-up is known. The phase change resulting from the displacement is due to a path-length difference $\zeta$ between parallel wavefronts, shown for a simple geometrical case in Figure 1.

This noise occurs due to displacement across the grating's corrugation with respect to the beam. Such displacement can appear from angular misalignment of the end-mirror projecting the beam across the grating, or the misalignment of the injection optics causing a similar effect, or physical displacement of the grating itself. Freise et al determined that, by assuming a sensitivity goal of $h = 10^{-23}/\sqrt{\text{Hz}}$, using a set of parameters reasonably consistent with Virgo [3] and a grating both typical and practical, a standard cavity required angular end mirror alignment of $\gamma < 2 \cdot 10^{-16}\,\text{rad}/\sqrt{\text{Hz}}$ whereas a three-port coupled grating cavity required $\gamma < 1 \cdot 10^{-21}\,\text{rad}/\sqrt{\text{Hz}}$. Hence five orders of magnitude improvement in the end turning mirror alignment is required for equivalent insensitivity to this noise source. In this article we consider the physical displacement of the grating which couples into phase noise.

## 3. Frequency domain model of phase noise coupling in a grating cavity

In order to analyse the coupling of the phase noise introduced by grating displacement to the output ports of a grating cavity quantitatively we performed a frequency domain analysis. By

Journal of Physics: Conference Series **228** (2010) 012022 doi:10.1088/1742-6596/228/1/012022

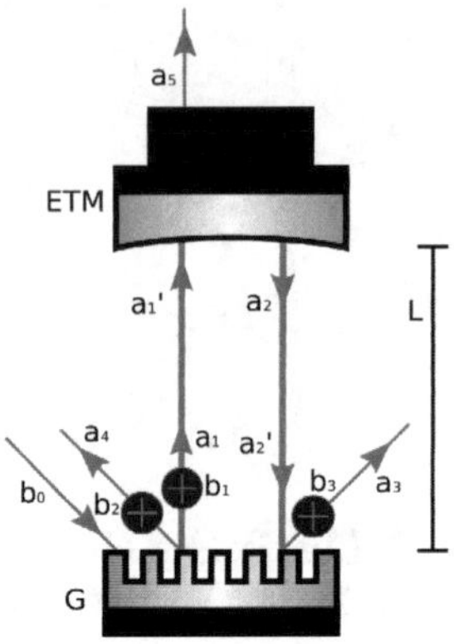


**Figure 2.** External input port $b_0$, internal input ports $b_1$, $b_2$, $b_3$ (marked by encircled addition symbols) and the transfer ports $a_1$, $a_1'$, $a_2$ and $a_2'$ in a grating cavity

carrying out this analysis for the effect of a potential gravitational-wave signal and the noise effect of lateral grating displacement we obtained the signal to noise ratio (SNR) at all three output ports of the grating cavity. Particularly important was to find the port with the highest SNR, a potentially strong driver on how to integrate such a cavity into interferometer configurations.

To compare the signal to noise ratio at the different output ports we performed the mathematical steps outlined here and discussed fully elsewhere [9]. First we determined the grating coupling relations using the symmetrical coupling matrices previously determined in [10]. Using these coupling relations we calculated both carrier light fields (shown blue in Figure 3) incident on the grating in terms of a single external input carrier light field. By definition, these carrier fields were unchanged in frequency from the input light field and therefore contained neither gravitational wave signal nor grating displacement noise.

These were then used to determine the amplitude of the sideband fields at the location of their creation. The sidebands were created by modulation, either from a gravitational wave or grating displacement. Again using the coupling relations we propagated these sidebands to the output ports and obtained their amplitude. Dividing the gravitational wave sidebands by the grating displacement noise sidebands obtained the signal to noise ratio.

### *3.1. Lateral grating displacement noise coupling in a grating cavity*

The complex amplitudes $a_3$, $a_4$ and $a_5$ associated with the upper noise sideband at each output consistent with Figure 2 are given by:

$$a_3 = ip_0\eta_1^2 e^{-2i\phi_1}\frac{\pi\Delta x}{d}\left(B_u - B_c\right), \quad (1)$$

$$a_4 = ip_0\eta_1^2 e^{-2i\phi_1}\frac{\pi\Delta x}{d}\left(B_u + B_c\right) + ip_0\eta_2 e^{-i\phi_2}\frac{2\pi\Delta x}{d}, \quad (2)$$

$$a_5 = -p_0\eta_1 e^{-i\phi_1}\frac{\pi\Delta x}{d}\frac{B_u}{\rho_2 e^{-ik_u L}}\tau_2. \quad (3)$$

Where $p_0$ is the complex amplitude for the carrier field input in the port $b_0$, $\rho_2$ is the reflectivity of the end-mirror, $\rho_0$ is the reflectivity of the diffraction grating in the special case of zeroth order diffraction normal to the surface of the grating, $\eta_1$ and $\eta_2$ are the first and second order diffraction efficiencies and $\phi_1$ and $\phi_2$ are the phase shifts associated with them, $d$ is the periodicity of the grating, $L$ is the cavity length and $\Delta x$ is the displacement of the grating. $B_c = D(\omega_c)\rho_2 e^{-2ik_c L}$

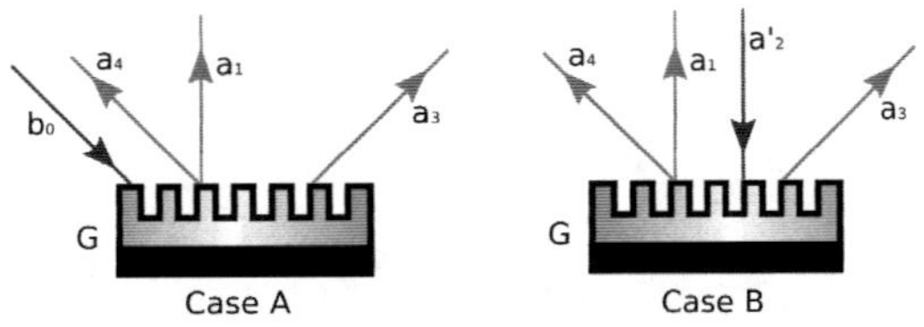


**Figure 3.** Input ($b_0$, $a_2'$, shown blue) and output ($a_1$, $a_3$, $a_4$, shown red) light field amplitudes at a phase modulating grating in second-order Littrow-configuration in case A, and zeroth-order-Littrow configuration in case B (labels consistent with Figure 2).

with $D(\omega_c) = 1/(1 - \rho_2\rho_0 e^{-2ik_cL})$, where $\omega_c$ is the absolute frequency of the input carrier light field and $k_c$ is the wavevector $(2\pi/\lambda)$ associated with it. $B_u$ can be similarly expressed except for its own absolute frequency given by $\omega_u = \omega_c + \omega_m$, the carrier frequency plus the frequency with which the grating is displaced. A similar result could be obtained for the lower sideband by subtracting rather than adding the modulation frequency, notationally replacing '$u$' with '$l$' in the subscript. Thus we computed the optical signal in all output ports generated by lateral motion of the grating.

*3.2. Signal coupling in a grating cavity*

The complex amplitudes $a_3$, $a_4$ and $a_5$ associated with the upper signal sideband induced by potential gravitational wave at each output consistent with Figure 2 are given by:

$$a_3 = i\rho_0\eta_1^2 e^{-2i\phi_1}\frac{2\pi\Delta z}{\lambda}B_cB_u, \tag{4}$$

$$a_4 = i\rho_0\eta_1^2 e^{-2i\phi_1}\frac{2\pi\Delta z}{\lambda}B_cB_u, \tag{5}$$

$$a_5 = -\rho_0\eta_1 e^{-i\phi_1}\frac{2\pi\Delta z}{\lambda}\frac{B_c}{\rho_2 e^{-ik_cL}}\frac{B_u}{\rho_2 e^{-ik_uL}}\tau_2. \tag{6}$$

Where $B_c$ and $B_u$ retain their prior meanings, excepting that $\omega_m$ will now be the frequency of mirror motion induced by the gravitational wave. The gravitational wave was assumed to propagate perpendicular to the cavity, and it's effect has been expressed in terms of an displacement of the mirror $\Delta z$ equivalent to our previously considered grating displacement, resulting in modulation index $m = 4\pi\Delta z/\lambda$. This effect appears at cavity internal input $b_1$ in Figure 2. Thus we computed the optical signal in all output ports due to a potential gravitational wave signal.

## 4. Ratio of signal to noise at grating cavity output ports

An appropriate figure of merit to evaluate the length-sensing performance of the different grating cavity output ports is the ratio of gravitational wave signal to lateral grating displacement noise (the signal to noise ratio). The SNR was derived by dividing the absolute field amplitude in the gravitational wave case by the absolute field amplitude in the grating lateral displacement case for each output. For a standard two-mirror cavity with equivalent finesse the SNR will be infinity as in principle two-mirror cavity input mirrors are insensitive to lateral displacement although in practice any suspension system has some coupling from lateral excitation to longitudinal displacement. Using Equations 1 through 3 and 4 through 6 we obtained the following signal to noise ratios at the output port of the equivalent subscript:

$$\mathrm{SNR}_3 = \Lambda_{\mathrm{cav}} \frac{B_u}{B_u - B_c}, \tag{7}$$

$$\mathrm{SNR}_4 = \Lambda_{\mathrm{cav}} \frac{B_u}{B_u + B_c + 2\eta_2/\eta_1^2 e^{-i\phi_2 + 2i\phi_1}}, \tag{8}$$

$$\mathrm{SNR}_5 = \Lambda_{\mathrm{cav}}/\rho_2 e^{-ik_c L}, \tag{9}$$

with

$$\Lambda_{\mathrm{cav}} = \frac{2d\Delta z B_c}{\lambda \Delta x}. \tag{10}$$

These equations reveal the differences between the ratio of gravitational wave signal to lateral grating displacement noise for the three cavity output ports. The SNR at the back-reflected ($a_3$) and forward-reflected ($a_4$) output ports contain the modulation frequency dependent term $B_u$ whilst at the transmitted ($a_5$) port the SNR is independent of the frequency of modulation. This frequency independence is somewhat counter-intuitive and only occurs because the displacement-noise and gravitational wave signal were compared at the same frequency. As can be seen from Equations 3 and 4 the transmitted output port has the same frequency response to both signal and noise.

It was found that the SNR for the forward-reflected ($a_4$) and back-reflected ($a_3$) output ports includes the SNR from inside the cavity $\Lambda_{\mathrm{cav}}$, multiplied by a fraction containing the different resonance factors for the carrier light field and the sidebands. The $\eta_2/\eta_1^2$ term in Equation 8 will be small despite the square in the denominator, as the $\eta_2$ diffraction efficiency is a loss term for this configuration and will be minimised by design, whereas $\eta_1$ is the much larger input coupling efficiency term. Apart from this term the only difference between Equations 7 and 8 is whether the resonance terms of the carrier and sideband in the denominator are added or subtracted. The important feature of these equations is that $B_u \approx B_c$ for modulation frequencies within the cavity bandwidth. This brings the denominator in Equation 7 close to zero, thus strongly increasing the SNR at the forward-reflected ($a_3$) port. There is therefore a partial cancellation of the phase noise introduced from lateral grating displacement at the forward-reflected port.

Physically this can be understood quite simply: The carrier light field enters the cavity, generating noise sidebands proportional to its intensity. It then resonates in the cavity and increases in intensity proportional to the cavity finesse. When it exits the cavity via the grating it generates noise sidebands proportional to its new, increased intensity. The input sidebands generated when the carrier entered the cavity will partially cancel with these output sidebands. The cancellation will be complete if the input sidebands also increased in intensity due to resonance in the cavity. How resonant these input sidebands are within the cavity depends on the finesse which sets the cavity's bandwidth and on how far they are from the perfect resonant frequency of the cavity (defined by the carrier frequency). Therefore we can say that the cancellation will be quite good for modulation frequencies within the cavity bandwidth.

From this it is clear that of the three expected sources of displacement noise: angular end-mirror misalignment, injection optic misalignment and lateral grating displacement only the latter two will benefit from this cancellation, as only they will generate both input and output sidebands. The end-mirror misalignment occurs within the cavity and hence generates only output sidebands. It might appear that misaligned injection pointing only generates input sidebands, but it has been shown that this also changes the alignment of the field circulating in the cavity [11] and hence also generates output sidebands. The end-mirror angular misalignment will remain a problem [8]; however, the large number of optical elements in the injection path, and that many of them are not suspended suggests the reduction of noise from this source will be important. In any-case suspension design will be eased if the transverse isolation requirement imposed by the grating can be weakened.

It is also necessary to discuss where the gravitational wave signal appears most strongly as given by Equations 1 through 6. It is useless to have good SNR versus the displacement noise source if the signal is not strongly present. For proposed interferometric gravitational wave detector layouts the end-mirror will be highly reflective ($\rho_2$ will be close to one) and therefore the complex field amplitudes for the potential gravitational wave signal will obey the relations $a_3 \gg a_5$, $a_4 \gg a_5$. There will be approximately half the signal in the forward-reflected port with good SNR, half will be in the back-reflected port with poor SNR and very little will be in the transmitted port.

In a broader context this result demonstrates that there is a small but useful advantage to be gained by utilising the forward-reflected output port of a three-port-coupled diffractive cavity for detection, especially to the exclusion of the back-reflected port if at all possible, although this will result in the loss of some signal. Due to the small amount of gravitational wave signal present there, the dumping of the transmitted port will have negligible effect, advantageous as many configurations will place this port at an end-station some kilometers from the central optics. In any-case it is likely this lost light can be used for monitoring purposes, e.g. in alignment control.

Conversely as both forward-reflected and back-reflected output ports will be co-located, and likely co-located at the central optics for both arm-cavities if a Michelson-like design is used it may be preferred to use both outputs, or to link the two cavities in some way. Such an idea was sketched by Drever [4] although we are aware of no formal noise-analysis of such a configuration, which anyway could not be complete with respect to noise without the work presented above and previously in [9].

### *4.1. Numerical result*

In this section we present a quantitative analysis for one example configuration of a three-port-coupled diffractive cavity. The figures selected were $L = 3\,\text{km}$, $\rho_2 = \sqrt{0.99995}$ (50 ppm power transmittance) and $\rho_0 = \sqrt{0.95}$ for a possible implementation of diffractive cavities in an Advanced Virgo-like interferometer [12]. Demanding that the $\mu = 1$ diffraction order serves as the input coupler imposes the grating design requirement $d \leq 2\lambda$ [13]. Grating phase relations impose the minimum possible value of $\eta_2 \geq 0.0127$. From [10] for minimum $\eta_2$ we find that $\eta_2 e^{-i\phi_2} = -0.0127$, and the cavity input parameters set $\eta_1 e^{-i\phi_1}$. Together these fix the last term in the denominator in Equation 8, and as expected it was found to be small compared to the other terms of the denominator. Setting the carrier light resonant in the cavity imposes $e^{-2ik_cL} = 1$ and the modulation frequency $\omega_m$ of the lateral grating displacement is chosen to be 10 Hz.

Our interest focuses on the relative magnitude of the frequency sidebands. As in Freise et al [8] we set $\Delta z/\Delta x = 1$ in order to compare potential gravitational-wave signal to the effect of grating displacement. Since we are dealing with complex field amplitudes for this analysis we take the absolute value of the ratios. The ratio contains a real part as well as an imaginary part, so we assumed that the a homodyne measurement can be performed, choosing the readout quadrature of each output port individually by adding the proper local oscillator. This yields the following result:

$$|\text{SNR}_3| = 3181, \tag{11}$$
$$|\text{SNR}_4| = 79, \tag{12}$$
$$|\text{SNR}_5| = 158. \tag{13}$$

In all output ports the SNR is greater than one, as the cavity naturally suppresses the displaced grating phase noise sidebands versus the signal sidebands by a factor of 158 as seen

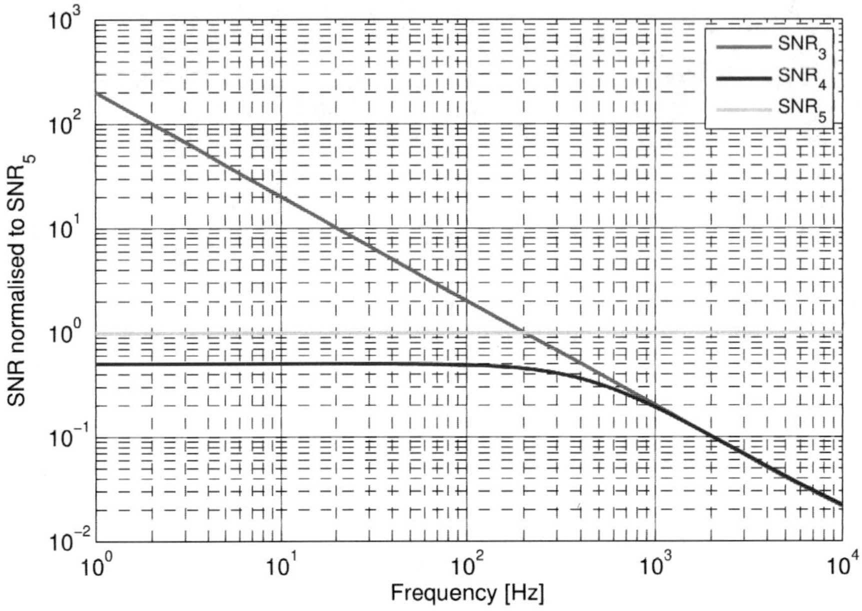


**Figure 4.** Signal to noise ratio of gravitational wave equivalent mirror displacement and lateral grating displacement at the three different output ports of a grating gravity. All traces are normalized to $\mathrm{SNR}_5$.

in the transmitted output port. This is because the displacement noise only appears when light enters the cavity, whereas the gravitational wave signal appears in every round trip. In the three-port-coupled case we have found a further factor of twenty improvement in the SNR can be obtained through the cancellation of input and output grating displacement sidebands in the forward-reflected ($a_3$) port compared to the transmitted ($a_5$) port.

In a next step, instead of a single frequency of interest, we consider modulation frequencies covering the full detection band of gravitational wave detectors. As $\mathrm{SNR}_3$ and $\mathrm{SNR}_4$ are frequency dependent, we normalise them to $\mathrm{SNR}_5$ and plot them over frequency to obtain Figure 4.

At low modulation frequencies where the sidebands are to a good approximation resonant in the cavity, imposing $B_u \approx B_c$ the forward-reflected port ($a_3$) has good cancellation between the input and output sidebands generated by the grating lateral displacement. This results in high SNR compared to the transmitted port ($a_5$). In the back-reflected port ($a_4$) summation occurs instead of cancellation and hence the SNR is lower than in the transmitted port. As the modulation frequency diverges from the cavity resonance, the sidebands do not resonate in either the gravitational-wave or grating lateral displacement case as $B_u$ trends to zero. Thus the sideband contribution generated when the carrier field exits the cavity (the $B_c$ terms in the denominator of Equations 7 and 8) dominates causing the SNR of the reflected ports to converge below the level of the transmitted port.

### *4.2. Suspension requirements*

The potential suspension requirements for a grating cavity used as an arm cavity within the planned Advanced Virgo detector provide a useful reference point for this work. Using a contemporary Advanced Virgo design sensitivity [14] we plot in Figure 5 the corresponding tolerable lateral grating motion for each of the three potential readout ports of the grating arm

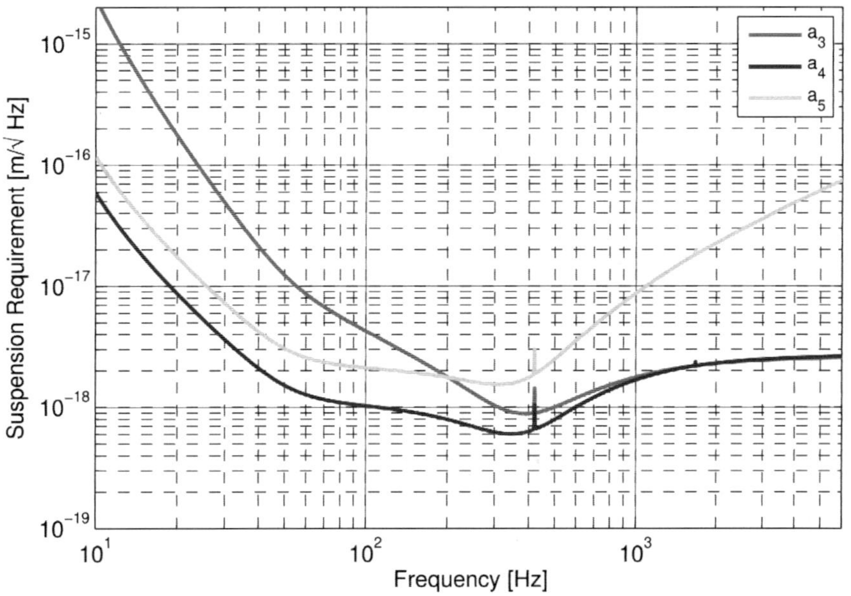


**Figure 5.** Suspension requirement for the maximum tolerable lateral grating displacement that is necessary to achieve the Advanced Virgo design sensitivity. The suspension required strongly depends on the readout port as well as the frequency of interest. At low frequencies the phase noise suppression in port $a_3$ allows significant relaxation of the required suspension isolation. The narrow peak around 410 Hz is due to a resonance in the Advanced Virgo design, not a grating effect.

cavity.

We have found that we can relax by a factor of twenty at 10 Hz (Figure 5) the suspension requirement for the the un-suppressed transmitted ($a_5$) port by utilizing the forward-reflected ($a_3$) port for the signal readout. As discussed above this suppression, because it relies on cancellation between input and output noise sidebands, only applies to the suspension of the diffractive optic itself (shown in Figure 5) and proportionally to the injection optics pointing. It does not, unfortunately, reduce the end-mirror alignment requirements calculated by Freise et al [8].

## 5. Summary and Outlook

Diffraction gratings are an important avenue of research because they allow all-reflective interferometer configurations. Apart from offering thermal noise improvements which may be especially relevant in mitigating effects made important when high power is used to reduce shot-noise contribution, these diffraction gratings may allow new configurations which have other advantageous properties without significantly changing the overall detector shape.

An additional noise source is lateral grating displacement relative to the beam causing path-length differences that couple phase-noise into the gravitational wave detection channel. From [8] the grating-cavity end-mirror angular alignment suspension requirement (five orders of magnitude higher than that required by a two-mirror cavity), and associated likely increases in injection optic pointing and lateral grating stability were a significant impetus against using grating cavities.

To determine the effect a frequency domain analysis was carried out to calculate the coupling of lateral grating displacement to the different output ports of a three-port-coupled grating cavity [9]. For the output port in forward-reflection of the grating we found a suppression of phase noise originating from lateral grating displacement over the transmitted port, resulting in a factor of twenty relaxation in the lateral displacement suspension requirement at 10 Hz. This will likely also apply to the injection optics pointing stability. This factor will increase with a wide cavity bandwidth and hence the noise suppression will be greater in cavities of lower finesse.

**Acknowledgments**
This work was supported by the Science and Technology Facilities Council (STFC), the European Gravitational Observatory (EGO) and the Deutsche Forschungsgemeinschaft (DFG) within the Sonderforschungsbereich (SFB) TR7.

## References

[1] K. X. Sun and R. L. Byer. All-reflective Michelson, Sagnac, and Fabry-Perot interferometers based on grating beam splitters. *Optics letters*, 23(8):567–569, April 1998.

[2] J. R. Smith and LIGO Scientific Collaboration. The path to the enhanced and advanced LIGO gravitational-wave detectors. *Classical and Quantum Gravity*, 26(11):114013–+, June 2009.

[3] F Acernese et al. Virgo status. *Classical and Quantum Gravity*, 25(18):184001 (9pp), 2008.

[4] R. W. P. Drever. Concepts for Extending the Ultimate Sensitivity of Interferometric Gravitational Wave Detectors Using Non-transmissive Optics with Diffractive or Holographic Coupling. In *Proceedings of the Seventh Marcel Grossman Meeting on recent developments in theoretical and experimental general relativity, gravitation, and relativistic field theories*, pages 1401–+, 1996.

[5] A. Bunkowski, O. Burmeister, P. Beyersdorf, K. Danzmann, R. Schnabel, T. Clausnitzer, E.-B. Kley, and A. Tünnermann. Low-loss grating for coupling to a high-finesse cavity. *Optics Letters*, 29:2342–2344, October 2004.

[6] A. Bunkowski, O. Burmeister, T. Clausnitzer, E.-B. Kley, A. Tünnermann, K. Danzmann, and R. Schnabel. Optical characterization of ultrahigh diffraction efficiency gratings. *Applied Optics*, 45:5795–5799, August 2006.

[7] M. D. Perry, R. D. Boyd, J. A. Britten, D. Decker, B. W. Shore, C. Shannon, and E. Shults. High-efficiency multilayer dielectric diffraction gratings. *Optics Letters*, 20:940–942, April 1995.

[8] A. Freise, A. Bunkowski, and R. Schnabel. Phase and alignment noise in grating interferometers. *New Journal of Physics*, 9:433–+, December 2007.

[9] J. Hallam, S. Chelkowski, A. Freise, S. Hild, B. Barr, K. A. Strain, O. Burmeister, and R. Schnabel. Coupling of lateral grating displacement to the output ports of a diffractive FabryPerot cavity. *Journal of Optics A: Pure and Applied Optics*, 11(8):085502–+, August 2009.

[10] A. Bunkowski, O. Burmeister, K. Danzmann, and R. Schnabel. Input-output relations for a three-port grating coupled fabry-perot cavity. *Opt. Lett.*, 30:1183–1185, 2005.

[11] E. Morrison, D. I. Robertson, H. Ward, and B. J. Meers. Experimental demonstration of an automatic alignment system for optical interferometers. *Appl. Opt.*, 33:5037–5040, August 1994.

[12] The Virgo Collaboration. Advanced virgo baseline design. Technical report, Virgo, 2009.

[13] T. Clausnitzer, E.-B. Kley, A. Tünnermann, A. Bunkowski, O. Burmeister, K. Danzmann, R. Schnabel, S. Gliech, and A. Duparré. Ultra low-loss low-efficiency diffraction gratings. *Optics Express, vol. 13, Issue 12, p.4370*, 13:4370–+, June 2005.

[14] S Hild, G Losurdo, and A Freise. Sensitivity options for the advanced virgo preliminary design. Technical report, Virgo, 2009.

# Quantum limit of different laser power stabilization schemes involving optical resonators

**P Kwee[1], B Willke[1], and K Danzmann[1]**
[1] Max-Planck-Institut für Gravitationsphysik (Albert-Einstein-Institut) and Leibniz Universität Hannover, 30167 Hannover, Germany

E-mail: patrick.kwee@aei.mpg.de

**Abstract.** Three different laser power stabilization schemes are compared: a traditional power stabilization, a traditional one with subsequent optical resonator, and a power stabilization with the novel optical ac coupling technique. The performance of the schemes is evaluated using the theoretical quantum limit and the power stability achieved considering technical limitations. The scheme with optical ac coupling is superior to the other ones especially at high laser power levels that will be used in future interferometric gravitational wave detectors.

## 1. Introduction

Laser power stabilization is important in many high precision experiments. In particular gravitational wave detectors (GWDs) require one of the most demanding power stabilities for the laser beam to be injected into the interferometer at frequencies between a few hertz and a few kilohertz. Commonly photodetectors measuring the power fluctuations of a beam sample are used in a feedback control loop to suppress the fluctuations by means of appropriate actuators. The quantum limit of this traditional power stabilization scheme is at least 6 dB above the relative quantum power noise of the original beam [1]. In contrast the quantum limits of stabilization schemes involving optical resonators [2] are significantly closer than 6 dB to this quantum noise, and therefore a better power stability can be achieved.

In this article we derive the quantum limit of three different active power stabilization schemes: a traditional power stabilization, a traditional power stabilization with subsequent noise filtering by a resonator, and a power stabilization utilizing the optical ac coupling technique [2]. The performance of these schemes is compared with respect to their theoretical quantum limit and by evaluating the achievable stability in two representative examples at the 10 mW and 100 W power level taking technical limitations into account.

The power stabilization with optical ac coupling achieves the best stability for frequencies above the bandwidth of the resonator required, especially when considering technical limitations that arise, among others, from the limited power detection capability of the photodetectors. At low laser power, the performance difference to the traditional scheme is small and in some cases the additional experimental complexity due to the optical ac coupling might not be justified. However, at the high laser power levels that will be used, e.g., in future GWDs, the optical ac coupling scheme applied, e.g., at the power recycling cavity is significantly superior.

© 2010 IOP Publishing Ltd

Journal of Physics: Conference Series **228** (2010) 012023 doi:10.1088/1742-6596/228/1/012023

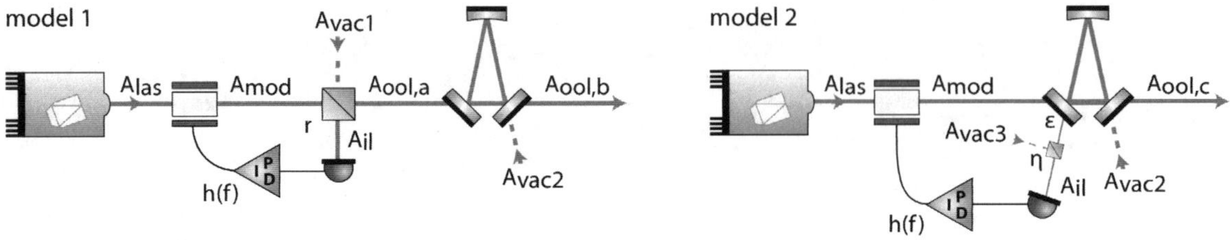


**Figure 1.** Models used to calculate the quantum noise limit of three different power stabilization schemes. Model 1 (left) showing the schemes A and B with OOL fields $A_{\mathrm{ool,a}}$ and $A_{\mathrm{ool,b}}$, model 2 showing scheme C with OOL field $A_{\mathrm{ool,c}}$.

## 2. Theory

Two models are used to calculate the quantum limit of three power stabilization schemes (Fig. 1). The field $A_{\mathrm{il}}$, called in-loop (IL) field, is used for the control loop sensor in order to compensate power fluctuations of the laser field $A_{\mathrm{las}}$ with an actuator. The required feedback loop is characterized by the frequency dependent loop gain $h(f)$. The fields $A_{\mathrm{ool,a}}$ to $A_{\mathrm{ool,c}}$ are the out-of-loop (OOL) fields of the three schemes A to C and are relevant for the actual control loop performance and for potential downstream experiments. The power reflectivity of the beam splitter $r$, the power transmission of the attenuator $\eta$ in front of the IL detector in model 2, and the impedance matching of the resonator $\epsilon$ (power fraction $\epsilon^2$ is reflected at the resonator) are important model parameters. Vacuum fluctuations $A_{\mathrm{vac}}$ couple into the system at open ports of the beam splitters and the resonator. Scheme A is a traditional power stabilization, scheme B is the same, but with subsequent resonator in the OOL beam, and scheme C is a stabilization with optical ac coupling [2].

The quantum limit of the stabilization schemes can be defined in terms of the relative power noise of the OOL beam. To determine this power noise, the spectral variance of the quantum mechanical photon number operator $N$ for the different fields $A_{\mathrm{ool,a}}$ to $A_{\mathrm{ool,c}}$ has to be calculated, as described in more detail, e.g., in [3]. We assume that all but the vacuum fields have a large average amplitude $\alpha$ and only small fluctuations $\delta A(t)$, so that $N$ can be expressed with linearized raising and lowering operators, $A^\dagger$ and $A$:

$$A^\dagger(t) = \alpha + \delta A^\dagger(t)\,, \qquad A(t) = \alpha + \delta A(t)\,, \qquad N = A^\dagger \cdot A \approx \alpha^2 + \alpha \cdot (\delta A^\dagger + \delta A)\,. \tag{1}$$

The equations for the operators $\delta A_{\mathrm{ool,a}}$ to $\delta A_{\mathrm{ool,c}}$ are deduced using the classical field equations and the canonical quantization. Assuming an optical resonator of high finesse, the following equations are yielded for model 1:

$$\begin{aligned} \delta A_{\mathrm{mod}} &= \delta A_{\mathrm{las}} + h(f)\delta A_{\mathrm{il}}\,, & \delta A_{\mathrm{il}} &= \sqrt{r}\,\delta A_{\mathrm{mod}} + \sqrt{1-r}\,\delta A_{\mathrm{vac1}}\,, \\ \delta A_{\mathrm{ool,a}} &= \sqrt{1-r}\,\delta A_{\mathrm{mod}} - \sqrt{r}\,\delta A_{\mathrm{vac1}}\,, & \delta A_{\mathrm{ool,b}} &= \frac{\mathrm{i}f\delta A_{\mathrm{vac2}} + \delta A_{\mathrm{ool,a}}}{1-\mathrm{i}f}\,, \end{aligned} \tag{2}$$

where $f$ is the Fourier frequency in units of the resonator bandwidth. With the control loop closed, the first two equations can be combined to the steady state solution

$$\delta A_{\mathrm{mod}} = \frac{\delta A_{\mathrm{las}} + \sqrt{1-r}\,h(f)\,\delta A_{\mathrm{vac1}}}{1-\sqrt{r}\,h(f)}\,. \tag{3}$$

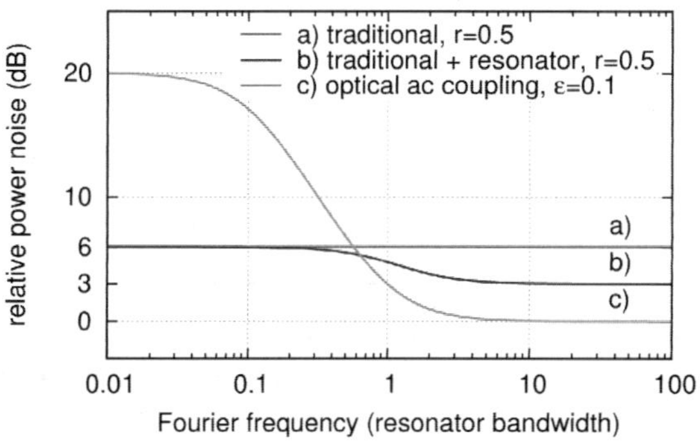


**Figure 2.** Quantum limit for the stabilization schemes. The limits are given in relation to the relative quantum noise of the original beam.

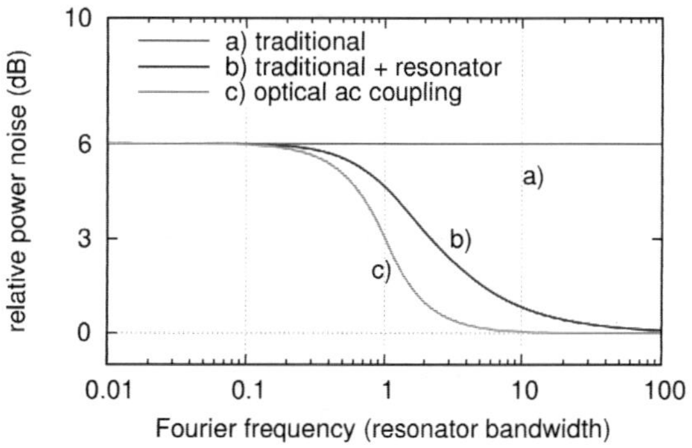


**Figure 3.** Minimal quantum limit for the three stabilization schemes, where the beam splitter reflectivity and the impedance matching were optimized for each Fourier frequency.

The double-sided power spectrum of the relative power fluctuations, $P_{\mathrm{s}}^2$, is the spectral variance of $N$ normalized with $\langle N\rangle^2$ where $\langle N\rangle$ is its expectation value, which is the average photon flow $\alpha^2$:

$$P^2_{\mathrm{s,ool,a}} = \frac{\mathrm{Var}(N_{\mathrm{ool,a}})}{\langle N_{\mathrm{ool,a}}\rangle^2} = \frac{\mathrm{Var}(\delta A^\dagger_{\mathrm{ool,a}} + \delta A_{\mathrm{ool,a}})}{\langle N_{\mathrm{ool,a}}\rangle} = \frac{(1-r)\mathrm{Var}(\delta A^\dagger_{\mathrm{las}} + \delta A_{\mathrm{las}}) + |h(f) - \sqrt{r}|^2}{\langle N_{\mathrm{ool,a}}\rangle \cdot |1-\sqrt{r}h(f)|^2}$$
$$\overset{h\to\infty}{=} \frac{1}{\langle N_{\mathrm{ool,a}}\rangle \cdot r} = \frac{1}{r\cdot(1-r)\alpha^2_{\mathrm{las}}}, \tag{4}$$

$$P^2_{\mathrm{s,ool,b}} \overset{h\to\infty}{=} \frac{rf^2+1}{r(1-r)(f^2+1)\alpha^2_{\mathrm{las}}}, \tag{5}$$

where $\alpha^2_{\mathrm{las}}$ is the average photon flow of the original laser beam. Furthermore, $\mathrm{Var}(\delta A^\dagger_{\mathrm{vac}} + \delta A_{\mathrm{vac}}) = 1$ due to the commutation rule $[A, A^\dagger] = 1$ (see e.g. [3]). A high loop-gain $h\to\infty$ is assumed, which is the case in most power stabilization experiments. The single-sided linear spectral density $\sqrt{2\cdot P_{\mathrm{s}}^2}$ is often used in experiments and is used later-on in the case study.

In general the quantum noise of the original laser beam $P_{\mathrm{s}}^2 = 1/\alpha^2_{\mathrm{las}}$ is the fundamental limit for all schemes. In scheme A the relative power noise of the OOL beam (Fig. 2 a) does not depend on the laser noise $\delta A_{\mathrm{las}}$ due to the high-gain feedback loop and is frequency independent. The noise is minimal $P^2_{\mathrm{s,min,a}} = 4/\alpha^2_{\mathrm{las}}$ for $r$=0.5 (Fig. 3 a) and thus is 6 dB above the quantum noise of the original beam ($1/\alpha^2_{\mathrm{las}}$) or 3 dB above its own quantum noise ($2/\alpha^2_{\mathrm{las}}$).

In scheme B the resonator has a high transmission at low frequencies and thus the resonator filtering effect is insignificant. The power noise is equal to the noise of a traditional stabilization $P^2_{\mathrm{s,ool,b}}(f\to 0) = P^2_{\mathrm{s,ool,a}}$. At high frequencies the reflectivity of the resonator and hence its filtering effect for noise at point $A_{\mathrm{ool,a}}$ increases. However, the reflectivity for the field $A_{\mathrm{vac2}}$ increases as well such that the power noise is at these frequencies equal to the quantum noise of the OOL beam $1/[(1-r)\alpha^2_{\mathrm{las}}]$, which in general is above the quantum noise of the original beam (Fig. 2 b).

The optimal reflectivity of the beam splitter $r_{\rm opt}(f)$ and the minimal power noise $P^2_{\rm s,min,b}(f)$ now depends on the frequency and is given by (Fig. 3 b)

$$r_{\rm opt}(f) = \frac{\sqrt{f^2+1}-1}{f^2}, \qquad P^2_{\rm s,min,b}(f) = \frac{f^4}{(f^2+1)\left(\sqrt{f^2+1}-1\right)^2 \alpha^2_{\rm las}}. \tag{6}$$

It should be noted that this minimal power noise cannot be achieved in the whole frequency band simultaneously since to our knowledge such a frequency dependent beam splitter does not exist. At low frequencies $r_{\rm opt}(f\to 0) = 0.5$ and at high frequencies $r_{\rm opt}(f\to\infty) \to 0$ yield the best stability.

The power noise of the OOL beam in model 2 is calculated in the same way as in model 1. The initial equations for model 2 are given by

$$\begin{aligned} \delta A_{\rm mod} &= \delta A_{\rm las} + h(f)\delta A_{\rm il}\,, \qquad \delta A_{\rm ool,c} = \frac{(-\epsilon + {\rm i}f)\delta A_{\rm vac2} + \sqrt{1-\epsilon^2}\,\delta A_{\rm mod}}{1-{\rm i}f}\,, \\ \delta A_{\rm il} &= \frac{(\epsilon + {\rm i}f)\delta A_{\rm mod} + \sqrt{1-\epsilon^2}\,\delta A_{\rm vac2}}{1-{\rm i}f} \cdot \sqrt{\eta} + \sqrt{1-\eta}\,\delta A_{\rm vac3}\,. \end{aligned} \tag{7}$$

The IL detector is optical ac coupled and the OOL beam is filtered in addition by the resonator. The power spectrum of the relative power noise of the OOL beam is given by

$$P^2_{\rm s,ool,c} = \frac{1+f^2+(1-\epsilon^2)(1-\eta)/\eta}{\epsilon^2+f^2}\,\frac{1}{(1-\epsilon^2)\alpha^2_{\rm las}}\,. \tag{8}$$

The attenuation controlled by $\eta$ in front of the IL detector is integrated in the model only to be able to take technical limitations into account later on. Thus for the rest of this section we assume $\eta$=1. For low frequencies the resonator has a high transmission and the imprinted noise of $A_{\rm il}$ is dominant at the OOL field $A_{\rm ool,c}$. For these frequencies the stabilization is comparable to a traditional power stabilization with a beam splitter reflection of $r$=$\epsilon^2$. The factor $\epsilon$ is determined by the impedance matching of the resonator. For small impedance mis-matches the OOL noise $P^2_{\rm s,ool,c}(\epsilon \ll 1) \approx 1/\epsilon^2\alpha^2_{\rm las}$ is about a factor of $1/\epsilon^2$ above the quantum noise of the original beam $A_{\rm las}$.

The reflection of the resonator increases for higher frequencies causing two effects: On the one hand the noise imprinted by the control loop onto $A_{\rm mod}$ is reduced. On the other hand the vacuum fluctuations $A_{\rm vac2}$ are mainly reflected by the resonator and dominate the power noise of $A_{\rm ool,c}$. For $f \gg 1$ the relative power noise of $A_{\rm ool,c}$ is only a factor of $1/(1-\epsilon^2)$ above the quantum noise of the original beam $P^2_{\rm s,ool,c}(f\to\infty) = 1/[(1-\epsilon^2)\alpha^2_{\rm las}]$. The power noise $P^2_{\rm s,ool,c}$ is shown in Fig. 2 c for $\epsilon$=0.1.

The quantum limit of scheme C could in principle be optimized for a specific frequency by altering the impedance matching $\epsilon$. The minimal noise $P^2_{\rm s,min,c}$ is reached for (Fig. 3 c)

$$\epsilon(f) = \begin{cases} \sqrt{\dfrac{1-f^2}{2}} & \text{for} \quad f<1 \\ \to 0 & \text{for} \quad f \geq 1 \end{cases}, \qquad P^2_{\rm s,min,c}(f) = \begin{cases} \dfrac{4}{1+f^2}\dfrac{1}{\alpha^2_{\rm las}} & \text{for} \quad f<1 \\ \dfrac{1+f^2}{f^2}\dfrac{1}{\alpha^2_{\rm las}} & \text{for} \quad f \geq 1 \end{cases}. \tag{9}$$

At low frequencies an impedance matching of $\epsilon^2$=0.5 is ideal. In this case and at these frequencies the resonator acts like a 50:50 beam splitter, which is as well optimal in the traditional power stabilization scheme. In contrast a nearly perfectly impedance matched resonator yields the best

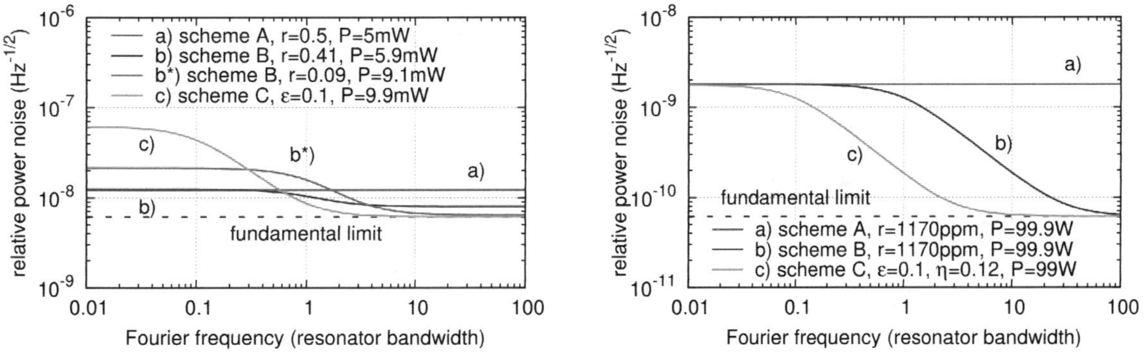


**Figure 4.** Out-of-loop achievable power noise for the different power stabilization schemes for laser beams of 10 mW (left) and 100 W (right). The model parameters chosen and the OOL available power are given.

stability at high frequencies and the minimal noise approaches the quantum noise of the original beam $P^2_{\mathrm{s,min,c}}(f\to\infty) = 1/\alpha^2_{\mathrm{las}}$.

All in all scheme C involving the optical ac coupling is superior compared to the other schemes at frequencies $f>1$ (Fig. 2 c). However, at low frequencies the traditional power stabilization scheme A and its modification, scheme B, achieve the lowest power noise.

A comparison of the minimal power noise (Fig. 3) shows that for a fixed Fourier frequency, the quantum limit of scheme C is at or below the quantum limit of the other schemes. As already mentioned this minimal power noise cannot be reached in the whole frequency band for scheme B and C since to our knowledge no beam splitter or resonator exists that allows to tune their reflectivity and impedance matching, respectively, with the required frequency dependence.

## 3. Case Study

So far the performance of the different power stabilization schemes were compared and evaluated using only the theoretical quantum limit. However, in real experiments several technical limitations have to be considered as well: The maximum photocurrent that can be detected with quantum-noise-limited noise performance at frequencies of interest for GWDs is about 100 mA. This corresponds to a beam power of 117 mW at 1064 nm wavelength and photodiode quantum efficiency of 1. For frequencies above 10 Hz this is a typical value in experiments [4, 5, 6]. Furthermore, in current experiments the impedance matching has to be $\epsilon\geq 0.1$ so that additional technical noise sources caused by non-resonant modes do not limit the achievable stability with optical ac coupling [2]. This corresponds to a maximal carrier reduction down to $\epsilon^2$=1%. In the following two realistic cases are described, where beams are stabilized for 10 mW ($\alpha^2_{\mathrm{las}} = 5.4\times10^{16}$ Hz) and 100 W ($\alpha^2_{\mathrm{las}} = 5.4\times10^{20}$ Hz).

The linear relative quantum noise of the laser beam with 10 mW power is $6.1\times10^{-9}\,\mathrm{Hz}^{-1/2}$. The quantum limits of the three schemes are shown in Fig. 4. For scheme A $r$=0.5 is chosen and scheme B is optimized once for $f$=1 and once for $f$=10. The lowest quantum limit in scheme C is achieved for $\epsilon\to 0$ for frequencies $f\geq 1$. However, to account for technical limitations a realistic impedance matching of $\epsilon$=0.1 was chosen, but the power noise still gets as close as 0.04 dB to the fundamental quantum limit.

At low frequencies $f<1$ the traditional power stabilization scheme A achieves the best stability, whereas at high frequencies $f>1$ the optical ac coupling scheme C achieves a better one. Scheme B gives comparable results to scheme C at very high frequencies $f>10$. All in all the OOL power noise differs only by at most 6 dB at high frequencies and thus the achieved stabilities are close to each other. However the 5 mW beam power, available for downstream experiments, is in scheme A significantly lower compared to 9.9 mW in scheme C. If a power noise difference of 6 dB and a beam power difference of a factor of up to 2 is acceptable for the downstream experiment, a traditional power stabilization should be used since it is less complex and no optical resonator is needed.

The linear relative quantum noise of a 100 W beam is $6.1\times10^{-11}\,\mathrm{Hz}^{-1/2}$. Due to the technical limit of the IL photodetector in scheme A and B a reflectivity of $r=1.17\times10^{-3}$ has to be chosen. Scheme C is limited by the photodetector and as well by the impedance matching. Since even with $\epsilon=0.1$ the power on the IL detector is too high, the power needs to be attenuated with $\eta=0.12$ (see Fig. 1, model 2). The quantum limits of the three schemes are shown in Fig. 4, and in all cases most of the laser power is available in the OOL beam (99 W to 99.9 W).

The differences between the schemes are more significant at the high laser power of 100 W. The best stability is achieved with scheme C. Scheme B is at most frequencies a factor of 10 worse, but reaches the same power noise as scheme C at very low and very high frequencies. In the whole frequency band the traditional power stabilization achieves the worst performance. At this laser power level the higher experimental complexity of a power stabilization with optical ac coupling compared to a traditional stabilization might well be justified by the requirement of the downstream experiment.

## 4. Discussion

The quantum limits of three different power stabilization schemes are calculated, compared, and evaluated. The optimal stabilization parameters are determined for two different laser powers, 10 mW and 100 W, taking technical limitations into account.

In conclusion, the power stabilization scheme with optical ac coupling can achieve a better quantum limited performance compared to the other schemes, especially compared to the traditional scheme, at frequencies above the resonator bandwidth.

Ground-based GWDs will require even higher laser powers than a few 100 W in the future in order to improve the high-frequency sensitivity limited by shot-noise. With increasing laser power, the power stability requirements will become more stringent as well. Especially at these laser power levels, a stabilization with optical ac coupling has important advantages in contrast to traditional stabilization schemes.

## Acknowledgments

This research is supported by the DFG and is part of QUEST (Centre for Quantum Engineering and Space-Time Research). The authors would like to thank Albrecht Rüdiger for his helpful comments during the preparation of this manuscript.

## References

[1] Taubman M S, Wiseman H, McClelland D E and Bachor H A 1995 *J. Opt. Soc. Am. B* **12** 1792–1800
[2] Kwee P, Willke B and Danzmann K 2009 *Appl. Opt.* **48** 5423–5431
[3] Bachor H A and Ralph T C 2004 *A Guide to Experiments in Quantum Optics* (WILEY-VCH) chap 8.3
[4] Rollins J, Ottaway D, Zucker M, Weiss R and Abbott R 2004 *Opt. Lett.* **29** 1876–1878
[5] Seifert F, Kwee P, Heurs M, Willke B and Danzmann K 2006 *Opt. Lett.* **31** 2000–2002
[6] Kwee P, Willke B and Danzmann K 2009 *Opt. Lett.* **34** 2912–2914

# Optomechanical correlations and signal self-amplification in interferometric measurements

**P.-F. Cohadon, P. Verlot, A. Tavernarakis, T. Briant, A. Heidmann**
Laboratoire Kastler Brossel, ENS, UPMC, CNRS,
Case 74, 4 place Jussieu, F75252 Paris Cedex 05, France

**Abstract.** Radiation pressure exerted by light in interferometric measurements is responsible for displacements of mirrors which appear as an additional back-action noise and limit the sensitivity of the measurement. We experimentally study these effects by monitoring in a very high-finesse optical cavity the displacements of a mirror with a sensitivity at the $10^{-20}\,\mathrm{m}/\sqrt{\mathrm{Hz}}$ level. This very high sensitivity is a step towards the observation of fundamental quantum effects of radiation pressure such as the standard quantum limit in interferometric measurements. We report the observation of optomechanical correlations between two optical beams sent into the same moving mirror cavity. We also observed a self-amplification of a signal, which is a consequence of dynamical back-action of radiation pressure in a detuned cavity, and may improve the interferometric measurement sensitivity beyond the standard quantum limit.

## 1. Introduction

In quantum mechanics, measurements are always related to a back-action on the measured system, which may limit the sensitivity. Interferometers are ruled by such limits: quantum effects of optomechanical coupling, the radiation-pressure coupling between a moving mirror and an incident light field, cause displacements of mirrors and lead to a standard quantum limit (SQL) for the measurement [1, 2, 3]. Such radiation-pressure effects are so weak that they haven't been experimentally demonstrated yet, although the future generations of gravitational-wave interferometers will most probably be confronted to these limits [4, 5].

Overcoming these limits was a major motivation for the quantum optics community, and number of quantum noise reduction schemes have been proposed which rely on the use of squeezed light sent into the interferometer [6, 7, 8], or on the use of the quantum correlations induced by radiation pressure between phase and intensity fluctuations [9]. Dynamical effects of radiation pressure play also an important role in detuned signal-recycled gravitational-wave interferometers, giving the possibility for resonant amplification of the gravitational-wave signal and for reshaping the noise curves beyond the standard quantum limit [10, 11].

We present in this paper an experiment devoted to the observation of quantum effects of radiation pressure and to the test of quantum-noise reduction schemes. Our experimental setup is based on a high-finesse Fabry-Perot cavity used to monitor the small displacements of a mirror. The experiment already achieved a sufficient sensitivity to analyze the internal thermal noise of the mirror at room temperature and to fully characterize the optomechanical properties of its internal vibration modes [12, 13, 14]. Recent progress in the experimental setup and the development of a dual optical injection system allowed us to observe various radiation-pressure effects in our experiment[15, 16]. We report the observation of radiation-pressure induced

© 2010 IOP Publishing Ltd

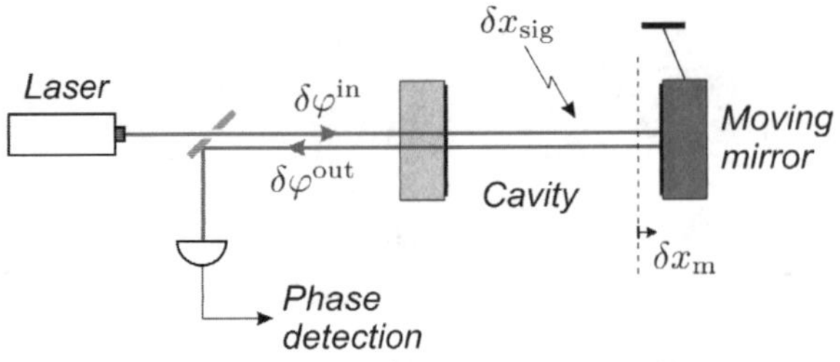


**Figure 1.** Principle of an optomechanical sensor based on a high-finesse cavity. Any displacement of the moving mirror or any optical length variation $\delta x_{\rm sig}$ of the cavity are detected through the phase shift $\delta\varphi^{\rm out}$ of the field reflected by the cavity. Quantum back-action noise corresponds to the spurious displacements of the moving mirror induced by the quantum fluctuations of the intracavity radiation pressure.

correlations between the intensity of an optical beam and the phase of a second beam, both sent into the same moving mirror cavity. We also observed a self-amplification of a signal, which may allow to overcome the standard quantum limit [17] as predicted in detuned signal-recycled gravitational-wave interferometers.

## 2. Radiation pressure and optomechanical coupling

The optomechanical coupling is the cross-coupling between a movable mirror and the light reflected upon it. Any displacement of the mirror induces a linear phase-shift of the reflected beam, providing a possibility to read out the displacement. On the other hand, radiation pressure appears as an additional force which perturbs the mirror motion. This corresponds to the back-action of the displacement measurement. When the mirror is part of a high-finesse optical cavity (Fig. 1), the optomechanical coupling is enhanced and the phase of the reflected field becomes very sensitive to mirror displacements. At resonance, the corresponding phase-shift $\delta\varphi^{\rm out}$ is given by

$$\delta\varphi^{\rm out} = \delta\varphi^{\rm in} + 8\frac{\mathcal{F}}{\lambda}\left(\delta x_{\rm sig} + \delta x_{\rm m}\right), \tag{1}$$

where $\lambda$ is the optical wavelength, $\mathcal{F}$ the cavity finesse, and $\delta\varphi^{\rm in}$ the phase fluctuations of the incident beam. The signal $\delta x_{\rm sig}$ can either be a physical displacement of the mirror, such as the response to an external force, or an apparent variation of the cavity length due for example to a gravitational wave. Fluctuations $\delta x_{\rm m}$ represent spurious displacements of the mirror, including thermal and radiation-pressure noises. The latter corresponds to the mechanical response of the mirror to the radiation pressure exerted by the intracavity field. In the framework of linear response theory [18], the mirror response $\delta x_{\rm rad}[\Omega]$ to the radiation-pressure fluctuations $\delta F_{\rm rad}[\Omega]$ at a frequency $\Omega$ is proportional to the mechanical susceptibility $\chi[\Omega]$ of the mirror:

$$\delta x_{\rm rad}[\Omega] = \chi[\Omega]\,\delta F_{\rm rad}[\Omega] = 2\hbar k\,\chi[\Omega]\,\delta I[\Omega], \tag{2}$$

where $k$ is the field wavevector, and $\delta I[\Omega]$ the intracavity intensity fluctuations.

Neglecting thermal and classical noises, the sensitivity of the measurement is ultimately limited by the sum of two quantum noises: the phase noise $\delta\varphi^{\rm in}$ of the incident light beam and the radiation-pressure noise $\delta x_{\rm rad}$. A compromise between the two leads to the so-called standard quantum limit [1, 2, 3, 4], which corresponds to a minimum measurable length variation $\delta x_{\rm sig}$ equal to $\sqrt{\hbar|\chi[\Omega]|}$. This limit has never been experimentally observed and a step towards its experimental demonstration would be the observation of the quantum radiation-pressure noise $\delta x_{\rm rad}$, which is responsible for correlations between the intensity and the phase of the light.

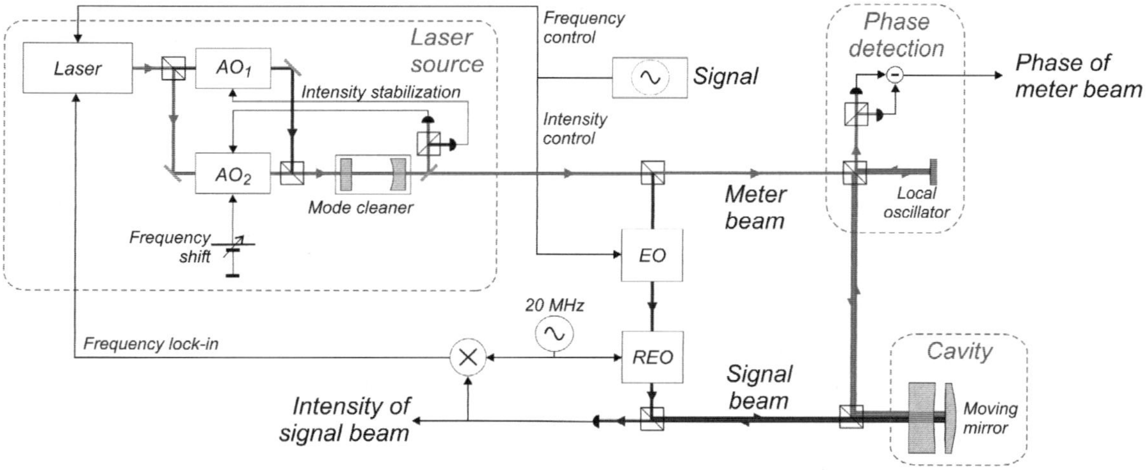


**Figure 2.** Experimental setup. A laser source provides two orthogonally polarized beams which are both sent into the moving mirror cavity. The signal beam is used to set the laser frequency at resonance via a Pound-Drever-Hall technique, while the meter beam can be detuned by an acousto-optic modulator ($AO_2$). The phase of the reflected meter beam is monitored by a homodyne detection, and the intensity of the signal beam by a photodiode. A signal generator provides either an intensity noise of the signal beam (via the electro-optic modulator EO) to test optomechanical correlations, or a modulation of the laser frequency to demonstrate the self-amplification of a signal.

To observe these optomechanical correlations, we send two beams in the cavity: the intensity fluctuations of the first, intense, signal beam drive the mirror into motion by radiation pressure, whereas the resulting position fluctuations $\delta x_{\mathrm{rad}}$ are monitored through the phase $\delta\varphi_m^{\mathrm{out}}$ of the second, weaker, meter beam. As the intensity fluctuations of the signal beam are unaltered by reflection upon the cavity and as far as the radiation pressure of the meter beam is negligible, the intensity-phase correlations observable between the two reflected beams provide a direct measurement of the optomechanical correlations. Also note that this measurement actually corresponds to a quantum non-demolition (QND) measurement of the intensity of the signal beam by the phase of the meter beam, as it has already been performed with nonlinear optical media [19], but here based on the optomechanical coupling [20, 21].

## 3. Experimental setup

Our experimental setup (Fig. 2) is based on a high-finesse cavity built with a cylindrical input mirror and a fused-silica plano-convex end mirror which exhibits gaussian internal vibration modes [14]. We work at frequencies close to one mechanical resonance with the following optomechanical characteristics, deduced from the observation of the thermal noise at room temperature: resonance frequency $\Omega_{\mathrm{m}}/2\pi = 1128$ kHz, mass $M = 72$ mg, and mechanical quality factor $Q = 760\,000$ when the cavity is operated in vacuum. The cavity is very short and compact (0.25 mm long) and we have obtained a cavity finesse up to $\mathcal{F} = 330\,000$.

A Titane-Sapphire laser working at 810 nm provides two cross-polarized signal and meter beams, both sent into the high finesse cavity. The laser is frequency stabilized on the resonance of the cavity via a Pound-Drever-Hall technique: the incident signal beam is phase-modulated at 20 MHz by a resonant electro-optical modulator (REO), and the resulting intensity modulation of the reflected beam provides the error signal. The meter beam can be arbitrarily detuned from the cavity resonance by an acousto-optic modulator ($AO_2$). Both beams are intensity stabilized

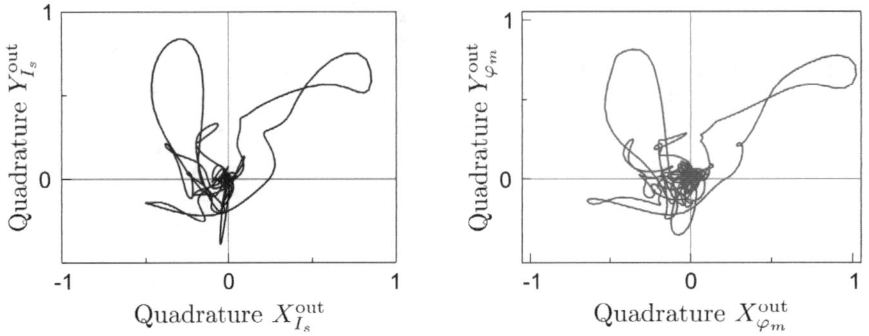


**Figure 3.** Phase-space trajectories of the intensity noise of the signal beam (left) and the phase noise of the meter beam (right).

and spatially filtered by a mode cleaner.

The phase fluctuations $\delta\varphi_m^{\mathrm{out}}$ of the reflected meter beam are monitored by a homodyne detection, with a local oscillator derived from the incident meter beam and phase-locked in order to detect the phase quadrature. With this setup, we obtain a shot-noise-limited sensitivity to mirror displacements of $2.7 \times 10^{-20}\,\mathrm{m}/\sqrt{\mathrm{Hz}}$ at frequencies above 200 kHz. Intensity fluctuations $\delta I_s^{\mathrm{out}}$ of the reflected signal beam are monitored by a high-efficiency photodiode. We have carefully eliminated unwanted optical reflections so that the optical rejection of the double-beam injection system is higher than 35 dB: the phase fluctuations of the meter beam are insulated from the intensity fluctuations of the signal beam in such a way that any observable effects of the signal beam are necessarily induced by intracavity radiation pressure.

## 4. Observation of optomechanical correlations

We have performed an experiment to demonstrate the optomechanical correlations induced by radiation pressure between the intensity of the signal beam and the phase of the meter beam. Since quantum radiation-pressure effects are dominated by thermal noise in our system at room temperature, we use a tiny classical intensity noise of the signal beam that mimics its quantum fluctuations but at a level at least 30 dB higher [15, 22]. This noise is produced by a wideband electro-optical modulator (EO in Fig. 2) driven by a digitized noise source. We generate a gaussian intensity noise centered at a frequency $\Omega_{\mathrm{c}}$ close to the mechanical resonance $\Omega_{\mathrm{m}}$,

$$\delta I_s^{\mathrm{in}}(t) = X_{I_s}^{\mathrm{in}}(t)\cos(\Omega_{\mathrm{c}}t) + Y_{I_s}^{\mathrm{in}}(t)\sin(\Omega_{\mathrm{c}}t)\,, \tag{3}$$

where the two slowly-varying quadratures $X_{I_s}^{\mathrm{in}}$ and $Y_{I_s}^{\mathrm{in}}$ are random gaussian functions.

The intensity of the reflected signal beam and the phase of the meter beam are independently acquired by two spectrum analysers, which directly extract the quadratures $X_{I_s}^{\mathrm{out}}(t)$, $Y_{I_s}^{\mathrm{out}}(t)$, $X_{\varphi_m}^{\mathrm{out}}(t)$, and $Y_{\varphi_m}^{\mathrm{out}}(t)$ of the reflected signal intensity and meter phase, respectively. Figure 3 presents the observed phase-space trajectories: clear correlations are obtained between the intensity noise of the signal beam (left) and the meter phase noise (right). Small discrepancies are visible between the two trajectories, mainly due to the thermal noise of the mirror which is 5 times smaller than the displacements induced by the radiation pressure. The optomechanical correlations can be made more quantitative by computing the correlation coefficient $C_{I_s,\varphi_m}$ defined as

$$C_{I_s,\varphi_m} = \frac{\left|\langle \delta I_s^{\mathrm{out}}\,\delta\varphi_m^{\mathrm{out}\star}\rangle\right|^2}{\langle|\delta I_s^{\mathrm{out}}|^2\rangle\,\langle|\delta\varphi_m^{\mathrm{out}}|^2\rangle}, \tag{4}$$

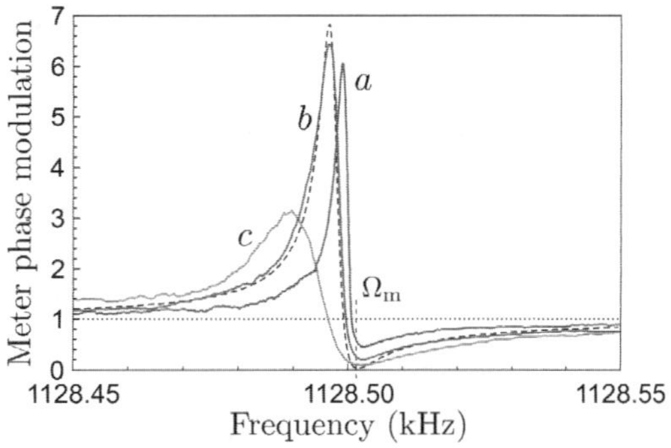


**Figure 4.** Modulation of the meter phase in response to a signal modulation swept around the mechanical resonance frequency $\Omega_{\mathrm{m}}$. Curves $a$ to $c$ correspond to increasing negative cavity detunings. The optomechanical sensor exhibits a self-amplification of the signal at a frequency close to the effective resonance of the mirror, down-shifted from $\Omega_{\mathrm{m}}$ by optical spring. The dashed curve is a fit of curve $b$ deduced from eq. (5).

where the brackets $\langle ... \rangle$ stand for a temporal average. We obtain a coefficient $C_{I_s,\varphi_m} \simeq 0.96$, in perfect agreement with the theoretical value expected when radiation-pressure and thermal noises are assumed to be preponderant in the measurement [16].

## 5. Observation of signal self-amplification

When the cavity is detuned with respect to the laser frequency, dynamical effects of radiation pressure are responsible for a change of the mechanical behavior of the moving mirror. As the intracavity intensity follows an Airy peak when the cavity length is changed, the displacements of the mirror are coupled to the intensity fluctuations in such a way that the radiation-pressure force becomes proportional to the displacement itself. This restoring force leads to an optical spring effect which changes the mechanical susceptibility of the mirror. The resulting effective susceptibility $\chi_{\mathrm{eff}}$ exhibits both a shift of the resonance frequency and a modification of its damping. Both effects have been observed in many experiments [23, 24, 25, 26, 27, 28], the optical damping being responsible for an efficient self-cooling of the mirror.

We demonstrate in this section that the radiation pressure is also responsible for a self-amplification of a signal, enabling the measurement sensitivity to beat the standard quantum limit. The signal is a modulation $\delta x_{\mathrm{sig}}$ of the optical length of the cavity, experimentally obtained through a modulation of the laser frequency (Fig. 2). In a detuned cavity, the radiation-pressure force becomes proportional to this length variation, leading to a mirror displacement $\delta x_{\mathrm{m}}$ proportional to $\delta x_{\mathrm{sig}}$. According to eq. (1), one can show [17, 29] that the phase $\delta\varphi_m^{\mathrm{out}}$ reproduces the total length variation equal to

$$\delta x_{\mathrm{m}}[\Omega] + \delta x_{\mathrm{sig}}[\Omega] = \frac{\chi_{\mathrm{eff}}[\Omega]}{\chi[\Omega]} \delta x_{\mathrm{sig}}[\Omega]. \tag{5}$$

Depending on the ratio between the initial and effective susceptibilities, one then gets either an amplification or a compensation of the signal by the mirror motion.

The signal beam is now only used to lock the meter beam frequency at a specific detuning with respect to the optical resonance of the cavity. We use a network analyzer to sweep the signal $\delta x_{\mathrm{sig}}$ around the mechanical resonance and to monitor the resulting phase modulation of the reflected meter beam. Curves in fig. 4 show the experimental results obtained for different detunings, together with a theoretical curve deduced from eq. (5). They clearly show a self-amplification of the signal near the effective mechanical resonance of $\chi_{\mathrm{eff}}$, which is down-shifted

Journal of Physics: Conference Series **228** (2010) 012024 doi:10.1088/1742-6596/228/1/012024

from $\Omega_m$ by the optical spring effect. We reach a self-amplification factor larger than 6 for curve *b*: dynamical back-action effects induce a motion $\delta x_m$ of the mirror in phase with the signal $\delta x_{sig}$, and with an amplitude larger than the signal itself.

## 6. Conclusion

We have presented an experiment based on the ultrasensitive measurement of the motion of a mirror in a high-finesse cavity. The mirror exhibits a mechanical resonance at a frequency around 1 MHz with a quality factor close to $10^6$. Our setup is then very sensitive to the intracavity radiation pressure and may be used to demonstrate quantum effects of optomechanical coupling.

We have observed the optomechanical correlations induced by radiation pressure between two light beams sent into the cavity. Such correlations are still at the classical level but similar correlations are expected in the quantum regime at low temperature. Inserting the cavity in a cryogenic environment should enable to detect these quantum correlations and hence demonstrate quantum radiation-pressure noise.

We have also observed a dynamical radiation-pressure effect corresponding to a self-amplification of a signal in a detuned cavity. Although our setup is currently limited by thermal noise, such a scheme has been shown to allow to overcome the standard quantum limit [17], as it is expected in detuned signal-recycled gravitational-wave interferometers. Using the parameters of our experiment and the self-amplification factor obtained so far, sensitivity may be improved beyond the standard quantum limit by more than 9 dB for frequencies close to the effective mechanical resonance of the mirror. Further experimental work is underway to reduce the thermal noise and to reach the corresponding sensitivity.

## References

[1] Caves C M 1981 *Phys. Rev. D* **23** 1693
[2] Jaekel M T and Reynaud S 1990 *Europhys. Lett.* **13** 301
[3] Braginsky V B and Khalili F Ya 1992 *Quantum measurement* (Cambridge: Cambridge University Press)
[4] Fritschel P 2003 *Proc. SPIE* **4856** 282
[5] The Advanced Virgo Team 2009 Virgo Internal Report VIR-089A-08
[6] Xiao M, Wu L and Kimble H J 1987 *Phys. Rev. Lett.* **59** 278
[7] Grangier P, Slusher R E, Yurke B and LaPorta A 1987 *Phys. Rev. Lett.* **59** 2153
[8] McKenzie K *et al* 2002 *Phys. Rev. Lett.* **88** 231102
[9] Kimble H J *et al* 2002 *Phys. Rev. D* **65** 022002
[10] Buonanno A and Chen Y 2001 *Phys. Rev. D* **64** 042006
[11] Buonanno A and Chen Y 2002 *Phys. Rev. D* **65** 042001
[12] Pinard M, Cohadon P-F, Briant T and Heidmann A 2000 *Phys. Rev. A* **63** 013808
[13] Briant T, Cohadon P-F, Pinard M and Heidmann A 2003 *Eur. Phys. J. D* **22** 131
[14] Briant T, Cohadon P-F, Heidmann A and Pinard M 2003 *Phys. Rev. A* **68** 033823
[15] Caniard T, Verlot P, Briant T, Cohadon P-F and Heidmann A 2007 *Phys. Rev. Lett.* **99** 110801
[16] Verlot P, Tavernarakis A, Briant T, Cohadon P-F and Heidmann A 2009 *Phys. Rev. Lett.* **102** 103601
[17] Arcizet O, Briant T, Heidmann A and Pinard M 2006 *Phys. Rev. A* **73** 033819
[18] Landau L and Lifshitz E 1958 *Course of Theoretical Physics: Statistical Physics* (New York: Pergamon)
[19] Grangier P, Levenson J A and Poizat J-P 1998 *Nature* **396** 537
[20] Jacobs K, Tombesi P, Collett M J and Walls D F 1994 *Phys. Rev. A* **49** 1961
[21] Heidmann A, Hadjar Y and Pinard M 1997 *Appl. Phys. B* **64** 173
[22] Mow-Lowry C M *et al* 2004 *Phys. Rev. Lett.* **92** 161102
[23] Sheard B S, Gray M B, Mow-Lowry C M, McClelland D E and Whitcomb S E 2004 *Phys. Rev. A* **69** 051801
[24] Arcizet O, Cohadon P-F, Briant T, Pinard M and Heidmann A 2006 *Nature* **444** 71
[25] Gigan S *et al* 2006 *Nature* **444** 67
[26] Corbitt T *et al* 2007 *Phys. Rev. Lett.* **99** 160801
[27] Schliesser A, Rivière O, Anetsberger G, Arcizet O and Kippenberg T J 2008 *Nature Physics* **4** 415
[28] Thompson J D *et al* 2008 *Nature* **452** 72
[29] Verlot P, Tavernarakis A, Briant T, Cohadon P-F and Heidmann A arXiv:0912.4085

# Study of three-mode parametric instability

**Fengchao Liang[1, 2], Chunnong Zhao[1], Slawomir Gras[1], Li Ju[1], D. G. Blair[1]**

[1] School of Physics, University of Western Australia, 35 Stirling Highway, Crawley, Western Australia 6009, Australia
[2] Key Laboratory of Optical System Advanced Manufacturing Technology, Changchun Institute of Optics, Fine Mechanics and Physics, Chinese Academy of Sciences, Changchun 130033, China

E-mail: fcliang@yahoo.cn

**Abstract**. The effect of parametric instability in advanced interferometric gravitational wave detectors is a potential problem for their proper operation. Great efforts have been made to study the onset of parametric instabilities and to find ways to control them. Here we present an experimental design for studying parametric instability in a 72 m cavity with suspended high quality fused silica mirrors. With 5 W input power and 20 kW circulation power inside the cavity, it is predicted that parametric instability will occur. The resonant condition of parametric instability can be met by thermally tuning the radius of curvature of a test mass. We will present simulation results of parametric gains for different radii of curvature of a test mass. The simulation results will provide the basis for designing the thermal tuning and observation parametric instability experiments. This will provide a test bed for studying parametric instability and its control for next generation detectors.

## 1. Introduction

Gravitational waves (GW) are a prediction of Einstein's general theory of relativity. The detectable gravitational wave signals on the Earth are extremely small in the order of $10^{-21}$ in strains. The first-generation ground-based laser interferometers including LIGO in USA [1] and VIRGO in Italy [2] are in operation. LCGT in Japan [3] and AIGO in Australia [4] are future detectors in the plan. The possibility to detect GW signal in the first generation detectors will be rare but it is possible. To increase the sensitivity to certain detection of GW, increasing the power stored in the Fabry-Perot cavity is necessary to reduce the quantum shot noise in addition to other improvements. However, extreme high optical power introduces some unavoidable problems, such as thermal lensing and parametric instabilities [5-7].

Parametric instability in Fabry–Perot cavity is the direct result of opto-acoustic interaction mediated by light radiation pressure force, where high circulating power, low optical and acoustic losses are necessary for high sensitivity, also could lead to the potential instability. If not controlled, parametric instability will cause the test masses' ultrasonic modes to ring at very large amplitudes until saturation [8]. To characterize the effect of the parametric instability, Braginsky et al. [9] obtained a dimensionless parametric gain $R$. The higher the parametric gain $R$, the faster a sensitive detector will be disrupted [10].

Various ways to control the instability has been proposed. Zhao et al proposed to thermally tune the radii of curvature (ROC) of a test mass of interferometer arm cavities in order to detune the high order mode resonance and achieve parametric instability suppression [8, 11, 12]. Gras, et al. proposed to use a ring damper (a high loss optical coating on test mass barrel) to reduce the test mass acoustic mode $Q$-factors with minimal thermal noise degradation [13]. Most recently, the group at MIT proposed to use resonant damper to damp the test mass acoustic modes, which has little effect on the

© 2010 IOP Publishing Ltd

test mass thermal noise if carefully designed [14]. The group at UWA proposed and demonstrated optical feedback control scheme by injecting an out-of-phase high order mode to destructively interfere with the high order mode inside the cavity.

In this paper, we will present an experimental design of a 72 m high finesse, high *Q*-factor suspended optical cavity for studying parametric instability. In this cavity we expect to observe self-sustained parametric instability by tuning the radius of curvature of a test mass. It will serve as a test bed for studying parametric instability and testing various control schemes. This paper is organized as follows: In the section 2, we will briefly review the theory of parametric instability. In section 3, we will show the configuration of the optical cavity. In section 4, we will show the simulation results of parametric gain as a function of test mass radius of curvature. At the end, we will give brief conclusions.

## 2. Review of parametric instability theory

In an optical cavity, the thermally driven test mass internal mode vibration will scatter the light incident on the test mass surface into upper and lower sidebands. We define the lower frequency sideband as anti-Stokes mode and the upper frequency sideband as the Stokes mode. Depending on the test mass internal mode shape on the surface, the scattering modes could have high overlap with cavity high order modes. If the scattering sidebands have the same frequencies as the cavity mode frequencies then these sidebands could be resonant inside the cavity and be amplified. The high overlap and the good frequency matching are necessary for the parametric instability.

We know that the frequency of a cavity mode is

$$\omega_{qmn} = \frac{\pi c}{L}\left[q + \frac{(m+n+1)}{\pi}\arccos(\pm\sqrt{g_1 g_2})\right] \tag{1}$$

So the frequency difference between fundamental mode $TEM_{00}$ (we assume it be Hermite-Gaussian $TEM_{00}$ mode in this paper) and Stokes mode $TEM_{mn}$ is

$$\omega_0 - \omega_s = \frac{\pi c}{L}\left[q_1 - \frac{(m+n)}{\pi}\arccos(\pm\sqrt{g_1 g_2})\right] \tag{2}$$

and the frequency difference between anti-Stokes mode and the fundamental mode is

$$\omega_{as} - \omega_0 = \frac{\pi c}{L}\left[q_{1a} + \frac{(m+n)}{\pi}\arccos(\pm\sqrt{g_1 g_2})\right] \tag{3}$$

where $q$, $q_1$ and $q_{1a}$ are longitudinal mode indices, $m$, $n$ are transversal indices, $g_1=1-L/R_1$, $g_2=1-L/R_2$ are $g$-factors, $\arccos(\pm\sqrt{g_1 g_2})$ is Guoy phase, $\omega_0$ is the frequency of fundamental optical mode, $\omega_s$ is the frequency of Stokes mode, $\omega_{as}$ is frequency of the anti-Stokes mode, $L$ is the length of the cavity, $R_1$ and $R_2$ are the radii of curvature of input test mass (ITM) and end test mass (ETM), respectively.

Braginsky *et al.* [5, 6] have shown that the effective parametric gain $R$ in a simple cavity is given by

$$R = \frac{4PQ_m}{McL\omega_m^2}\left(\frac{Q_1\Lambda_1}{1+(\Delta\omega_1/\delta_1)^2} - \frac{Q_{1a}\Lambda_{1a}}{1+(\Delta\omega_{1a}/\delta_{1a})^2}\right) \tag{4}$$

When the parametric gain $R$ exceeds unity, the acoustic mode will be excited by scattered higher order optical mode. Here $P$ is the circulating power inside the cavity, $Q_1$ and $Q_{1a}$ are the quality factors of the Stokes and anti-Stokes modes, $Q_m$ is the quality factor of the test mass, $\delta_{1(a)}=\omega_{1(a)}/2Q_{1(a)}$ is relaxation rate, $M$ is the mass of a test mass, $\Delta\omega_1=\omega_0-\omega_s-\omega_m$ and $\Delta\omega_{1a}=\omega_{as}-\omega_0-\omega_m$ are the possible detuning from the ideal resonance case, and $\Lambda_1$ and $\Lambda_{1a}$ are the spatial overlap factors between optical mode shape and acoustic normal mode shape. The overlap factors are defined as

$$\Lambda_1 = \frac{V(\int f_0(\vec{r}_\perp) f_1(\vec{r}_\perp) u_z d\vec{r}_\perp)^2}{\int |f_0|^2 d\vec{r}_\perp \int |f_1|^2 d\vec{r}_\perp \int |\vec{u}|^2 dV}, \quad \Lambda_{1a} = \frac{V(\int f_0(\vec{r}_\perp) f_{1a}(\vec{r}_\perp) u_z d\vec{r}_\perp)^2}{\int |f_0|^2 d\vec{r}_\perp \int |f_{1a}|^2 d\vec{r}_\perp \int |\vec{u}|^2 dV} \tag{5}$$

where $f_0$ and $f_{1(a)}$ describe the optical field distribution over the mirror surface for the fundamental and Stokes (anti-Stokes) modes, respectively, $\vec{u}$ is the spatial displacement vector for the mechanical mode, $u_z$ is the component of $\vec{u}$ that normal to the mirror surface, the integrals $\int d\vec{r}_\perp$ and $\int dV$

correspond to integration over the mirror surface and the mirror volume, separately. We can see from equation (4) that the parametric gain depends on the frequency detuning, overlapping factors, the $Q$-factors of the test mass acoustic mode and the cavity high order mode. The parametric instability will occur if the parametric gain is greater than unity.

## 3. Optical cavity design

In order to observe the parametric instability with the laser power that we currently have, we designed an optical cavity that its optical modes have the maximum overlap factor and tunable frequency difference between cavity fundamental mode and the required high order mode to bring them to simultaneous resonance. The second consideration of the design is to use test mass acoustic modes with the simplest mode structure that will simplify the identification of the acoustic mode from the observation.

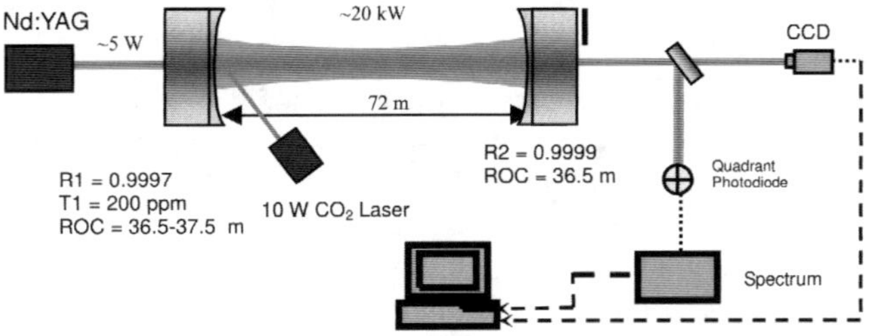


**Figure 1.** Schematic diagram of three-mode interactions in a simple Fabry-Perot cavity design for parametric instability investigation.

Considering above, we chose the $TEM_{01}$ and the $TEM_{10}$ mode as the candidate optical modes for studying the parametric instability. We will show in the following section that the modes have also large overlap factors with a few test mass acoustic modes from the Finite Element Modeling (FEM). These acoustic mode frequencies are around 155 to 200 kHz. From equation (4), in order to make the scattering mode resonance in the cavity, the frequency difference between $TEM_{00}$ and $TEM_{01}/TEM_{10}$ modes have to be close to the acoustic mode frequency. For the existing vacuum envelope of an optical cavity length ~72 m, we will need the cavity $g$-factors to be ~ 0.95 to match the cavity frequency spacing to the acoustic mode frequencies. In order for these frequencies to be exactly match we will use the $CO_2$ laser heating the test mass to tune the cavity mode spacing *in situ*.

**Table 1**. Parameters of optical cavity

| | ITM | ETM |
|---|---|---|
| Radius of curvature (m) | 36.5~37.5 | 36.5 |
| Material | Fused silica | Fused silica |
| Reflectivity | 0.9997 | 0.9999 |
| Transmissivity (ppm) | 200 | <100 |
| Cavity length (m) | 72 | |
| Wavelength (μm) | 1.064 | |
| Beam waist radius (mm) | 10 | |
| Circling power (kw) | 20 | |
| Finesse | 15000 | |

The layout of the optical system for investigation of three-mode parametric instability is shown in figure 1. A simple Fabry-Perot optical cavity consists of two fused silica mirrors suspended 72 m apart in a vacuum system. The single-frequency neodymium-doped yttrium aluminum garnet (Nd:YAG) laser is phase locked and mode matched to the $TEM_{00}$ mode of the optical cavity, providing about 5W at the ITM. The nominal finesse of the optical cavity is 15000, giving an

Journal of Physics: Conference Series **228** (2010) 012025 doi:10.1088/1742-6596/228/1/012025

intracavity $TEM_{00}$ power of about 20 kW. With 10 W $CO_2$ laser thermal tuning of the ROC of ITM, parametric gain $R$ can be changed. If $R$ is greater than 1, parametric instability should occur. Detailed parameters of optical cavity design are showed in table 1.

## 4. Suspension design and FEM simulation

The test masses will be suspended from the control mass using 25μm niobium ribbon removable modular suspension elements designed by Lee [15]. Such suspension can be removed and exchanged with suspensions of different geometry without the need to change the test mass itself. This will be convenient for us to test the different suspensions. Another advantage is that this method avoids the need for permanent bonding of suspensions to the test mass. Problems associated with bonding, in particular the stresses resulting from the different thermal expansion coefficients of bonded objects, are avoided. Figure 2(a), (b) and (c) show the suspension design of fused silica test mass and a modular suspension peg. To avoid the introduction of thermal noise due to the slip-stick phenomena at the point where the suspension element makes contact with the test mass, it is important that well defined high pressure contact points are created. To achieve this, a 'pin-in-hole' interface has been designed, where pins connected to the suspension element slot into small holes machined into the optic. It is solely the gravitational force acting on the test mass that induces the pressure on the contact points. There are two suspension holes at each flat side of the test mass in an equatorial position. Two holes on each side assure good control of mirror positioning without bonding. Low mechanical loss ribbons and cantilevers suspend the test mass in the vibration isolated vacuum tank. However, the drilled holes will introduce high surface loss since they have not been polished. The high pressure contact between the surface of suspension holes and suspension cantilever could be a source of excess localized loss from friction. These extra losses will lower the $Q$ factor of the test mass. We will simulate this effect of the extra surface loss to the test mass mode $Q$-factor using FEM and prove that this effect will not stop us to see the parametric instability with designed experimental parameters.

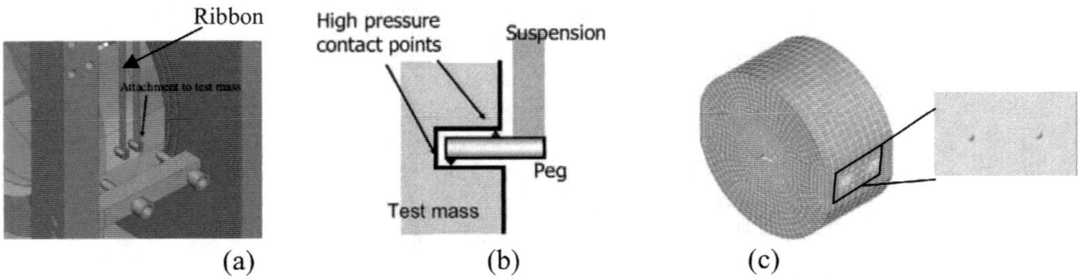


**Figure 2.** (a) Cantilever-ribbon suspension, (b) suspension peg, (c) Fused silica test mass (Φ100×50) with flats and holes (Φ3 ×4 ) for cantilever-ribbon suspension.

Figure 2(b) shows the model of a test mass created in the FEM software package, ANSYS®. The dimension of the test mass is 100 mm in diameter and 50 mm in thickness, suspension holes dimension is 3 mm in diameter and 4 mm long. We use Shell93 element to model the high loss holes' surface with the assumption that the surface loss is concentrated on the 100 μm thickness of surface. We use Solid95 element to model the bulk test mass. Test mass material parameters are as follows, Young's modulus is 73 GPa, Poisson ratio is 0.17 and the density is 2202 kg/m$^3$. After meshing, there are 14505 elements for bulk test mass and holes' surface in the model.

When considering localized loss, the $Q$ factors of the test mass will be different from that of without localized loss, so as the parametric gain $R$. To include the high surface loss when predicting parametric gain, three types of losses, $\phi_{bulk}$, $\phi_{surf}$ and $\phi_{TM}$ are defined, where $\phi_{bulk}$ corresponds to bulk structural loss, $\phi_{surf}$ associates with the surface loss of the suspension holes, $\phi_{TM}$ is the total loss of the test mass. Loss angle and $Q$ factors of test mass at different frequency can be calculated by following formulas [16, 17].

$$\phi_{bulk}(f) = 6.5\times(\frac{S}{V}) + 7.6\times10^{-12}\times f^{0.77} \qquad (6)$$

$$\phi_{TM} = \phi_{bulk}(f) + \phi_{surf} \times \sum_i E_i / \sum_j E_j \tag{7}$$

$$Q_m(f) = 1/\phi_{TM}(f) \tag{8}$$

where $S$ is the area of test mass, $V$ is the volume of the test mass, $f$ is the frequency, $\sum_i E_i / \sum_j E_j$ is the energy dilution factor depending on the individual mode structure, $E_i$ is strain energy of no. $i$ element of high loss surface of suspension holes, $E_j$ is the strain energy of no. $j$ element of bulk test mass, $Q_m(f)$ is the quality factor of the test mass at frequency $f$.

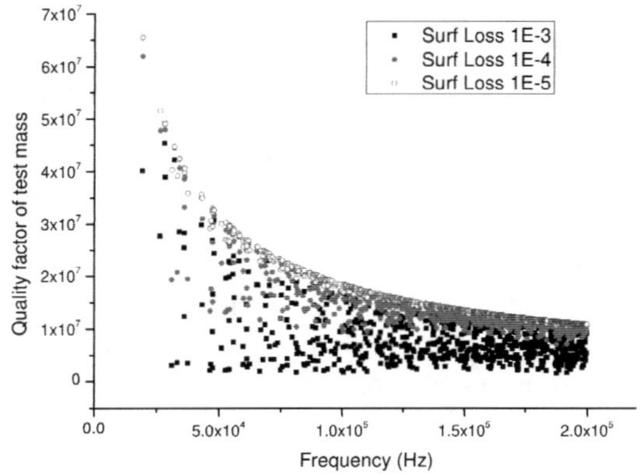


**Figure 3.** Frequency dependant $Q$ factors of test mass considering different suspension holes' surface loss in the range of $10^{-5}$ to $10^{-3}$.

The resonant frequency and the mode shape of each acoustic mode of the test mass have been obtained using FEM. $Q$ factors of the test mass with the assumption of the holes surface loss in the range of $10^{-5}$ to $10^{-3}$ are shown in figure 3. It can be seen from figure 3 that $Q$ factors decrease along with the increasing of the frequency because the fused silica's mechanical loss is frequency depended. The $Q$ factors decrease with increased suspension hole surface losses and the changes depend on the mode frequency because the different mode shape has different coupling effect to the suspension holes' surfaces. This agrees with ideas of test mass barrel coating for suppressing parametric instability with the minimum thermal noise penalty [13].

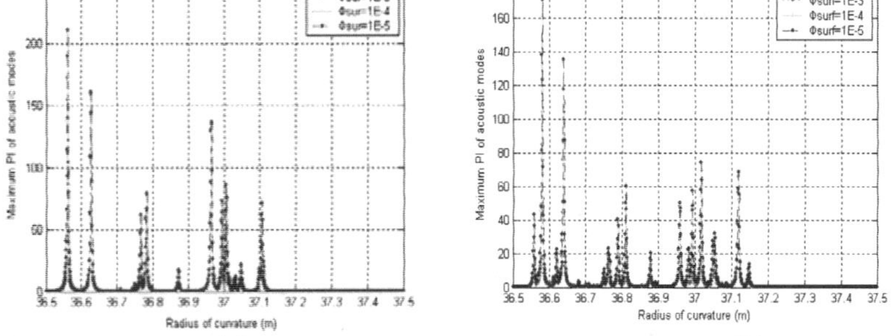


**Figure 4.** Maximum parametric gain vs. ROC of ITM (a) for $TEM_{01}$; (b) for $TEM_{10}$.

To match the cavity mode spacing to the test mass acoustic mode frequencies and then to obtain the maximum parametric gain, a $CO_2$ laser will be used to thermally tune the ROC of the ITM from 36.5 to 37.5 m, which will scan through various cavity mode spacing and then various parametric gain peaks. In order to obtain high precision of parametric instability gain in the simulation, a step of 1 mm is selected for tuning the ROC of ITM in the simulation. During this tuning course, Gouy phase for the optical cavity ranges from 2.81 to 2.91 radian, and the frequency distance between fundamental optical mode and Stokes mode, $\omega_0-\omega_s$, ranges from 155 kHz to 219 kHz. So test mass' acoustic modes with resonant frequency within this range should be able to get small detuning when interacting with the other two optical modes. If there is also a great overlapping factor between the acoustic mode shape and Stokes mode shape, the parametric gain may be greater than unity and parametric oscillation instability may occur. So when evaluating parametric gain, we put emphasis on acoustic modes with resonant frequency between 155 kHz to 200 kHz.

We calculated the parametric gain for each acoustic mode as a function of ROC of ITM for the designed optical cavity. Our analysis assumes that the fundamental optical mode is $TEM_{00}$ mode and higher order mode is $TEM_{01}$ and $TEM_{10}$. Acoustic modes are modeled using ANSYS® finite element modeling package with 14505 meshing elements. The fundamental optical mode and Stokes mode, the overlapping factors and parametric gains are evaluated in MATLAB®. Figure 4 shows the maximum parametric gain of each acoustic mode versus the ROC of the ITM. Figure 4 (a) is for $TEM_{01}$ mode (b) is for $TEM_{10}$ mode while including the suspension holes' high surface loss. It can be seen from figure 4 that a little change of ROC of the ITM can result in dramatic change of parametric gain value. There are more than 10 acoustic modes that could interact with the other two optical modes result in parametric instability, i.e., the acoustic modes can scatter $TEM_{00}$ into $TEM_{01}$ or $TEM_{10}$ mode with small detuning and high overlap factors. But through tuning the ROC of ITM, parametric gain can be suppressed much lower until less than 1. So in principle, thermally tuning of ROC of ITM can scan through the maximum parametric gain regime that will be used to investigate parametric instability effect. Two typical instable modes examples and their relative parameters are showed in figure 5.

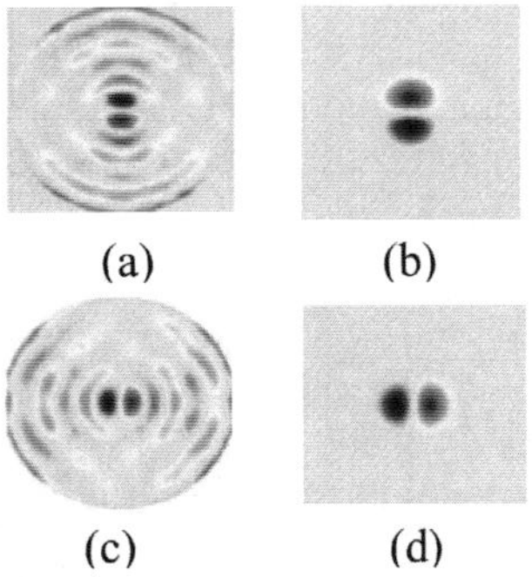

**Figure 5.** Typical acoustic and optical mode shapes, (a) ROC = 37.107 m, $\omega_m$=196.69 kHz, $\Lambda$ = 2.1608, $R$ = 60.891 m, $\Delta\omega_1$=31.844 Hz, $Q_m$= 9.377×10$^7$, $\Phi_{surf}$ = 10$^{-3}$; (b) the field distribution of the cavity $TEM_{01}$ mode showing high overlap of the acoustic mode and optical mode structures; (c) ROC = 37.118 m, $\omega_m$ = 197.35 kHz, $\Lambda$ = 2.0864, $R$ = 65.672 m, $\Delta\omega_1$ = 37.509Hz, $Q_m$=10.551×10$^7$, $\Phi_{surf}$ = 10$^{-3}$; (d) the field distribution of the $TEM_{10}$ showing high overlap of the acoustic mode and optical mode structures.

## 5. Conclusion

We presented an experimental design of a 72 m high finesse, high $Q$-factor suspended optical cavity for studying parametric instability. By tuning the ROC of the ITM the simulation shows that there are about 10 potentially unstable acoustic modes in the frequency range of 155 to 200 kHz with parametric gains between 1 and 250 if the test mass mechanical $Q$-factor ~1 to 7×10$^7$. Even if the actual test mass $Q$-factor is ~10$^6$ we can still see some modes of parametric gain greater than unity.

This experimental setup will serve as a test bed for studying parametric instability and testing various control schemes.

**Acknowledgment**

This is a project of Australian Consortium for Gravitational Astronomy, supported by Australian Research Council and Western Australia government research centre of excellence scheme.

**References**

[1] Abramovici, et al., Science 256 (1992) 325.
[2] F. Acernese, et al., Class. Quantum Grav. 23 (2006) S635.
[3] K. Kuroda, Class. Quantum Grav. 23 (2006) S215.
[4] D.E. McClelland, et al., Class. Quantum Grav. 23 (2006) S41.
[5] V.B. Braginsky, S.E. Strigin, S.P. Vyatchanin, Phys. Lett. A287, 331, (2001).
[6] V.B. Braginsky, S.E. Strigin, S.P. Vyatchanin, Phys. Lett. A305, 111, (2002).

[7] W. Kells and E. D'Ambrosio, Phys. Lett. A299, 326, (2002).
[8] L. Ju, C. Zhao, S. Gras, J. Degallaix, D.G. Blair, et al., Phys. Lett. A 355 (2006) 419.
[9] V. B. Braginsky and S. P. Vyatchanin, Phys. Lett. A 293, 228, (2002).
[10] D.E. McClelland, et al., Class. Quantum Grav. 23 (2006) S41.
[11] C. Zhao, L. Ju, J. Degallaix, S. Gras, D.G. Blair, Phys. Rev. Lett. 94 (2005) 121102.
[12] L. Ju, S. Gras, C. Zhao, J. Degallaix, D.G. Blair, Phys. Lett. A 354 (2006) 360.
[13] S. Gras, D.G. Blair, L. Ju. Physics Letters A 372 (2008) 1348–1356.
[14] Evans M et al.. 2008 LSC/Virgo Collaboration Meeting, LIGO-G080541-00-Z.
[15] B. Lee, Advanced test mass suspensions and electrostatic control for AIGO, PhD thesis, University of Western Australia, 2007.
[16] S.D. Penn, A. Ageev, D. Busby, G.M. Harry, et al., Physics Letters A 352 (2006) 3–6.
[17] Gregory M Harry, Andri M Gretarsson, et al., Class. Quantum Grav. 19 (2002) 897–917.

Journal of Physics: Conference Series **228** (2010) 012026 doi:10.1088/1742-6596/228/1/012026

# Noise cancellation properties of displacement noise free interferometer

**Shuichi Sato**[1], **Seiji Kawamura**[2], **Atsushi Nishizawa**[2] **and Yanbei Chen**[3]

[1]Hosei University, 3-7-2, kajinocho, Koganei, Tokyo, 184-8584, Japan
[2]National Astronomical Observatory of Japan, 2-21-1 Osawa, Mitaka, Tokyo, 181-8588, Japan
[3]California Institute of Technology, Pasadena, California 91125, USA

E-mail: sato.shuichi@k.hosei.ac.jp

**Abstract.** We have demonstrated the practical feasibility of a displacement- and frequency-noise-free laser interferometer (DFI) by partially implementing a recently proposed optical con guration using bi-directional Mach-Zehnder interferometers (MZIs). The noise cancellation e ciency was evaluated by comparing the displacement noise spectrum of the MZIs and the DFI, demonstrating up to 50 dB of noise cancellation. In addition, the possible extension of DFI as QND device is explored.

## 1. Introduction

Gravitational waves (GWs) and test mass displacements have di erent e ects on an interferometer. The idea of displacement noise (and frequency noise) free interferometry (DFI) is based on this fact and uses a network of interferometers to exploit it [1, 2]. DFI was suggested as a potential interferometer topology for third generation GW antennas in 2004. It can in principle cancel the e ects displacements of optics (and laser frequency noise) completely in the whole frequency band: this is a frequency independent feature. The distinguishing aspect of this property is that DFI cancellation does not depend on the driving force (disturbance) on the optics. DFI can cancel the optic's motion driven by any kind of disturbances, even including quantum e ects from the light such as quantum radiation pressure noise. This can allow an ultimate sensitivity limited only by the quantum shot-noise. The key challenge is that DFI should have nite response to GW signal at the same time. The e ect of GW and test mass displacement are indistinguishable in the low frequency limit, which means that DFI will certainly cancel the GW signal out to some extent in some frequency bands. The key task is to realize a DFI topology with reasonable sensitivity to GWs in the frequency range of our interest, which is still a challenge to work on.

After the invention of a practical topology for DFI using conventional laser interferometry [3], several proof-of-principle experiments were performed to demonstrate the fundamental feature of DFI [4, 5]. In these experiments, the characteristics of noise cancellation and a response to GW signals are con rmed to be as expected using arti cial excitations simulating the displacement of optics and GW signals.

In this article, we focus on the cancellation properties of naturally existing displacement noise in the frequency domain by measuring the interferometer sensitivity spectrum. A signi cant

© 2010 IOP Publishing Ltd

suppression of displacement noise (presumably seismic disturbances) was observed in the low frequency band, but there still remain some practical noise contributions that cannot be cancelled out with DFI mechanism. These may come from an "asymmetry" of the instrument in a broad sense, which can easily degrade the cancellation e ciency of the DFI. However, once these technical noises have been overcome, the DFI will be limited only by the quantum shot noise. Thus, DFI will have no equivalent of the concept of "standard quantum limit (SQL)" for conventional free mass laser interferometers, and that implies the capability of free mass laser interferometry to go far beyond the SQL. Thus, the idea to use DFI as a QND device is discussed in this article, together with the current status of the experiment.

## 2. DFI experiments

### *2.1. Proof-of-principle experiments*

Practical optical topologies of DFI using conventional laser interferometry in two and three dimensions have been proposed recently [3]. There are many possible con gurations to build a DFI, but the proposed ones are particularly simple. In these con gurations, the conventional, equal-arm Mach-Zehnder interferometer was used as a building block to eliminate laser frequency noise. Four such Mach-Zehnder interferometers were combined, in such a way that they form two pairs of counter propagating Mach-Zehnder interferometers. Within each pair, the 2 Mach-Zehnders share the same beam splitters and folding mirrors; subtraction of their outputs eliminates displacement noise from motions of the folding mirrors. The two pairs share the same beam splitters, which allows the elimination of beam splitter displacement noise. It should be mentioned that DFI con gurations work only for nonzero frequencies, because at DC gravitational waves are indistinguishable from relative mirror motions. We started to characterize DFI instruments with a series of proof-of-principle experiments using partial and full three-dimensional DFI topologies [4, 5]. The goal of these experiments was to con rm that DFI topologies work as GW detectors, while being insensitive to displacement noise. The scale of the interferometers was about a meter, so the optimal frequency for maximum response is around a few hundreds of MHz. This is because the frequency depends mainly on the scale of the instruments. The displacement noise and the GW signal were simulated using EOM phase modulators with a bandwidth up to several hundreds of MHz. The transfer function from simulated source to the DFI output gives the cancellation property for displacement noise and the response to simulated-GW signals. Thus, these experiments were performed with an arti cial excitation in a relatively high frequency region.

### *2.2. Demonstration experiments*

Following the completion of a series of proof-of-principle experiments, we had moved to a demonstration experiment. Our next interest was to operate a DFI in the terrestrial detector band to see the characteristics of DFI. This is because there are natural displacement noises (mainly driven by seismic and acoustic disturbances) to be suppressed by the DFI mechanism, and because this experiment directly connected the extension of DFI to a QND device, which will be described in a later section 3.

The setup is as simple as in one of the previous experiments, but we now focus on observing a natural displacement noise and its suppression. Thus we now use the noise spectrum instead of the transfer function to assess the performance. The goal of this experiment is to see the suppression of displacement noise to achieve quantum shot noise levels down to the 1 Hz region, in combination with the elimination of other technical noises that are not within the scope of DFI.

For this experiment we have used a partial-DFI composed of a single pair of counter propagating MZIs to demonstrate the cancellation properties of the DFI. The practical experimental setup is shown in Figure 1, which is very similar to that of a previous setup.

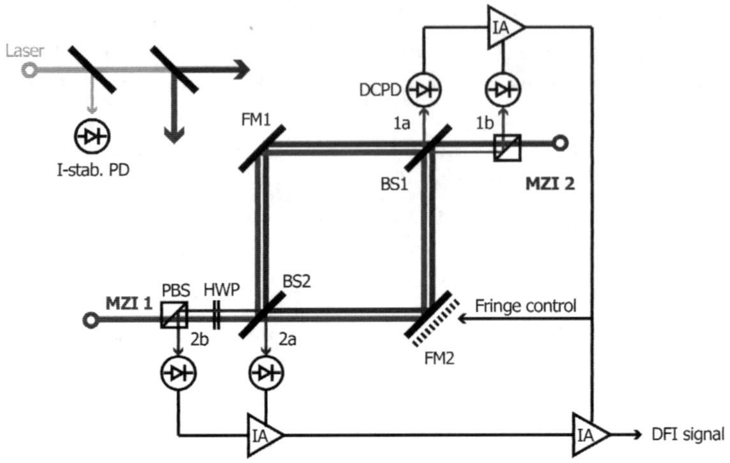


**Figure 1.** The experimental setup for the practical demonstration of DFI. This is one pair of bi-directional MZIs, which is part of a 3D DFI, to con rm the cancellation e ciency for both folding mirrors and the beam splitters.

The laser source is a commercial solid-state Nd:YAG laser (Innolight Mephius) yielding 500 mW at 1064 nm. The output beam is split into two, and each beam enters a MZI after passing through polarizing beamsplitter (PBS), which allow the detection of light exiting the counter propagating MZI. The light paths of both MZIs are carefully adjusted and superposed on each other so that the two MZIs can exactly share the displacement noise. With this setup, the displacement e ects from the two folding mirrors should cancel out, while that of the beam splitters will not, because this setup is only one building block of a full DFI. However, some suppression of displacement noise can be expected depending on the scale of the instrument and the frequency of interest. Thus, the arm length of the MZIs a ects the cancellation e ciency of the displacement noise; in this experiment, the arm lengths were chosen to be about $l = 10$ cm, to ensure the ideal balance up to $l\omega/c \simeq 160$ dB around $\omega = 1$ Hz.

Both output ports of MZI 1 (output 1a and 1b) were monitored with two identical dc detectors (DCPD) to provide a di erential error signal; this signal was fed back to a PZT-actuated folding mirror (FM2) after appropriate ltering to give a mid-fringe locking control. The control bandwidth was very low, around a few hundred Hz. Once the fringe of the MZI 1 is controlled, that of MZI 2 is also automatically controlled because the two MZIs share common optical paths. The other two output ports (output 2a and 2b) were equipped with dc detectors to monitor the di erential optical path length variations a ecting the MZI 2 sensitivity. The di erential error signals from MZI 1 and MZI 2 were subtracted with an instrumentation ampli er to produce the DFI output signal, which was then monitored with a dynamic signal analyzer (Agilent 35670A). The DFI features of the bi-directional MZI setup were con rmed by comparing the spectra of the single MZIs and DFI in the frequency domain.

The results for displacement-noise suppression are shown in Figure 2. The sensitivities of the MZIs are limited by shot noise above 1 kHz corresponding to tens of mW of power on the beam splitter, while presumable seismic noise is dominant at lower frequencies. Proving this, the DFI sensitivity showed a signi cant reduction of structured noise, which a cancellation of up to 50 dB. The DFI output was shot noise limited down to around 200 Hz, however, there remained some technical noise which the DFI mechanism cannot suppress. It is inferred that any "asymmetry" of the instruments in a broad sense, i.e. optical, mechanical and electronic imbalance, can easily degrade the performance of the DFI. We so far believe that this excess

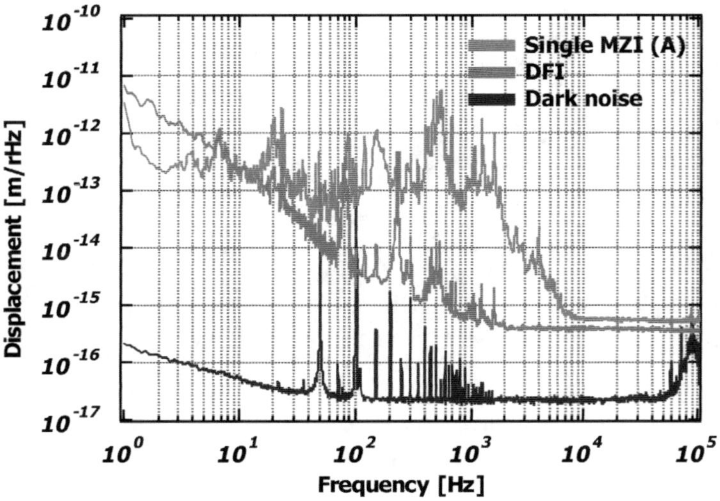


**Figure 2.** The displacement sensitivity spectra for both single MZI and the DFI, together with the dark-noise level of the instrument.

noise came from such an asymmetry and is thus not a fundamental problem. Therefore making the instrument more symmetric is expected to reduce this excess noise.

## 3. Extension of DFI to a QND device

DFI is essentially capable of eliminating the displacement information of optics, which is a frequency independent and driving force independent feature. Any noise that appears as motion of optics, even if it arises from the quantum nature of light, will thus be cancelled in principle. Therefore, once technical noise sources have been eliminated, the DFI should be limited only by the quantum shot noise at all frequencies. Thus, DFI will have no analogue to the "standard quantum limit (SQL)" concept for conventional free mass laser interferometers, and this implies that free mass laser interferometers can have a sensitivity far below the SQL. At the same time, a DFI should have a reasonable response to GWs in the frequency band of interest, such that a DFI could be an excellent GW detector. However, the sensitivity to GWs depends on the exact DFI con guration, which is still a challenge to work on, and is beyond the scope of this article. So here we focus on a noise reduction aspect: the possibility to demonstrate sub-SQL measurement of the displacement using DFI.

A sketch of the sensitivity of MZIs is shown in Figure 3, assuming that they are limited by quantum shot noise and radiation pressure noise. When the setup described here is extended to an input-equivalent laser power of 100 W and 10 mg free mass, the sensitivity curve will touch the SQL level of $10^{\ 16}\,\mathrm{m}/\sqrt{\mathrm{Hz}}$ at 10 Hz. Although there is still room to improve the instrument con guration, this is already not so far from a practical design using current technologies.

## 4. Summary

We studied the practical features of DFI using a single pair of counter-propagating Mach-Zehnder interferometers to see the elimination of folding mirror and beam splitter displacement noise. The fundamental features of DFI were con rmed by a sensitivity spectrum measurement. The maximum attained suppression of displacement noise was 50 dB in a particular frequency band; it is believed that some asymmetry in the instruments was the limiting factor for the cancellation. In addition, the possibility of using DFI as a QND device was discussed. 100 W input laser power

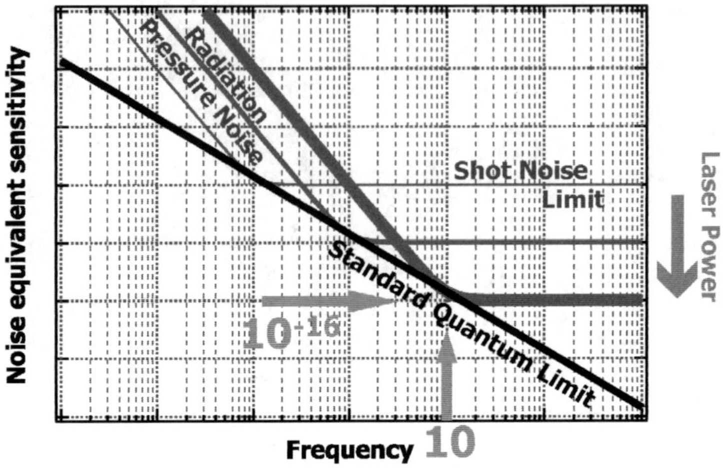


**Figure 3.** A sketch of quantum-noise-limited sensitivity of conventional MZIs. With 100 W-equivalent input power and free optics weighing only 10 mg, the sensitivity touches the 10 $^{16}$ m/rHz QSL level at 10 Hz. If this sensitivity is reasonably attainable, DFI will be able to show the capability of sub-SQL displacement measurement below 10 Hz.

and 10 mg free mass with the current setup can move the SQL up to an attainable level with current technologies, but there is still room for improving the con guration.

## Acknowledgments

This research was partially supported by the japanese Ministry of Education, Culture, Sports, Science and Technology under Grant-in-Aid for Scienti c Research (B), No. 18340070, 2006, and was also partially supported by the U.S. National Science Foundation under Cooperative Agreement No. PHY-0107417.

## References

[1] Kawamura S and Chen Y 2004 *Phys. Rev. Lett.* 93, 211103
[2] Chen Y and Kawamura S 2006 *Phys. Rev. Lett.* 96, 231102
[3] Chen Y *et al.* 2006 *Phys. Rev. Lett.* 97, 151103
[4] Sato S *et al.* 2007 *Phys. Rev. Lett.* 98, 141101
[5] Kokeyama K *et al.* 2009 *Phys. Rev. Lett.* 103, 171101

# Towards a Suspension Platform Interferometer for the AEI 10 m Prototype Interferometer

**K Dahl[1], A Bertolini[1], M Born[1], Y Chen[4], D Gering[2], S Goßler[1], C Gräf[2], G Heinzel[1], S Hild[3], F Kawazoe[2], O Kranz[1], G Kühn[2], H Lück[1], K Mossavi[2], R Schnabel[1], K Somiya[4], K A Strain[3], J R Taylor[2], A Wanner[1], T Westphal[1], B Willke[2], and K Danzmann[1,2]**

[1]Leibniz Universität Hannover, D-30167 Hannover, Germany
[2]Albert-Einstein-Institut, Max-Planck-Institut für Gravitationsphysik, D-30167 Hannover, Germany
[3]University of Glasgow, Glasgow, G12 8QQ, United Kingdom
[4]California Institute of Technology, Theoretical Astrophysics 130-33, Pasadena, California 91125, USA

E-mail: `katrin.dahl@aei.mpg.de`

**Abstract.** Currently, the AEI 10 m Prototype is being set up at the Albert Einstein Institute in Hannover, Germany. The Suspension Platform Interferometer (SPI) will be an additional interferometer set up inside the vacuum envelope of the AEI 10 m Prototype. It will interferometrically link the three suspended in-vacuum tables. The inter-table distance will be 11.65 m. The SPI will measure and stabilise the relative motions between these tables for all degrees of freedom, except roll around the optical axis. In this way, all tables can be regarded as one large platform. The design goal is $100\,\mathrm{pm}/\sqrt{\mathrm{Hz}}$ differential distance stability between 10 mHz and 100 Hz.

## 1. Introduction

Currently, the AEI 10 m Prototype is being set up at the Albert Einstein Institute in Hannover, Germany. As with other prototypes in the gravitational wave community, the purpose of this prototype is to prove and develop new techniques for future generations of gravitational wave detectors. Furthermore, the AEI 10 m Prototype [1] will probe at (and later beyond) the standard quantum limit (SQL) of interferometry and will be a testbed for geodesy and LISA [2] related experiments.

The AEI 10 m Prototype is housed inside an L-shaped, ultra-high vacuum envelope of 11.65 m arm length (measured from tank centre to tank centre). Each of the three vacuum tanks will contain a seismically isolated optical table of dimension 1.75 m by 1.75 m. The table design is a modified version of the HAM-SAS table tested at LASTI [3]. This design provides excellent passive and independent seismic isolation at the resonant frequencies of the optics suspension system in both vertical and horizontal directions by use of geometric anti-spring filters [4] and inverted pendulums [5]. However, the optical tables will still move slightly relative to each other.

This residual differential motion of the tables is too large for the experiments planned for GRACE follow-on, to be conducted in the AEI 10 m Prototype facility. For GRACE follow-on tests, two of the tables will act as satellites. The goal of these tests is to show that the GRACE

© 2010 IOP Publishing Ltd

follow-on interferometer can resolve changes in the satellite distance smaller than $2.5\,\mathrm{nm}/\sqrt{\mathrm{Hz}}$ between 10 mHz to 100 mHz, increasing with $1/f$ when going from 10 mHz to 1 mHz [6]. To demonstrate this, the prototype tables will be moved by a well-defined amount. It is obvious that the relative motion between the tables has to be smaller than the GRACE follow-on goal. This sets the requirement on the residual relative motion between the in-vacuum optical tables to $100\,\mathrm{pm}/\sqrt{\mathrm{Hz}}$ between 10 mHz to 100 mHz, increasing with $1/f$ when going from 10 mHz to 1 mHz and $10\,\mathrm{nrad}/\sqrt{\mathrm{Hz}}$ for angular deviation in the same frequency range. To realise this, all three in-vacuum tables will be linked interferometrically by a so-called Suspension Platform Interferometer (SPI). The SPI will measure the residual table motions and apply the appropriate feedback to stabilise the table distances via coil-magnet actuators.

The SPI can be beneficial for other projects in the AEI 10 m Prototype. For example, reducing residual motion of suspended mirrors in the quantum-measurement experiment [1] would help lowering the required actuator power and thus its electronic noise. Hence, we set the requirement of the SPI to be $100\,\mathrm{pm}/\sqrt{\mathrm{Hz}}$ relative displacement noise up to 100 Hz. Due to this constraint, and the fact that we can benefit from in-house knowledge gained during LISA Pathfinder development [7], we decided to choose a heterodyne Mach-Zehnder interferometer configuration for the Suspension Platform Interferometer.

The idea of stabilising relative displacements is not new within the gravitational wave community. It was first proposed by Ronald Drever in 1987 [8]. He suggested to set up an ancillary interferometer above the measurement interferometer (MIFO). This so-called suspension point interferometer is locked to the injected laser. It will reduce the residual motions of the mirrors and thus, ease the lock acquisition of the MIFO. This idea was experimentally tested by Yoichi Aso, who demonstrated 40 dB noise reduction of a Fabry-Perot interferometer below 1 Hz, down to 0.1 Hz [9]. The experiment coming closest to the AEI 10 m Prototype SPI is the metrology testbed depicted in [10]. In this experiment the relative longitudinal and yaw motions between two hexapods separated by 1 m were measured by use of three homodyne Michelson interferometers. The longitudinal displacement was stabilised down to $1\,\mathrm{nm}/\sqrt{\mathrm{Hz}}$ at 1 mHz.

## 2. Optical layout of the Suspension Platform Interferometer

The optical layout of the SPI (see Fig. 1) consists of two parts: the modulation bench and the measurement bench. The modulation bench containing the laser is located outside the vacuum system. Since the mismatch of the arm lengths in each Mach-Zehnder MIFO is about 22 m the free running laser frequency needs to be stabilised. In order not to let frequency noise limit the SPI sensitivity, we are aiming to reduce the position noise due to frequency fluctuations to $10\,\mathrm{pm}/\sqrt{\mathrm{Hz}}$ at 10 mHz, resulting in a frequency stability better than $128\,\mathrm{Hz}/\sqrt{\mathrm{Hz}}$ at 10 mHz. We chose an iodine-stabilised Nd:YAG laser with a wavelength of 1064 nm, which is commercially available [11]. On the modulation bench the laser light is split into two paths and modulated at the heterodyne frequency by two AOMs. The heterodyne frequency has been chosen to be around 20 kHz in order to provide a control bandwidth up to 100 Hz. The light is then coupled into two 20 m long polarisation-maintaining single-mode optical fibres. These are fed through into the vacuum system towards the measurement bench. In order to not be limited by thermal expansion, all optics on the measurement bench will be bonded [12] onto a baseplate made of the ultra-low expansion material Clearceram®-Z HS, with a coefficient of thermal expansion of $(0.0 \pm 0.2) \cdot 10^{-7}\,\mathrm{K}^{-1}$ [13].

The measurement bench consists of three interferometers: the reference interferometer (RIFO), and two MIFOs. The RIFO is used to cancel all noise individually picked up along each path starting from the very first beam splitter (labelled BS in Fig. 1) on the modulation bench, ending at the recombination beam splitter (labelled RBS in Fig. 1) of the RIFO. Furthermore, the RIFO defines the point to which the phase information gained at the MIFOs is compared.

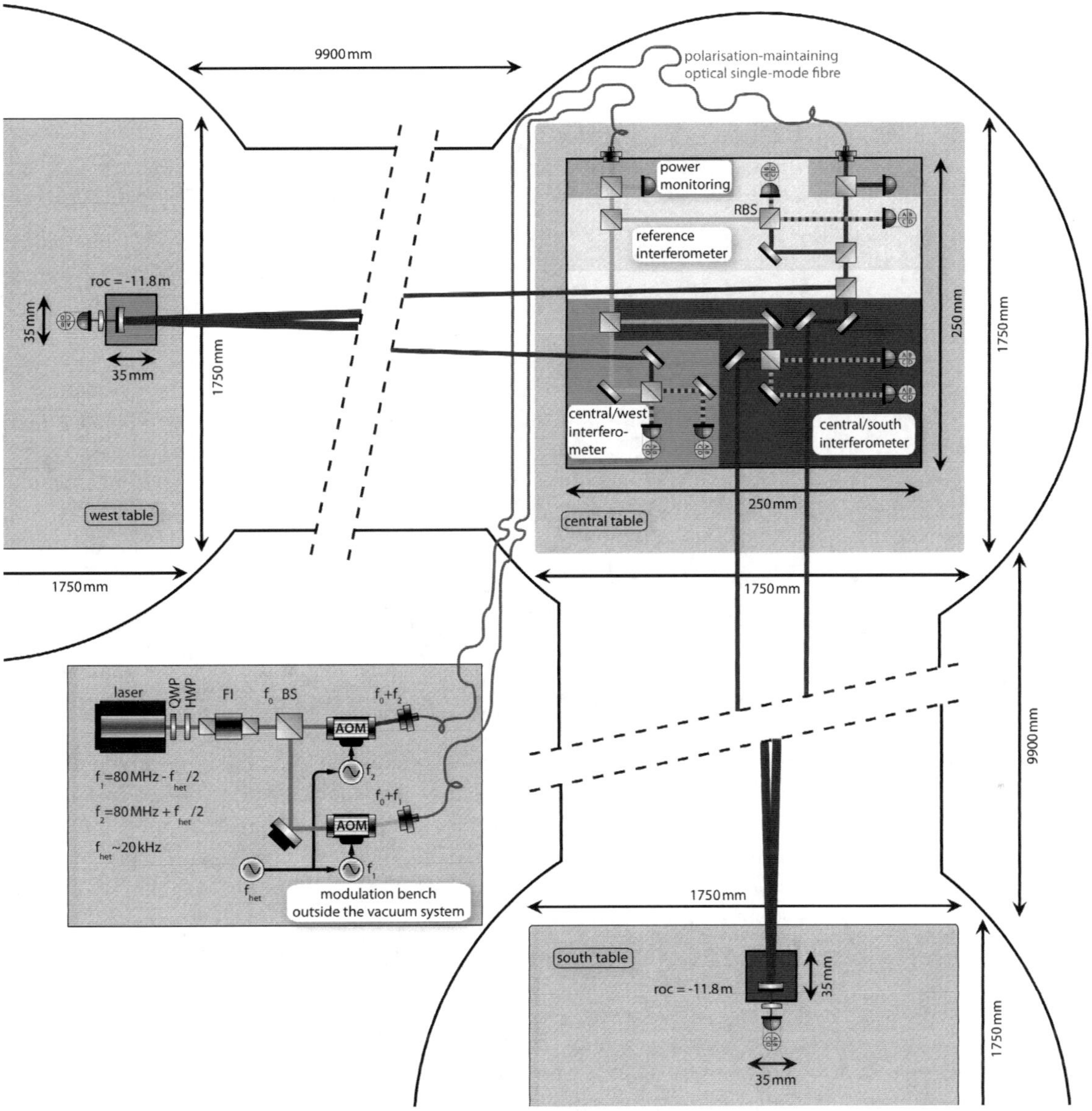


**Figure 1.** Schematic setup of the SPI. The reference beam is drawn in orange, the measurement beam in blue. For illustration purposes, position and size of the in-vacuum baseplates do not correspond to their real positions. Due to the small footprint of the baseplate (and overall height of 50 mm) it is still possible to place e.g. the suspended beam splitter of the SQL interferometer in the centre of the central table.
AOM: acousto-optical modulator; BS, RBS: beam splitter, $f_0$: laser frequency; $f_1$, $f_2$: modulation frequencies, $f_{het}$: heterodyne frequency, FI: Faraday isolator, HWP: half-wave plate, QWP: quarter-wave plate, roc: radius of curvature

One of the two MIFOs measures the motion between the west and central table, the other MIFO detects the relative motion between the south and central table. It can be seen that the RIFO is located only on the central table, whereas each MIFO has one mirror mounted on the south or west table, respectively. The position change of these end mirrors is then individually measured. The light is detected by quadrant photodiodes (QPD). Therefore, we will be able to monitor all three degrees of freedom of translation, as well as pitch and yaw. However, the sensitivity is not identical for all degrees of freedom.

In order to be able to separate linearly dependent signals, e.g. a vertical motion of a table that can be reproduced by a combination of pitch rotation and longitudinal translation; the curved end mirrors on the south and west tables will be partially transmissive. The light coming from the central table can be directly monitored on a QPD placed behind these end mirrors. Thus, we will have an additional — independent — source of information for signal separation.

## 3. Signal processing

The photo-currents for each QPD quadrant will be fed into a phasemeter outside the vacuum, which was originally developed for LISA Pathfinder [14]. The photo-currents are converted into voltages, the signals are then digitised, and a single-bin discrete Fourier transform is performed at the heterodyne frequency. The output values of the phasemeter for each channel are DC, real and imaginary parts of the complex amplitude of the photodiode signal at the heterodyne frequency. The argument of this complex amplitude is the phase of the signal. The phase values refer to the $\sim 20\,\mathrm{kHz}$ signal produced by the optical beat with respect to the phase of an electrical reference signal. The DC value is the light power of the two interfering beams, averaged over all QPD quadrants. The proper combination of this phase and DC information leads to four different types of values for each QPD: contrast, longitudinal phase information, differential wave-front sensing (DWS) and DC alignment signals. The contrast is a measure of how well measurement and reference laser beams interfere at the recombination beam splitter. The longitudinal phase indicates how far the tables have moved towards or away from each other. The relative angle between the wavefronts of the two interfering beams is given by the DWS signals. For the horizontal plane they are obtained by comparing the phase values of the left and right QPD sides, for the vertical plane by evaluating the phases for the upper and lower sides of the QPD. The DC alignment signals are derived in a way similar to the DWS signals, except that the evaluated signals are DC signals instead of phases. The DC alignment signals quantify how well the laser beam centre corresponds with the centre of the QPD. DC alignment signals are less sensitive than DWS signals. However, they are useful over a larger dynamic range than the DWS signals. DC alignment and DWS signals, as well as longitudinal phase information, will be used to derive error signals for the feedback control of the optical benches via coil actuators.

The phasemeter output port is an enhanced parallel port (EPP) which is currently connected to a personal computer. At a later stage, the phasemeter will be connected to a realtime Control and Data System (CDS) developed for the Advanced LIGO project [15] and adopted for the AEI 10 m control. For several reasons, it is not possible to connect the phasemeter directly to the CDS. First of all, the CDS has no parallel input port. Secondly, the phasemeter will be located as close as possible to the vacuum system, whereas the CDS is (due to infrastructural reasons) 10 m to 20 m away. However, the EPP cable length is limited to approximately 4 m. Therefore, a phasemeter interface based on a microcontroller board with an EPP input port and an ethernet output port is currently under development. Furthermore, an ethernet input port to the CDS will be implemented.

## 4. Status, conclusion and outlook

The SPI modulation bench has already been assembled. The measurement bench has been set up as a table-top test setup on a metal breadboard. In this particular test setup the arm length mismatch of the MIFO was about 3 m. Motions of the south and west tables were mimicked by a piezo motor-driven mirror-mount placed on two motor-driven linear stages. In this way pitch, yaw, and longitudinal, as well as transversal motions of the end tables could be simulated. Data evaluation showed that these motions are clearly distinguishable from each other.

One of the next steps towards an SPI for the AEI 10 m Prototype will be to bond the components onto the measurement bench. The in-vacuum tables are expected to be installed in spring 2010. Following the commissioning of the first two tables, the SPI will be installed and start to stabilise their relative motion.

## Acknowledgments

The authors would like to thank the AEI LISA and LISA Pathfinder group for fruitful discussions and support as well as the Excellence Cluster QUEST (Centre for Quantum Engineering and Space-Time Research) for financial support.

## References

[1] Goßler S *et al.* 2010 *accepted for publication in Class Quantum Grav*
[2] Danzmann K and Rüdiger A 2003 *Class Quantum Grav* **20** 1
[3] Stochino A *et al.* 2009 *Nucl Instrum Meth A* **598** 737
[4] Stochino A, DeSalvo R, Huang Y and Sannibale V 2007 *Nucl Instrum Meth A* **580** 1559
[5] Takamori A *et al.* 2007 *Nucl Instrum Meth A* **582** 683
[6] Dehne M, Guzmán Cervantes F, Sheard B, Heinzel G and Danzmann K 2009 *J Phys Conf Ser* **154** 012023
[7] Heinzel G *et al.* 2005 *Class Quantum Grav* **22** 149
[8] Drever R W P 1987 *LIGO Document* **T870001-00-R**
[9] Aso Y, Ando M, Kawabe K, Otsuka S and Tsubono K 2004 *Phys Lett A* **327** 1
[10] Numata K and Camp J 2008 *Appl Opt* **47** 6832
[11] http: //innolight.de/pdf/laser_accessories.pdf
[12] Elliffe E J, Bogenstahl J, Deshpande A, Hough J, Killow C, Reid S, Robertson D, Rowan S, Ward H and Cagnoli G 2005 *Class Quantum Grav* **22** 257
[13] www.oharacorp.com/pdf/ohara_optimax_ccz_study_oft.pdf
[14] Heinzel G *et al.* 2004 *Class Quantum Grav* **21** 581
[15] Shoemaker D 2009 *LIGO Document* **M060056-v1**

Journal of Physics: Conference Series **228** (2010) 012028 doi:10.1088/1742-6596/228/1/012028

# Designs of the frequency reference cavity for the AEI 10 m Prototype Interferometer

**Fumiko Kawazoe[1], John Robert Taylor[1], Alessandro Bertolini[2], Michael Born[2], Yanbei Chen[3], Katrin Dahl[2], Daniel Gering[1], Stefan Goßler[2], Christian Gräf[1], Gerhard Heinzel[1], Stefan Hild[4], Oliver Kranz[2], Gerrit Kühn[2], Harald Lück[2], Kasem Mossavi[1], Roman Schnabel[2], Kentaro Somiya[3], Kenneth Strain[4], Alexander Wanner[2], Tobias Westphal[2], Benno Willke[2], and Karsten Danzmann[1,2]**

[1] Max-Planck-Institut für Gravitationsphysik (Albert-Einstein-Institut), 30167 Hannover, Germany

[2]Leibniz Universität Hannover, 30167 Hannover, Germany

[3] California Institute of Technology, Pasadena, CA 91125 USA

[4] University of Glasgow, Glasgow, G12 8QQ, United Kingdom

E-mail: `fumiko.kawazoe@aei.mpg.de`

**Abstract.** The AEI 10 m Prototype is in its designing phase and will provide a test-bed for very sensitive interferometric experiments, such as the sub-SQL interferometer. It will test new techniques to reach – and even surpass – the Standard Quantum Limit. The experience and knowledge that can be gained from this experiment can be applied to large-scale interferometric gravitational detectors to improve the detector sensitivities. In order for the sub-SQL interferometer to achieve the required sensitivity all limiting noise sources need to be suppressed sufficiently. Noise sources can include seismic noise, thermal noise, and laser noise; laser frequency noise will be the main focus of this document. The laser frequency noise will be suppressed to a level of $10^{-4}\,\mathrm{Hz}/\sqrt{\mathrm{Hz}}$ at 20 Hz dropping to below $10^{-6}\,\mathrm{Hz}/\sqrt{\mathrm{Hz}}$ at 1 kHz. The proposed design to suppress the laser frequency noise with a ring cavity is described in this paper.

## 1. Introduction

The second generation of interferometric gravitational wave detectors such as Advanced LIGO[1], Advanced VIRGO[2], LCGT[3], and GEO HF[4], have been proposed, and research and design work has been started. The design upgrade will enable the sensitivity of these detectors to approach the Standard Quantum Limit (SQL). However, studies and research on SQL with prototype interferometers is necessary to fully understand new techniques to reach and surpass the limit, as well as to obtain experience that can be applied to large-scale detectors.

The AEI 10 m Prototype project began in 2005 with the aim of providing a test-bed for very sensitive interferometric experiments. One of these is the sub-SQL interferometer, which is in the initial designing phase. It aims to reach and go beyond the SQL with a Fabry-Pérot Michelson interferometer. In order to realize this goal, all the classical noise sources must be suppressed well below the SQL level at the frequencies of interest($\sim$ 20 Hz to 1 kHz). Laser frequency noise

© 2010 IOP Publishing Ltd

would deteriorate the sensitivity if not suppressed properly. The AEI Prototype will use a laser frequency stabilization system composed of a high-finesse triangular cavity (called 'the reference cavity' throughout this document). The sub-SQL interferometer will use Fabry-Pérot cavities with very light-weight (100 g) mirrors which are susceptible to radiation pressure noise, hence, the arm cavities do not provide a good frequency reference. In the following sections, the current design of the laser frequency stabilization system is described.

## 2. Laser frequency stabilization system

The mirrors of the reference cavity are suspended from a triple pendulum system. It will be placed on optical tables whose relative positions are controlled by a Suspension Platform Interferometer [5], inside the vacuum tanks, and will provide a frequency stabilized laser to the Sub-SQL interferometer input.

### *2.1. Sensitivity goal*

In order to reach the SQL all the classical noise sources need to be suppressed so that their levels are well below that of the SQL. In the experiment the designed sum of all the classical noise sources is below the SQL level by about a factor of 2 to 4, ranging from $\sim$ 50 Hz to $\sim$ 400 Hz. With the assumed common-mode rejection ratio of 1 %, this sets the maximum allowable level of the frequency noise in the input beam to the interferometer to be $10^{-3}\,\mathrm{Hz}/\sqrt{\mathrm{Hz}}$ at 20 Hz dropping to below $10^{-5}\,\mathrm{Hz}/\sqrt{\mathrm{Hz}}$ at 1 kHz. This corresponds to a reference-cavity length-noise of $10^{-16}$ to $< 10^{-18}\,\mathrm{m}/\sqrt{\mathrm{Hz}}$, from $\sim$ 20 Hz to 1 kHz. Therefore the laser frequency stabilization system must stabilize the laser frequency down to this level. This maximum allowable noise level is shown in Fig. 1, along with the target for the noise, which includes a safety margin (i.e. its level is an order of magnitude lower than the maximum). The noise level is calculated with carefully chosen suspension design parameters as well as optical design parameters.

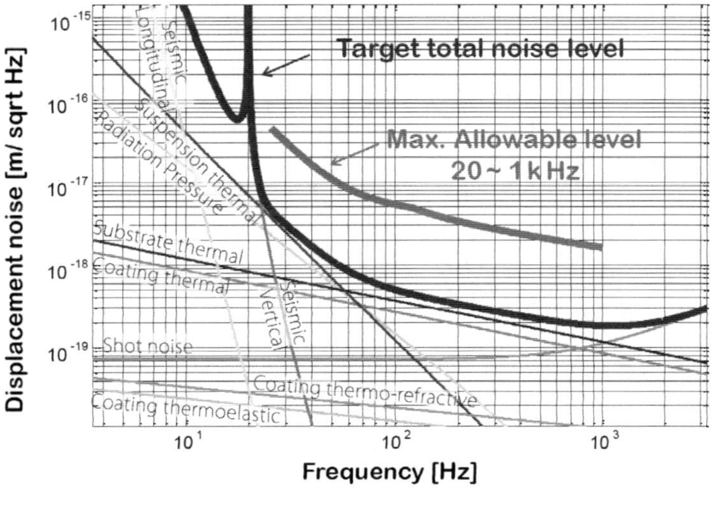


**Figure 1.** Required frequency stabilization level expressed in terms of equivalent displacement noise.

### *2.2. Suspension design*

The mirrors for the frequency reference cavity must be sufficiently seismically isolated, as their residual motion sets limits on the achievable laser frequency stability, as shown in Fig. 1.

The measured seismic noise in the laboratory is approximately $10^{-7}\,\mathrm{m}/\sqrt{\mathrm{Hz}}$ at 1 Hz, scaling with $f^{-2}$ at higher frequencies; while the required noise performance is better than $\sim 10^{-18}\,\mathrm{m}/\sqrt{\mathrm{Hz}}$ at 50 Hz.

To reduce the contribution from seismic noise, all optics are first placed on an optical table that also functions as a passive pre-isolation stage. This table is based on the HAM-SAS (Horizontal Access Module - Seismic Attenuation System) table [6, 7] developed for Advanced LIGO, and above several Hz will provide at least 60 dB and 70 dB attenuation in horizontal and vertical directions, respectively.

Additional seismic isolation is required to meet the design goal however. The mirrors are further isolated by a triple-stage suspension system – suspended from a rigid enclosure constructed from extruded aluminium struts – with the mirror comprising the final pendulum mass. The stage directly above the mirror is an aluminium cylinder of the same mass and outer dimensions. The uppermost stage is a rectangular piece, designed to hold combined position sensing, eddy-current damping and position feedback assemblies – used to align the suspension system and damp the main pendulum modes, over six degrees of freedom.

To alleviate any coupling from vertical to horizontal seismic motion, the wires suspending the upper mass from the suspension enclosure, and those suspending the intermediate mass from the upper mass, are further isolated by attaching them to maraging steel cantilever blades. This results in a combined spring system more compliant in the vertical direction than the wires alone, thus lowering the vertical resonant frequencies.

In order to reduce the thermal noise contribution from the suspension system, the suspension elements are high carbon-content stainless-steel wires with a high breaking stress and relatively low mechanical loss. The wires holding the mirror mass will be extremely thin (28 $\mu$m radius) thus keeping the wire's vertical bounce mode as low as possible ($\sim$ 20 Hz) and the first lateral violin resonance as high as possible (> 400Hz) while still loading the wire to less than a third of the breaking point.

The modelled seismic and thermal noise spectra for the reference cavity suspension systems can be seen in Fig. 2.

### *2.3. Optical design*

The residual cavity length fluctuations cause frequency fluctuations in the output beam with an amplitude inversely proportional to the cavity length. Therefore it is beneficial to have a long cavity length, and the design utilizes two optical tables which are separated by $\sim$ 11 m in such a way that two mirrors (input and output couplers) will be placed on one table, and the curved end mirror will be placed on the other table.

The circulating power is optimized so that the radiation pressure noise and shot noise are well below the required noise level, with the mirror mass being fixed to be 850 g due to load/space restrictions for the optical tables. The input power and finesse are optimized to meet the circulating power restriction, and to optimize the shot noise limited performance.

The cavity needs to be non-degenerate because higher order spatial-mode contents will disturb the auto-alignment signals. Assuming there will be $\sim$ 5 % of higher order spatial-mode contributions in the input beam, it is sufficient to suppress them by a factor of $\sim$ 1000 by using the cavity. A beam waist size (the waist is located between the two couplers) of 2.4 mm is large enough such that the coating Brownian thermal noise (with its linear spectral density being inversely proportional to the beam size) contribution is small enough. Simultaneously, it is small enough that the optically induced torque, as well as the power fluctuation due to angular fluctuation, are negligible. The cavity $g$-factor (which provides an indication of the optical stability; here, the condition of $0 < \mathrm{g} < 1$ means the reference cavity is optically stable) is calculated to be 0.67 to best suit the beam size and the cavity length.

The design parameters are summarized in Table 1.

Journal of Physics: Conference Series **228** (2010) 012028 doi:10.1088/1742-6596/228/1/012028

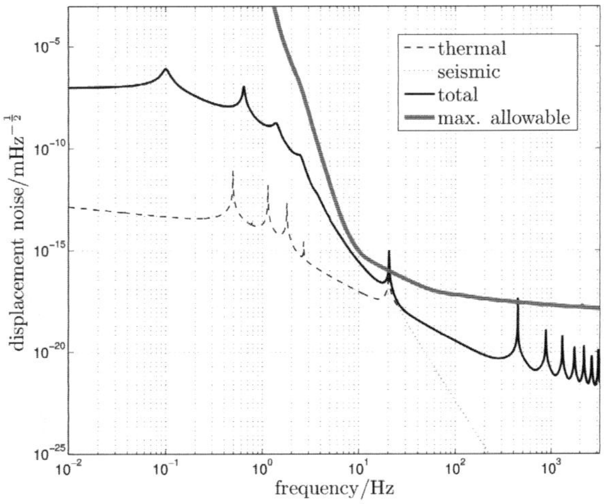


**Figure 2.** Simulated noise performance of the reference cavity mirror suspensions. The total noise from both seismic and thermal contributions is seen to be less than the frequency reference maximum allowable level at all points, except those where the wire vertical and lateral violin resonances lie.

**Table 1.** Optical design.

| | |
|---|---|
| Round trip length | 24.6 m |
| Input power | 0.133 W |
| Finesse | 7305 |
| Circulating power | 232 W |
| Higher order spatial-mode suppression | >1000 |
| Cavity $g$-factor | 0.67 |
| Beam waist size | 2.43 mm |

### *2.4. Control of the system*

In principle the laser frequency needs to be stabilized in the frequency range of interest, although in reality frequency noise at other frequencies could also couple into the target frequency region and deteriorate the performance of the system. Thus, control must be implemented over all the relevant frequency regions. The following frequency stabilization scheme utilizes three components; the main laser, the reference cavity, and the auxiliary laser. The main laser is a 2 W InnoLight Mephisto non-planar monolithic ring oscillator (NPRO), which is amplified by a four head Nd:$YVO_4$ laser amplifier to a power level of 35 W.

Below $\sim 4$ Hz, the iodine line stabilized Nd:YAG auxiliary laser (iodine laser from here on), used for the Suspension Platform Interferometer provides a better frequency reference than the reference cavity eigenmode, with its frequency noise at a level of 50 Hz/$\sqrt{\text{Hz}}$ between $\sim 0.5$ Hz to $\sim$4 Hz, increasing very roughly with $f^{-0.62}$ when going from $\sim 0.5$ Hz to $10^{-4}$ Hz. Above $\sim 4$ Hz the cavity eigenmode will provide the most stable frequency reference. Therefore below $\sim 4$ Hz, the eigenmode of the reference cavity is controlled via a coil-magnet actuator system to follow the iodine laser frequency while the NPRO is controlled to follow the cavity eigenmode from DC to several tens of kHz via temperature control loop, piezoelectric transducer control loop, and a phase-correcting pockels cell loop. An open loop gain of $\sim 10^7$ is required above 20 Hz. A Pound-Drever-Hall (PDH) scheme is used for the control. The control topology is sketched in Fig. 3. The 35 W main laser is split into two beams. One of them is phase modulated with an Electro-optic Modulator (EOM) to create phase modulation sidebands at a modulation frequency of 8 MHz, which do not resonate inside the reference cavity. A control signal is applied to the main laser, and to a phase corrector EOM. The other beam is coupled with the iodine laser that is phase modulated with an Acousto-optic modulator (AOM). A control signal is applied to the coil-magnet actuator of one of the mirrors of the frequency reference cavity. The figure does not include the auto-alignment system, which also uses the same modulation sidebands created by the EOM to align the reference cavity. The control parameters are shown in Table 2. The modulation index is optimized for the shot noise performance of the reference cavity.

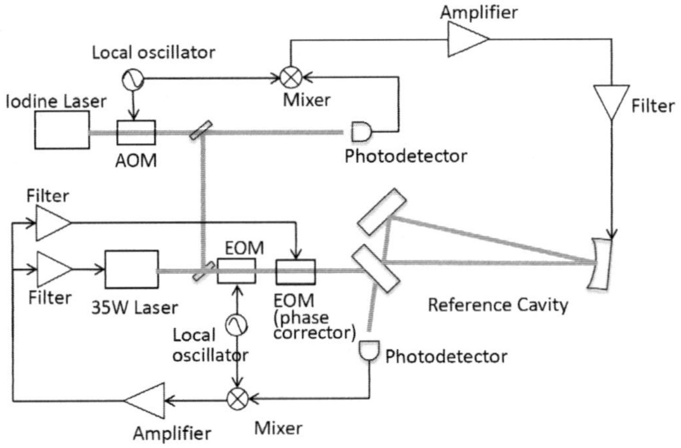


**Figure 3.** Control topology. It consists of two control loops. One is for controlling the frequency of the beam that goes into the reference cavity from DC to several tens of kHz, with its reference being the eigenmode of the reference cavity. The feedback signal is applied to the 35 W main laser and to the phase corrector EOM at lower and higher frequencies, respectively. The second loop is for controlling the length of the reference cavity below $\sim$ 4 Hz to follow the frequency of the iodine laser.

**Table 2.** Control parameters.

| | |
|---|---|
| Cavity FSR | 12.1867 MHz |
| Modulation frequency | 8 MHz |
| Modulation index | 0.6 |

## 3. Summary

The laser frequency stabilization system is required for the successful operation of the sub-SQL interferometer. The system will utilize a high-finesse optical cavity with suspended mirrors of which the suspension design, the optical design, and the control design have been developed.

With the frequency stabilization system the laser frequency noise will be suppressed to a level of $10^{-4}\,\mathrm{Hz}/\sqrt{\mathrm{Hz}}$ at 20 Hz dropping to below $10^{-6}\,\mathrm{Hz}/\sqrt{\mathrm{Hz}}$ at 1 kHz.

## References

[1] Fritschel P 2003 *Proc. SPIE Int. Soc. Opt. Eng.* **4856** 282
[2] Acernese F *et al.* 2008 *Class. Quantum Grav.* **25** 184001
[3] Kuroda K *et al.* 2006 *Class. Quantum Grav.* **23** S215–S221
[4] Lück H *et al.* 2009 The transition to GEO HF(Submitted to JPCS as proceedings of the 8th Amaldi conference)
[5] Dahl K *et al.* 2009 Towards a Suspension Platform Interferometer for the AEI 10 m Prototype Interferometer(Submitted to JPCS as proceedings of the 8th Amaldi conference)
[6] Stochino A *et al.* 2009 *Nucl. Instr. and Meth. A* **598** 737
[7] Goßler S *et al.* 2009 The AEI 10 m Prototype Interferometer(Submitted to CQG as proceedings of the 8th Amaldi conference)

Journal of Physics: Conference Series **228** (2010) 012029 doi:10.1088/1742-6596/228/1/012029

# Building blocks for future detectors: Silicon test masses and 1550 nm laser light

**R Schnabel[1], M Britzger[1], F Brückner[2], O Burmeister[1], K Danzmann[1], J Dück[1], T Eberle[1], D Friedrich[1], H Lück[1], M Mehmet[1], R Nawrodt[3], S Steinlechner[1], and B Willke[1]**

[1] Albert-Einstein-Institut, Max-Planck-Institut für Gravitationsphysik, Institut für Gravitationsphysik der Leibniz Universität Hannover, Callinstr. 38, 30167 Hannover, Germany
[2] Institut für Angewandte Physik, Friedrich-Schiller-Universität Jena, Max-Wien-Platz 1, 07743 Jena, Germany
[3] Institute for Gravitational Research, University of Glasgow, G12 8QQ Glasgow, Scotland

E-mail: `roman.schnabel@aei.mpg.de`

**Abstract.** Current interferometric gravitational wave detectors use the combination of quasi-monochromatic, continuous-wave laser light at 1064 nm and fused silica test masses at room temperature. Detectors of the third generation, such as the Einstein-Telescope, will involve a considerable sensitivity increase. The combination of 1550 nm laser radiation and crystalline silicon test masses at low temperatures might be important ingredients in order to achieve the sensitivity goal. Here we compare some properties of the fused silica and silicon test mass materials relevant for decreasing the thermal noise in future detectors as well as the recent technology achievements in the preparation of laser radiation at 1064 nm and 1550 nm relevant for decreasing the quantum noise. We conclude that silicon test masses and 1550 nm laser light have the potential to form the future building blocks of gravitational wave detection.

## 1. Introduction

The measuring principle of interferometric gravitational wave detectors is based upon laser beams that monitor changes in the distances between quasi-free falling test masses [1, 2, 3]. A high sensitivity to gravitational waves require a great number of cutting edge-technologies. The test masses need to be suspended as multiple pendulums in order to mechanically decouple them from the environment. They should have masses of several tens of kilogramms in order to reduce their susceptibility to radiation pressure forces and they should have a very high reflectivity for the laser wavelength used. The laser beams must show very low fluctuations in all their mode parameters like polarisation, spatial intensity distribution, frequency and power. In this contribution we focus on the thermal noise of the test masses, i.e. the thermally excited motions of the test mass surfaces with respect to the center of masses, and on the shot-noise, i.e. the photon counting noise of the photo-electric detection of the laser light. We are in particular interested in potential noise reductions that might be achievable if the current fused silica test mass material is replaced by crystalline silicon, as previously considered in Refs. [18, 19, 20], and the laser wavelength changed from 1064 nm to 1550 nm.

© 2010 IOP Publishing Ltd

## 2. Fused silica and crystalline silicon

According to the equipartition theorem, thermal noise can be reduced by decreasing the temperature. Additionally, as decribed by the fluctuation-dissipation theorem [4, 5], the thermal noise at Fourier frequencies far away from the system's mechanical resonances can be reduced by increasing the mechanical quality factors (Q-factors), if the mechanical dissipation associated with the sample under study is spatially homogeneous. The higher the Q-factor, the more thermal noise is transferred from the off-resonant spectrum into the resonances. The detection band of gravitational wave detectors is below the fundamental internal resonance frequency of test mass mirrors, which implies that a high Q-factor is very useful and therefore an essential requirement for detector test masses.

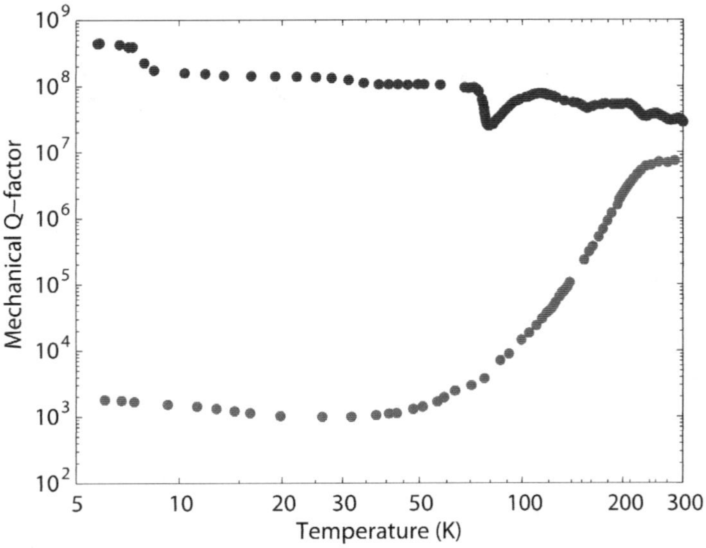


**Figure 1.** Mechanical quality factors of crystalline silicon (top, blue) and fused silica (bottom, red) versus temperature. The data were measured at the Friedrich-Schiller-Universität Jena on a 11.6 kHz mode of a fused silica and on a 14.6 kHz mode of a silicon(100) substrat. Both substrates were cylindrical, were 12 mm thick and had a diameter of 3" [15].

Current interferometric detectors are operated at room temperature with test masses made from fused silica. Fused silica offers high Q-factors of at least $2 \times 10^8$ at room temperature [6]. However, future gravitational wave detectors require even higher Q-factors and, most likely, also an operation of the test masses at temperatures far below room temperature. Unfortunately, Q-factors of fused silica samples significantly decrease with decreasing temperatures [8]-[14]. This effect is also clearly found in the measurement data of Fig. 1 (bottom trace, red dots) [15, 16]. Crystalline silicon, on the other hand, generally shows high Q-factors that even increases with lower temperature (top trace, blue dots in Fig. 1). A Q-factor of even more than $10^9$ at cryogenic temperatures was observed in [17]. The high Q-factors of silicon, which are also available at room temperature, resulted in proposals that this material might be a good test mass material for gravitational wave detectors as early as in 1991 [18, 19, 20].

An obvious problem with silicon is its rather high optical absorption at the currently used laser wavelength of 1064 nm. The absorption coefficient of silicon at 1064 nm is about $10\,\mathrm{cm}^{-1}$ [21], see Fig. 2. In contrast fused silica may show absorption at 1064 nm as low as $2.5\times10^{-7}\mathrm{cm}^{-1}$,

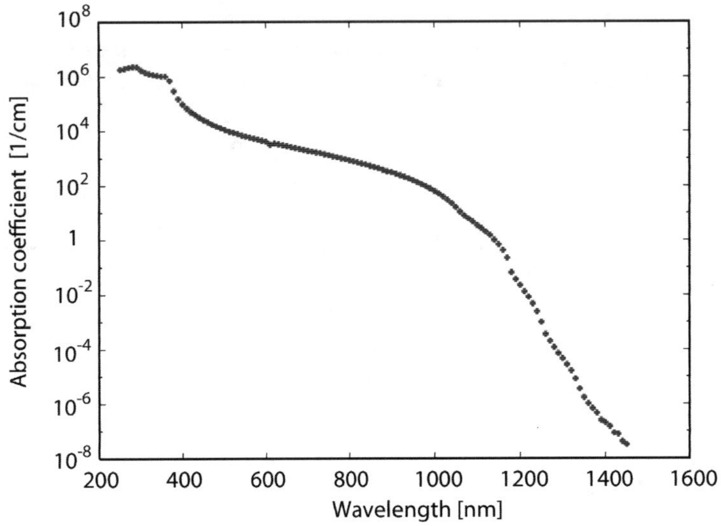


**Figure 2.** Measured optical absorption coefficients of silicon. The data, taken from Ref. [21], suggest an absorption of the order of $10^{-8}$cm$^{-1}$ at wavelengths around 1550 nm. Above 1200 nm the data are based on sensitive photocurrent measurements on high efficiency silicon solar cells. In this wavelength regime actual absorptions might be higher due to impurities.

such as the Heraeus Suprasil 3001. Test mass mirrors in gravitational wave detectors also serve as transmissive optical coupling mirrors to the Fabry-Perot arm resonators. Therefore, the test mass materials need to have extremely low absorption in order to minimize thermal lensing and photon absorption induced thermo-refractive noise [22]. In order to be able to use non-transparent materials like silicon in combination with 1064 nm radiation, all-reflective schemes have been proposed and demonstrated [23, 24, 25, 26]. Such schemes are based on diffractive couplers and could, in principle, solve this problem.

However, a signficant and actually dominating source of thermal noise arises from the high-reflectivity dielectric multi-layer coating stacks on the test mass mirrors. Such coatings consist of a several micrometer thick stack of up to 40 alternating layers of low- and high-refractive index materials. Recent research has shown that in particular the typically used high-index material tantala ($Ta_2O_5$) is responsible for the dominant thermal noise contribution of the coating [27]. At a laser wavelength of 1064 nm the coating thermal noise problem is very difficult to solve because of a lack of high index materials with high mechanical quality and low optical absorption at that wavelength. Doping of $Ta_2O_5$ with $TiO_2$ has also been investigated and a reduction of the mechanical loss by a factor of 1.5 was observed at room temperature [28]. If a wavelength is used at which silicon is highly transparent, not only can the effects of thermal lensing and the photon absorption induced thermo-refractive noise be minimized, but also the coating thermal noise can be reduced. Since silicon has a rather high index of refraction of 3.48 it could serve as an ideal high-index material in a conventional dielectric multi-layer stack. A combination of silicon with silica would generate the same high reflectivity as a combination of tantala and silica but with less layers. One inch optics with such coating stacks are already successfully used in our experiments as steering mirrors. A precise measurement of the reflectivities achieved is in preparation. The data from Ref. [21] as presented in Fig. 2 shows that at wavelengths above 1064 nm the absorption of silicon is drastically reduced. At the telecommunication wavelength of 1550 nm the absorption might be below $10^{-8}$cm$^{-1}$. Since the measurement of such low absorptions is difficult reliable data in this wavelength region is still missing.

Journal of Physics: Conference Series **228** (2010) 012029 doi:10.1088/1742-6596/228/1/012029

Recently, a coating-free, monolithic high-reflectivity surface of crystalline silicon was proposed [29] and fabricated [30]. A reflectivity of 99.8% was achieved for a silicon surface without adding any material by etching a specific structure into the crystalline silicon surface. In principle this method also works at other wavelengths and other materials. However, again the high refraction index of silicon is very important for designing an error-tolerant high-reflection surface and the combination of silicon and 1550 nm seems to be an excellent choice.

Of course several parameters of a material are important to judge its suitability as a gravitational wave test mass mirror. This contribution does not aim for a comprehensive survey. Only some other promissing properties of silicon should be mentioned here. Silicon has a vanishing linear thermal expansion coefficient at temperatures of 125 K and around 20 K [31]. At these temperatures the thermo-elastic noise of silicon is therefore minimal.

Another positive feature of silicon is its rather high thermal conductivity of about 2 W/(cm K) at room temperature. At 125 K and at 20 K the thermal conductivities are even higher, of about 10 W/(cm K) [32]. These high values would help in cooling the test mass mirrors down to cryogenic temperatures and in avoiding the thermal lensing problem. The values for fused silica are considerably lower, 0.015 W/(cm K) at room tempeature, less than 0.01 W/(cm K) at 125 K and 0.001 W/(cm K) at 20 K [33].

## 3. Laser wavelengths of 1064 nm and 1550 nm

Interferometric gravitational wave detectors require ultra-stable high-power laser radiation in order to reach high measurement sensitivities. The higher the laser power, the lower is the photo-electric counting noise in relation to the measurement signal, i.e. the signal-normalized shot-noise. After many years of research and developement of solid-state laser-systems, now ultra-stable single-mode laser radiation of more than 200 W at the wavelength of 1064 nm is available [35]. A laser-system of such power has been developed for the Advanced LIGO detector at the Laser Zentrum Hannover and the Albert-Einstein-Institut Hannover [34]. Systems with powers from 10 W to 35 W have been operated in current gravitational wave detectors and a lot of expertise at 1064 nm is therefore available.

In view of future detector generations the laser wavelength of 1064 nm also offers a rather mature technology for generating squeezed states of light. In addition to the high-power laser light input, future detector generations will, most likely, also have a (dim) squeezed light input into the (almost dark) signal out-put port of the interferometer. This additional light input reduces ('squeezes') the measurement's photon counting statistics. Most importantly, with squeezed light the shot-noise is reduced without a laser power increase, i.e. without a thermal load increase inside the interferometer. Since even the best mirror test mass materials have a residual absorption, this property is highly valuable for the operation at low temperatures. Suspended test masses are highly decoupled from the environment and heating due to light absorption is not compatible with a low temperature operation.

Squeezed light sources at 1064 nm were tested in many table-top experiments and at the suspended 40m-prototype at the California Institute of Technology [36, 37, 38]. For the first time, squeezed shot-noise in the audio detection band of gravitational wave detectors was demonstrated [39, 40], a detector compatible control scheme developed [41] and the strongest squeezing effect ever observed [42, 43].

At 1550 nm, research and developement of laser light sources for gravitational wave detectors have just begun. Erbium-doped fibre lasers are commercially available from a number of companies and provide up to 2 W of continuous-wave single-mode laser radiation around 1550 nm. Research and developement in the area of high-power fiber amplifiers is currently under way at the Laser Zentrum Hannover and the Albert-Einstein-Institut (AEI) supported by QUEST (Quantum Engineering and Space Time Research), the centre of excellence at the Leibniz Universität Hannover. We expect that single-mode output powers of the order of 100 W

should be achievable at 1550 nm within the next years.
Strongly squeezed light at 1550 nm was recently realized at the AEI Hannover [44]. An improved result, squeezing of 6.4 dB below the shot-noise is presented in this contribution, see Fig. 3. Here, trace (a) represents the noise power (the variance) of the shot-noise of a laser beam with a certain power. Trace (b) shows the squeezed noise of a laser beam with the same power. Here the phase angle of the squeezed field quadrature is controlled, which is the setting one would use for a measurement sensitivity improvement. The set-up used to generate squeezed light at 1550 nm is in direct analogy to the previously used set-ups at 1064 nm. Starting from a few tenths of a Watt of laser radiation at the fundamental wavelength (here 1550 nm) laser radiation at the second harmonic frequency is produced that subsequently pumps sub-threshold optical parametric oscillation in a second-order nonlinear crystal. The down-converted field, again at fundamental frequency, shows the squeezed photon statistics at sub-shot-noise levels. Clearly, the expertise gained at 1064 nm can directly be transferred in order to optimize the squeezed light sources at 1550 nm, eventually reaching the same performances already demonstrated at 1064 nm.

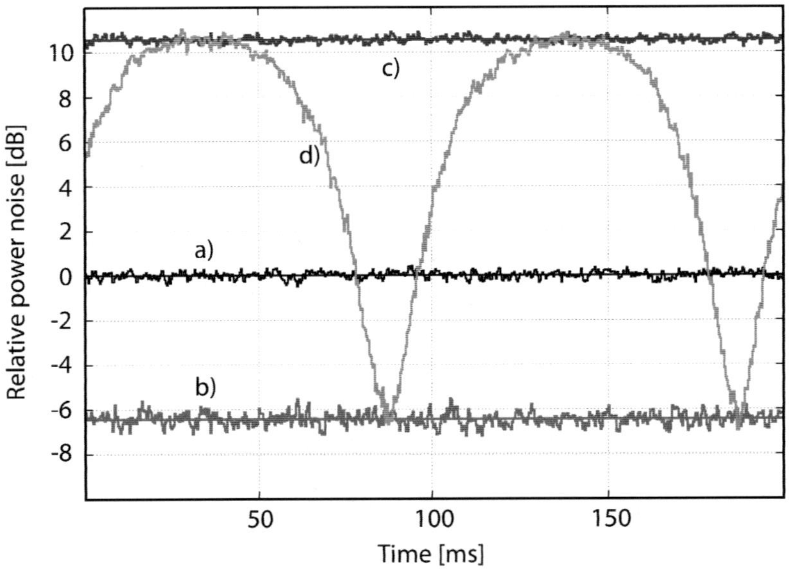


**Figure 3.** Noise powers at a Fourier frequency of 5 MHz, normalized to the laser beam's shot noise as shown in trace (a). Trace (b) represents the observation of 6.4 dB squeezed shot noise, and trace (c) the observed anti-squeezing. In both traces the phase angle of the squeezed light beam was controlled to be fixed. Trace (d) shows the noise power when the phase angle was scanned.

## 4. Conclusion

The combination of silicon as the test mass material and 1550 nm as the laser wavelength has a high potential to provide the building blocks of future generations of gravitational wave detectors like the European Einstein-Telescope [45]. Silicon has excellent well-known mechanical properties at low temperatures, in particular a high mechanical Q-factor, high thermal conductivity, and a vanishing thermal expansion coefficient at around 20 K and 125 K. Additionally, silicon most likely shows very low optical absorption at 1550 nm of the order $10^{-8}\text{cm}^{-1}$. The high index of refraction of silicon might be very useful to solve the coating noise problem, for wavelengths around 1550 nm. A laser material for this wavelength region already exists, and the developement of ultra-stable high-power lasers and state of the art squeezed light sources at 1550 nm should be possible. We also would like to mention here that the transition to a longer wavelength reduces Rayleigh scattering and relaxes tolerance requirements for the generation of high-reflectivity test mass surfaces. Last but not least, silicon industry as well as telecommunication industry already offer a great availability of ultra-pure silicon crystals and laser optical components at 1550 nm, which certainly would ease a transition from fused silica and 1064 nm to crystalline silicon and 1550 nm laser light.

## Acknowledgments

We thank Raffaele Flaminio for useful discussion. This work was supported by the Centre for Quantum Engineering and Space-Time Research QUEST and the Deutsche Forschungsgemeinschaft (DFG) within the Sonderforschungsbereich TR7. The research leading to these results has also received funding from the European Community's Seventh Framework Programme (FP7/2007-2013) under grant agreement N. 211743 (Einstein Telescope Design Study).

## References

[1] Saulson P R 1994 *Fundamentals of Interferometric Gravitational Wave Detectors* (Singapore: World Scientific)
[2] Rowan S and Hough J 2000 *Living Rev. Relativity* **3** http://www.livingreviews.org/lrr-2000-3
[3] Aufmuth P and Danzmann K 2005 *New J. Phys.* **7** 202 1
[4] Callen H B and Welton T A 1951 *Phys. Rev.* **83** 34
[5] Callen H B and Greene R F 1952 *Phys. Rev.* **86** 702
[6] Ageev A, Palmer B C, De Felice A, Penn S D and Saulson P R 2004 *Class. Quantum Grav.* **21** 3887
[7] Braginsky V B, Gorodetsky M L and Vyatchanin S P 1999 *Phys. Lett. A* **264** 1
[8] Marx J W and Sivertsen J.M. 1953 *J. Appl. Phys.* **24** 81
[9] McSkimin H J 1953 *J. Appl. Phys.* **24** 988
[10] Fine M E, Van Duyne H and Kenney N T 1954 *J. Appl. Phys.* **25** 402
[11] Anderson O L and Bommel H E 1955 *J. Amer. Ceram. Soc.* **38** 125
[12] Tielbürger D, Merz R, Ehrenfels R and Hunklinger S 1992 *Phys. Rev. B* **45** 2750
[13] Vacher R, Courtens E and Foret M 2005 *Phys. Rev. B* **72** 214205
[14] Travasso F *et al.* 2007 *Europhys. Lett.* **80** 50008
[15] Schwarz C *et al.* 2009 Proceedings of ICEC22-ICMC2008, edited by Chang H-M *et al.*, 978-89-957138-2-2
[16] Nawrodt R *et al.* 2008 *Journal of Physics: Conference Series* **122**, 012008
[17] McGuigan D F *et al.* 1978 *J. Low Temp. Phys.* **30** 621
[18] Winkler W, Danzmann K, Rüdiger A, Schilling R 1991, *Phys. Rev. A* **44** 7022
[19] Rowan S *et al.* 2003 in: M. Cruise, P. Saulson (Eds.), *Gravitational Wave Detection, Proceedings of SPIE*, SPIE, Wellingham, WA, p. 292
[20] Rowan S, Hough J, Crooks D 2005, *Phys. Lett. A* **347** 25
[21] Green M and Keevers M 1995, *Progress in Photovoltaic research & Appl.* **3** 189
[22] Braginsky V B, Gorodetsky M L, and Vyatchanin S P 2000 *Phys. Lett. A* **271**, 303
[23] Drever R W P 1995, in *Proceedings of the Seventh Marcel Grossman Meeting on General Relativity*, M. Keiser and R. T. Jantzen, eds. (World Scientific)
[24] Sun K-X and Byer R L 1997 *Opt. Lett.* **23**, 567
[25] Bunkowski A, Burmeister O, Beyersdorf P, Danzmann K, Schnabel R, Clausnitzer T, Kley E B and Tünnermann A 2004 *Opt. Lett.* **29** 2342
[26] Bunkowski A *et al.* 2006 *Appl. Opt.* **45** 5795
[27] Penn S D et al. 2003 *Class. Quant. Grav.* **20** 2917
[28] Harry G M *et al.* 2006 *Appl. Opt.* **45,** 1569
[29] Brückner F *et al.* 2008 *Opt. Lett.* **33**, 264
[30] Brückner F, Friedrich D, Clausnitzer T, Britzger M, Burmeister O, Danzmann K, Kley E-B, Tünnermann A and Schnabel R 2009 *submitted*
[31] Okada Y and Tokumaru Y J 1984 *Appl. Phys.* **56** 2 314
[32] Glassbrenner C J and Slack G A 1964 *Phys. Rev.* **134** 4A A1058
[33] Damon D H 1973 *Phys. Rev. B* **8** 5860
[34] Willke B *et al.* 2008 *Class. Quantum Grav.* **25**, 114040
[35] Frede M, Wilhelm R, Kracht D, and Fallnich C 2005 *Opt. Express* **13** 7516

[36] McKenzie K, Shaddock D A, McClelland D E, Buchler B C, and Lam P K 2002 *Phys. Rev. Lett.* **88** 231102
[37] Vahlbruch H, Chelkowski S, Hage B, Franzen A, Danzmann K, and Schnabel R 2005 *Phys. Rev. Lett.* **95** 211102
[38] Goda K *et al.* 2008 *Nat. Phys.* **4** 472
[39] McKenzie K, Grosse N, Bowen W P, Whitcomb S E, Gray M B, McClelland D E, and Lam P K 2004 *Phys. Rev. Lett.* **93** 161105
[40] Vahlbruch H, Chelkowski S, Danzmann K, and Schnabel R 2007 *New J. Phys.* **9** 371
[41] Vahlbruch H *et al.* 2006 *Phys. Rev. Lett.* **97** 011101
[42] Vahlbruch H *et al.* 2008 *Phys. Rev. Lett.* **100** 33602
[43] Mehmet M, Vahlbruch H, Lastzka N, Danzmann K, and Schnabel R. 2009 *submitted* arXiv:0909.5386
[44] Mehmet M *et al.* 2009 *Opt. Lett.* **34** 7
[45] http://www.et-gw.eu/

# Preliminary results on the cryogenic payload for the 3rd generation g.w. interferometers.

**F. Basti[1], F. Frasconi[3], M. Granata[1,2], E. Majorana[2] , V. Moscatelli[2], L.Naticchioni[1,2], M. Perciballi[2], P. Puppo[2], P. Rapagnani[1,2], F. Ricci[1,2]**

[1]*Department of Physics, University of Rome La Sapienza*
[2]*I.N.F.N. Sezione di Roma , 00185 Roma, Italy*
[3]*I.N.F.N. Sezione di Pisa, 56127 Pisa, Italy*

E-mail: paola.puppo@roma1.infn.it

**Abstract.** Thermal Noise is one of the limiting factor in the low and intermediate frequency range of the interferometric Gravitational Wave (GW) detectors. For beating this limit, a full scale last stage suspension (payload) prototype has been designed and built. Together with a mirror made of silicon, it has been cooled down at low temperature. Suspending the mirror from a cradle system, identical to that one used in the present VIRGO payload, the cooling strategy, the thermal behavior as well as the system mechanical response have been deeply studied. In this paper, the results obtained during the first cooling of the tested prototype are reported.

## 1. Introduction

Current GW antennas are designed to achieve the first direct detections and their proposed upgrades will look at typical GW sources distant of up to 100 $Mpc$. For effective Gravitational Wave Astronomy the sensitivity of future advanced detectors must increase ten-fold. The main noise limitation to the current and upgraded detectors is the Brownian thermal noise of the test masses. This limit can be overcome by cooling at cryogenic temperature the suspended mirror [1].

The use of the cryo-techniques have several other advantages:

- the thermo-elastic and thermo-refractive noises of the coated mirror decrease;
- the thermal expansion rate decreases at low temperature reducing drastically the thermal lensing effect induced by the high light power stored in the optical cavities of the interferometer;
- the mechanical loss of the payload will decrease significantly at low temperature if suitable materials are chosen.

On the other hand, for an effective use of low temperature in a GW detector we need to design the last stage suspension system in such a way

- to transmit the refrigeration power to the mirror without worsening its position control performance;
- to keep the residual vibration of the mirror due to the cryo system well below the noise sensitivity curve of the detector;

© 2010 IOP Publishing Ltd

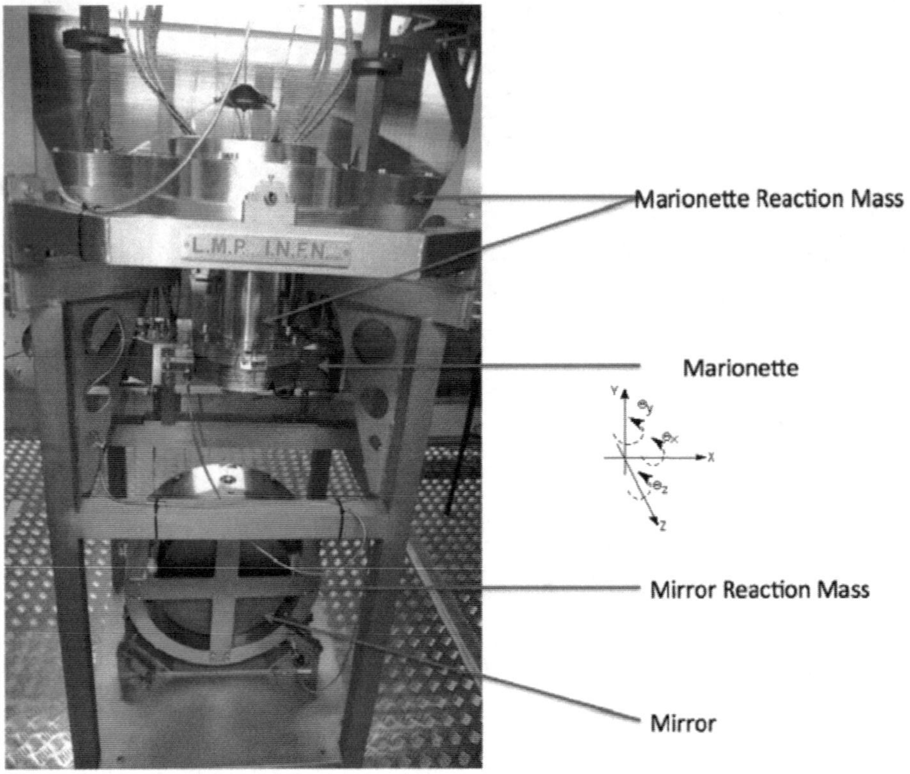


**Figure 1.** The cryogenic payload in its assembling frame.

In the following sections we present a first attempt to design a cryogenic payload and the preliminary results obtained during its first cooling up to $\sim 25\ K$.

## 2. The payload and its cooling system

The payload prototype we tested (figure 1) is composed by four main elements: a fake mirror, its reaction mass, the marionette and a fourth element, the marionette reaction mass (MRM). The marionette is hung to the aluminum flange of the inner vacuum chamber of the cryostat by means of the main Ti-6Al-4V suspension rod having a diameter of 4 $mm$. The MRM is suspended by three Ti-6Al-4V wires (1.5 $mm$ in diameter) inclined by an angle of $58^o$, which are connected, to the main suspension rod by means a of junction box. This last element, having the shape of a chinese hat, is conceived to rotate the MRM holding the actuation coils in front of the permanent magnets attached to the marionette wings.

The mirror is suspended to the marionette by means of two copper-beryllium ribbons wrapped around it. The 23 $kg$ mirror is a silicon crystal plate, 350 $mm$ diameter and 100 $mm$ thick. We choose this material because of its high thermal conductivity, extremely low mechanical loss at low temperature and very low thermal expansion (zero below 17 $K$).

The reaction mass (RM) is suspended to the marionette by two steel wires (0.6 $mm$) wrapped around it. The thermal links are made by electrolytic copper bundles 3 $mm$ thick and 30 $mm$ large. They are attached from one end to the aluminum flange of the inner vacuum chamber of the cryostat and on the other end to copper plates of the marionette where the clamps of the mirror ribbons are located. The cryogenic payload temperature has been monitored on different point within the experimental chamber by using calibrated diode sensors from Lake Shore. The

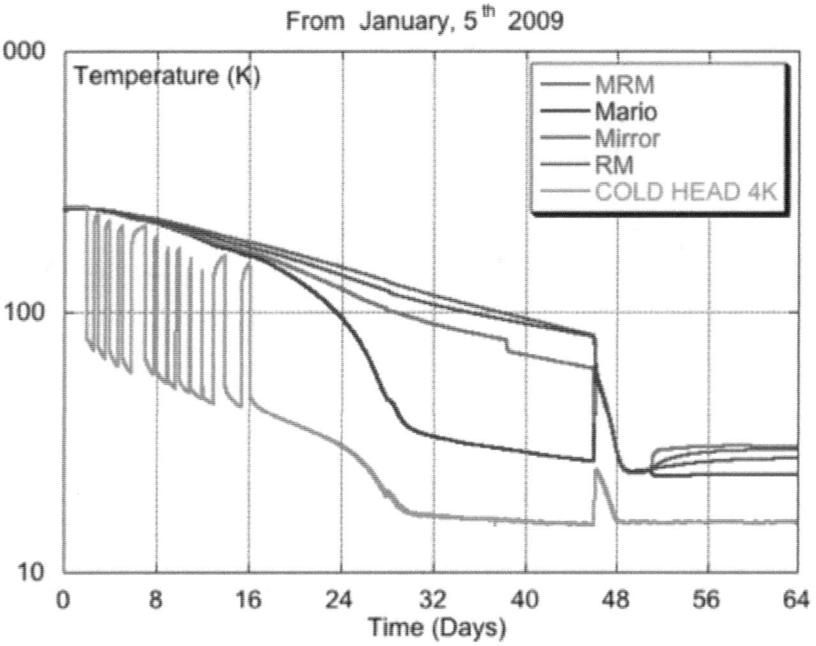


**Figure 2.** The temperature of the mirror, MRM, RM and marionette vs. time.

total mass ~300 kg of the payload is hosted in the inner vacuum chamber of a cryostat. The cooling system is a large cryostat installed at the European Gravitational Observatory (EGO) in Cascina. The dewar is equipped with two pulse-tube (PT) refrigerator. The pulse-tube is a closed loop cryo cooler invented by Alexander Mikulin in 1984. He modified a device known as the "basic pulse tube" and created a new class of cryo-cooler, the Orifice Pulse Tube cryocooler. In our case the two refrigerators are commercial devices made by the CRYOMECH Company: the first one is a double stage PT410 while the second one is a single stage PT60. The cold head of the PT410 is attached to the external side of the aluminum vacuum flange of the inner chamber of the cryostat, while the PT60 cold head is connected to the intermediate thermal shield of the cryostat. In the figure 2 we show the plot of temperature versus time during the cooling. We notice the abrupt change of the temperature of the PT410 cold head due to failures of the cooling water system of the PT compressor experienced during the first $16^{th}$ days of the run. Moreover the sharp temperature decrease of the all the payload elements the $46^{th}$ day of the cooling is due to the payload vacuum chamber He gas ling, operated to speed-up the cooling. Once a stationary condition was achieved, the $51^{st}$ day we removed the gas exchange to perform the vibration measurements. The final mirror temperature was 30 $K$ while the cold head is at 15 $K$. In the stationary condition the heat transfer trough the suspension of the payload element results to be low: along the four $CuBe_3$ wires of the mirror the heat power transmission was 1.9 $mW$, while along the four Steel c70 wires of the recoil sass we have 1 $mW$. During the test most of the thermal power (more than 10W at $\sim$ 20 $K$) was lost in the cooling process of the whole cryostat and this implies that a significant improvement of the thermal super insulation of the EGO cryostat is needed.

**Table 1.**

| Frequency at room temperature | Frequency at cryogenic temperature | Finite element simulation (no thermal links) | Mode identification |
|---|---|---|---|
| | 26.40 mHz | 54.00 mHz | $\theta_y$ |
| 93.75 mHz | 71.00 mHz | 87.00 mHz | $\theta_y$ |
| 0.460 Hz | 0.495 Hz | 0.467 Hz | $z$ |
| 0.470 Hz | 0.510 Hz | 0.477 Hz | $x$ |
| 0.560 Hz | 0.580 Hz | 0.590 Hz | $\theta_z$ |
| 0.554 Hz | 0.643 Hz | 0.653 Hz | $x$ |
| 0.660 Hz | 0.754 Hz | 0.680 Hz | $z$ |
| 0.750 Hz | 0.763 Hz | 0.710 Hz | $x$ |
| 0.924 Hz | 0.974 Hz | 0.740 Hz | $z$ |
| 1.050 Hz | 1.021 Hz | 0.893 Hz | $\theta_x$ |
| 1.087 Hz | 1.092 Hz | 0.995 Hz | $z$ |
| 1.126 Hz | 1.122 Hz | 1.013 Hz | $x$ |
| 1.485 Hz | 1.163 Hz | 1.024 Hz | $\theta_y$ |
| 1.650 Hz | 1.228 Hz | 1.455 Hz | $\theta_y$ |
| 1.724 Hz | 2.230 Hz | 2.136 Hz | $\theta_z$ |
| 1.850 Hz | 2.288 Hz | 2.271 Hz | $\theta_x$ |
| 2.040 Hz | 2.323 Hz | 3.972 Hz | $\theta_x$ |

## 3. The payload characterization at room and low temperature

We measured the mechanical response of the system at room and cryogenic temperature using two optical fiber bundle sensors set in front of the reaction mass and of the marionette. Moreover, we had the possibility to use as a velocity sensor also the e.m. actuators of the mirror. They are made by a 1 $T$ magnet set on the mirror border and a 250 turns coil mounted in front of it on the reaction mass. In this case, we amplified the output signal of the coil by means of the low noise amplifier PAR 51133 set in cascade to a select amplifier PAR 189. We measured the low frequency Fast Fourier spectra (FFT) of the ber-bundle sensor output set in front the reaction mass with the payload at room temperature and, then, once the stationary condition was achieved, at cryogenic temperature. In the low temperature spectra we noticed stray peaks at 1.4 $Hz$, 2.8 $Hz$ and 2 $Hz$, which are due to the standing wave of the two PT cryo cooler vibration. The other frequencies of the peaks measured during this first cooling have been compared with the output of the finite element mechanical simulation of the complete payload based on the ANSYS software. The simulation permits us to identify for each frequency the main degree of freedom associated to the corresponding normal mode. In the table we report the frequencies measured at room and low temperature and its mode identification. The d.o.f. in the mode identication column are referred to the reference framework drawn in figure 1 These results contribute to highlight the design strategy of the cryo suspension and they provide the crucial framework to start the study of the related control issues. Thanks to the good agreement between the simulation and the experimental results, we are confident to be able to optimize the payload parameters for maximizing the thermal contacts with the mirror and minimizing the thermal noise budget.

## 4. Conclusions

We have designed the first prototype of the last stage suspension system for mirrors of the $3^{th}$ generation of gravitational wave antennas with the aim to study the cooling and control strategy for the optical elements of interferometer. The system was cooled for the first time at low temperature. The preliminary results have been compared with the finite element simulation. Useful information concerning the change with temperature of the system dynamic behavior have been derived. The results will impact on mirror suspension control and, also, will reasonably require specific attenuation techniques of pulse-tube vibration.

## Acknowledgments

The research was partially supported by the European Commission under the Framework Program 7 (FP7, Grant Agreement 211743) and by the Italian minister of education under the financial program Prin 2007 (Grant prot. 2007NXMBHP_002).

## References

[1] `http://www.icrr.u-tokyo.ac.jp/gr/home/gre.html`

# A thermal noise model for a branched system of harmonic oscillators.

**Paola Puppo**

INFN Roma, P.le A. Moro 2, 00185, Roma, Italy.

E-mail: paola.puppo@roma1.infn.it

**Abstract.** We have calculated the thermal noise of a branched system of oscillators through the normal mode representation. This model describes well the mechanical behavior of the last stage suspension system like in the Virgo interferometer and is consistent with the predictions coming from the fluctuation-dissipation theorem. Moreover, the developed formalism can be useful to infer informations on the mechanical quantities of the uncoupled elements of the suspension and on the suspension thermal noise predictions for a third generation gravitational interferometer like the Einstein Telescope (ET).

## 1. Introduction

It is well understood that the pendulum thermal noise limits the sensitivity curve of a ground based interferometeric g.w. antenna in the low frequency range and the prediction of its noise spectrum is important for a proper design of the detector. For this reason the use of low dissipative materials is a consolidated choice to improve the sensitivity of the enhanced and advanced interferometers [3, 4, 5]. Another challenging choice is to go at cryogenic temperatures pushing down to 1 Hz the sensitivity bandwidth of the third generation interferometers like ET [2]. The recent new calculation of the mirror thermal noise [1] for the Virgo-like detectors has shown that as soon as we are dealing with low losses mirror pendulum, the upper suspension losses starts to contribute via its recoil and must be included in the computation. This paper is meant to give an overview of the thermal noise for a Virgo-like suspension focusing on the normal-mode expansion treatment. A similar treatise was already performed for double systems by P. Rapagnani [8] for the study of the electromechanical transducers in the resonant g.w. antennas and by E. Majorana and Y. Ogawa [7] for a double suspension. The results are equivalent to the calculation performed using the Fluctuation-Dissipation theorem (FDT), however the developed formalism can be useful for the noise estimation of a cryogenic suspension at a thermally steady state but with the stages at different temperatures [10].

## 2. The branched model

A Virgo-like last stage suspension is a cascade of three pendula [6]. To the first pendulum (the marionette) the mirror and the recoil mass are hung as branches. The system is equivalent to a combination of three harmonic oscillators (see figure 1). This model can be used to calculate the thermal noise of the pendulum, and also of the vertical degrees of freedom by properly defining

© 2010 IOP Publishing Ltd

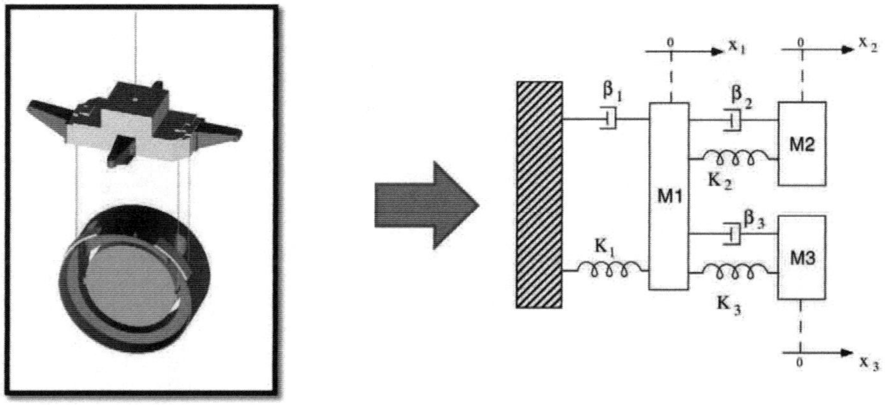


**Figure 1.** A Virgo-like last stage suspension is a cascade of three pendula. To the first pendulum (the marionette, $M_1$) the mirror $M_2$ and the recoil mass $M_3$ are hung as branches.

the uncoupled frequencies as follows:

$$
\begin{array}{ll}
\text{Pendulum modes} & \\
\omega_1^2 = \left(\omega_{g1}^2 + \omega_{w1}^2\right)/\mu_t;\ \mu_t = M_1/M_T & \text{Vertical modes} \\
\omega_i^2 = \omega_{gi}^2 + \omega_{wi}^2; \quad i = 2,3 & \omega_{v1}^2 = (2\pi 0.4)^2/\mu_t \\
\omega_{gi} = \left(\dfrac{g}{L_i}\right)^{1/2} \quad \omega_{wi} = \left(\dfrac{k_{el}^i}{M_i}\right)^{1/2} & \omega_{vi}^2 = \dfrac{\left(\frac{\pi r_{wi}^2 Y_i}{L_i}\right)}{M_i/4}; \quad i = 2,3 \\
k_{el}^i = \dfrac{n_i f_i \left(\Lambda_i Y_i I_2^i\right)^{1/2}}{2L_i^2}; \ I_2^i = \dfrac{\pi}{4} r_{wi}^2 &
\end{array}
\tag{1}
$$

where $Y_i, r_{wi}, L_i$ are the Young modulus, the radius and the length of the suspension wires, $\Lambda_i$ is the wires tension, $M_T = M_1 + M_2 + M_3$ and $f_i, n_i$ the number of flexural points and the number of suspension wires for each mass.

The mechanical losses are defined by the mechanical quality factors $Q_i$ and $Q_{iv}$ of the oscillators describing the viscous dissipation mechanisms and by the structural loss angles which are the immaginary parts of the oscillators elastic constants $k_j = k_j'(1 + i\phi_j)$. Thus the overall loss angles are defined as follows:

$$
\begin{array}{ll}
\text{Pendulum modes} & \\
\Phi_1(\omega) = \left(\phi_1(\omega) + \dfrac{\omega}{\omega_1} Q_1^{-1}\right)/(\mu_t)^{1/2} & \text{Vertical modes} \\
& \Phi_{iv}(\omega) = \left(\phi_{iv}(\omega) + \dfrac{\omega}{\omega_{iv}} Q_{vi}^{-1}\right); \ i = 1,2,3 \\
\Phi_i(\omega) = \left(\phi_i(\omega) + \dfrac{\omega}{\omega_i} Q_i^{-1}\right); \ i = 2,3 &
\end{array}
\tag{2}
$$

The equations of motion can be found starting from the Lagrangian and the dissipation functions:

$$
\begin{aligned}
\mathcal{L} &= \frac{1}{2}M_1\dot{x}_1^2 + \frac{1}{2}M_2\dot{x}_2^2 + \frac{1}{2}M_3\dot{x}_3^2 - \frac{1}{2}M_1\omega_1^2 x_1^2 - \frac{1}{2}M_2\omega_2^2(x_1 - x_2)^2 - \frac{1}{2}M_3\omega_3^2(x_1 - x_3)^2 \\
E_d &= \frac{1}{2}M_1\frac{\omega_1}{Q_1}\dot{x}_1^2 + \frac{1}{2}M_2\frac{\omega_2}{Q_2}(\dot{x}_1 - \dot{x}_2)^2 + \frac{1}{2}M_3\frac{\omega_3}{Q_3}(\dot{x}_1 - \dot{x}_3)^2
\end{aligned}
\tag{3}
$$

and, from the Lagrange equations $\frac{d}{dt}\frac{\partial \mathcal{L}}{\partial \dot{x}_i} = \frac{\partial \mathcal{L}}{\partial x_i} - \frac{\partial E_d}{\partial \dot{x}_i}$ in the frequency space, they can be written in matricial form as:

$$\hat{Z}\begin{pmatrix} \dot{X}_1 \\ \dot{X}_2 \\ \dot{X}_3 \end{pmatrix} = \begin{pmatrix} \mathrm{F_{thI}} \\ \mathrm{F_{thII}} \\ \mathrm{F_{thIII}} \end{pmatrix} \tag{4}$$

where $\hat{Z}$ is the mechanical impedance of the system. The stochastic generalized forces in the right side can be consistently related to the uncoupled stochastic terms as:

$$\begin{aligned} f_{th_I} &= f_{th_1} - f_{th_2} - f_{th_3};\, f_{th_{II}} = f_{th_2};\, f_{th_{III}} = f_{th_3} \\ \langle {f_{thi}}^2 \rangle &= \frac{k_b T_i M_i}{\tau_{0i}};\ i = 1,2,3 \end{aligned} \tag{5}$$

where $T_i$ are the temperatures and $\tau_{0i}$ the decay times of each oscillator. Using the impedance matrix, the FDT leads to the thermal noise of the mirror in the case of homogeneous temperature ($T_1 = T_2 = T_3 = T$) [1]:

$$X_{th_2}(\omega)^2 = \frac{4k_b T}{\omega^2}\mathrm{Re}\{(\hat{Z}(\omega)^{-1})_{22}\} \tag{6}$$

this result can be used when the whole mechanical system is at temperature $T$, but it does not deal with the case of different temperature oscillators.

## 3. Normal mode treatment

In the following we will illustrate the main steps of the normal mode treatment for a triple oscillator like that one sketched in figure 1. The Lagrangian (3) can be expressed as a sum of independent quadratic terms obtained by diagonalizing the characteristic matrix of the motion equations. The eigenvectors of such a matrix are coordinates of modes $(y_+, y_o, y_-)$ obtaining:

$$\mathcal{L} = \frac{1}{2}m_-\dot{y}_-^2 + \frac{1}{2}m_0\dot{y}_0^2 + \frac{1}{2}m_+\dot{y}_+^2 - \frac{1}{2}m_-\omega_-^2 y_-^2 - \frac{1}{2}m_0\omega_0^2 y_0^2 - \frac{1}{2}m_+\omega_+^2 y_+^2 \tag{7}$$

while the dissipation function is:

$$E_{dn} = \frac{1}{2}m_+\frac{\omega_+}{Q_+}\dot{y}_+^2 + \frac{1}{2}m_0\frac{\omega_0}{Q_0}\dot{y}_0^2 + \frac{1}{2}m_-\frac{\omega_-}{Q_-}\dot{y}_-^2 + Cross(m_i, \omega_i, Q_i, \dot{y}_+\dot{y}_-, \dot{y}_+\dot{y}_0, \dot{y}_0\dot{y}_-) \tag{8}$$

The coupled quality factors can be found by imposing the equivalence between the dissipation function expressed in terms of the normal modes and the coupled system one, their full expression is:

$$\begin{aligned} Q_- &= \omega_- m_- \left[\frac{\omega_1 M_1}{Q_1} + \frac{\omega_2 M_2}{Q_2}\left(\frac{\omega_-^2}{\omega_-^2 - \omega_2^2}\right)^2 + \frac{\omega_3 M_3}{Q_3}\left(\frac{\omega_-^2}{\omega_-^2 - \omega_3^2}\right)^2\right]^{-1} \\ Q_+ &= \omega_+ m_+ \left[\frac{\omega_1 M_1}{Q_1} + \frac{\omega_2 M_2}{Q_2}\left(\frac{\omega_+^2}{\omega_+^2 - \omega_2^2}\right)^2 + \frac{\omega_3 M_3}{Q_3}\left(\frac{\omega+}{\omega_+^2 - \omega_3^2}\right)^2\right]^{-1} \\ Q_0 &= \omega_0 m_0 \left[\frac{\omega_1 M_1}{Q_1}\left(1 - \frac{\omega_0^2}{\omega_3^2}\right)^2 + \frac{\omega_3 M_3}{Q_3}\frac{\omega_0^4}{\omega_3^4} - \frac{\omega_2 M_2}{Q_2}\left(\frac{\omega_1^2\omega_3^2 + \omega_0^4 - \omega_0^2\left(\omega_1^2 + (1+\mu_{31})\,\omega_3^2\right)}{\mu_{21}\omega_2^2\omega_3^2}\right)\right]^{-1} \end{aligned} \tag{9}$$

the found formulas for the quality factors fully agrees with the calculations made in the paper [9]. We notice that in general, there is a cross-term $Cross(m_i, \omega_i, Q_i, \dot{y}_+\dot{y}_-, \dot{y}_+\dot{y}_0, \dot{y}_0\dot{y}_-)$ in the dissipative function which can be neglected for systems with high Qs but is not null. For this reason $(y_+, y_o, y_-)$ are called quasi-normal modes. The relationships between the physical and the normal coordinates are related to the uncoupled frequencies and mass by means of the equations:

$$\begin{pmatrix} Y_- \\ Y_0 \\ Y_+ \end{pmatrix} = \hat{\Lambda}^{-1} \begin{pmatrix} X_1 \\ X_2 \\ X_3 \end{pmatrix} \tag{10}$$

where the diagonalization matrix is:

$$\hat{\Lambda} = \begin{pmatrix} \lambda_{11} & \lambda_{12} & \lambda_{13} \\ \lambda_{21} & \lambda_{22} & \lambda_{23} \\ \lambda_{31} & \lambda_{32} & \lambda_{33} \end{pmatrix} = \begin{pmatrix} 1 & \frac{\omega_3^2-\omega_0^2}{\omega_3^2} & 1 \\ \frac{\omega_2^2}{\omega_2^2-\omega_-^2} & -\frac{\mu_{31}\omega_3^2}{\mu_{21}\omega_2^2} + \frac{\omega_1^2+\mu_{21}\omega_2^2+\mu_{31}\omega_3^2-\omega_0^2}{\mu_{21}\omega_2^2}\left(1-\frac{\omega_0^2}{\omega_3^2}\right) & \frac{\omega_2^2}{\omega_2^2-\omega_+^2} \\ \frac{\omega_3^2}{\omega_3^2-\omega_-^2} & 1 & \frac{\omega_3^2}{\omega_3^2-\omega_+^2} \end{pmatrix} \tag{11}$$

and $\omega_-, \omega_0, \omega_+$ are the modal frequencies of the system. Within this formalism we can express the mirror coordinate $X_2$ in terms on the normal eigenvectors and find its thermal noise by computing the power spectrum. Thus remembering that the stochastic thermal forces expressed in the equation (5) are uncorrelated we can write:

$$\begin{gathered} \left\langle X_{th2}(\omega)^2 \right\rangle = \left\langle |\lambda_{21} Y_-(\omega) + \lambda_{22} Y_0(\omega) + \lambda_{23} Y_+(\omega)|^2 \right\rangle \\ \Downarrow \\ \left\langle X_{th2}(\omega)^2 \right\rangle = \left\langle {f_{th1}}^2 \right\rangle |T_{n1}(\omega)|^2 + \left\langle {f_{th2}}^2 \right\rangle |T_{n2}(\omega)|^2 + \left\langle {f_{th3}}^2 \right\rangle |T_{n3}(\omega)|^2 \end{gathered} \tag{12}$$

where $T_{ni}(\omega), i = 1, 2, 3$ are generalized transfer functions depending on the uncoupled mechanical parameters of the pendulum. Their complete expression can be found on the reference [1]. The relation (12) differs considerably by the naïve treatment of the thermal noise in which the calculation is performed by a simple quadratic sum of the modal frequencies thermal noise:

$$\left\langle X_{th2}^{naive}(\omega)^2 \right\rangle = {\lambda_{21}}^2 |T_-(\omega)|^2 \left\langle {F_{th-}}^2 \right\rangle + {\lambda_{22}}^2 |T_0(\omega)|^2 \left\langle {F_{th0}}^2 \right\rangle + {\lambda_{23}}^2 |T_+(\omega)|^2 \left\langle {F_{th+}}^2 \right\rangle \tag{13}$$

The main difference is the presence of a cross-term due to the correlation between the normal forces that in the equation (13) are erroneously neglected. In figure 2 the general difference between naïve and the modal curves is shown. We notice that the off-resonance zone at high frequencies is quite different, and in some cases, the cross-term give rise to a helpful cancellation, which reduces the thermal noise in the zone of the sensitivity. The result with the modal treatment is fully equivalent to the FDT calculation shown in the equation (6) at homogenous temperature and can be useful for the thermal noise computation of systems at a thermal steady state but with different temperatures of the stages.

## 4. Numerical estimations

As an example we have computed the thermal noise of a Virgo-like last stage suspension with a very low Q marionette stage. The result is sketched in figure 3 where it is compared with the simple mirror pendulum. In presence of low dissipative mirror suspensions, the contributions of the other last stage suspension elements cannot be neglected. In particular the marionetta mechanical losses give a non negligible effect via its recoil, in the off-resonance high-freq. range. The modal formalism agrees with the FDT approach [1].

The developed formalism can be useful to infer informations on the mechanical quality factor of the uncoupled pendula using the formulas (9) and consequently helpful for their optimization.

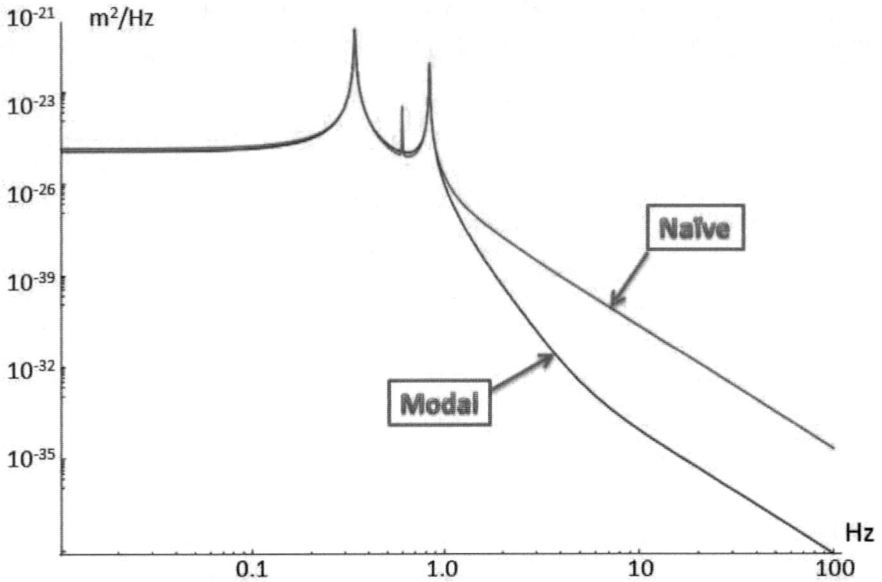


**Figure 2.** Thermal noise prediction of for the Virgo Branched pendulum calculated with the Normal Mode method, compared with the naïve calculation.

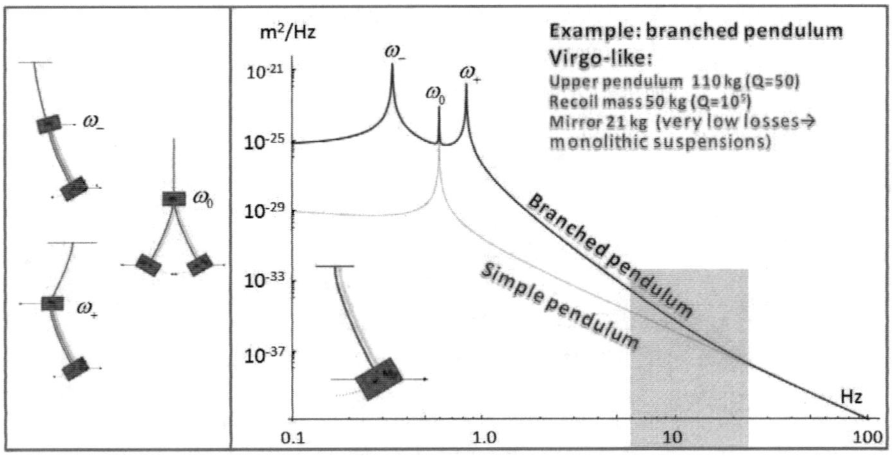


**Figure 3.** The thermal noise spectrum of a Virgo-like last stage suspension with a very low Q marionette stage compared with the single mirror thermal noise. The light blue rectangle indicates the frequency zone at which the marionetta losses change the curve with respect to the single pendulum one. This difference can affect the sensitivity at low frequency and consequently lower the expected average sight for the BBH signals.

As an example we have calculated the quality factors of the uncoupled pendula using the measurements on the monolithic payload prototype installed in air, the results are shown in the Table 1. The fitted values can be used to predict the quality factors of the suspension coupled to the Superattenuator chain in Virgo and to have the right prediction of the overall sensitivity curve of the detector[11].

**Table 1.** From the measured quality factors of the mirror last stage suspension the uncoupled losses can be fitted using the formula (9)

| Coupled Measured Q's | | Modal frequencies | | Fitted uncoupled Q's | | Coupled Q's from the fit | |
|---|---|---|---|---|---|---|---|
| $Q_-$ | 600 | $\nu_-$ | 0.421 $Hz$ | $Q_{mario}$ | 1000 | $Q_-$ | 500 |
| $Q_0$ | 160 | $\nu_0$ | 0.598 $Hz$ | $Q_{mirror}$ | 200 | $Q_0$ | 230 |
| $Q_+$ | 220 | $\nu_+$ | 0.842 $Hz$ | $Q_{rm}$ | 300 | $Q_+$ | 260 |
| $Q_{-vert}$ | 950 | $\nu_{-vert}$ | 5.316 $Hz$ | $Q_{mariov}$ | 900 | $Q_{-vert}$ | 960 |
| $Q_{0vert}$ | 900 | $\nu_{0vert}$ | 8.746 $Hz$ | $Q_{mirrorv}$ | 1300 | $Q_{0vert}$ | 900 |
| $Q_{+vert}$ | 950 | $\nu_{+vert}$ | 19.95 $Hz$ | $Q_{rmv}$ | 1500 | $Q_{+vert}$ | 970 |

## 5. Conclusions

We have calculated the thermal noise of a branched system of oscillators through the normal mode representation. This model describes well the mechanical behavior of the last stage suspension system like in the Virgo interferometer and is consistent with the predictions coming by the Fluctuation-Dissipation Theorem. In the evaluation of a multimodal system thermal noise it is important to take into account the cross-correlation term among the various modes which is neglected in a naïve calculation. In presence of low dissipative mirror suspensions, the contributions of the other suspension elements to thermal noise of the mirror cannot be neglected and a new thermal noise estimation must be done by including the viscous and structural dissipations of the marionette and recoil mass pendulum. The developed formalism can be useful to estimate the thermal noise of a system in a thermally steady state and with the stages at different temperatures. This study can be important for an evaluation of the thermal noise of the suspensions in a 3rd generation g.w. interferometers [12].

## Acknowledgments

The research was partially supported by the European Commission under the Framework Programme 7 (FP7, Grant Agreement 211743) and by the Italian minister of education under the financial program Prin 2007 (Grant prot. 2007NXMBHP_002).

## References

[1] Piergiovanni F, Punturo M and Puppo P 2009 *Virgo Document* available at `https://tds.ego-gw.it/ql/?c=2199`

[2] Punturo M et al 2009 Third generation of gravitational gave gbservatories and their science reach *Classical and Quantum Gravity: Proceedings of 8th Edoardo Amaldi Conference*

[3] Flaminio R, Freise A, Gennai A, Hello P, La Penna P, Losurdo G, Lueck H, Man N, Masserot A, Mours B, Punturo M, Spallici A and Viceré 2005 *Advanced Virgo white paper*, Virgo Note VIR-NOT-DIR-1390-304.

[4] `http://www.ligo.caltech.edu`

[5] `http://geo600.aei.mpg.de`

[6] Bernardini A, Majorana E ,Puppo P ,Rapagnani P , Ricci F, Testi G 1999 Suspension last stages for the mirrors of the Virgo interferometric gravitational wave antenna *Rev. Sci. Instr.* **70** 8 3463.

[7] Majorana E and Ogawa Y 1997 Mechanical thermal noise in coupled oscillators *Physics Letters A* **233** 162-168.

[8] Rapagnani P 1982 Development and Test at T=4.2 K of a Capacitive Resonant Transducer for Cryogenic Gravitational Wave Antennas *Il Nuovo Cimento* July-August

[9] Majorana E and Tricarico P 1993 Decay Times of an N-Normal-Mode System *Il Nuovo Cimento* September

[10] Basti F, Frasconi F, Granata M, Majorana E , Moscatelli V, Naticchioni L, Perciballi M, Puppo P ,Rapagnani P, Ricci F, *J. Phys.: Conf. Series* **same issue as this**.

[11] Puppo P 2009 *Virgo+MS sensitivity curve Virgo Document* available at `https://tds.ego-gw.it/ql/?c=6957`
[12] Puppo P 2009 Suspension thermal noise for ET ET-023-09 (2009)

# Potential mechanical loss mechanisms in bulk materials for future gravitational wave detectors

**D Heinert[1], A Grib[1,2], K Haughian[3], J Hough[3], S Kroker[4], P Murray[3], R Nawrodt[1,3], S Rowan[3], C Schwarz[1], P Seidel[1], A Tünnermann[4]**

[1] Institut für Festkörperphysik, Friedrich-Schiller-Universität, 07743 Jena, Germany
[2] Physics Department, Kharkiv V. N. Karazin National University, 61077 Kharkiv, Ukraine
[3] Institute for Gravitational Research, Department of Physics and Astronomy, Glasgow University, G12 8QQ Glasgow, UK
[4] Institut für Angewandte Physik, Albert-Einstein-Str. 15, 07745 Jena, Germany

E-mail: daniel.heinert@uni-jena.de

**Abstract.**
Low mechanical loss materials are needed to further decrease thermal noise in upcoming gravitational wave detectors. We present an analysis of the contribution of Akhieser and thermoelastic damping on the experimental results of resonant mechanical loss measurements. The combination of both processes allows the fit of the experimental data of quartz in the low temperature region (10 K to 25 K). A fully anisotropic numerical calculation over a wide temperature range (10 K to 300 K) reveals, that thermoelastic damping is not a dominant noise source in bulk silicon samples. The anisotropic numerical calculation is sucessfully applied to the estimate of thermoelastic noise of an advanced LIGO sized silicon test mass.

## 1. Introduction

The detection of gravitational waves demands a very high precision length measurement. Even strong gravitational wave sources will just cause an effect in the order of $10^{-19}$ m on earth. In modern interferometric gravitational wave detectors thermal noise significantly limits the sensitivity. Especially in the most sensitive frequency range from 10 Hz - 200 Hz thermally driven random vibrations of the surface of the optical substrates become important. Besides the classical Brownian noise [1, 2] other noise mechanisms have been identified to contribute to the total thermal noise: thermoelastic and photothermal noise [3] as well as thermo-refractive noise [4, 5]. All noise processes can equivalently be described by a corresponding process consuming energy. The link between noise and loss is given by the fluctuation-dissipation-theorem [6]. Consequently knowing the dissipative properties, i.e. the temperature and frequency behaviour of all loss processes, of a system allows the computation of its thermal noise.

In the case of a gravitational wave detector the displacement noise of the surface is connected with the dissipation of mechanical energy within the material. Therefore, the mechanical loss of different substrate materials has been studied in the past, e.g. [7, 8, 9, 10, 11], to understand the origin of the mechanical dissipation. All these studies rely on ring down measurements of cylindrical bulk samples at their resonances. The estimated characteristic ring down time is finally used as a measure for the mechanical loss.

© 2010 IOP Publishing Ltd

The results of these measurements consist of a superposition of many different internal loss processes. In order to understand or even minimise these different loss contributions it is necessary to separate them within the measured data sets. Therefore a subsequent analysis aims at the identification of the different mechanical loss processes within the material. Fundamental intrinsic loss processes in bulk materials are thermoelastic damping and Akhieser damping, which is the result of phonon-phonon-interaction. We will focus on a detailed description of these fundamental loss processes within this paper. Due to the small height of the cylindrical samples an isotropic plate theory [12] is applied for evaluating the effect of thermoelastic damping. Furthermore, a finite element analysis is developed to extend the theory of thermoelastic loss to anisotropic samples. This treatment allows an easy application to arbitrarily shaped test masses as well and helps identifying the loss sources in the ring-down experiments. For silicon we examplarily compare our approach to the isotropic plate model.

The second part of the paper applies the developed code to the calculation of thermoelastic noise of a gravitational wave detector bulk material having anisotropic material properties.

## 2. Mechanical loss processes

### *2.1. Thermoelastic damping*

A test sample under vibration exhibits regions being compressed and regions being expanded during a vibrational cycle. Accordingly, the described volume change is linked to a temperature change by means of the coefficient of thermal expansion. The arising temperature gradient within the sample leads to a heat flow and, consequently, increases the entropy of the system [3] which leads to dissipation. This dissipation process causes the vibrational amplitude to decrease with time and is known as thermoelastic damping (TED). A non-vanishing coefficient of thermal expansion only occurs in a material if the interatomic potential has anharmonic contributions and thus TED is classified as a nonlinear effect of the lattice.

In order to calculate the TED numerically the mode shape of the resonant vibration of the substrate is required. With the knowledge of the mode shape one can estimate the volume change within the sample under deformation. This volume change is directly linked to the stress tensor $\sigma$. Quantitatively the temperature distribution $T$ within the sample is obtained by using the equation of heat transport (see e.g. [13]):

$$C_p\dot{T} - \lambda_{ij}\frac{\partial T}{\partial x_i \partial x_j} = -\alpha_{ij}\dot{\sigma}_{ij}T_0 \ . \tag{1}$$

In this equation $\lambda$ and $\alpha$ represent the tensors of thermal conductivity and thermal expansion respectively. Furthermore $C_p$ is the specific heat per unit volume and $T_0$ the average temperature of the sample. In Eq. (1) the Einstein convention is used, i.e. to sum over double indices.

Summing the work done by the new thermal field against the original stress field finally gives the dissipated energy $\Delta E$ per oscillation cycle [14, 15]. By using the complex notation for harmonic processes one obtains:

$$\Delta E = \pi \int_V \alpha_{ij}\hat{\sigma}_{ij}\Im\left(\hat{T}\right) dV \ , \tag{2}$$

where $\Im(\hat{T})$ is the imaginary part of the temperature amplitude $\hat{T}$ ($T = \hat{T}e^{i\omega t}$) which is taken as the solution from Eq. (1). This part of the temperature change is phase shifted compared to the mechanical stress. The mechanical loss $\phi$ is then given by:

$$\phi = \frac{1}{2\pi}\frac{\Delta E}{E} \tag{3}$$

with the total vibrational energy $E$.

The presented scheme allows the full anisotropic treatment of thermoelastic damping in solids.

### *2.2. Akhieser damping*

The second important contribution to the mechanical damping in bulk samples results from the interaction of acoustic waves with thermal phonons. Thermal phonons carry the thermal energy of the sample and obey a Planck distribution in thermal equilibrium. In 1939 Akhieser [16] developed a theory for this effect, where he treated the acoustic wave as a homogeneous stress background for the thermal phonons. In the field of resonant loss measurements the homogeneous stress background is provided by the excited eigenmode of the sample. This approximation requires that the lifetime of thermal phonons $\tau$ is short compared to the period $1/f$ of the acoustic vibration. For all materials and temperatures considered in this paper this condition ($f\tau \ll 1$) is well satisfied. At a frequency of 25 kHz and at a temperature of 10 K $f\tau$ reveals to be $6 \times 10^{-3}$ for silicon and $2.2 \times 10^{-3}$ for quartz.

Mechanical stress leads to a change of the phonon energies in the crystal via the Grüneisenparameter $\gamma$. In general each phonon branch shows a distinct parameter and therefore a distinct energy shift. Consequently, the whole phonon ensemble is brought out of thermal equilibrium by an external stress. The thermodynamical equilibrium state is characterized by the maximum of entropy. Any process reestablishing equilibrium, here the collision between phonons, will, therefore, increase the entropy and, thus, dissipate energy. The speed of phonon redistribution is limited by the phonon lifetime $\tau$ as the typical time between interactions.

Again Akhieser damping is a nonlinear effect, because without having anharmonic contributions in the interatomic potential phonon-phonon-interaction would be forbidden.

Akhieser losses $\phi_{ph}$ for the bulk vibrating solid are given by the following expression [17]

$$\phi_{ph} = \frac{C_p T \gamma^2}{\rho v_s^2} \frac{\omega\tau}{1+\omega^2\tau^2} \ . \tag{4}$$

In this model the Akhieser loss $\phi_{ph}$ depends on the sound velocity $v_s$, the angular freqency of the acoustic wave $\omega$ and the density $\rho$ of the material.

## 3. Numerical estimation and application to experimental data

### *3.1. Crystalline quartz*

Akhieser and thermoelastic damping have been applied to the measured loss values of a crystalline quartz sample. The sample has a cylindrical shape (z-axis parallel to the cylinder axis) with a diameter of 76.2 mm and a height of 12 mm. The mechanical loss above 25 K is dominated by processes which can be described in the model of double well potentials [18]. Former investigations identified interstitial aluminum together with alkali atoms to cause mechanical loss in quartz [19] at temperatures above 25 K. For the defect model does not satisfactorily explain the low temperature data, one has to take further mechanical loss processes into consideration.

In this paper we concentrate on the temperature region below 25 K to exclude the influence of defects and to explain the behaviour with a combined model of TED and Akhieser damping. For a detailed description of this model see [20]. Due to the small height of the sample compared to its diameter we used the approach of Bishop and Kinra [12] to estimate TED in a plate. Mechanical losses due to Akhieser damping were calculated with the use of Eq. (4) into which we inserted the material properties of quartz from the literature [21].

The theoretical values using both TED and Akhieser damping are compared to the measured loss values in Fig. 1 for an eigenmode at f = 11.7 kHz. It is seen that we obtained a good quantitative agreement between experiment and theory for the quite reasonable value of the Grüneisen parameter $\gamma = 2$. This is an indication that the low temperature loss in quartz can exclusively be described by these two mechanisms.

At the experimental frequency of 11.7 kHz the contribution of Akhieser losses is negligibly small (see Fig. 1). However, when the eigenfrequencies exceed a few MHz, the contribution of Akhieser losses will become dominating.

Journal of Physics: Conference Series **228** (2010) 012032 doi:10.1088/1742-6596/228/1/012032

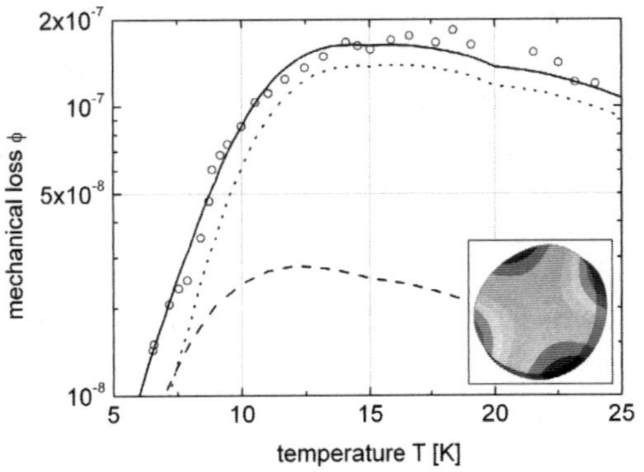


**Figure 1.** Measured mechanical loss (○) of a Ø 76.2 mm × 12 mm quartz sample in z-cut at f = 11.7 kHz. Calculated mechanical loss (——) as the sum of TED (······) and phonon damping (- - - -) as described in the text. In the lower right corner the deformation of the substrate in the cylinder axis is shown.

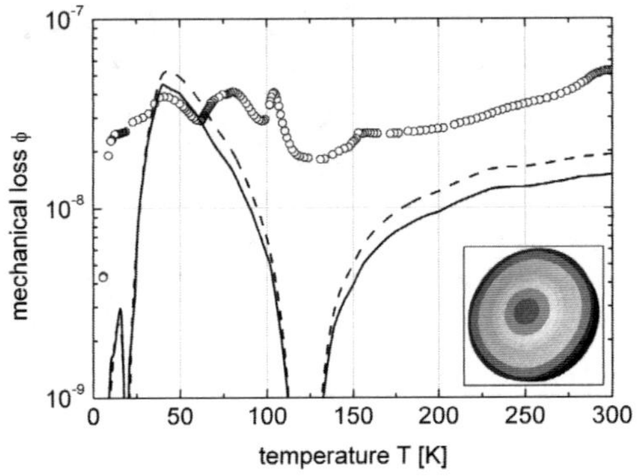


**Figure 2.** Measured mechanical loss (○) of a Ø 76.2 mm × 12 mm silicon sample in (100) direction at f = 24.1 kHz. The data are compared to TED obtained by a isotropic model (——) and our anisotropic approach (- - - -). In the lower right corner the deformation of the substrate in the cylinder axis is shown.

*3.2. Crystalline silicon*

We also investigated silicon, the most promising candidate for future cryogenic gravitational wave detectors [22, 23, 10]. The cylindrical sample has the same geometry as the quartz sample (Ø 76.2 mm × 12 mm) and was oriented along the cylinder axis in (100) direction. Again TED in silicon was at first calculated with the isotropic plate theory described in [12]. We used the FE software COMSOL [24] for a completely numerical and anisotropic calculation of TED. Consequently, the results of both models are compared in Fig. 2. Thermophysical properties of silicon are taken from the references [21, 25].

For the given silicon resonance we encounter deviations between the anisotropic and isotropic model of up to 20%. However, TED alone is not capable of explaining the measured loss data over the whole temperature range. The results show, that TED significantly affects mechanical loss in our bulk-like sample only at a small temperature region around 40 K. There the calculated dissipation even exceeds the experimental values. The origin of this discrepancy could be found in the deviation of the real sample parameters from the table values used in the calculation.

The Akhieser damping strongly depends on the thermal conductivity (see [17]) which is in turn strongly dependent on the impurity concentration. Unfortunately, for the sample, used in this paper, this parameter is not known to a degree which would justify any estimate on Akhieser damping. Further investigation of the effect of phonon interaction on damping in silicon is on the way.

## 4. Thermal noise due to TED in a gravitational wave detector

The effect of TED was first outlined in 1937 by Zener for vibrating reeds [26]. In 1999 the importance of TED as a serious noise contribution in gravitational wave detectors was pointed out in [3]. This behaviour is especially relevant for crystalline test masses as sapphire and silicon which are proposed to be used in future detectors [27, 28]. A refined model was developed in order to cover finite test mass sizes and non-adiabatic conditions [29]. These conditions become

Journal of Physics: Conference Series **228** (2010) 012032 doi:10.1088/1742-6596/228/1/012032

important when small beams (e.g. in laser stabilization cavities), high thermal conductivity materials or low frequencies are used.

Using Levin's theorem [30] the resulting thermoelastic noise of the mirror is obtained by the calculation of the losses due to a virtual mechanical pressure applied to the front surface of the sample. The spatial distribution of pressure along the surface $p(\vec{r})$ of the sample has to mimic the intensity distribution of the laser beam:

$$p(\vec{r}) = \frac{F_0}{r_0^2} \exp\left(-\frac{r^2}{r_0^2}\right) , \tag{5}$$

where $F_0$ is an arbitrarily chosen force amplitude of the excitation. The beam radius $r_0$ indicates the point of the 1/e decay of the laser intensity.

The spectral noise power density $S_x^2(f)$ due to surface fluctuations then results from [30]:

$$S_x^2(f) = \frac{2k_B T}{\pi^2 f^2} \frac{W_{diss}}{F_0^2} . \tag{6}$$

In Eq. (6) $W_{diss}$ describes the energy dissipated per cycle due to all relevant loss mechanisms. In order to obtain thermoelastic noise the dissipation due to TED needs to be considered. Applying the pressure $p(\vec{r})$ in the quasistatic approximation allows the calculation of $W_{diss}$ with the FE software COMSOL. Fig. 3 compares the full anisotropic FEA with the results obtained by using the analytical calculation in [29, Eq. (20)].

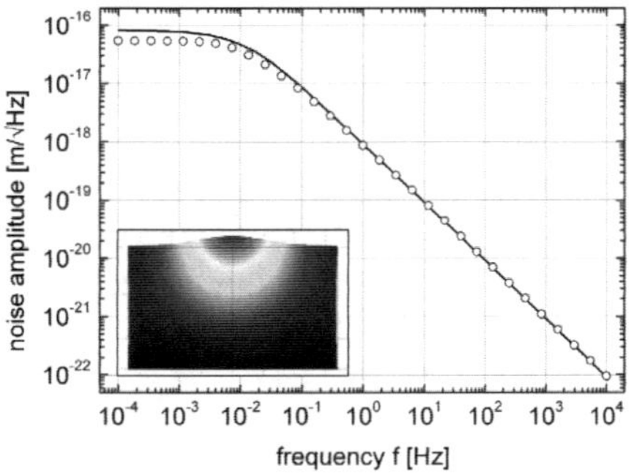


**Figure 3.** Numerical estimate (○) of thermoelastic noise in an isotropically modelled silicon substrate with advLIGO geometry (Ø 340 mm × 220 mm) in comparison to the analytical treatment given in [29] for the same substrate (——). For both calculations a beam waist radius of $w_0 = \sqrt{2} r_0 = 60$ mm and a temperature T = 300 K were used. The full anisotropic treatment leads to the same result as the isotropic one. Thus, the plot is omitted. The inset shows the deformation due to the Gaussian pressure profile.

At first we compared the thermoelastic noise level given by the analytical and our numerical theory for an isotropic body of finite size (advLIGO geometry). Both results exhibit a good agreement although the analytical calculation is strictly valid for infinite test masses. This effect of finite sized geometry explains the encountered difference below 5 % over the frequency range of 1 Hz to 10 kHz between the models. An analytic expression for the effect of finite sizes [31] on isotropic test masses also explains changes in the order of 5 %. The only major discrepancy occurs at low frequencies where the difference in the boundary conditions become important.

Our analysis showed that introducing anisotropy into the problem only leads to a minor change in the noise level of less than one per cent. Thus, in the quasistatic model far away from any resonances, for silicon the effect of anisotropy on thermoelastic noise is negligible. Close to the resonances this assumption breaks down and a full anisotropic treatment is needed. Although this is not necessary for the optical components (where the resonances are always well outside the detection band) this might become important for the suspension elements having resonances inside the operation frequency band.

## 5. Conclusion

Considering both Bömmels model for phonon damping and the model of Bishop and Kinra for thermoelastic damping in plates allows the interpretation of the low temperature behaviour of mechanical losses in quartz. We showed a quantitative agreement with our measurement for the temperature interval from around 10 K to 25 K. In contrast, the repeated analysis for a bulk silicon sample revealed that thermoelastic damping is not to be the dominating factor within the experimental temperature interval from 5 K to 300 K. TED alone does not explain the measured data. Therefore we expect other loss mechanisms to contribute essentially to the mechanical loss within the considered temperature region.

Finally the comparison of our noise calculation to an analytical approach proved the reliability of our numerical code. This code allows to include arbitrary sample geometries as well as the anisotropic character of the material. Its application on bulk silicon shows, that the influence of the anisotropy on thermoelastic noise can be neglected for this material for typical mirror geometries.

Nevertheless, in the field of mechanical loss measurements on samples of varying size the effect of anisotropy is important in order to seperate different loss mechanisms. Thus, the developed code is a valueable tool to include anisotropy into the mechanical loss data analysis as well as the thermal noise calculation for small structures.

## Acknowledgments

This work was in part supported by the German science foundation DFG under contract SFB TR7.

## References

[1] Einstein A 1906 *Ann. Phys.* **19** 371–81
[2] Gonzalez G I and Saulson P R 1994 *L. Acoust. Soc. Am.* **96** 207–12
[3] Braginsky V B, Gorodetsky M L and Vyatchanin S P 1999 *Phys. Lett. A* **264** 1–10
[4] Braginsky V B, Gorodetsky M L and Vyatchanin S P 2000 *Phys. Lett. A* **271** 303–7
[5] Evans M, Ballmer S, Fejer M, Fritschel P, Harry G and Ogin G 2008 *Phys. Rev. D* **78** 102003
[6] Callen H B and Welton T A 1951 *Phys. Rev.* **83 (1)** 34–40
[7] McGuigan D F et al 1978 *Journal of Low Temperature Physics* **30** 621–9
[8] Crooks D R M et al 2004 *Class. Quantum Grav.* **21** S1059–65
[9] Amico P et al 2002 *Class. Quantum Grav.* **19** 1663–8
[10] Nawrodt R et al 2008 *J. Phys.: Conf. Ser.* **122** 012008
[11] Penn S D et al 2006 *Phys. Lett. A* **352** 3–6
[12] Bishop J E and Kinra V K 1997 *Int. J. Solids Structures* **34** 1075–92
[13] Norris A N 2005 *Q. J. Mech. Appl. Math.* **58 (1)** 143–63
[14] Lifshitz R and Roukes M L 2000 *Phys. Rev. B* **61** 5600–9
[15] Fejer M M et al 2003 *Phys. Rev. D* **70** 082003
[16] Akhieser A 1939 *J. Phys.* **1 (4)** 277–87
[17] Braginsky V B, Mitrofanov V P and Panov V I 1985 *Systems with Small Dissipation* (Chicago Press)
[18] Zimmer A et al 2007 *Rev. Sci. Instrum.* **78** 063905
[19] Fraser D B 1964 *J. Appl. Phys.* **35 (10)** 2913–8
[20] Grib A et al 2009 *J. Appl. Phys.* **accepted**
[21] Touloukian Y S 1970 *Thermophysical properties of matter* (IFI/Plenum, New York-Washington)
[22] Winkler W, Rüdiger A, Schillig R, Strain K A and Danzmann K 1994 *Optics Communications* **112** 245–52
[23] Rowan S, Hough J and Crooks D R M 2005 *Phys. Lett. A* **347** 25–32
[24] URL `www.comsol.com`
[25] Hull R, ed 1999 *Properties of Crystalline Silicon* (INSPEC, the Institution of Electrical Engineers, London)
[26] Zener C 1937 *Phys. Rev.* **52** 230–5
[27] Kuroda K 2006 *Class. Quantum Grav.* **23** S215–21
[28] URL `www.et-gw.eu`
[29] Cerdonio M, Conti L, Heidmann A and Pinard M 2001 *Phys. Rev. D* **63** 082003
[30] Levin Y 1998 *Phys. Rev. D* **57 (2)** 659–63
[31] Liu Y T and Thorne K S 2000 *Phys. Rev. D* **62** 122002

# Anomalous low frequency dissipation processes in metal springs

**Riccardo DeSalvo[a], Arianna DiCintio[a,b], Fabio Marchesoni[c], Abhik Bhawal[d]**

a LIGO Laboratory, California Institute of Technology, Pasadena, CA, 91125 USA
b Dipartimento di Fisica, Università degli studi di Roma "Sapienza", 00185 Roma, Italy
c INFN-VIRGO Project Dipartimento di Fisica, Universita' di Perugia, 06100 Perugia, Italy
d Arcadia High School, Arcadia, CA, USA
E-mail: riccardo@caltech.edu

**Abstract**. The dissipation processes of leaf springs used in seismic isolation chains of Gravitational Wave detectors have been studied. A low frequency phase transition from viscous-like to fractal-like dissipation, controlled by Self Organized Criticality of dislocations, was observed. The new understandings suggest different best practices for the operations of the seismic isolation chains of the second generation of Gravitational Wave observatories and require new techniques and materials for the third generation.

## 1. Introduction.

Measurement on Maraging steel blades used in seismic attenuators of Gravitational Wave interferometers [1-5], coupled with an old prediction on self organization of dislocations, lead to novel understanding of low frequency losses in metals, with unpredictable, avalanche dominated, loss mechanisms and mechanical noise.

It is commonly accepted that dislocation motion inside metal grains (crystals) plays an important role in dissipation [6,7].

The dislocation model of dissipation (considering the motion of many independent dislocations) leads to a description of viscous-like dissipation, or a loss angle one, which can be easily mathematically treated (damped oscillator) and lead to predictable results [8, 9].

Neither is completely satisfactory, especially when dealing with equilibrium point hysteresis. As a result dissipation studies are full of exceptions from the rule, and ad hoc explanations [10], often with no basis on microscopic behavior understanding.

Mostly the models do not explain hysteresis [11] (which should not be confused with plasticity, in which the material deforms permanently) and low frequency instabilities (sudden collapses [12], random walk [13], excess losses, history dependent mechanical noise [14]).

To get some deeper understanding, one needs better understanding of the collective processes at the microscopic level.

Dislocations are ubiquitous and isotropic in polycrystalline materials, a large number of dislocations are required by the process of rapid freezing of a liquid into a solid, due to the material's thermal expansion coefficient, to reconnect separate parts of growth of a crystal during the rather chaotic freezing process. Crystalline defects (not all line dislocations) appearing as often as every 200 lattice spacings [8, 15], are required to generate the electron scattering necessary to explain the measured metal resistivity in pure metals. Many of these defects are dislocations of various kind.

© 2010 IOP Publishing Ltd

Line dislocations take different forms. One of the simpler and most common ones is the edge dislocation, that appears when there is a missing half plane of atoms in the crystal. For simplicity we limit the reasoning to edge dislocations, although screw and other kinds of dislocations can participate as well. While we picture line dislocation as straight, they have no reason to be straight, neither at birth, nor as a result of movements; line dislocations form arches and bends.

Dislocations are deviations from the crystal, which is the state of minimal energy. Therefore dislocations are energetically expensive (annealing dislocations in a material is an exothermal process).

Forcing two dislocations to overlap would cause additional energy expense. Therefore line dislocation generally display short range repulsion force (except for combinations that may lead to annihilation processes).

Atoms in a metal lattice are tied by covalent bonds, therefore missing atomic bonds along a dislocation can switch to the neighboring atoms at virtually no energetic cost. If there is an external stress applied to the material, say a compression gradient, there is an energetic advantage for the dislocation to move along that gradient. Although quantum and thermal fluctuations may cause bond switching in either direction, the net result is always in the direction of lower energy states and therefore of energy dissipation.

Additionally, dislocations, being the expression of local deformations of the crystal, carry both strain and stress, which means that the slow accumulation of dislocations in response to a stress gradient results in a material deformation and appearance of internal forces, either permanent (if grain cleavage, creep [16,17], acoustic emission [18], or other irreversible processes occur) or recoverable by simply reversing the force (hysteresis).

Although the individual bond switching in a plane happens almost instantaneously, a sequence of switches must repeat in a "zipper"-fashion along the dislocation line for actual dislocation movement, which takes considerable time. This makes that dislocations respond sluggishly to applied forces, thus leading to dissipation.

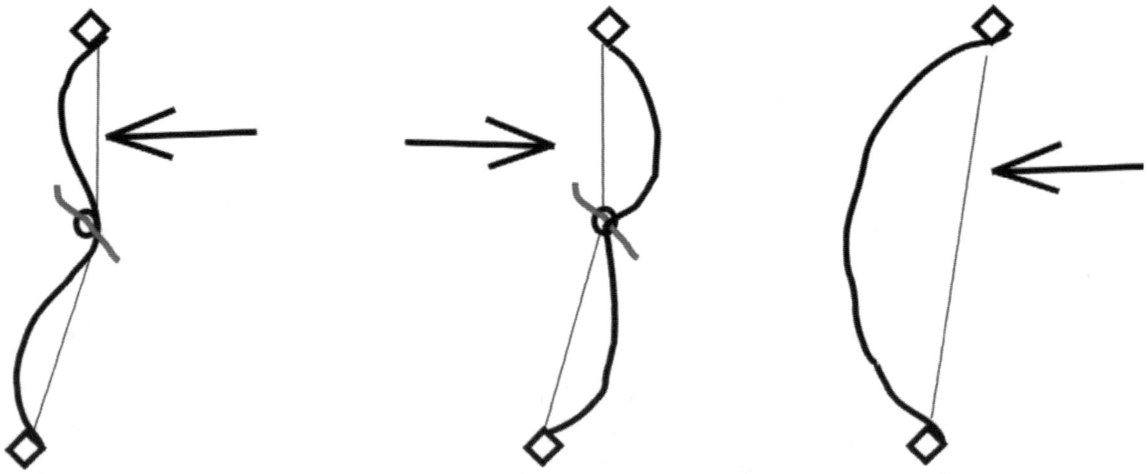

**Figure 1:** Schematics of a dislocation constrained between two anchoring points (diamonds), responding to alternating stress (the arrows). The dislocation arches until the energy gain from drifting along the stress equals to the energy necessary to stretch it longer. In the left and centre panel the dislocation is further constrained by entanglement (circle) with another dislocation. In the right panel the other dislocation is removed and the dislocation movement covers much more surface, thus dissipating more energy; it also takes longer time to move farther. Entangled dislocations lead to lower dissipation losses.

The zipping propagates along the dislocation length, as long as the pressure gradient offers energetic advantage. A zipping event moves a dislocation by an atomic spacing; it also allows the start of additional zipping events on the same dislocation. A sequence of many zipping processes is needed to allow a dislocation to glide over any appreciable distance. If a dislocation is anchored in one or more points it will deform into an arching profile until the energy cost of stretching the dislocation length exceeds the advantage of following the stress gradient (figure 1.). Dislocations in an anchoring-defect-poor grain respond to a varying stress gradient with longer drifts, resulting in longer arches, which absorb more energy from the external excitation.

If the stress field is cyclic (as in a macroscopic oscillation or a vibration) the dislocation's motion is also cyclic, but because dislocation movement happens with a delay with respect to the applied force, it will dissipate energy. The result is either a viscous or a loss angle dissipation regime if the oscillation period is shorter or longer than the average arch response time.

These well-known behaviors hypothesize independent movement of dislocations.

The system becomes much more interesting when dynamic and static interactions between dislocations are considered, resulting in non-linear, unpredictable, and sometimes even preposterous effects.

When a large number of dislocations respond to a stress gradient, they tend to pile up against fixed defects. Because they repel and cannot cross, they entangle to form a chaotic and meta-stable lattice permeating the ordered crystal lattice. Because dislocations carry stress and strain, the dislocation lattice contributes to the mechanical properties of the sample with its own stiffness and its own internal stresses and strains. When piled up and entangled, dislocations are blocked at many points along their length, and can respond to oscillating stresses only with short arches, which being short respond more rapidly, thus dissipating little energy.

If a sufficiently large number of dislocations is introduced (work hardening) the dislocation lattice can even become permanently entangled, and the material becomes harder and more elastic.

If there are only few dislocations and blocking defects (as in the annealed state of a pure metal grain), dislocations can glide under stress across the entire crystal volume, and pile up at the edge of it, until the accumulated stress exceeds the material ultimate strength and grain cleavage ensues, with acoustic emission [19], crystal cleavage, and creep; this also happens if the applied stress is sufficient to allow dislocations to jump blocking defects [16,17].

In the intermediate case, a sufficient number of dislocations moving under a pressure gradient can easily entangle. If one waits for a time sufficient to allow all dislocations to settle in a tangled pileup, the material can be expected to become slightly more elastic (slightly larger Young's modulus) and less lossy (entangled dislocations either cannot move or move less in response to variable stress gradients and therefore dissipate less energy).

The lattice of entangled dislocations is not stable though. Reversing the stress field in absence of irreversible events (like crystal cleavage for example) eventually results in disentanglement. Disentanglement is not immediate, though, and can require non-negligible stress reversal. If disentanglement is induced the excess stiffness contributed by the chaotic lattice disappears, and the disentangled dislocations become mobilized in the form of avalanches, and available for larger dissipation.

Avalanching and its effects are non local. The impact of a whipping dislocation can activate a nearby one. Similarly the stress field of a dislocation avalanche can trigger the motion of other avalanches at longer range. The entire sample can eventually be interested by spreading avalanches and temporarily change its physical properties.

A regime where dislocation can pile up and get entangled is described by the statistics of systems controlled by Self Organized Criticality [20,21] (SOC). Such a behavior for dislocations was foreseen by [22-24] and observed in our laboratory. When in SOC controlled regime, the character of material internal dissipation changes dramatically.

In a SOC-controlled regime entangled dislocations act collectively, i.e. they their entanglement is stable below a critical "slope" and above that they move in avalanches of unpredictable amplitudes and timing that return the system below critical slope but in a different equilibrium state. The dislocation under varying stresses act like sand in a box that is tilted until the sand's critical slope is reached and the sand starts flowing. The avalanching process is scale-free, in the sense that any event size can happen, between a single dislocation shifting over a small distance, and the dislocation lattice of the entire macroscopic sample rearranging dramatically.

Therefore one cannot expect anymore a smooth reaction from the spring in response to movements sufficiently slow to allow time for entanglement.

If the macroscopic movements are sufficiently fast, many smaller avalanches will happen at the same time, and merge, masking the effects of individual larger avalanches happening at low frequency, and an apparently smooth behavior is restored.

If an oscillatory motion faster than the avalanche growth time is applied, no avalanche have time to fully grow before the stress field reverses direction and aborts it. Because the avalanches have exponential growth, i.e. the bulk of action of an avalanche happens at its very end, the energy dissipated by the avalanching process should drop off dramatically for oscillations with period shorter than the avalanching characteristic time. Note that, because the dislocation avalanches can spread over the entire elastic sample, this characteristic time depends on the macroscopic dimensions of the object, as well as the material characteristics and its stress conditions.

## 2. Expected unexpected behaviours in GAS springs and other low frequency oscillators

Although quantitative predictions would require unyielding simulations, extending from the microscopic to the macroscopic level, several qualitative predictions can be made from the theory of SOC of dislocations. Many of these "predictions" were made with hindsight, taking advantage of many puzzling experimental observations accumulated over years by one of the authors and his collaborators. The theory of SOC of dislocations can also explain many unexplained or ill-explained effects observed in other experiments and precision instruments.

The rest of this section is a listing of some "predictions", while the next section lists the corresponding observations, either from our own measurements, or from effects or behaviors reported in literature.

2.1. Geometric Anti Springs (GAS) are mechanical oscillators that can be tuned to very low frequency [25,26] by progressively neutralizing their restoring forces by means of radial blade compression and then by means of ElectroMagnetic Anti Springs (EMAS) [27]. At sufficiently low frequency the stable crystalline lattice elasticity can be completely neutralized and the oscillator may remain stable only because of the metastable elasticity contributed by the entangled dislocation lattice. In such a state, if dislocation movement is triggered (by an external excitation of by a continuing slowly changing stress, like a thermal drift), the component of the restoring force provided by the entangled dislocations can temporarily disappear and the oscillator can run-off from its equilibrium position. In absence of slow stress variations, like thermal drifts for GAS springs and tilt drifts for IPs, the metastable equilibrium may last indefinitely even with excessively small restoring forces.

2.2. A similar effect should also be observed in Inverted Pendulum (IP), when their restoring force is reduced by means of progressive top loading they should fall in arbitrary direction [11,28].

The run-off events in both GAS and IP should be reversible by simply returning the system to an equilibrium point and holding them there for the time necessary for dislocation entanglement.

2.3. The initial phase of the run-off, when only a small number of individual and localized avalanches is involved, the acceleration of a run-off should to be "bumpy", reflecting the effect of individual, non predictable, avalanches.

The same should be observed in ringdowns of oscillators close to their instability limit.

2.4. A sequence of small excitations combined with a slowly changing stress can induce many small avalanches before the critical slope is reached. While a slow thermal drift can reach the critical slope, and result in a catastrophic run-off in a quiet environment, in noisy conditions it can produce a series of bumpy motion and less likely run-off.

2.5. If slow motion of the oscillator is induced, with period longer than avalanching characteristic times, dislocation avalanches that are mobilized by the motion are expected to re-entangle in a different configuration. When the force inducing the motion is removed the oscillator's equilibrium point must be changed by an amount that depends, on the relative strength of the residual "true" and stable restoring force from the crystal lattice, and the changeable contribution provided by the entangled dislocation lattice. The hysteresis is expected to reach dramatic levels when the oscillator is tuned close to the instability (i.e. the residual restoring forces are dominated by the entangled dislocation contribution).

Multi stability can be observed in these conditions if forces of different amplitudes and directions are applied.

The dislocation landscape can be altered by a sequence of fast stresses, even orthogonal to the oscillator motion, thus inducing random changes of equilibrium point.

2.6. The amount of hysteresis should depend on the past motion or stress history, if the applied force causes the oscillator to go in directions just explored, the dislocations are already entangled in the correct location and no further hysteresis should be observed.

2.7. An oscillator tuned to high frequency with period shorter than the characteristic avalanche growth time should show much lower dissipation than when tuned with longer periods, thus affecting for example the Quality factor of the system.

2.8. Forced oscillations of a metal flexure cause dislocation disentanglement, which can be expected to grow with larger induced motion amplitudes. The thus disentangled dislocations should then result in a reduction of the fraction of Young's modulus contributed by the entangled dislocation lattice, and in an enhanced viscous-like dissipation.

2.9. When an oscillator is tuned or moves to a frequency sufficiently low to allow sizable entanglement, and is subject to excitations sufficient to generate dislocation avalanches, it cannot be expected to respond like a harmonic oscillator, and its transfer function must be affected by 1/f response characteristic of avalanching systems.

2.10. Precision instruments based on mechanical oscillators (like tiltmeters, seismometers, torsional pendulums and accelerometers) should present low noise performance only after a sufficient dislocation settling time, and only in absence of slow excitations, like thermal drifts and tilts, or large excursions or stresses even in orthogonal vibrations, that can drive parts of the system to its "critical" slope and generate dislocation avalanches. These instruments cannot be expected to produce low noise operation during and after large excitations like earthquakes.

## 3. Observational behaviour and results

For lack of time only some of our experimental observations were illustrated in the conference presentation, and for lack of space even fewer of them are reported here, please refer to [29] and to the forthcoming longer paper [30] for more detail.

3.1. GAS springs were observed to spontaneously run-off either up or down (figure 2. Left panel). In particularly quiet conditions, under vacuum and by night, the HAM SAS system remained stable at 10 mHz tuning until an external perturbation was applied [31].

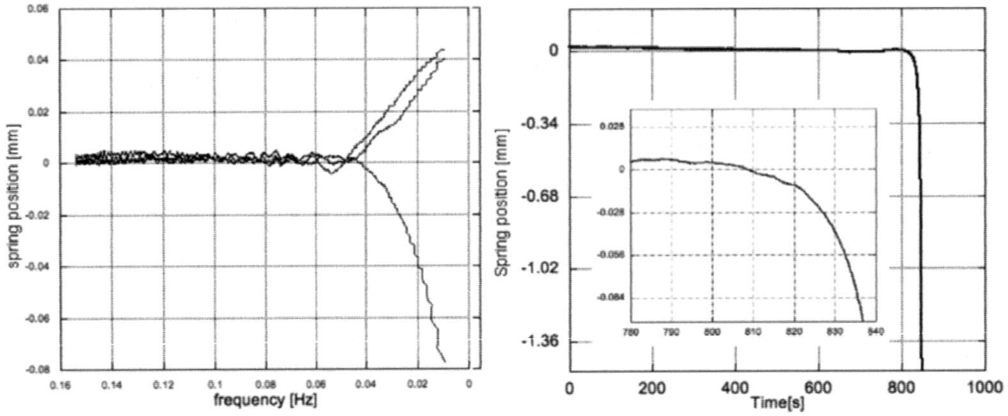


**Figure 2:** Example of GAS filter run-offs. In the left panel the resonant frequency of the filter is tuned to lower and lower frequency, until it destabilizes and runs off to the end stops. Note that run-off in an ideal spring should occur only at frequency equal to zero, while run-off is observed for frequencies as high as 200 mHz. The run-off of the 65 kg mass happens indifferently up or down (ruling out creep or material failing effects). In the right panel a run-off in a filter tuned at fixed resonant frequency, happens after several hours of night operation. Note, in the zoomed insert, the very slow start of the event, with an irregular and bumpy motion signalling the effects of individual avalanches, followed by a progressive acceleration.

3.2. IP were similarly observed to fall in arbitrary direction, sometimes after waiting times as long as many hours. In both IP and GAS filters the equilibrium can be easily restored.

3.3. Runoff and very slow frequency oscillations are indeed bumpy and of unpredictable shape (figure 2. Insert of right panel).

3.4. GAS filters were observed to show jumping equilibrium point and less run-off during the noisy day time and run-off more easily after long, quiet periods during the nights.

3.5. Hysteresis was observed to grow very rapidly in GAS filters below 0.2 Hz, much more than what could be expected by the lower stiffness (Figure 3.). Slow forces can drive the system to almost arbitrary equilibrium points [32] over a wide range of positions. Bad-weather-induced perturbations were observed to induce a slow equilibrium point random walk in the Virgo Superattenuators Inverted Pendula [13].

3.6. While excitations of alternate sign show alternate hysteresis of approximately the same amplitude, repeated excitations of the same sign produce no additional hysteresis (Figure 3.).

3.7. The Quality factor of GAS filters was observed to increase dramatically above a threshold of 0.2 Hz [29,30]. Below 0.2 Hz it behaves quadratically, as expected from time independent (completely developed avalanches) dissipation processes.

Journal of Physics: Conference Series **228** (2010) 012033 doi:10.1088/1742-6596/228/1/012033

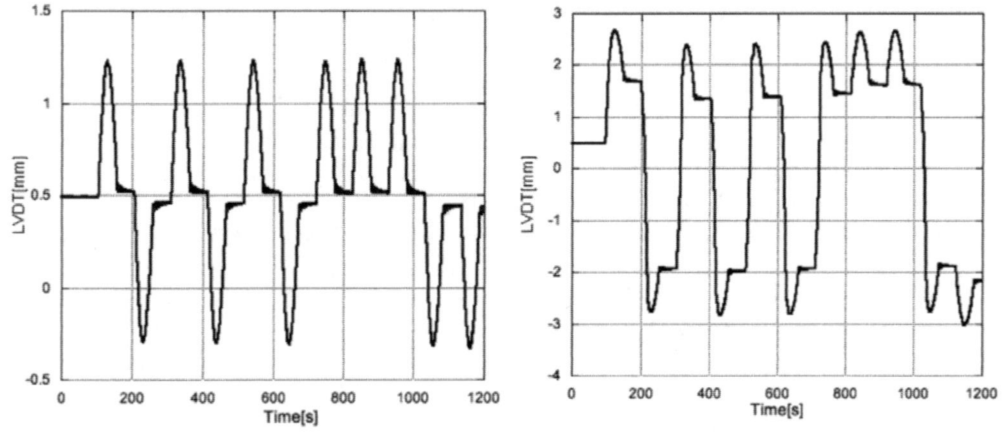


**Figure 3:** Measurement of hysteresis at 0.21 and 0.15 Hz GAS filter tune, responding to the same half-sine excitation. The restoring force differs by less than a factor of two, while the measured hysteresis grows by more than an order of magnitude, marking the phase transition between viscous and avalanche dominated dissipation. Note how the leading edge of the response in the left panel has the same curvature of the sine, while in the right panel the leading edge has opposite curvature as if the system was about to run-off.

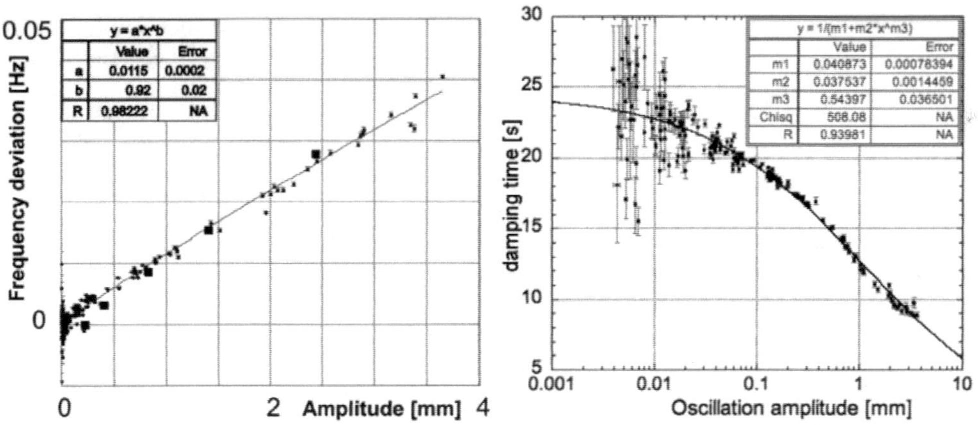


**Figure 4:** Left panel, frequency deficit as function of oscillation amplitude (19% in frequency for oscillation amplitude of 4 mm, 34% in elastic constant k). The dots were evaluated along ringdown measurements, the bigger squares in swept sine measurements. Right panel, change of damping constant as function of amplitude, the dissipation is more than doubled for amplitude oscillations of 4 mm. Both measurements taken at 0.21 Hz (without EMAS).

3.8. The oscillation frequency of GAS filters is lower than the frequency measured for vanishing amplitude by a small amount (several percent) roughly linear with the oscillation amplitude, while the dissipation was observed to more than double with a growth rate compatible with the square root of the oscillation amplitude (Figure 4.). These effects were observed both in our measurements of swept sine excitation and ringdowns of GAS springs.

3.9. The 1/f transfer function was observed by [33,34] in a GAS filter operated at 100 mHz. It remains to observe if, at lower excitation level, the transfer function resume its ordinary 1/f2 slope as would be expected for oscillations small enough not to generate critical slope and not to trigger avalanches.

3.10. Development of at least three generations of mechanical tiltmeters ended in disappointing performance [35,36], thousands of times the level that could be extrapolated from higher frequency behavior, growing at least with 1/f at the low frequency side. Additionally, it is well known that high precision seismometers need to settle for a "thermalization" time of many hours, which is much longer than the thermalization time constants that one could expect from the heath capacitance and conductivity of their structure. Careful choice of the flexure material and its treatments can minimize this effect but never fully eliminate it. Torsion pendula reach their thermal noise performance only after settling time of hours, to be compared with the wire thermalization time of seconds or less [37]. In view of our understandings, the cause of the settling times is more likely connected to the settling of dislocations in their flexures and along the wires. Similarly it is well known that seismometers subject to thermal drifts are more noisy. Finally, attempts to use low noise, co-located seismometers to measure the actual displacement generated by an earthquake through integration of their signal, are reported to give inconsistent values and even directions.

## 4. Conclusions

Hysteresis effects and excess low frequency noise were a nagging and long studied issue in the development of the SAS systems for Virgo, LIGO, TAMA and for the HAM chambers for Ad-LIGO. After more than 10 years of studying, these effects were linked to the theory of SOC of dislocation. While work is still ongoing, better practices are slowly emerging from the new understandings, which include different engineering, different materials, and sometimes simply different procedures.

Most of the measurements leading to these results were performed with precipitation wire Maraging steel, but hysteretic behavior foreseen by SOC was observed also in work hardened piano wire [38] and other metals (Copper Beryllium, Nispan-C, et c.). SOC effects are thought to be ubiquitous in low frequency behavior of metal flexures.

During the dislocation rearrangement leading to hysteresis, springs and flexures can be expected to be noisy. Rearrangement of dislocations may take substantial time, even hours of settling time after perturbation in high sensitivity seismometers or tilt pendulum. To reach quiet spring behavior, the dislocation rearrangement in the flexures must lead to a "flat" dislocation landscape. Dislocation landscape smoothing happens naturally (a SOC controlled system always settles below its "critical" slope). The corresponding instrument settling time can likely be reduced, and the flattening of the dislocation landscape can be improved, and driven well below the critical slope, by inducing gentle oscillations around the true equilibrium point and slowing tapering them down to zero amplitude.

Fluctuation dissipation estimations of thermal noise are valid only in absence of avalanching, and therefore they are valid only when the dislocation landscape is well below its critical "slope", i.e. after a sufficiently long settling time, in absence of external mechanical perturbations and of any other slow stress variations. All sources susceptible to alter the dislocation landscape slope must be avoided (which include thermal drifts, slowly changing forces, physical tilt variations, vibrations and occasionally stronger excitations of flexures, even in orthogonal degrees of freedom, and any other stimuli that we may not have identified yet) if minimal noise conditions are desired. While any sustained external perturbation (like seismic noise, thermal drifts and other slow moving forces) will produce avalanches, it is possible that, at the lower stages of the GW detectors seismic attenuation chains, the residual seismic and mechanical excitation may be insufficient to induce avalanching, and that the full design attenuation power and thermal noise performance may be recovered. Thermal stabilization of a small fraction of a degree should be required. An example of behavior to be avoided is to compensate for earth's tides by pushing on the mirror itself. The slow flexure bending close to the test mass would generate avalanching noise especially at the extremities of the tidal motion range. Tidal compensation should be applied as far upstream as possible.

We also suggest that any oscillator expected to operate at low noise should be operated around its "true" equilibrium point, i.e. the one determined by the elasticity of the crystalline structure alone, in absence of the contribution from the entangled dislocations lattice. Tests have shown that this state can be reached through induced oscillations of slowly decreasing amplitudes [32]. If the system does not have sufficient Q-factor to do it by simple ringdown, these oscillations must be forced [30]. Old practices based on hysteresis effects, like finding the working tilt of a suspended mirror by over-tilting it by hand and then slowly releasing it, were effective but inductive of a metastable equilibrium determined by large dislocation pile-up and likely source of noisy conditions. These practices should be avoided, the required tilt of the mirror should be determined by moving weights up the chain, after a suitable dislocation landscape relaxation routine.

Materials free of dislocations (like glasses) or with frozen dislocations (like ceramics) can be expected to be free of SOC noise. Such SOC noise free materials will be needed for the seismic isolation of the third generation, low frequency Gravitational Wave detectors.

## 5. Acknowledgments

We gratefully thank the National Science Foundation, cooperative agreement number PHY-0823459. This document has LIGO number LIGO-P0900176.

## References

[1] LIGO project, www.ligo.caltech.edu
[2] B.Abbott et al, Detector description and performance for the coincidence observations between LIGO and GEO, Nuclear Instruments and Methods in Physics Research A 517, 154-179, (2004)
[3] VIRGO project, www.virgo.infn.it
[4] VIRGO description, www.ego-gw.it
[5] GEO600 project, http://geo600.aei.mpg.de/
[6] J.M. Ziman, Models of disorder, CUP Cambridge UK, (1979)
[7] J.P. Hirth and J. Lothe, Theory of Dislocations, McGraw-Hill New York, (1968)
[8] A.S.Nowick BS. Berry, Anelastic Relaxation in Crystalline Solids, Academic New York, (1972)
[9] Granato A V and Lucke K 1966 Physical Acoustics, edited by Mason and Thurston Academic New York IVA 225
[10] Y. L. Huang, Dissipation mechanisms in pendulums and their implications for gravitational wave interferometers, Rev. Sci. Instrum., Vol. 69, No. 2, February 1998
[11] A.L. Kimball, D.E. Lowell, "Internal friction in solids", Physics review, 30, 948-960, 1927
[12] P. Saulson et al, The inverted pendulum as a probe of anelasticity, Rev. Sci. Instrum. 65, (1994), 182-91
[13] P. Ruggi, VIRGO project, private comm. 2005
[14] Virtually all high precision seismometer installation manuals recommend settling times longer than typical thermalization time constant, in these periods glitches are often observed; presumably these long times are necessary to allow for dislocation avalanche settling.
[15] Ad Lagendijk, et al., Fifty years of Anderson localization, Physics Today, Volume 62, Issue 8, August 2009, pp. 24-29
[16] M. Beccaria, V. Rubino, et al., The creep problem in the Virgo suspensions: a possible solution using Maraging steel., Nucl. Instr. Meth. Phys. Res. 404 (2–3) (1998), 455-469.
[17] N. Virdone, et al., Extended-time-scale creep measurement on Maraging cantilever blade springs, Nuclear Instruments and Methods in Physics Research A 593 (2008) 597– 607
[18] R. DeSalvo, Non-stochastic noise in gravitational wave detectors, in: Proceedings of the 2nd Edoardo Amaldi Conference on Gravitational Waves, CERN, Switzerland, 1–4 July 1997, World Scientific Publishing Co., P.O. Box 128, Farrer Road, Singapore 912805, pp. 228–239, LIGO Document No. P970036-00-D, available at /http://admdbsrv.ligo.caltech.edu/dcc
[19] S. Braccini, et al., The maraging-steel blades of the Virgo super attenuator, Meas. Sci. Technol. **11** (2000) 467–476
[20] P. Bak et al, Self-organized criticality: An explanation of the 1/f noise, Phys. Rev. Lett. 59, 381-84, (1987)
[21] P. Bak et al, Self-organized criticality, Phys. Rev. A 38, 364-74, (1988)

[22] F. Marchesoni and M. Patriarca, Self-Organized Criticality in Dislocation Networks, Phys. Rev. Lett. 72, 4101, (1994)
[23] F. Marchesoni et al, Damping in low-frequency mechanical pendulums, Phys. Lett. A 187, 359, (1994)
[24] G. Cagnoli, et al., "On dislocation damping at low frequencies", Phil. Mag. A 68, p. 865, 1993
[25] R. De Salvo, Passive, Nonlinear, Mechanical Structures for Seismic Attenuation, J. Comput. Nonlinear Dynam. 2, 290-98, (2007)
[26] G. Cella et al, Monolithic Geometric Anti-Spring Blades, Nucl. Instrum. Methods Phys. Res. A 540, 502-519, (2005)
[27] M. Mantovani and R. De Salvo, One hertz seismic attenuation for low frequency gravitational waves interferometers, Nucl. Instrum. and Meth. 554, 546-54, (2004)
[28] A. Takamori, et al., Inverted pendulum as low-frequency pre-isolation for advanced gravitational wave detectors, Nuclear Instruments and Methods in Physics Research A 582 (2007) 683–692
[29] A. DiCintio et al., Dislocation movement and hysteresis in Maraging blades Class. Quantum Grav. 26, 204018 (12pp) (2009)
[30] R. DeSalvo et al., Observations of a new dissipation regime in metals, to be published LIGO DCC number P0900216, and a longer paer on "Dissipation processes in metal springs" being written.
[31] A. Stochino et al, Improvement of the seismic noise attenuation performance of the Monolithic Geometric Anti-Spring lters for gravitational wave interferometric detectors, Nucl. Instrum. Methods Phys. Res. A 580, 1559-64, (2007)
[32] R. De Salvo et al., Study of Quality Factor and Hysteresis Associated with the State-of-the-art Passive Seismic Isolation System for Gravitational Wave Interferometric Detectors, Nuclear Instr. and Meth. A, 538, 526-537, (2005)
[33] M. Mantovani, and Riccardo DeSalvo, "One hertz seismic attenuation for low frequency gravitational waves interferometers", Nucl. Instr. and Meth. vol. 554, no1-3, 2005,pp. 546-554
[34] A. Stochino et al, The SAS Seismic Attenuation System for the Advanced LIGO Gravitational Wave Interferometric Detectors, Nucl. Instrum. Methods Phys. Res. A, 598, 737-753, (2008)
[35] C.G. Speake, The design and application of a novel high-frequency tiltmeter, Rev. Sci. Instrum. 61 (5), 1500-1503, May 1990
[36] A.N. Luiten, et al., "Ground tilt seismic spectrum measured with a new high sensitivity rotational accelerometer", Rev. Sci. Instrum. 68 (4), 1889-1893, May 1997
[37] J. Gerlach, Private communication, (2009)
[38] J. Greenhalgh and I. Wilmut, LIGO Technical Report LIGO DCC T080033-01-K, (2008)

Journal of Physics: Conference Series **228** (2010) 012034 doi:10.1088/1742-6596/228/1/012034

# Towards time domain finite element analysis of gravity gradient noise

**M.G. Beker, J.F.J. van den Brand, E. Hennes, D.S. Rabeling**

Nikhef, National Institute for Subatomic Physics, Science Park 105, 1098 XG Amsterdam, The Netherlands

E-mail: mbeker@nikhef.nl

**Abstract.** Gravity gradient noise generated by seismic displacements constitute a limiting factor for the sensitivity of ground based gravitational wave detectors at frequencies below 10 Hz. We present a finite element framework to calculate the soil response to various excitations. The accompanying gravity gradients as a result of the seismic displacement field can then be evaluated. The framework is first shown to accurately model seismic waves in homogenous media. Calculations of the gravity gradient noise are then shown to be in agreement with previous analytical results. Finally results of gravity gradient noise from a single pulse excitation of a homogenous medium are discussed.

## 1. Introduction

Advances in gravitational wave (GW) detection have, throughout the past 20 years, overcome many technical obstacles and suppressed a vast number of instrumental noise sources. At this moment first generation interferometric detectors have reached their design sensitivities and second generation detectors are poised to upgrade these sensitivities by a factor of 10 within the next decade [1] [2] [3]. Now research is starting to explore possibilities beyond the second generation detectors. A third generation of detectors will once again improve sensitivities by a factor of 10 over a broad frequency spectrum [4]. A particular challenge will be to detect signals at frequencies below 10 Hz. It is expected that sensitivity at these frequencies will be limited by fluctuations in the local gravitational field as a result of fluctuating density variations in the surrounding soil [5] [6] [7]. These field fluctuations couple directly to the interferometer test masses and are known as gravity gradient (or Newtonian) noise (GGN). A full understanding of seismic activity, seismic wave characteristics and how these affect GW detectors through gravity gradients is imperative to achieve the desired sensitivity. Sources of seismic noise include earthquakes, atmospheric and oceanic disturbances and cultural noise [8]. The latter is a result of human activity and infrastructure such as footsteps, traffic and machinery. Analytical descriptions of seismic activity fall short when considering a non-homogeneous medium or complex geologies. Here we present a time domain finite element (FE) framework to model seismic activity and discuss the impact of GGN on GW detectors.

## 2. Ambient ground motion and gravity gradient noise

As a result of seismic disturbances a displacement wave field is produced that is governed by the elasto-dynamic equations. The wave field in a homogeneous elastic medium can be expressed as

© 2010 IOP Publishing Ltd

a combination of plane body waves [9]. Two types of body waves exist, pressure (P) and shear (S) waves [10]. In the case of P-waves the movement of a ground particles is parallel to the direction of wave propagation. For S-waves the particle motion is perpendicular to the direction of the wave. The characteristics of seismic waves can be described by the ground properties, parameterized by the Young's modulus, $E$, the density, $\rho$, and the Poisson ratio, $\nu$, describing the relationship between shear and strain forces. The wave velocities are then given by

$$c_P = \sqrt{\frac{E(1-\nu)}{(1-2\nu)(1+\nu)\rho}}, \quad \text{and} \quad c_S = \sqrt{\frac{E}{2(1+\nu)\rho}}, \tag{1}$$

for P and S-waves respectively. Typical values for clay-like soils range from 500 - 1000 m/s for $c_P$, and 250 - 600 m/s for $c_S$. In a medium that is bounded by another medium, such as air, or is comprised of layers, surface and head waves also exist. Head waves emerge in stratified media where modes propagating along an interface, cause energy to radiate into the low velocity zone. Surface waves are typically referred to as Rayleigh and Love waves. Love waves involve particle motion parallel to the surface and transverse to the direction of propagation. They produce no density variations and therefore have no effect on gravity gradients [6]. Rayleigh waves are polarised perpendicular to the surface and vanish with depth. Solving the wave equation for harmonic Rayleigh waves results in the following displacement fields [11]

$$\begin{aligned} \xi_x &= iA(k_R e^{-\kappa_P z} - \zeta \kappa_S e^{-\kappa_S z}) e^{i(k_R x - \omega t)}, \\ \xi_z &= -A(\kappa_P e^{-\kappa_P z} - \zeta k_R e^{-\kappa_S z}) e^{i(k_R x - \omega t)}, \end{aligned} \tag{2}$$

where $A$ is an arbitrary amplitude, $t$ denotes time, $\omega$ denotes the angular frequency, $z$ is the depth, $k_R$ is the wave number of the Rayleigh wave, and $\kappa_S = \sqrt{k_R^2 - k_S^2}$ and $\kappa_P = \sqrt{k_R^2 - k_P^2}$ are decay factors related to the shear and pressure wave numbers. Finally, $\zeta = \sqrt{\kappa_P/\kappa_S}$. The horizontal Rayleigh wave speed, $c_R$, is slightly lower than the S-wave speed and when expressed in units of $c_S$ is purely a function of $\nu$. It can be found through $c_R/c_S = \chi$ where $\chi$ is the real root, in the range $0 < \chi < 1$ of the equation [11]

$$\chi^6 - 8\chi^4 + 8\left(\frac{2-\nu}{1-\nu}\right)\chi^2 - \frac{8}{1-\nu} = 0. \tag{3}$$

The above describes harmonic Rayleigh waves propagating far from the source. In reality, excitation of a medium results in a combination of all the different wave fields; body and surface.

As waves propagate through the medium, their amplitude decreases. This attenuation can be attributed to two factors; material and geometric damping. Geometrical damping is a result of energy spreading over an increasing area. The frequency dependent material damping involves energy lost due to friction. Seismic wave attenuation for homogeneous media, can be described by [12]

$$A_2 = A_1 \left(\frac{r_1}{r_2}\right)^n e^{-\frac{\pi \eta f}{c}(r_2 - r_1)}, \tag{4}$$

where $A_1$ and $A_2$ are the wave amplitudes at distance $r_1$ and $r_2$ from the source, $n$ is the geometric damping coefficient, $f$ is the frequency and $c$ the propagation speed of the wave. The material damping is represented by the loss factor $\eta$. The geometric damping coefficient can be determined analytically by assessing the type of wave involved and the source type. For radial surface waves $n = 1/2$ while radial body waves within the medium decay with $n = 1$. For this paper $\eta$ is set to zero.

For a given distribution of masses, which can be described by the mass density function $\rho(\mathbf{r}, t)$, the acceleration experienced by a test mass (i.e. an interferometer mirror) located at $\mathbf{y}$

can be written as

$$\mathbf{a}(\mathbf{y},t) = G\int_V \rho(\mathbf{r},t)\frac{\mathbf{r}'}{|\mathbf{r}'|^3}dV, \tag{5}$$

where $\mathbf{r}$ is the position of the mass volume $dV$ and $\mathbf{r}' = \mathbf{r} - \mathbf{y}$. In our FE analysis the acceleration is the summation of the contributions from each node $i$ with mass $m_i$ located at $\mathbf{r}_i$. The acceleration at the test mass is given by

$$\mathbf{a} = \sum_i \mathbf{a}_i = \sum_i Gm_i\frac{\mathbf{r}'}{|\mathbf{r}'|^3}, \tag{6}$$

with $G$ the universal gravitational constant. When a seismic disturbance is present, the nodes suffer a displacement denoted by $\xi_i(\mathbf{r},t)$. The gravity gradient acceleration due to these displacements is given by

$$\mathbf{a}^{NN}(\mathbf{y},t) = \sum_i(\nabla\otimes\mathbf{a}_i)^T\xi_i(\mathbf{r},t). \tag{7}$$

Note that in the gravity gradient calculations presented here, the mass associated with each node is assumed to be constant. The corresponding analytical expression can be obtained by substituting $m_i \rightarrow \rho dV$ and evaluating the resulting integral.

## 3. Finite element ground motion simulations

Simulation of the ground displacement fields was done with the FE software package *Comsol* [13]. In the FE framework we subdivide a 3D continuum into small hexahedral elements. Within each element the relevant physical parameters, like displacement and stress, are approximated by spline functions of arbitrary order. To confirm the ground motion response calculated by a FE model, comparison is made with analytical solutions described in the previous paragraph.

The FE displacement wave fields for one wavelength of a Rayleigh wave propagating in the $x$ direction have been plotted in Fig. 1. It is interesting to note the sign change in the particle movement in the $x$ direction. This occurs at a depth of about $0.2\lambda_R$. In Fig. 1b the maximum amplitude in both the $x$ and $z$ displacements are plotted, the two are separated by a phase difference of $\pi/2$. These are in good agreement with the analytical solutions of Eq. (2).

A homogenous half-space was simulated by creating a half-sphere model with no reflection of waves incident to the spherical boundary. In view of symmetry the model could be further simplified to a quarter half-space with symmetric boundary conditions on the vertical surfaces. A single vertical excitation force was applied uniformly within a circular area at the origin with a time dependent factor given by $F(t) = A\sin^2(\pi t/T_e)$ for $0 \leq t \leq T_e$ where $T_e$ is the excitation period. The amplitude scaling factor, $A$, was adjusted to create a vertical displacement at the excitation point of 1 $\mu$m. The model has parameters $E$ = 10 GPa, $\rho$ = 2.0 g/cm$^3$, $\nu$ = 0.25 and a radius of 2.2 km. This results in wave speeds of $c_P$ = 800 m/s, $c_S$ = 462 m/s and $c_R$ = 420 m/s. To check the consistency of the model with analytical expectations the arrival times and geometric damping of the wave fields can be studied. No material damping was implemented in this model. Fig. 2a shows the FE results of seismic displacements along with expected arrival times at a location on the surface and at a depth of 800 m. Note the phase difference of $\pi/2$ between the $x$ and $z$ displacements of the Rayleigh wave. The rms wave amplitude with increasing distance from the source across the surface and within the medium are plotted in Fig. 2b. As expected from Eq. (4) the wave attenuation is proportional to $1/\sqrt{r}$ along the surface and $1/r$ within the medium.

## 4. Comparison with previous analytical GGN results

The calculation of GGN via FE models was validated by creating simple rectangular homogenous half-space models, equivalent to those discussed by Saulson for a surface detector [5]. The

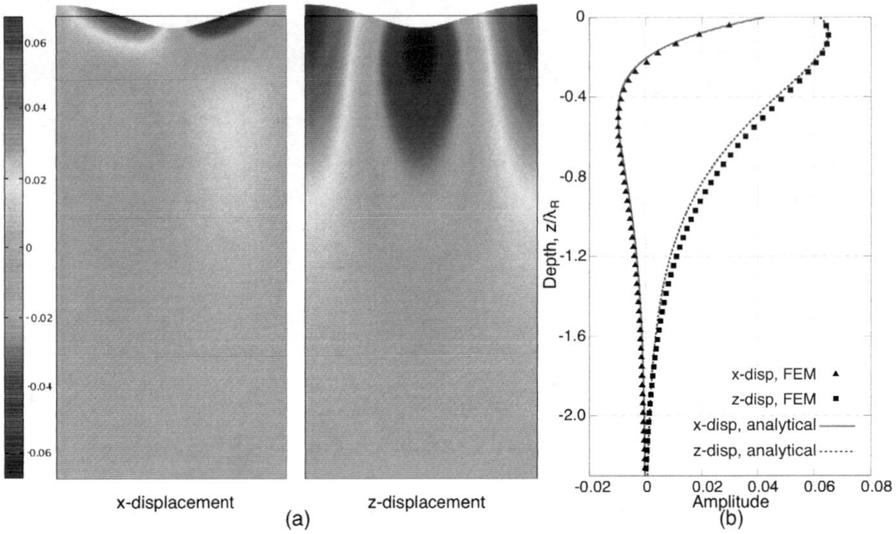


**Figure 1.** (a) Rayleigh wave displacement fields for a wave propagating in the $x$ direction. (b) FE and analytical calculations of Rayleigh wave displacements. The $x$ and $z$ components attenuate differently with depth. The $x$ component changes sign at a depth of about 0.2 $\lambda_R$.

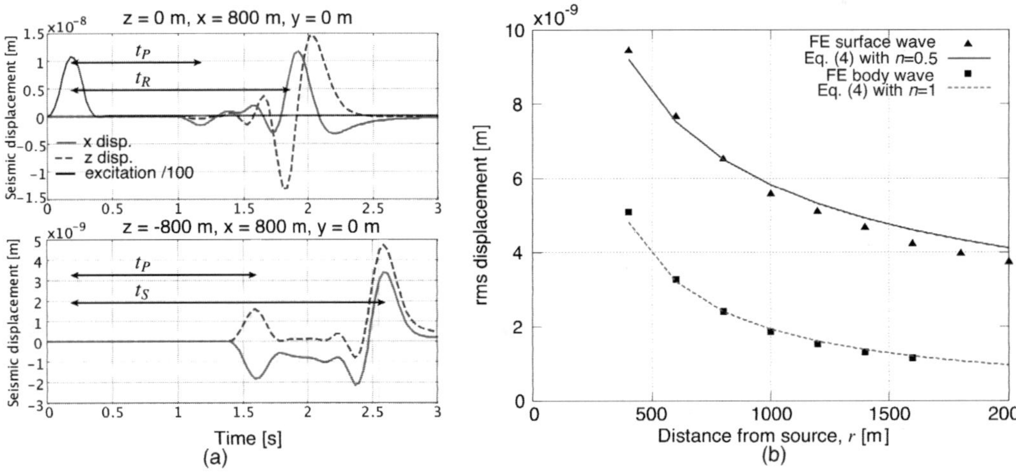


**Figure 2.** (a) FE ground displacement measurements at a surface and subterranean location after a pulse excitation. P, S and Rayleigh wave arrival times are indicated. (b) The FE results for the rms amplitude of the surface and body waves with increasing distance from the source. The geometric damping contribution to Eq. (4) is plotted for comparison.

isotropic, elastic half-space with $\rho = 1.8$ g/cm$^3$, $\nu = 0.33$, $c_P = 440$ m/s and $c_S = 220$ m/s was excited on one boundary to yield plane harmonic pressure waves scaled to a flat ambient seismic noise spectrum of 1 nm/$\sqrt{\text{Hz}}$ between 1 and 10 Hz and the subsequent nodal displacements were recorded as a function of time. Boundary conditions were set such that no reflections occurred and seismic waves were continuous. With this input spectrum the gravity gradient displacement noise amplitude at the interferometer output was calculated and the result is shown in Fig. 3b. The FE results are compared with the analytic results of Saulson and Hughes and Thorne [5][6].

To facilitate comparison, an integral cut-off radius equal to that used in Saulson's analysis ($r_{\text{cutoff}} = \lambda/4$) was employed in the summation process. Fig. 3 shows that good agreement is obtained. To assess the effect of this cut-off the above model was calculated analytically with Eq. (5). Removing the cut-off leads to an increase of GGN by about a factor 2 (see Fig. 3). The FE results approach those of the analytic expression in the limit that $r_{\text{cutoff}}$ decreases to zero.

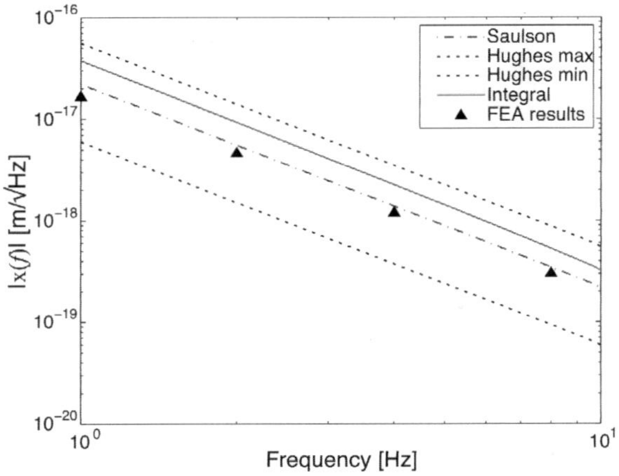


**Figure 3.** FE calculation of the Newtonian displacement noise amplitude for a surface detector. For comparison the results of Saulson, Hughes and Thorne, and the analytic integral are shown.

## 5. GGN results from pulse excitations on a homogeneous medium

The pulse excitations and the half-sphere model described in section 3, were used to investigate FE gravity gradient modeling (see Fig. 4). The nodal displacements were recorded as a function of time and the GGN was calculated at various depths on a vertical line at a distance $\lambda_P = 800$ m from the $z$-axis. In order to artificially separate the contributions of the surface and body waves to the GGN acceleration the nodes with a depth less than 200 m were summed separately to those deeper that 200 m. The respective surface and body contributions were combined to give a total acceleration. Note that for times shortly after the excitation, this distinction is not precise. The results for a test mass at the surface and a test mass at a depth of $\lambda_P$ are shown in Fig. 4. Only the GGN acceleration in the horizontal direction is shown since it has the largest effect on the performance of an interferometer. The expected arrival times of the different waves are indicated in the figures and show that the Rayleigh wave dominates the GGN contribution of a detector on the surface. At a depth of $\lambda_P$ the arrival of the S and P-waves can clearly be distinguished, the S-wave producing a larger contribution, attributed to the larger seismic displacements seen in Fig. 2a. The GGN contribution of a wave is initially negative as the wave approaches then changes sign as the wave passes by the test mass. It is interesting to note that the sign change of the surface contributions between a surface and underground detector. This is due to the sign change of the horizontal component of the Rayleigh wave for depths larger than $0.2\lambda_R \approx 80$ m. The figure also shows that GGN builds up before any seismic disturbance actually reaches the test mass and for short times is only dependent on surface contributions.

## 6. Conclusions

A time domain finite element framework has been presented that successfully models seismic motion. The displacement fields of a Rayleigh wave are accurately modelled. By investigating pulse excitations of a homogenous half-space it is shown that all wave types are excited. These propagate from the source as predicted by theoretical models and adhere to the geometrical damping laws. Gravity gradient accelerations from a homogeneous half-space are calculated and show agreement with previous analytical results. Finally results are presented

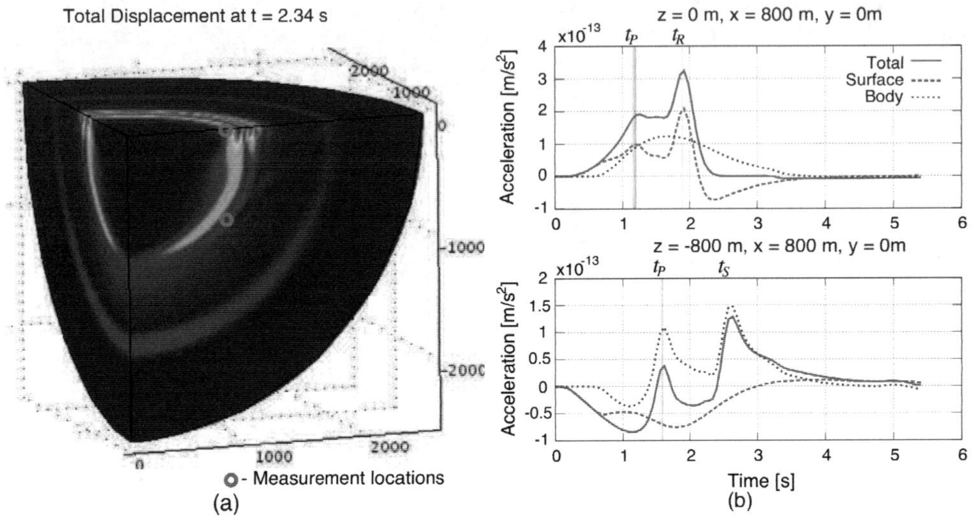


**Figure 4.** (a) Total displacement for a time domain simulation at 2.34 seconds after a 1 $\mu$m pulse excitation at the center of the half-sphere. (b) Time domain evolution of GGN acceleration at a surface ($z$=0 m) and underground ($z$=-800 m) test mass. Only the horizontal component of the GGN acceleration is shown. Arrival times of Rayleigh, S and P-waves are also indicated.

of GGN accelerations for the pulse excitation on a homogeneous half-sphere. This analysis indicates strong contributions from Rayleigh waves for a surface detector and body waves for a subterranean detector. This paper has shown that a FE framework can be an effective tool in investigating gravity gradients for future, possibly subterranean, gravitational wave detectors. This allows further research into quantitive effects of seismic motion on GGN and implications of geology and infrastructure.

## Acknowledgments

This work has been performed with the support of the European Commission under the Framework Programme 7 (FP7) Capacities, project Einstein Telescope design study (Grant Agreement 211743), http://www.et-gw.eu/. This work is part of the research programme of the Foundation for Fundamental Research on Matter (FOM), which is financially supported by the Netherlands Organisation for Scientific Research (NWO).

## References

[1] Acernese F *et al.* 2008 *Class. Quantum Grav.* **25** 184001
[2] Sutton P 2008 *J. Phys.: Conf. Ser.* **110** 062024
[3] Danzmann K 1995 *Proc. First E. Amaldi Conf. Grav. Wave Exp. (Rome)*
[4] Hild S *et al.* A 2008 Pushing towards the ET sensitivity using 'conventional' technology `arXiv.org:0810.0604`
[5] Saulson P 1984 *Phys. Rev. D* **30** 732–736
[6] Hughes S and Thorne K 1998 *Phys. Rev. D* **58** 122002
[7] Beccaria M *et al.* 1998 *Class. Quantum Grav.* **15** 3339–3362
[8] Peterson J 1993 *U.S. Department of Interior Geological Survey* Open-File Report 93-322
[9] Schevenels M 2007 Ph.D. thesis Katholieke Universiteit Leuven
[10] Achenbach J 1973 *Wave Propagation in Elastic Solids* (Amsterdam: North-Holland) pp 187–194
[11] Hassan W and Nagy P B 1998 *J. Acoust. Soc. Am.* **104** 3107
[12] Kim D S and Lee J S 2000 *Soil Dyn. and Earthq. Eng.* **19** 115–126
[13] Comsol Multiphysics version 3.5a URL `www.comsol.com`

Journal of Physics: Conference Series **228** (2010) 012035 doi:10.1088/1742-6596/228/1/012035

# Tunable mechanical monolithic horizontal sensor with high Q for low frequency seismic noise measurement

**Fausto Acernese[1,2], Rosario De Rosa[3,2], Fabio Garufi[3,2], Gerardo Giordano[1,2], Rocco Romano[1,2], Fabrizio Barone[1,2]**

[1] Università degli Studi di Salerno, Fisciano (SA), Italia
[2] INFN sez. di Napoli, Napoli, Italia
[3] Università di Napoli Federico II, Napoli, Italia

E-mail: `fabrizio.barone@na.infn.it`

**Abstract.** A tunable mechanical horizontal monolithic seismometer/accelerometer, developed for applications in the fields of geophysics and interferometric detection of gravitational waves of second and third generation, is described. The large measurement band ($10^{-3} \div 10\,Hz$) with sensitivities of $\approx 10^{-12}\,m/\sqrt{Hz}$, as seismometer, and better than $10^{-11}\,m/s^2/\sqrt{Hz}$, as accelerometer, have been obtained with an optimised mechanical design and the introduction of a very sensitive laser interferometric optical readout, the latter aimed also to ensure a very good immunity to environmental noises. Prototypes of seismometers are operational in selected sites both to acquire seismic data for scientific analysis of seismic noise and to collect all the useful information to understand their performances in the very low frequency band ($10^{-6} \div 10^{-3}\,Hz$).

## 1. Introduction

The structure of the tunable mechanical horizontal monolithic sensor is based on the Folded Pendulum (hereafter FP), called also *Watt-linkage*, a suspension system developed in 1962 [1], recently applied as ultra-low frequency pendulum resonator for vibration isolation in interferometric detectors of gravitational waves [2]. Based on the FP basic scheme, single-axis monolithic accelerometers were developed as sensors in the control system for advanced seismic attenuators [3]. Recently, tunable monolithic seismometer/accelerometers of small size have been developed at the University of Salerno with extremely soft flexures at the pendulum's hinges coupled to the innovative application of laser optics readout techniques, that have largely improved the sensor sensitivity in the low frequency band and increased its immunity to environmental noises [4, 5]. Special tuning procedures have been also developed to decrease the sensor natural frequency to values as low as $\approx 70\,mHz$, allowing its use both as seismometer and accelerometer of small dimensions for seafloors or boreholes [4]. On the basis of these good scientific results, we have developed a low-noise high-resolution horizontal monolithic FP sensor for geophysical application aimed to explore the low frequency band of the seismic spectrum, both as a stand-alone sensor or as part of large and geographically distributed seismic networks.

## 2. Theoretical Model

The FP dynamics is theoretically well described by the simplified Lagrangian model developed by J.Liu et al. [2], very useful to understand and discuss the FP global dynamics together its

© 2010 IOP Publishing Ltd

Journal of Physics: Conference Series **228** (2010) 012035 doi:10.1088/1742-6596/228/1/012035

main characteristics and expected performances. The FP mechanical scheme, shown in Figure 1, consists of two vertical beams, a pendulum of length $l_p$ and mass $m_p$ and an inverted pendulum of length $l_{ip}$ and mass $m_{ip}$, linked to a central rigid mass, $m_c$. The distance between the FP central mass pivot points is fixed and equal to $l_d$. In the model of Liu et al. [2] the two vertical beams are modeled with equivalent concentrated masses positioned in their geometrical centre, that is $l_{b_1} = l_p/2$ for the pendulum and $l_{b_2} = l_{ip}/2$ for the inverted pendulum, approximations well justified by the mechanical design of our FP prototypes. The central mass, $m_c$, is modeled, instead, with two equivalent concentrated masses, $m_{c_1}$ and $m_{c_2}$ ($m_c = m_{c_1} + m_{c_2}$), positioned at the pivot points of the central mass at distances $l_{c_1}$ and $l_{c_2}$, measured with respect to the pivot points of the pendulum arm and of the inverted pendulum, respectively. The addition of a tuning mass, $m_t$, at a distance $l_t$ from the pendulum-central-mass pivot point, changes the values of the equivalent masses $m_{c_1}$ and $m_{c_2}$, increased by fractions of the tuning mass, function of its position, $l_t$, according to the relations

$$m_{c1_{new}} = m_{c1_{old}} + m_t\left(1 - \frac{l_t}{l_d}\right) \qquad m_{c2_{new}} = m_{c2_{old}} + m_t\left(\tfrac{l_t}{l_d}\right) \tag{1}$$

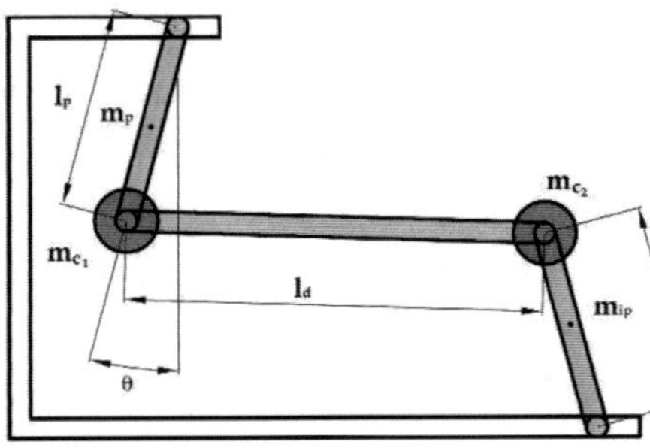


**Figure 1.** FP mechanical scheme.

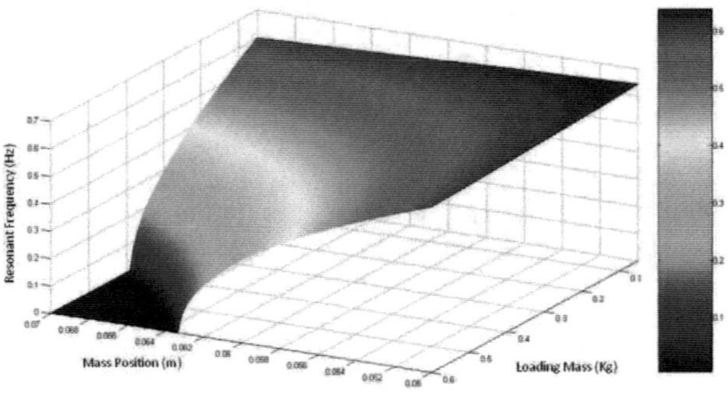


**Figure 2.** $f_o$ vs. the tuning mass value, $m_t$, and the position, $l_t$, for the FP prototype.

Assuming equal lenght of the two vertical beams, $l_b = l_{b_1} = l_{b_2}$, and equal distance of the pivot points of the central mass measured $l_c = l_{c_1} = l_{c_2}$, hypothesis satisfied by our FP monolithic prototypes, the FP resonance frequency, $f_o$, is expressed by [4, 5]

$$f_o = \frac{\omega_o}{2\pi} = \frac{1}{2\pi}\sqrt{\frac{(m_{b_1} - m_{b_2})\frac{gl_b}{2l_c^2} + (m_{c_1} - m_{c_2})\frac{g}{l_c} + \frac{k_\theta}{l_c^2}}{(m_{b_1} + m_{b_2})\frac{l_b^2}{3l_c^2} + (m_{c_1} + m_{c_2})}} = \frac{1}{2\pi}\sqrt{\frac{K_{eq}}{M_{eq}}} \tag{2}$$

where $\omega_o$ is the resonance angular frequency, $K_{eq}$ is the equivalent stiffness constant, sum of the equivalent gravitational linear stiffness constant, $K_{g_{eq}}$, and of the equivalent elastic constant, $K_{e_{eq}}$, given by

$$K_{g_{eq}} = (m_{b_1} - m_{b_2})\frac{gl_b}{2l_c^2} + (m_{c_1} - m_{c_2})\frac{g}{l_c} \qquad K_{e_{eq}} = \frac{k_\theta}{l_c^2} \tag{3}$$

while $M_{eq}$ is the equivalent mass

$$M_{eq} = (m_{b_1} + m_{b_2})\frac{l_b^2}{3l_c^2} + (m_{c_1} + m_{c_2}) \tag{4}$$

Another important FP design parameter is the tuning sensitivity, $S_{f_o}$, relevant to guarantee a comfortable and stable tuning of $f_o$. An analytic expression for $S_{f_o}$ is easily obtained deriving Equation 2 with respect to the tuning mass position, $l_t$, that is

$$S_{f_o} = \frac{df_o}{dl_t} = \frac{g}{2\pi l_c l_d} \frac{m_t}{\sqrt{M_{eq}(m_t) K_{eq}}} \tag{5}$$

Equation 5 shows that the FP sensitivity is function of the value of the tuning mass, $m_t$. The heavier is the tuning mass, the smaller are its displacements necessary to tune $f_o$. This relation must be taken into full account during the FP design phase. In Figure 2, the FP resonance frequency is plotted as function of the value and position of the tuning mass for the FP prototype. The frequency plateau shows the FP instability region.

The FP theoretical transfer functions, describing its dynamics as seismometer and as accelerometer are easily obtained using the simplified Lagrangian model developed by J.Liu et al. [2], but do not include the dissipation effects, that are measurable and are the real limitation to the FP performances. To improve the FP model we introduced a global dissipation term in the lagrangian in the form of the global quality factor $Q(\omega_o)$ [4, 5]. The dependance of the quality factor on the FP resonance frequency, $\omega_o$, has been experimentally demonstrated and will be discussed in the following section. Therefore, defining the coordinates of the pendulum frame (fixed to the ground) as $x_g$ and the coordinate of the FP central mass as $x_c$ (see Figure 1), then the FP transfer function as seismometer is [4, 5]

$$\frac{x_c(\omega) - x_g(\omega)}{x_g(\omega)} = \frac{(1 - A_c)\omega^2}{-\omega^2 + i\frac{\omega_o}{Q(\omega_o)}\omega + \omega_o^2} \tag{6}$$

A FP accelerometer is, instead, implemented with a force feed-back control, with a generated feed-back force proportional to the ground acceleration, experimentally obtained when the current used to drive the actuator coil is proportional to the ground acceleration. Thus defining the ground acceleration as $a_g(\omega) = \omega^2 x_g(\omega)$, then, with the same reasonment done before, the FP transfer function as accelerometer is [4, 5]

$$\frac{x_c(\omega) - x_g(\omega)}{a_g(\omega)} = \frac{(1 - A_c)}{-\omega^2 + i\frac{\omega_o}{Q(\omega_o)}\omega + \omega_o^2} \tag{7}$$

where

$$A_c = \frac{\left(\frac{l_b}{3l_c} - \frac{1}{2}\right)(m_{b_1} - m_{b_2})}{M_{eq}} \tag{8}$$

is the parameter related to the centre of percussion effects [2].

## 3. Test of the monolithic prototype

All the tests were performed on an Aluminium prototype (mod. 08F_100_AL1), shown in Figure 3, shaped with precision machining and electric discharge machining (EDM) from a $134 \times 134 \times 40\,mm$ block of metal (Alloy 7075-T6). The four flex joints, the sensor most critical parts, have an elliptical profile with $100\,\mu m$ minimum thickness and ellipticity ratio of $\epsilon = 16/5$. This shape ensures robustness and long-term durability to the mechanics [4, 5]. The pendula arms ($71.5\,mm$ length and spaced by $102\,mm$) minimize the mass and the moment of inertia, without reducing rigidity and symmetry. The values of the masses of the pendulum arm, of the inverted pendulum arm and of the central mass are $m_p \approx 40\,g$, $m_{ip} \approx 50\,g$ and $m_c \approx 600\,g$, respectively. The FP frequency tunability was obtained as described in [4], machining a large recess in the central mass to hosting suitable shaped and positioned tuning masses.

**Figure 3.** Mechanical monolithic Folded Pendulum with interferometric optical reaodut.

The gaps between the central mass-arms and arms-frame are $1\,mm$ large, much larger than the ones of previous versions of the FP sensor [4], for two reasons: to increase the FP dynamics and the quality factor for applications in air.

The width of the gaps is an important design parameter for a FP seismometer, because it determines its mechanical dynamics. In fact, according to the simplified lagrangian model [2] the lower is the FP natural frequency, the lower is the restoring force to external perturbations and the amplitude of the seismic signal that saturates the FP. Therefore, the width of the gaps are directly evaluted as function of the chosen FP natural resonance frequency and of the expected seismic signal maximum amplitude. The width of the gap is not a problem for a FP accelerometer, being the central mass forced in its rest position by a force feed-back control.

The performances of the monolithic FP strongly depend on the quality factor, $Q$, largely determined by the damping effect of the air in the gaps, and, therefore, function of their width and of the air pressure. It is also well known that $Q$ decreases together with the decrease of the FP natural frequency, $f_o$, so that the knowledge of this function, even if based only on experimental data, is necessary to predict the FP behaviour and to understand its limits in the low frequency band. On the other hand, it was demonstrated in [4] that the same FP resonance frequency can be obtained with tuning masses of different weight suitably positioned. Therefore, we expected to find a dependance of $Q$ on the value and position of the tuning mass, $m_t$, so that $Q = Q(f_o, m_t)$. Although this relationship is largely dependent on the design, structure and material of FP sensor, nevertheless an empirical knowledge would be very useful to predict the global behaviour of $Q$ at different resonance frequencies.

For this task we performed a set of measurements in air, using four different tuning masses of values $m_t = 100\,g, 240\,g, 350\,g, 500\,g$. We performed the tuning procedure with each tuning mass, changing the resonance frequency and measuring the quality factor. We then fitted the measured points for each value of the tuning mass, $m_t$. In Figure 4 all the measurements and the fitting curves are shown. This figure shows that the heavier is the value of the tuning mass, the higher is the value of $Q$. Moreover, the dependance of the $Q$ on the frequency, $f_o$, is linear, with a slope increasing at the increasing of the tuning mass.

We performed linear fittings ($Q$ vs. $f_o$) (Figure 4) to understand the FP real limits at low frequency, although we were aware that at very low frequencies the function is parabolic ($Q = a \cdot f_o^2$). These fittings demonstrate that the prototyype cannot be used in air, already at $f_o = 10\,mHz$. We then performed three sets of measurements, tuning the FP at three different resonance frequencies, that are $300\,mHz$, $370\,mHz$, $450\,mHz$, measuring the FP $Q$ at different values of pressure, $p$. The measurements, reported in Figure 5, show that $Q$ largely increases in the pressure range $0.1\,mBar \div 1\,mBar$, suggesting the need of a deeper study in vacuum, to get rid of the gas damping, extending the Q measurements towards the low frequencies, down

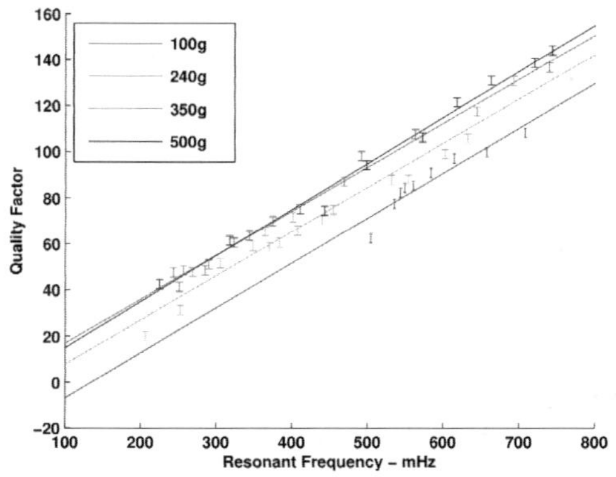


**Figure 4.** $Q$ vs. $f_o$ in air for $m_t = 100\,g,\ 240\,g,\ 350\,g,\ 500\,g$.

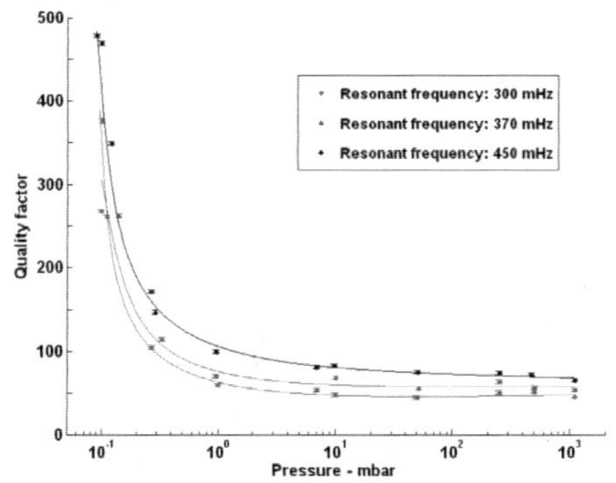


**Figure 5.** $Q$ vs. $p$ for $m_t = 120\,g$.

to tents of mHz resonance frequencies, in order to understand how long the expected square dependence of Q vs frequency lasts or new dissipation mechanisms come into play.

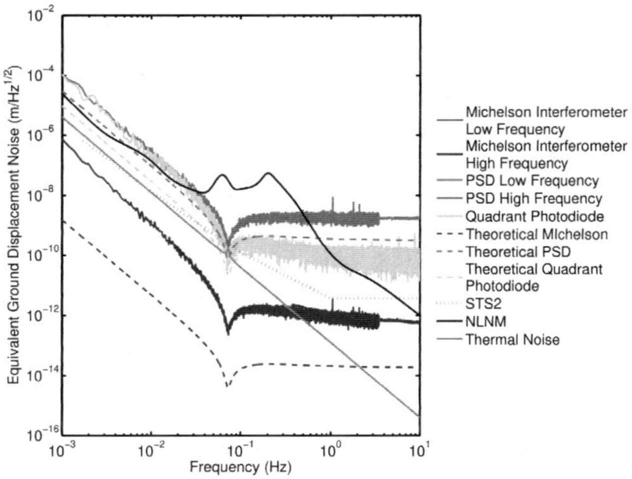


**Figure 6.** Theoretical and experimental sensitivity curves of the monolithic FP seismometer with optical lever and laser interferometric readouts at a natural frequency $f_o = 70\,mHz$ and temperature $T = 300\,K$.

Figure 6 shows the sensitivity curves of the FP seismometer, obtained with optical levers (PSD and quadrant photodiodes) and interferometer readouts. The theoretical sensitivity curves, evaluated according to the model described in [4] are reported for comparison, together with the sensitivity of the STS-2 by Streckeisen [6] and the Peterson New Low Noise Model (NLNM), that is the Minimum Earth Noise [7]. Note the sensitivity of $\approx 10^{-12}\,m/\sqrt{Hz}$ in the band $10^{-3} \div 10\,Hz$. At the present, two monolithic horizontal sensors are located in a blind-ended (side) tunnel 2000 ft deep in the Homestake (South Dakota, USA) mine hosting the Deep Underground Science and Engineering Laboratory (DUSEL) for the seismic characterization of the Homestake site in the frequency band $10^{-4} \div 30\,Hz$.

Figure 7 shows the sensitivity curves of the FP accelerometer, obtained with optical levers (PSD and quadrant photodiodes) and interferometer readouts, together with the sensitivity of the STS-2 by Streckeisen. Note the sensitivity better than $10^{-11}\,m/s^2/\sqrt{Hz}$ in the band $10^{-1} \div 10\,Hz$, obtained with the interferometric readout.

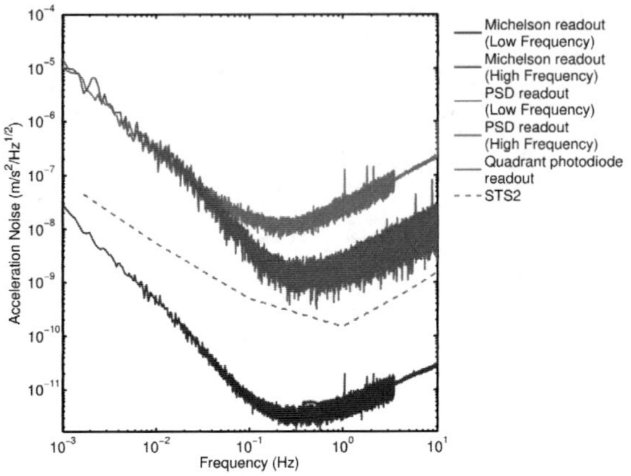


**Figure 7.** Sensitivity curves of the monolithic FP accelerometer with optical levers and laser interferometric readouts.

## 4. Conclusions and future developments

We described a mechanical horizontal monolithic sensor developed at the University of Salerno. The instrument is a monolithic tunable folded pendulum, very sensitive in the low-frequency seismic noise band ($10^{-3} \div 10\,Hz$), with very good immunity to environmental noises. The tunability of the resonance frequency and the integrated laser optical readout are its main characteristics. The interferometric readout has largely improved the sensitivity of FP seismometer that is now $\approx 10^{-12}\,m/\sqrt{Hz}$ in the band $10^{-1} \div 10\,Hz$. The sensitivity of the FP accelerometer is instead already better than $10^{-11}\,m/s^2/\sqrt{Hz}$ in the band $10^{-1} \div 10\,Hz$. Prototypes of monolithic seismometers are operational in selected sites both to acquire seismic data for scientific analysis of seismic noise and to collect all the useful information to understand their performances in the very low frequency band ($10^{-6} \div 10^{-3}\,Hz$).

## Acknowledgments

We acknowledge Galli & Morelli Factory and, in particular, Mr. Carlo Galli for the useful suggestions and for the technical support in the development and implementation of mechanical part of the monolithic FP sensor.

This work is part of the PRIN 2007 Research Project *Low frequency sensitivity in underground cryogenic gravitational wave interferometers* (Project 2007NXMBHP), partly supported by the Ministero dell'Istruzione, dell'Università e della Ricerca (MIUR), Italy.

## 5. References

[1] E. S. Fergusson, *US Nat. Museum Bull.*, **228**, 185 (1962).
[2] J. Liu, L. Ju, D.G. Blair, *Phys. Lett. A*, **228**, 243-249 (1997).
[3] A. Bertolini et al., *Nucl. Instr. and Meth.*, **556**, 616-623 (2006).
[4] F. Acernese, R. De Rosa, G. Giordano, R. Romano, F. Barone, *Rev. Sci. Instrum.*, **79**, 074501, doi:10.1063/1.2943415 (2008).
[5] F. Acernese, R. De Rosa, G. Giordano, R. Romano, F. Barone, Proc. SPIE Vol. 7292, SPIE, Bellingham, WA, 69320K, 72922J, doi: 10.1117/12.814106 (2009).
[6] Y. Nakayama et al., Proc. of IWAA2004, CERN, Geneva, 4-7 October 2004.
[7] J. Berger, P. Davis, *2005 IRIS 5-Year Proposal*, 38 (2005).

Journal of Physics: Conference Series **228** (2010) 012036 doi:10.1088/1742-6596/228/1/012036

# Long term seismic noise acquisition and analysis in the Homestake mine with tunable monolithic sensors

**Fausto Acernese[1,2], Rosario De Rosa[3,2], Riccardo De Salvo[4], Fabio Garufi[3,2], Gerardo Giordano[1,2], Jan Harms[5], Vuk Mandic[5], Angelo Sajeva[6], Thomas Trancynger[7], Fabrizio Barone[1,2]**

[1] Università degli Studi di Salerno, Fisciano (SA), Italia
[2] INFN sez. di Napoli, Napoli, Italia
[3] Università di Napoli Federico II, Napoli, Italia
[4] California Institute of Technology, Pasadena, CA, U.S.A.
[5] University of Minnesota, Twin Cities, MN, U.S.A.
[6] Università di Pisa, Pisa, Italia
[7] South Dakota Science and Technology Authority, Lead, SD, U.S.A.

E-mail: `fabrizio.barone@na.infn.it`

**Abstract.** In this paper we present and discuss the scientific data taken by two mechanical monolithic horizontal sensor prototypes located at the $2000\,ft$ deep level in the Homestake mine (South Dakota, USA). The main goal of this experiment was to provide a preliminary characterization of the Homestake site in the frequency band $10^{-4} \div 30\,Hz$, necessary to evaluate the feasibility of underground gravitational-wave interferometers sensitive at $1\,Hz$ and below, and to test the sensor prototypes in the band $(10^{-6} \div 30\,Hz)$. The first results show the good performances of the sensors, being they able both to detect the Peterson Low Noise Spectrum and, in the very low frequency region, the peaks due to Earth tides, although configured with a limited sensitivity (resonance frequency of $\approx 300\,mHz$ and optical lever readout).

## 1. Introduction

The Homestake mine is the largest known iron-formation-hosted gold-ore body, marked by a rather complicated stratification and history of folding events and metamorphosis of igneous and sedimentary rocks. It lies far from the oceans in the Lead-Deadwood Dome of the Black Hills (South Dakota, USA) and has the deepest reaching tunnels in North America ($8000\,ft$) which provides an optimal stage for monitoring the seismic noise with a three dimensional network of seismometers, but also a very good testbed for seismic sensor prototypes.

Recently, the Homestake mine was chosen to host the Deep Underground Science and Engineering Laboratory (DUSEL). For this task, in August 2008 a first series of seismic measurements was performed at different underground levels in the Homestake mine, the $300\,ft$, the $800\,ft$ and $2000\,ft$ levels, intended as a starting point for a more exhaustive survey [1].

In December 2008, the group of Applied Physics of the University of Salerno installed at Homestake two tunable monolithic seismometers in thermally insulating enclosures onto concrete slabs connected to the bedrock and behind a sound-proofing wall at the $2000\,ft$ level.

In this paper the monolithic seismometers, their installation at $2000\,ft$ and the DAQ system architecture are described. Then, the first scientific data acquired are presented and discussed.

© 2010 IOP Publishing Ltd

## 2. The Monolithic Sensor

The structure of the tunable mechanical horizontal monolithic sensor is based on the Folded Pendulum (hereafter FP), called also *Watt-linkage*, a classical suspension system developed in 1962 [2], recently rediscovered for application in gravitational wave research as ultra-low frequency pendulum resonators for vibration isolation in interferometric detectors of gravitational waves [3].

Based on the FP basic scheme, single-axis monolithic accelerometers have been developed as sensors in the control system for advanced seismic attenuators [4]. The progress in precision micro-machining, that has allowed the construction of tunable sensors of small size with extremely soft flexures at the pendulum's hinges, and the innovative application of laser optics readout techniques have largely improved the their sensitivity in the low frequency band and increased their immunity to environmental noises [5, 6]. Special tuning procedures have been also developed to decrease the sensor natural frequency (to values as low as $\approx 70\,mHz$) and dimensions, allowing its use both as seismometer and accelerometer for seafloors or boreholes [5].

On the basis of these scientific results, a low-noise high-resolution horizontal monolithic FP sensor was developed at the University of Salerno for geophysical applications aimed to explore the low frequency band of the seismic spectrum, both as a stand-alone sensor or as part of large and geographically distributed seismic networks.

The theoretical FP transfer function, describing the FP dynamics as seismometer and/or accelerometer, can be easily obtained using the simplified Lagrangian model developed by J.Liu et al. [3]. This model, based on the mechanical scheme shown in Figure 1, describes only the FP basic dynamics, but it is anyway very useful to understand the FP main characteristics and expected performances.

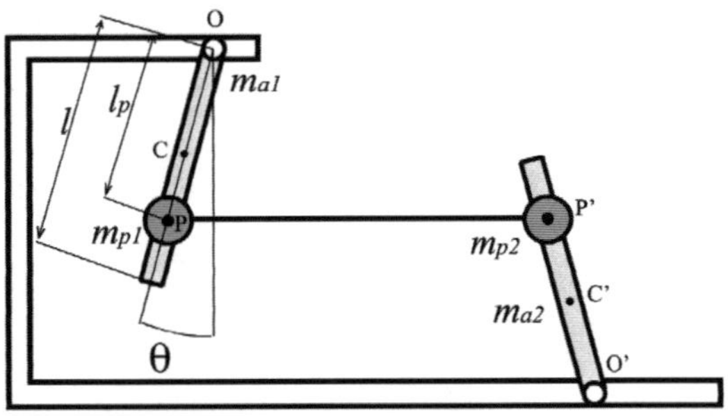


**Figure 1.** Folded Pendulum mechanical scheme.

**Figure 2.** Mechanical monolithic horizontal seismometer prototype (mod. 08F_100_AL1).

In particular, defining the coordinate of the pendulum frame (fixed to the ground) as $x_g$ and the coordinate of the FP central mass as $x_c$ (Figure 1), then the FP transfer function is

$$\frac{x_c(\omega)}{x_g(\omega)} = \frac{\omega_o^2 - A_c\omega^2}{\omega_o^2 - \omega^2} = 1 + \frac{(1 - A_c)\omega^2}{\omega_o^2 - \omega^2} \tag{1}$$

where $A_c$ is the parameter related to the center of percussion effects [3] and

$$\omega_o = \sqrt{\frac{(m_{a1} - m_{a2})\frac{gl}{2l_p^2} + (m_{p1} - m_{p2})\frac{g}{l_p} + \frac{k_\theta}{l_p^2}}{(m_{a1} + m_{a2})\frac{l^2}{3l_p^2} + (m_{p1} + m_{p2})}} \tag{2}$$

is the FP angular resonance frequency. The mass displacement transfer function with respect to the ground displacement can be obtained as

$$\frac{x_c(s) - x_g(s)}{x_g(s)} = \frac{(A_c - 1)s^2}{s^2 + \frac{\omega_o}{Q(\omega_o)}s + \omega_o^2} \tag{3}$$

where $Q(\omega_o)$ is the global quality factor, that includes the dissipative effects.

The configuration of the monolithic FP as seismometer is straightforward: the readout output signal $(x_p(\omega) - x_s(\omega))$ is acquired, calibrated and processed according to Equation 3.

The sensor prototype (mod. 08F_100_AL1) is made of Aluminum, shaped with precision machining and electric discharge machining (EDM) from a $134 \times 134 \times 40\,mm$ block of metal (Alloy 7075-T6) (Figure 2). The four flex joints have an elliptical profile with $100\,\mu m$ minimum thickness and ellipticity ratio of $\epsilon = 16/5$. This shape ensures robustness and long-term durability to the mechanics [5, 6]. The pendula arms ($71.5\,mm$ length and spaced by $102\,mm$) minimize mass and moment of inertia, without reducing rigidity and symmetry. The values of the masses of the pendulum arm, of the inverted pendulum arm and of the central mass are $m_p \approx 40\,g$, $m_{ip} \approx 50\,g$ and $m_c \approx 600\,g$, respectively. The FP frequency tunability was obtained as described in [5], machining a large recess in the central mass to hosting suitable shaped and positioned tuning masses. The gaps between central mass-arms and arms-frame are $1\,mm$ large to increase the FP mechanical dynamics and the quality factor for applications in air.

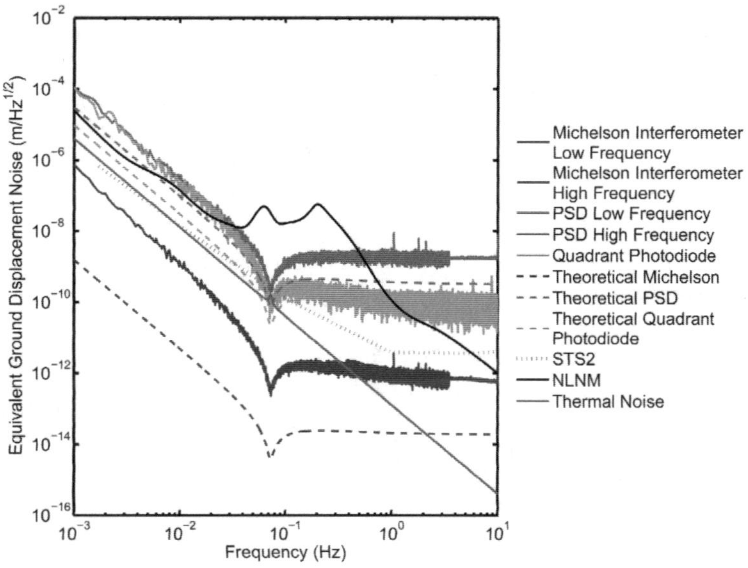


**Figure 3.** Theoretical and experimental sensitivity curves of the monolithic FP seismometer.

We measured the three sensitivity curves corresponding to optical levers (with PSD and quadrant photodiodes) and interferometer readouts. In Figure 3 the best theoretical and experimental sensitivities curves at $T = 300\,K$ and $f_o = 70\,mHz$ in air and without thermal stabilization are shown.

To compare the sensitivity of the FP seismometer with other geophysical instruments, we reported in Figure 3 also the sensitivity of the STS-2 by Streckeisen, that represents the state-of-art of the low frequency seismic sensors [7]. Figure 3 shows that the monolithic FP sensor has already a sensitivity comparable or better (with interferometric readout) than the STS-2 in the band $10^{-3} \div 10\,Hz$. We reported also the Peterson New Low Noise Model (NLNM) [8], that is the Minimum Earth Noise evaluated from a collection of seismic data from 75 sites located around the world, that describes the minimum level of earth noise: noise levels below this are never - or extremely rarely - observed.

## 3. The $2000\,ft$ Seismic Station at Homestake

The $2000\,ft$ station was built in August 2008 on the concrete platform of an old charging station located on the bedrock, having carefully checked that any audible source of dropping water were eliminated as good as possible. It was obtained sectioning a tunnel and sealing it from two sides with double layered walls. Many instruments were located in the station inside a specially designed insulated box. A separate hut at some distance to the instrument area hosts the data acquisition PCs and the DC power supply [1].

The environmental conditions at the $2000\,ft$ level are stable around $20^o\,C$ with an rms of about $0.2^o\,C$. Sensors are installed close to the seismometers to monitor the most important environmental quantities: temperature, pressure, humidity, sound and magnetic field. Humidity monitoring is relevant because the humidity levels may reach 100% and condensation of water on the electronics could start soon, although the insulation boxes are good enough to bring humidity down to $80\% - 90\%$ also in presence of 100% humidity. Pressure and sound monitoring is important because these quantities may show significant correlation with seismic data since both could generate rock vibrations or may even directly act on seismometers despite all efforts to isolate the instruments.

The data-acquisition system includes the configuration of the local network and remote access to PCs. The acquisition system is based on the PCI 6289 card from National Instruments$^{TM}$. It contains a 18 bit ADC and an internal amplifier with a maximum amplification of 100. The input range is limited to $\pm 10\,V$. The Data Acquisition is managed by Labview$^{TM}$programs which also generate analog-output signals to initiate mass-centering procedures of seismometers, if necessary. The data are first stored in ASCII files as $128\,s$ records sampled at $100\,Hz$. Then the data is made available to the outside world by an ftp server while the timing signals are provided to the acquisition system through optical fibers.

The sequence of installation of the monolithic sensor within the insulated box of the $2000\,ft$ station is shown Figure 4.

**Figure 4.** Installation of the monolithic sensor within the insulated box of the $2000\,ft$ station.

## 4. Experimental Results

In this section we present some of the preliminary results of the first month of data acquisition of FP monolithic sensor prototypes. In particular, the power spectral density of ground displacement at $2000\,ft$ underground level (compared with the theoretical expected FP noise, the simulated horizontal semidiurnal Earth tides [9] and the Peterson's New Low Noise Model) is shown in Figure 5. What is relevant to note is its large measurement band $(10^{-5} \div 30\,Hz)$

and the good sensitivity, although the latter is largely limited by our technical choice of using an optical lever readout, surely more robust than an interferometric one but less sensitive, and a safe resonance frequency ($300\,mHz$) for this first long unattended test in the Homestake mine. Tuning the FP sensor at a lower resonance frequency would simply translate the sensitivity curve towards the low frequency region, enlarging the low frequency band. Nevertheless, even with this not optimal configuration, the instruments is able to detect part of the Peterson Low Noise Spectrum and, in the very low frequency region, the peaks due to Earth tides (Figure 5).

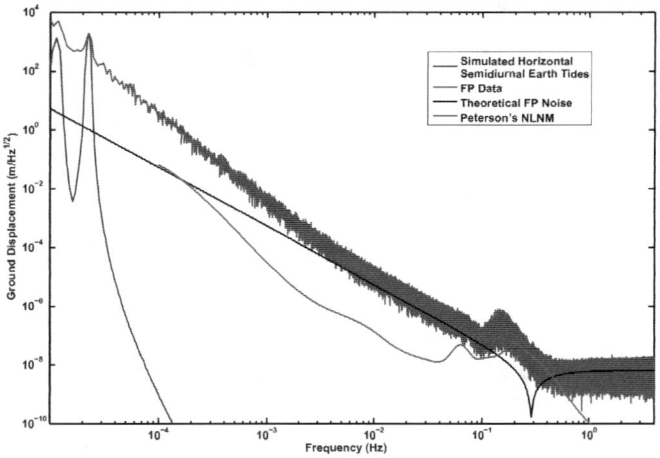


**Figure 5.** Power Spectral Density of ground displacement at 2000 ft underground level (compared with theoretical FP noise, simulated horizontal semidiurnal Earth tides and Peterson's New Low Noise Model).

The apparent limitation in the low frequency band is simply due to the fact that the power spectral density has been evaluated only for one month of data. Longer runs will allow the exploration of lower frequency regions. In fact, the FP monolithic sensor is an open loop sensor, so that there are no limitations coming from the feed-back control, as it happens in the majority of the commercial instruments used for seismic noise acquisition. The limitations are mainly due to the electronic noise of the readout system, to the thermal noise of the mechanical joints and to the effects of the air for sensors not operating in vacuum.

For completeness, the spectrogram $500\,h$ of data of ground displacement as recorded by our seismometer is shown in Figure 6. This figure shows the very resonance frequency of the instrument and its ability of measuring teleseimic earthquakes. Of course, longer data acquisitions are necessary for a more accurate analysis of the performances, although these preliminary results are already very promising.

## 5. Conclusions and future developments

Two FP monolithic sensors are installed into the $2000\,ft$ station in the Homestake mine since December 2008, equipped with a National Instruments$^{TM}$(32 channels - 18 bit ADC) data acquisition system, a network access through optical fibre cables and an environmental monitoring system.

The FP sensor sensitivity is close to the theoretical one in the band ($10^{-5} \div 30\,Hz$) for the resonance frequency of about $300\,mHz$ and for the optical lever readout configuration. Better sensitivities could be obtained reducing the resonance frequency of the instrument and using more sensitive optical readouts, but it was preferred to priviledge the instrument robustness

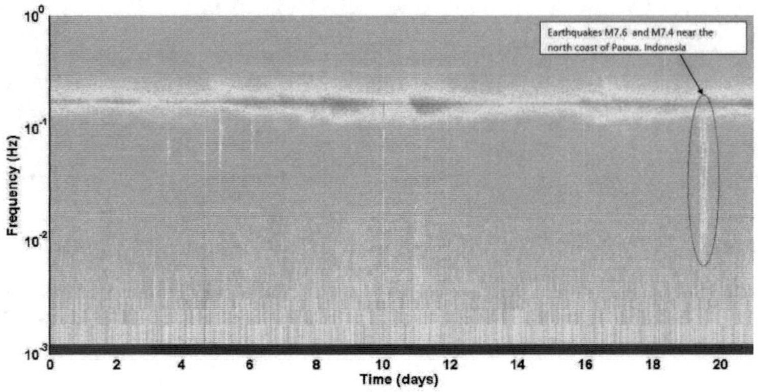


**Figure 6.** Spectrogram of ground displacement data of 500 $h$ as recorded by our seismometer. Two teleseismic earthquakes near north coast of Papua, Indonesia, are shown (January, 03, 2009, 19:43:50 UTC time and Saturday, January 03, 2009, 22:33:40 UTC time).

for this first unattended measurement in the Homestake mine. Nevertheless, even with this not optimal configuration, the instruments is able to detect part of the Peterson Low Noise Spectrum and, in the very low frequency region, the peaks due to Earth tides.

These data, together with all the auxiliary technical information obtained with this test, will be used to organize a new experiment for the low frequency seismic characterization of the Homestake site with an improved version and configuration of these sensors to be positioned at different levels and orientations.

**Acknowledgments**

We acknowledge Galli & Morelli Factory and, in particular, Mr. Carlo Galli for the useful suggestions and for the technical support in the development and implementation of mechanical part of the monolithic FP sensor.

This work is part of the PRIN 2007 Research Project *Low frequency sensitivity in underground cryogenic gravitational wave interferometers* (Project 2007NXMBHP), partly supported by the Ministero dell'Istruzione, dell'Università e della Ricerca (MIUR), Italy.

## 6. References

[1] F. Acernese, R. DeSalvo, J. Harms, A. Sajeva, L. Shaffner, T. Trancynger, LIGO T080344-00-R, Caltech, USA (2008).
[2] E. S. Fergusson, *US Nat. Museum Bull.*, **228**, 185 (1962).
[3] J. Liu, L. Ju, D.G. Blair, *Phys. Lett. A*, **228**, 243-249 (1997).
[4] A. Bertolini et al., *Nucl. Instr. and Meth.*, **556**, 616-623 (2006).
[5] F. Acernese, R. De Rosa, G. Giordano, R. Romano, F. Barone, *Rev. Sci. Instrum.*, **79**, 074501, doi:10.1063/1.2943415 (2008).
[6] F. Acernese, R. De Rosa, G. Giordano, R. Romano, F. Barone, Proc. SPIE Vol. 7292, SPIE, Bellingham, WA, 69320K, 72922J, doi: 10.1117/12.814106 (2009).
[7] Y. Nakayama et al., Proc. of IWAA2004, CERN, Geneva, 4-7 October 2004.
[8] J. Berger, P. Davis, *2005 IRIS 5-Year Proposal*, 38 (2005).
[9] Wilhelm et al., *Tidal Phenomena*, Vol.66 of *Lectures Notes in Earth Science*, Springer, USA (1997).

Journal of Physics: Conference Series **228** (2010) 012037 doi:10.1088/1742-6596/228/1/012037

# The 2 Degrees of Freedom Facility in Firenze for the study of Weak Forces

**L Marconi**[1,2], **R Stanga**[2,3], **M Lorenzini**[2,3], **C Grimani**[2,4], **M Bassan**[3,5], **G Pucacco**[3,5], **L Di Fiore**[6], **R De Rosa**[6,7], **F Garufi**[6,7], **L Milano**[6,7]

[1] Dipartimento di Fisica, Università degli Studi di Roma Tor Vergata, Roma, Italy.
[2] Istituto Nazionale di Fisica Nucleare, Firenze, Italy.
[3] Dipartimento di Astronomia e Scienza dello Spazio, Università degli Studi di Firenze, Firenze, Italy.
[4] Istituto di Fisica, Università degli Studi di Urbino, Urbino, Italy.
[5] Istituto Nazionale di Fisica Nucleare, Roma Tor Vergata, Italy.
[6] Istituto Nazionale di Fisica Nucleare, Napoli, Italy.
[7] Università Federico II, Napoli, Italy.

E-mail: marconi@fi.infn.it

**Abstract.**The LISA test-mass (TM) is sensitive to weak forces along all 6 Degrees of Freedom (DoFs). Extensive ground testing is required in order to evaluate the influence of cross-talks of read-outs and actuators operating on different DoFs. To best represent the flight conditions, we developed in Firenze a facility with 2 soft DoFs. Using this facility we measure the forces and stiffnesses acting simultaneously along the 2 soft DoFs, and, more specifically, we will be able to debug residual couplings between the TM and the capacitive position sensor that reads the TM position, and to measure actuation cross talks with closed feedback loop. The facility is now ready, and here we report on the commissioning tests, and on the first measurements.

## 1. Introduction

The LISA requirement for the residual acceleration noise in the free-falling frame of the test masses M is [1]:

$$\frac{S_f^{1/2}}{M} \leq 3\cdot 10^{-15}\left[1+\left(\frac{f}{0.003Hz}\right)^2\right]\frac{m}{s^2\sqrt{Hz}}$$

for $0.1mHz < f < 0.1Hz$. A drag-free operation mode is necessary, with minimal coupling between test mass and spacecraft, so that the above requirements can be met. A capacitive sensor system (GRS, *Gravitational Reference Sensor)* [2] is used to provide the relative position input to the contol loop that is closed on micro-thrusters that force the spacecraft to follow the test-mass. In the case of a single-mass/single-axis high gain control loop, the residual closed-loop test mass acceleration [1], is (parameters are indicates in figure 1):

$$a_n \approx \frac{f_{str}}{m} + \omega_p^2\left(x_n + \frac{F_{S/C}}{M\omega_{DF}^2}\right)$$

© 2010 IOP Publishing Ltd

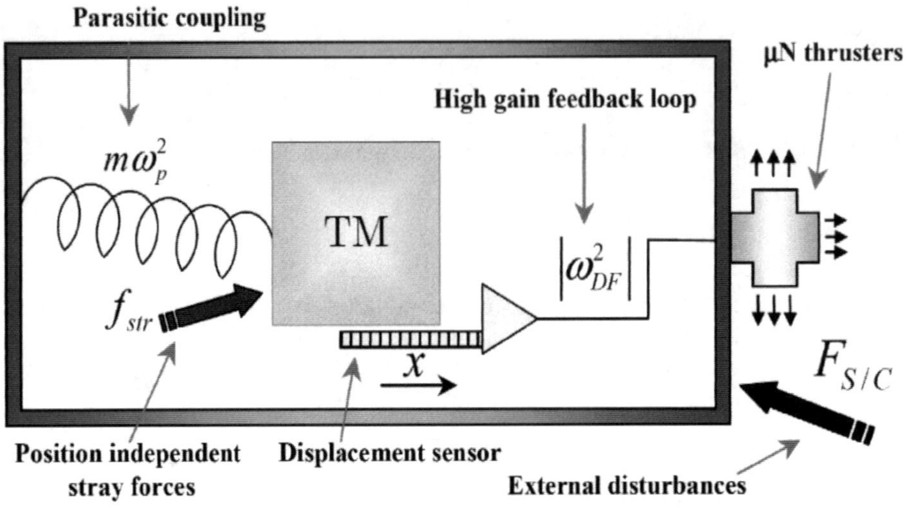


**Figure 1.** A sketch of the interaction between the spacecraft and the test mass.

Ground tests are required to study the residual weak forces that may couple the Test Mass to the GRS. The challenge for these tests is to cancel gravity, in order to simulate a free fall condition along as many DoFs as possible. Up to now, very good results have been obtained with a torsion pendulum, that only allows to study one single DoF [3], [4].

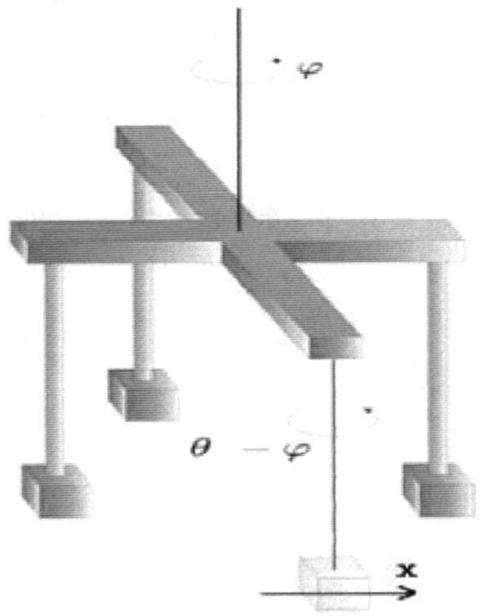


**Figure 2.** Schematic of the roto-translational pendulum.

**Figure 3.** The crossbar.

## 2. The Firenze facility

Our two DoF facility (the scheme is in figure 2 and 3) will better represent the flight conditions where the test mass is sensitive to force along all 6 DoFs [5], [6]. The facility will measure the forces and stiffnesses simultaneously acting along different soft DoFs. The advantages with respect to a single DoF test bench are a more effective identification and debug of spurious effects and the possibility to test actuation cross talk with closed feedback loop. In particular, it allows us to measure the residual disturbance along one DoF when we close the control loop on the other one. More details in [7], [8].

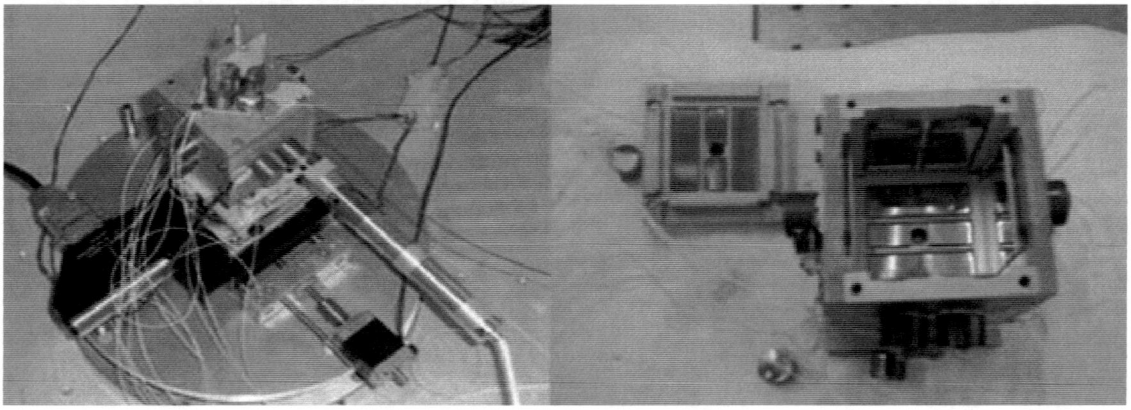

**Figure 4.** The capacitive sensor (GRS).

## 3. Sensing and actuation via capacitive coupling

The capacitive sensing and feedback electronics was designed building on the experience of analogous devices previously developed for similar, single degree of freedom applications [3]. Each of two identical channels reads out (and actuates) a pair of capacitive sensors (figure 4) placed along the x axis of the GRS on opposite sides of the Test Mass. Sum and differences of these two channels outputs provide information about translation along x axis and rotation ($\theta$-$\varphi$) as defined in figure 2. The signals are modulated at about 100kHz (in green the injection electrodes, figure 5), and then read and demodulated with lock-in amplifiers. The capacitors are also used as electrostatic actuators, to close feedback position loops on the TM. We have recently implemented a second version of this electronics, where the actuation and sensing are separated on different boards, in a modular design. With this set up, hosted in a NIM crate, we can add readout and feedback channels as we need.

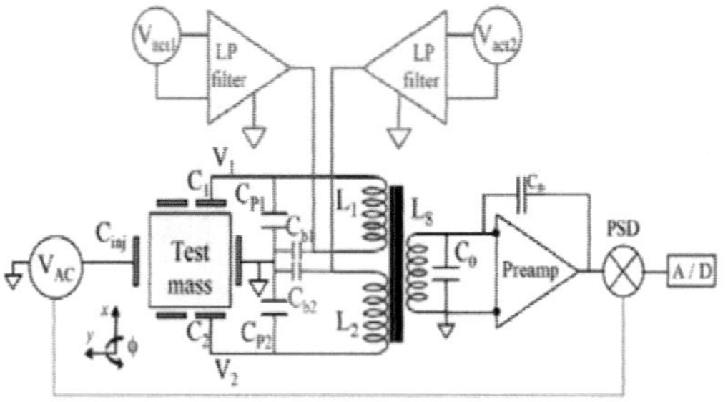


**Figure 5.** The capacitive sensor (GRS).

## 4. Closing the feedback loop on position

The output signals are filtered by the lock-in amplifiers, and then are fed to a commercial 18 bits ADC board in a PC. A Labview routine generates a feedback signal, which is converted by a DAC, which provides the input to an analog board, where is amplified, modulated at 200 Hz, and then sent to the same pair of sensing capacitors, to obtain an electrostatic damping of the oscillations, and hold the test mass in place. The frequency response of servo is generated with a PID network:

$$G(s) = G_1 \frac{1}{\omega_1 + s}\left(s^2 + \omega_0^2 + \gamma_G s\right)\left(\omega_{lockin} + s\right)\frac{1}{s^2 + \omega_2^2 + \gamma_2 s}$$

where $G_1$ is the gain of the loop, $\omega_1$ is the cut frequency of the filter, $\omega_0$, $\gamma_G$ are the resonant frequency and the damping coefficient of the rotational pendulum, $\omega_{lockin}$ is the cut frequency of the output of the lockin and $\omega_2$, $\gamma_2$ are the resonant frequency and the damping coefficient of the a further filter to stop noise from the high frequency (see figure 6 for the scheme).

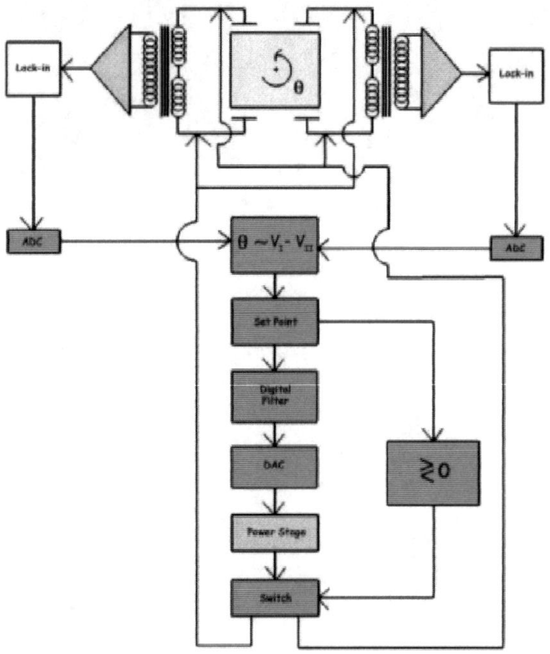


**Figure 6.** The block diagram of the feedback loop.

In the figures 7,8,9 it are shown the outcome of some of the first tests, made with the translational degree of freedom blocked.

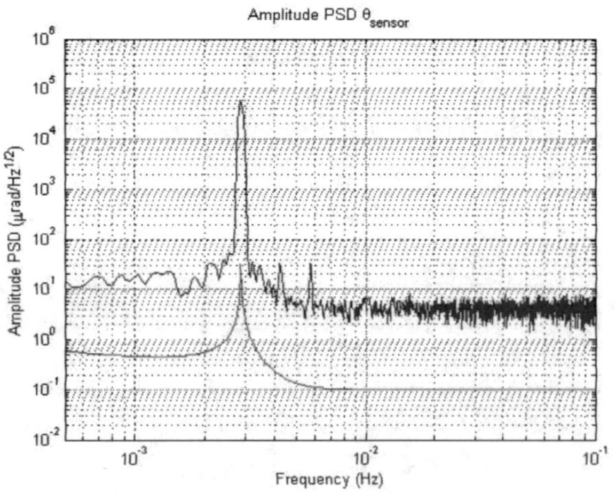


**Figure 7.** Measurement of the resonance frequency of the (θ-φ) pendulum. The red line is the sum of the thermal noise of the pendulum and the capacitive read-out noise.

We are presently engaged in the task of defining and implementing an algorithm that will allow us to acquire control of both Dofs simultaneously. The read out noise limited sensitivity is better than 2 $nm/\sqrt{Hz}$ down to 50 *mHz* [7]. It translates to a sensitivity in force of $10^{-13}$ $N/\sqrt{Hz}$, for the translational DoF, and of $2.5\times10^{-14}$ $N/\sqrt{Hz}$, for the rotational DoF, and it is adequate to reach the thermal noise limit of the two pendulum stages; at lower frequencies the noise is higher, with a consistent contribution from the residual motion of the test mass.

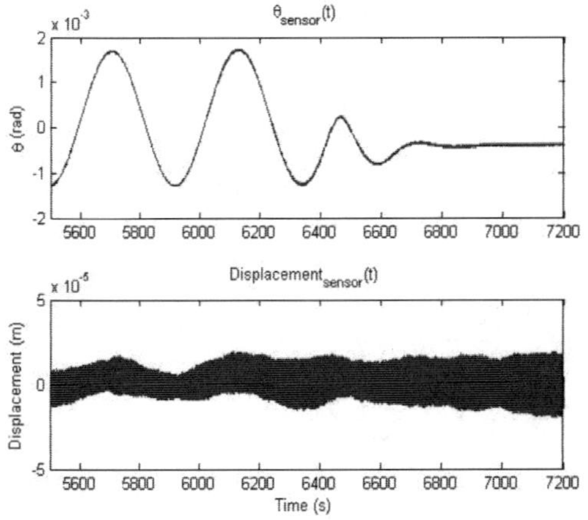


**Figure 8.** Closing the loop we damp only the rotation but not the translation (pendulum motion).

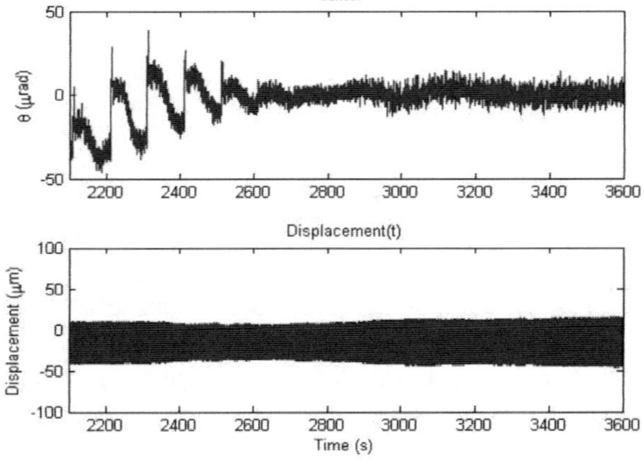


**Figure 9.** The feedback implemented after centering the capacitive sensor using the motorized rotation stage.

## 5. Optical position sensing: autocollimator and optical lever

A great flexibility can be obtained using second position readout, independent from the hardware through which the feedback is applied. Cross correlation of optical and electrostatic readout have proven effective in reducing the readout noise [4]. In addition to the capacitive readout, two optical readout schemes will also be implemented, one for each degree of freedom.

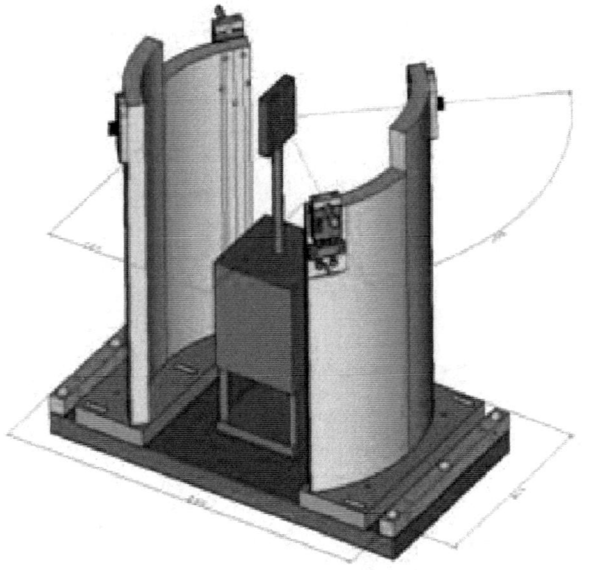

**Figure 10.** 3D scheme of the optical lever read-out.

The first is based on a commercial autocollimator (Elcomat Vario 300). It will read the rotation of the mirror attached to the crossbar, and it will be used to measure the angle φ, with a sensitivity of about 3 x $10^{-8}$ rad/$Hz^{1/2}$ [7] A second optical readout will soon be implemented on the facility (figure 10), and it will be used to measure rotation and displacement of the mirror attached to the TM. It will consist of two channels, that will allow to measure the displacement and the rotation (θ-φ) of the TM [9].

## 6. Conclusion

We have tested with satisfaction the capabilities of our two degrees of freedom facility, with one degree of freedom blocked. The next steps will be: to substitute the present TM/capacitive sensor system with a TM and a capacitive sensor more closely similar to the flight model, which will also allow an easier management of the translational degree of freedom; then, to unblock the second degree of freedom and test the performances of the entire facility; and, finally, to substitute the electronics with boards similar to the flight electronics. With that, we will have a test bench that will simulate free fall on ground, at least along two DoF's.

## 7. Reference

[1] Vitale S et al, "LISA and its in flight test precursor SMART-2", *Nuclear Physics B,* vol. 110, pp. 210, 2002 and references therin

[2] Dolesi R et al, "Gravitational sensor for LISA and its technology demonstration mission", *Class. And Quant. Grav.,* vol. 20, pp. S99, 2003

[3] M.Hueller, "Geodesic motion of LISA test masses: development and testing of drag-free position sensor*", PhD degree thesis,* Trento 2003

[4] Hueller H, Cavalleri A, Dolesi R, Vitale S and Weber W J, "Torsion Pendulum Facility for ground testing of gravitational sensors for LISA", *Class. And Quant. Grav.,* vol. 19, pp. 1757, 2002

[5] C.D.Hoyle et al, "4-Mass Pendulum for ground testing of LISA displacement", in *Proc. Marcel Grossman Meeting,* 2003

[6] Y.Su et al, "New test of the universality free fall", *Phys. Rev.,* vol. D50, pp. 3614, 1994

[7] R. Stanga et al, "Double Degree of Freedom pendulum facility for the study of weak forces", in *JPCS,* vol 154, 012032

[8] G.Bagni, C.Grimani, L.Marconi, R.Stanga, F.Vetrano, A.Viceré, "Ground Based Test for LISA and LISA Pathfinder", in *Proc. NSS/MIC IEEE,* Rome 2004

[9] Di Fiore L et al, *Class. And Quant. Grav.,* vol. 21, pp. S261, 2004

8th Edoardo Amaldi Conference on Gravitational Waves IOP Publishing
Journal of Physics: Conference Series **228** (2010) 012038 doi:10.1088/1742-6596/228/1/012038

# The magnetic diagnostics subsystem of the *LISA* Technology Package

**M Diaz-Aguiló[1,2], E García–Berro[1,2], A Lobo[2,3], N Mateos[2,3] and J Sanjuán[2]**

[1]Departament de Física Aplicada, Universitat Politècnica de Catalunya, c/Esteve Terrades, 5, 08860 Castelldefels, Spain

[2]Institut d'Estudis Espacials de Catalunya, c/Gran Capità 2–4, Edif. Nexus 104, 08034 Barcelona, Spain

[3]Institut de Ciències de l'Espai (CSIC), Campus UAB, Facultat de Ciències, Torre C-5, 08193 Bellaterra, Spain

E-mail: `marc.diaz.aguilo@fa.upc.edu`

**Abstract.** The Magnetic Diagnostics Subsystem of the *LISA* Technology Package (*LTP*) on board the *LISA* Pathfinder (*LPF*) spacecraft includes a set of four tri-axial fluxgate magnetometers, intended to measure with high precision the magnetic field at the positions they occupy. However, their readouts do not provide a direct measurement of the magnetic field at the positions of the test masses. Therefore, an interpolation method must be implemented to obtain this information. However, such interpolation process faces serious difficulties. Indeed, the size of the interpolation region is excessive for a linear interpolation to be reliable, and the number of magnetometer channels does not provide sufficient data to go beyond that poor approximation. Recent research points to a possible alternative to address the magnetic interpolation problem by means of neural network algorithms. The key point of this approach is the ability neural networks have to learn from suitable training data representing the magnetic field behaviour. Despite the large distance to the test masses and the insufficient magnetic readings, artificial neural networks are able to significantly reduce the estimation error to acceptable levels. The learning efficiency can be best improved by making use of data obtained from on-ground measurements prior to mission launch in all relevant satellite locations and under real operation conditions. Reliable information on that appears to be essential for a meaningful assessment of magnetic noise in the *LTP*.

## 1. Introduction

*LISA* Pathfinder (*LPF*) is a science and technology demonstrator mission, led by programmed by the European Space Agency (*ESA*), within its *LISA* (Laser Interferometer Space Antenna) mission activities [1]. *LISA* (Laser Interferometer Space Antenna) is a joint *ESA-NASA* mission which will be the first low frequency (milli-Hz) gravitational wave detector, and also the first space-borne gravitational wave observatory. *LPF*'s payload includes the *LISA* Technology Package (*LTP*), and this will be the highest sensitivity geodesic explorer flown to date. The *LTP* is designed to measure relative accelerations between two test masses (TMs) in nominal free fall (geodesic motion) with a noise budget of

$$S^{1/2}_{\delta a,LPF}(\omega) \leq 3\times 10^{-14}\left[1+\left(\frac{\omega/2\pi}{3\ \mathrm{mHz}}\right)^2\right]\frac{\mathrm{m\ s^{-2}}}{\sqrt{\mathrm{Hz}}} \tag{1}$$

© 2010 IOP Publishing Ltd

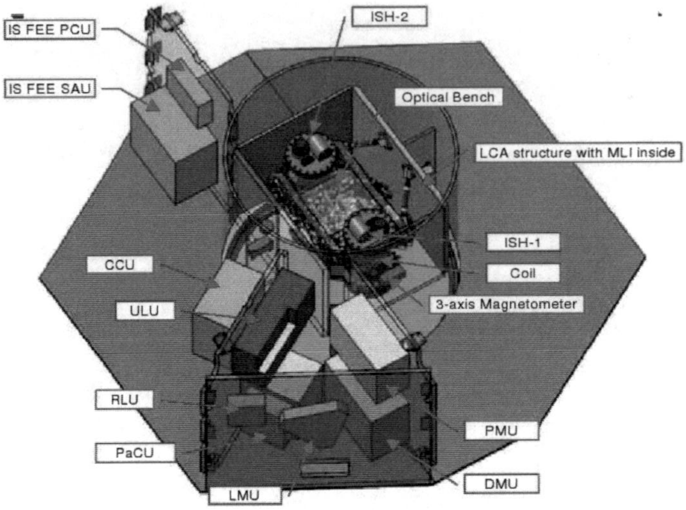


**Figure 1.** View of the *LPF* science-craft. The *LCA* is in the center, surrounded by a double cylindrical shield. Outside it, a number of electronic boxes are represented, most of which are sources of magnetic field. The four magnetometers are the grey little boxes labelled with the 3-axis Magnetometer label and are mounted on the outer cylindrical shell.

in the frequency band between 1 mHz and 30 mHz [2].

Noise in the *LTP* arises as a consequence of various disturbances, mainly generated within the spacecraft itself, which limit the performance of the instrument. A number of these disturbances are monitored and dealt with by means of suitable devices, which form the so-called Diagnostics Subsystem [3]. In *LPF*, this includes thermal and magnetic diagnostics, plus the radiation monitor, which provides counting and spectral information on ionizing particles hitting the spacecraft. The magnetic diagnostics system will be the subject of our attention here.

One of the most important functions of the *LTP* magnetic diagnostics is the determination of the magnetic field and its gradient at the positions of the TMs. For this, it includes a set of four tri-axial fluxgate magnetometers, intended to measure with high precision the magnetic field at the positions they occupy in the spacecraft — see figure 1. However, their readouts do not provide a direct measurement of the magnetic field at the positions where the TMs are, and an interpolation method must therefore be implemented to calculate it. In the circumstances we face, this is a difficult problem, mostly because the magnetometer layout is such that they are too distant from the locations of the TMs compared with the typical distances between magnetic sources on the satellite. It is important to solve this problem as magnetic noise can be as high as 40 % of the total budget [2] given in Eq. (1), and hence it must be properly quantified.

## 2. Description of the problem

Magnetic noise in the *LTP* is allowed to be a significant fraction of the total mission acceleration noise because 40 % of the total noise ( $3\times10^{-14}\,\mathrm{m\,s^{-2}\,Hz^{-1/2}}$ ) can be apportioned to magnetism, see Eq. (1). This noise occurs because the residual magnetisation and susceptibility of the TMs couple to the surrounding magnetic field. The magnetic field and its gradient randomly fluctuate in the regions occupied by the TMs, thus resulting in a randomly fluctuating force:

$$\delta\mathbf{F} = \left\langle \left[ \left( \mathbf{M} + \frac{\chi}{\mu_0}\mathbf{B} \right) \cdot \delta\boldsymbol{\nabla} \right] \mathbf{B} + \frac{\chi}{\mu_0}\,[\delta\mathbf{B}\cdot\boldsymbol{\nabla}]\,\mathbf{B} \right\rangle V \tag{2}$$

In this expression $\mathbf{B}$ is the magnetic field in the TM, $\boldsymbol{\nabla}B$ stands for the magnetic field gradients, $\chi$ its magnetic susceptibility ($\chi = 2.5 \cdot 10^{-5}$), $\mathbf{M}$ its density of magnetic moment (magnetisation, with a nominal value of $2 \cdot 10^{-4}\,\mathrm{A/m}$), $V$ is the volume of the TMs ($V = 9.7336 \cdot 10^{-5}\,\mathrm{m}^3$), $\mu_0$ is the vacuum magnetic constant, $4\pi \times 10^{-7}\,\mathrm{m\,kg\,s^{-2}\,A^{-2}}$), and $\langle \cdots \rangle$ indicates TM volume average of the enclosed quantity. Finally, $\delta\mathbf{B}$ represents the fluctuation of the magnetic field, and $\delta\boldsymbol{\nabla}$ stands for the fluctuation of the gradients.

Quantitative assessment of magnetic noise in the *LTP* clearly requires real-time monitoring of the magnetic field, which in *LPF* is done by means of a set of four tri-axial fluxgate magnetometers (12 magnetic channels) [4]. These devices have a high-permeability magnetic core, so they must be kept somewhat far from the TMs. The TMs are located at $x = \pm 0.1880\,\mathrm{m}$ ($y = 0$), and the magnetometers are located at the following positions: magnetometer $M_1$ is located at $x = -0.0758\,\mathrm{m}$ and $y = 0.3694\,\mathrm{m}$, magnetometer $M_3$ is at $x = 0.0758\,\mathrm{m}$ and $y = -0.3694\,\mathrm{m}$, whereas magnetometers $M_2$ and $M_4$ are along the x-axis but at positions $x = \pm 0.3765\,\mathrm{m}$ ($y = 0$). Therefore, the minimum distance of a magnetomer to the position of the TM is $\sim 20\,\mathrm{cm}$ (all TMs and magnetometers are place within the same $z-$plane). The consequence of this is that the measured field does not directly give that at the positions of the TMs, as is needed. Hence, a procedure to estimate the field at these positions, based on the data from the magnetometers, must be established.

As previously mentioned, the sources of magnetic field are essentially electronics boxes plus a few genuinely magnetic components inside the spacecraft. The interplanetary magnetic field is orders of magnitude weaker, hence of little relevance to the effects considered here, and solar panel effects will not be considered. There are no sources of magnetic field inside the *LTP* Core Assembly (*LCA*), all being placed outside its walls but within the spacecraft. Therefore, the magnetic field is smaller towards the centre of the *LCA* than it is in its periphery, where the magnetometers take measurements. The number of sources identified by Astrium (prime contractor of the mission) is around 40, and these can be modelled as point magnetic dipoles [5]. In the following sections two different interpolation schemes for the Magnetic Diagnostics System of *LISA* Pathfinder will be presented and their performance compared.

## 3. Classical interpolation theory

The *LCA* region is treated as a vacuum, with a magnetic field of zero divergence and curl, $\boldsymbol{\nabla}\cdot\mathbf{B}(\mathbf{x},t) = 0$ and $\boldsymbol{\nabla} \times \mathbf{B}(\mathbf{x},t) = 0$. Hence, a scalar function $\Psi(\mathbf{x},t)$ exists such that

$$\mathbf{B}(\mathbf{x},t) = \boldsymbol{\nabla}\Psi(\mathbf{x},t) \quad \text{with} \quad \nabla^2\Psi(\mathbf{x},t) = 0$$

The solution to this equation can be expressed as an orthogonal series of the form

$$\Psi(\mathbf{x},t) = \sum_{l=0}^{\infty} \sum_{m=-l}^{l} M_{lm}(t)\, r^l\, Y_{lm}(\mathbf{n})$$

where $r \equiv |\mathbf{x}|$ and $\mathbf{n} \equiv \mathbf{x}/r$ are the spherical coordinates of the field point $\mathbf{x}$. Finally, the multipole coefficients, $M_{lm}(t)$, depend on the sources of magnetic field, and boundary conditions. The interpolated magnetic field is given by cutting off the series after the terms with $l = l_{\rm max} \equiv L$, or

$$\mathbf{B}_{\rm estim}(\mathbf{x},t) = \sum_{l=1}^{L} \sum_{m=-l}^{l} M_{lm}(t)\, \boldsymbol{\nabla}[r^l\, Y_{lm}(\mathbf{n})]$$

The number of terms in this sum is $N(L) = \sum_{l=1}^{L}(2l+1) = L(L+2)$ which obviously equals the number of multipole coefficients needed to evaluate the sum. For example, we have $N(2) = 8$

**Table 1.** Averaged estimation errors in the components of the magnetic field and of its modulus. They are reported in relative percentage.

| TM | $\varepsilon_{B_x}$ [%] | $\varepsilon_{B_y}$ [%] | $\varepsilon_{B_z}$ [%] | $\varepsilon_{\|B\|}$ [%] |
|---|---|---|---|---|
| $TM_1$ | 496 | 323 | 349 | 79 |
| $TM_2$ | 645 | 539 | 352 | 68 |

and $N(3)=15$. On the other hand, the number of magnetometer data channels is 12 — three channels per magnetometer, as the devices are tri-axial. This means we cannot extend push the series beyond the quadrupole ($l=2$) terms. Since we have 12 data channels we have some redundancy in determining the first eight $M_{lm}(t)$ coefficients up to $l=2$, but we do not have sufficient information to evaluate the next seven octupole terms.

## 4. Results of the classical interpolation method

The most salient features of the numerical analysis of the above mentioned method can be briefly summarised. First, we find that magnetic field estimation errors are very variable, ranging from a few percent to more than 1 000% with only slight changes in the dipoles' configuration and, secondly, these huge uncertainties happen in an utterly random and unpredictable way. A summary of the numerical results is presented in table 1. The conclusion is quite simple: the intrinsic linear character of the interpolation scheme is not capable of reliably reproducing the field structure inside the *LCA* and hence at the positions of the TMs. In addition to this chaotic behaviour not being predictable, the average error is also unacceptably large. The ultimate reason for such poor performances is the small number of magnetometers as well as the positions they occupy: four magnetometers only allow for a field multipole expansion up to quadrupole terms, which means that the field values at the TMs are just linearly interpolated between magnetometer readouts at the boundary of the *LCA*. Further, the magnetometers are closer to the magnetic field sources than they are to the TMs, which prevents resolution of the spatial field structure details inside the *LCA* using only linear terms in the space coordinates.

## 5. A novel approach: neural network algorithms

Search for an alternative approach to the classical interpolation scheme is imperative, otherwise the information provided by the magnetometers will not be useful for the main goal of the *LTP* magnetic diagnostics system, i.e., to quantify the contribution of the magnetic noise to the total system noise. Here we present some promising results obtained using a completely different methodology, neural networks [6]. Artificial neural networks are made up of interconnecting artificial neurons (programming constructs that mimic the properties of biological neurons) that have the capacity to learn from processing data. Neural networks are often used in solving nonlinear classification and regression tasks by learning from data, hence are worth investigation for the present problem [7]. There are four sets of tasks which need to be implemented when solving a problem using artificial neural networks: (i) Neuron model selection, (ii) model and architecture selection, (iii) learning paradigm and learning algorithm selection and (iv) performance assessment. In this paper we will focus on tasks (i), (ii) and (iv), the third one is beyond the scope of this paper.

The neuron is the basic unit of any neural network. It collects the inputs from all other neurons connected to it and computes a weighted sum of all the signals injected into it, generally adding a bias as well. If we represent the inputs by a vector $\mathbf{x}\equiv(x_1,\ldots,x_n)$, and the weights

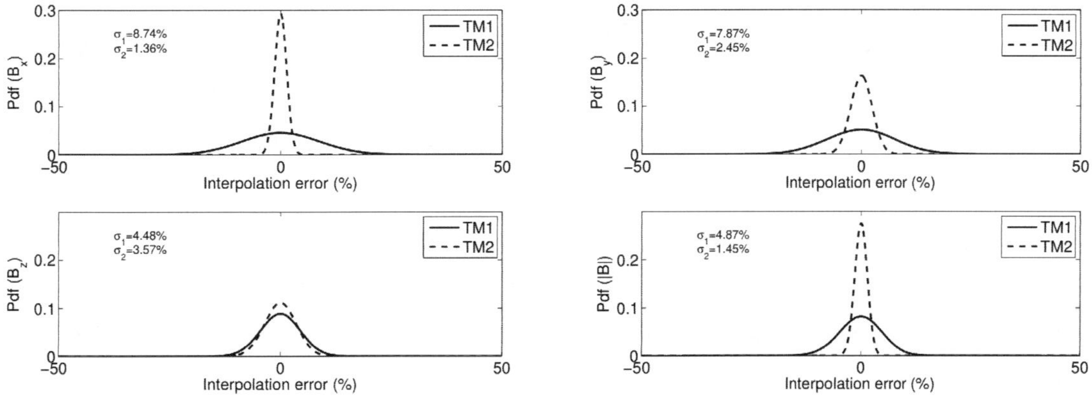


**Figure 2.** Field estimation performance for $B_x$, $B_y$, $B_z$ and $|B|$ using artificial neural networks. Note that the estimation errors are now very small and that the distributions of errors have zero mean errors and standard deviations below 10%, which represents an increase of one order of magnitude in the estimation performance.

by a $\mathbf{w} \equiv (x_1, \ldots, w_n)$ then this operation consists of calculating the sum

$$\Sigma = w_0 + \sum_{k=1}^{n} w_k x_k \equiv w_0 + \mathbf{w}^T \mathbf{x} \quad (3)$$

where the superindex $T$ stands for transpose matrix. In this case, $\mathbf{w}^T$ is a row vector while $\mathbf{x}$ is a column vector, so that $\mathbf{w}^T\mathbf{x}$ is the scalar product of $\mathbf{w}$ and $\mathbf{x}$. A term $w_0$ is added to form the most general linear function of the vector argument $\mathbf{x}$; it is called the bias.

Artificial neural networks are software or hardware models inspired by the structure and behaviour of biological systems, and they are created by a set of neurons distributed in layers. There are many different types of neural networks in use today, but the architecture of a so-called feed-forward network, where each layer of neurons is linked with the next by means of a set of weights, is the most commonly used, and will also be used here. The data streams coming from the magnetometers will be considered the system inputs, while magnetic field results and their gradients at the positions of the TMs will be the system outputs.

## 6. Implementation and results

The implementation of the neural network algorithm for this specific application has been performed according to the following characteristics. The implemented architecture is a feed-forward network implemented with only one hidden layer (15 neurons in the hidden layer). The input vector is characterized by the 12 onboard magnetometers channels and the output vector by the field and gradient values at the positions of the TMs.

Training and testing have been done based on different field realisations. Training has been performed using a backpropagation algorithm on batches of 1 000 examples, i.e., each *example* will consist in the magnetic field at the magnetometers' positions, plus the magnetic field and gradient at the TMs positions, all of them corresponding to a given configuration of the 37 dipole sources [5]. Two different batches of examples, each one including 1 000 realisations of a possible magnetic environment, have been generated. The first batch has been used as the training set for a neural network with 12 inputs (3 inputs for each of the 4 vector magnetometers) and 16 outputs representing the field information at the position of the two TMs (3 field plus 5 gradient components per TM). Note that only 5 of the 9 gradient components $\partial B_i/\partial x_j$ are independent.

This is because $\partial B_i/\partial x_j$ is a traceless and symmetric matrix. The second batch has been used for validation to assess the performance of the network in front of magnetometers readings not used during the training procedure.

Figure 2 shows the distribution of relative percentage errors of the estimated components of the magnetic field at the positions of each TM. The plots are based on the results of the 1 000 validation runs previously described. As can be observed, the order of magnitude of the errors of the estimated fields and gradients are now below $\sim 15\%$. This represents a reduction of estimation errors of more than one order of magnitude in comparison with the multipole expansion method, which is a substantial improvement for the magnetic diagnostics for *LISA* Pathfinder.

## 7. Conclusions

The research presented here allows to draw the following conclusions. First, neural network algorithms represent a significant improvement of the quality of the magnetic field and gradient interpolation in *LISA* PathFinder. We find that where classical interpolation methods produce very erratic results with very large (over 500%) average estimation errors, neural network interpolation algorithms achieve zero mean errors with 10% standard deviation. However, we stress that these results require a careful choice of the training data set and of the training algorithm. Hence, we foresee that real measured data will be required to perform a proper training on the actual spacecraft magnetic configuration. Nevertheless, the results obtained so far are promising, although they have been obtained for DC field and gradient estimation. Current research is underway for estimation in the presence of magnetic field drifts and fluctuations and will be presented elsewhere.

## Acknowledgments

Support for this work came from grants ESP2004-01647 and AYA08–04211–C02–01 of the Spanish Ministry of Education and Science. JS acknowledges a grant from MICINN. Part of this work was also supported by the AGAUR.

## References

[1] The *LISA* International Team 2008 ESA-NASA Report LISA-SCRD-Iss5-Rev1 (2008).
[2] Danzmann K, Vitale S et *al. Science Requirements and Top-level Architecture Definition for the LISA Technology Package (LTP) on Board LISA Pathfinder (SMART-2). LTPA-UTN-ScRD-3.1.* 30-06-2005.
[3] Araújo H *et al. J. of Phys. Conf. Ser.* **66**, 012003 (2007).
[4] Cañizares P *et al.* Class. & Quantum Grav. **26**, 094005 (2009).
[5] Private communication from David Wealthy (Astrium Stevenage, UK).
[6] Kecman V *Learning and soft computing* The MIT Press, 1st edition (2001).
[7] Serpico C and Visone C *IEEE Trans. on Magnetics*, **34** 623-628 (1998).

8th Edoardo Amaldi Conference on Gravitational Waves IOP Publishing
Journal of Physics: Conference Series **228** (2010) 012039 doi:10.1088/1742-6596/228/1/012039

# Proton irradiation test on the flight model radiation monitor for LISA Pathfinder

**I Mateos[1], A Lobo[1], J Ramos-Castro[2], J Sanjuán[1], M Diaz-Aguiló[1,3], PJ Wass[4] and C Grimani[5]**

[1] Institut de Ciències de l'Espai (CSIC-IEEC), Ed. Nexus, Gran Capità 2-4, 08034 Barcelona, Spain
[2] Departament d'Enginyeria Electrònica, Universitat Politècnica de Catalunya (UPC), Campus Nord, Ed. C4, Jordi Girona 1-3, 08034 Barcelona, Spain
[3]Departament de Física Aplicada, Universitat Politècnica de Catalunya (UPC), Campus Nord, Ed. B4/B5, Jordi Girona 1-3, 08034 Barcelona, Spain
[4] Dipartimento di Fisica, Università di Trento and INFN Gruppo collegato di Trento, via Sommarive 14, 38050 Povo (TN), Italy
[5]Istituto di Fisica Università degli Studi di Urbino "Carlo Bo", Urbino (PU) and Istituto a Nazionale di Fisica Nucleare, Florence, Italy

E-mail: mateos@ice.csic.es

**Abstract.** The design of the Radiation Monitor in the LISA Technology Package on board LISA Pathfinder is based on two silicon PIN diodes, placed parallel to each other in a telescopic configuration. One of them will be able to record spectral information of the particle hitting the diode. A test campaign for the Flight Model Radiation Monitor is proposed to verify its performance. This paper shows the results obtained with a simulated flight model geometry using GEANT4, to be compared with the real data that will be obtained in a proton irradiation facility.

## 1. Introduction

LISA Pathfinder (LPF) [1] is a technological mission, whose main goal is to verify the ability to set two proof masses in purely gravitational free fall to a certain level of accuracy. This idea is reflected in the differential acceleration noise requirements between two Test Masses (TMs) for the LISA Technology Package (LTP)

$$S^{1/2}_{\delta a,\mathrm{LTP}}(\omega) \leq 3\times 10^{-14}\left[1+\left(\frac{\omega/2\pi}{3\ \mathrm{mHz}}\right)^2\right]\ \mathrm{ms}^{-2}\mathrm{Hz}^{-1/2} \tag{1}$$

in the frequency band between 1 mHz and 30 mHz. This noise is the result of various disturbances which limit the performance of the instrumentation on-board.

LPF will be stationed in a Lissajous orbit around Lagrange point L1. There, the spacecraft will be exposed to ionising radiation coming from the Galaxy and from the Sun. Some of these charged particles will be stopped by the spacecraft structure surrounding the TMs, while others will make it to the TMs. The latter are particles having energies above a threshold of about 100 MeV/nucleon [2]. The deposition of excess charge in the TMs is of course a random process

© 2010 IOP Publishing Ltd

which results in acceleration noise because the TM position sensor is capacitive [1]. The Radiation Monitor (RM) aboard LPF is intended to complement the measurement of the electrostatic charging rate of the TMs, in order to correlate the flux of energetic particles (detected by the RM) with the longer term charging rates observed in the TMs by means of the TM position sensor.

The RM is based on two silicon PIN diodes placed parallel to each other in a telescopic configuration. Both diodes can count single events, and one of them measures the primary energy of the particle, inferred from the energy deposited, when the events are detected in coincidence in both PIN diodes [4]. The expected deposited energy in the diodes (with added Gaussian noise to mimic the electronic noise) can be predicted through simulation.

A test campaign for the Flight Model (FM) RM is proposed to verify its performance. A similar test was done with a prototype of the RM, which probed its design. Test results gave the go ahead to the FM manufacturing, but the latter presents a few structural differences with respect to the prototype. The FM test is thus required, although it will be less comprehensive, in particular to prevent damage by excess radiation. The results obtained with a simulated model geometry using the GEANT4 toolkit will be compared with the real data that will be obtained in the proton irradiation facility at the Paul Scherrer Institute, in Switzerland.

## 2. Test and simulation

The test plan is designed to verify the performance of the FM RM. The simulation should asses whether or not:

(i) The shield thickness is sufficient to stop soft protons with energy of less than 70 MeV.
(ii) The dynamic range of the Front-End Electronics (FEE) + Analog-to-digital Converter (ADC) is set properly to accommodate the smallest (50 keV) and the largest (5 MeV) energy deposition in the Si PIN diodes.
(iii) Angular acceptance for coincidence detection complies with the required value.

The initial beam for the Proton Irradiation Facility (PIF) has an energy of 254 MeV, and is guided to the experimental area passing through a set of movable energy degraders. These copper degraders can be used to select the energy of the protons finally sent to the target (from 250 MeV down to 70 MeV). The model geometry simulates the beam degrader plates, the aluminium box housing of the FM RM, the shield containing two Printed Circuit Boards (PCBs), the diodes mounted on ceramic substrates and the two PCBs where the electronic circuitry is placed. Figure 1 shows the model geometry simulated with GEANT4.

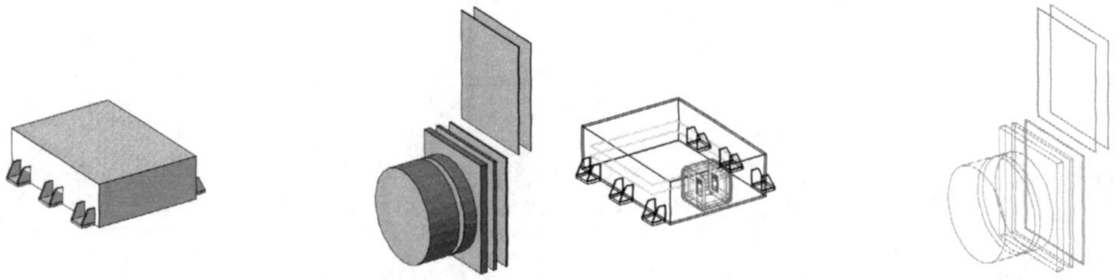

**Figure 1.** Beam test simulation setup in GEANT4. Left side: solid volume. Right side: clear volume showing the FM RM shield, PIN diodes and PCBs. Exchangeable degraders to select beam energies are also shown.

Several parameters of the simulation need to be tuned to improve confidence in the results. Initial theoretical values of the experimental setup are shown in brackets in the list below:

- Initial mean energy of the Beam: 255.27 MeV with an energy spread of 3.556 MeV (instead of a 254 MeV monoenergetic beam).
- Nominal thickness of the diode: 280 $\mu$m (instead of 300 $\mu$m).

## 3. Beam simulations

The beam is simulated as a parallel beam with a Gaussian energy profile and a flat radial profile with a diameter of 40 mm, wide enough to irradiate the diode uniformly —see figure 2. The full beam degradation was simulated in an attempt to make the calculation as representative as possible. The simulation results of the beam energy profiles after it passes through the set of movable degraders, are shown in figure 3.

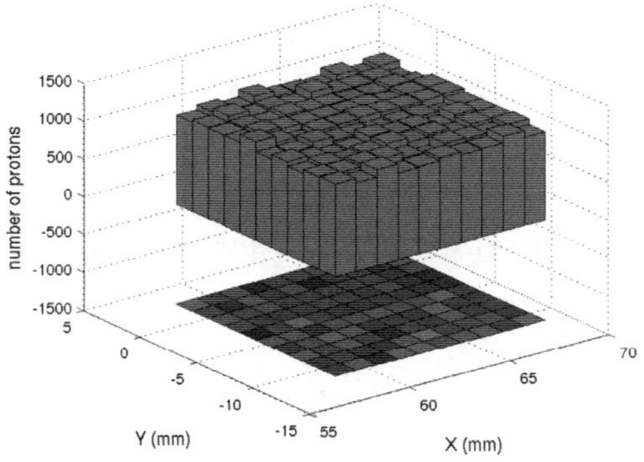


**Figure 2.** Uniform irradiation after exposure of the FM RM to $10^6$ high energy protons.

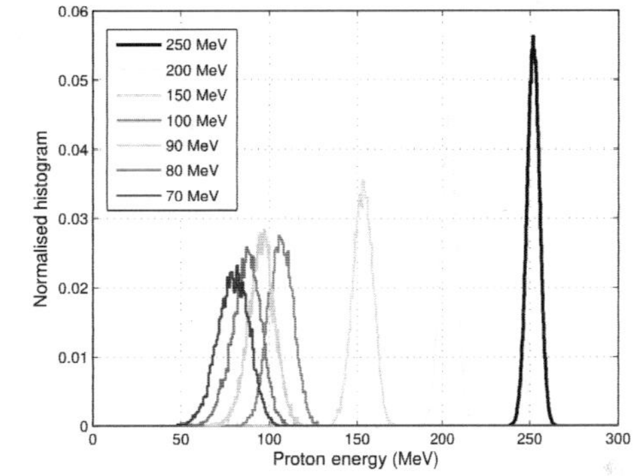


**Figure 3.** Simulation of the energy profiles of the final beam between 70 MeV and 250 MeV.

Based on these simulations, the properties of the energy spread of the beam ($\sigma$) for each energy setting ($\overline{E}_{\text{simulated}}$) are given in table 1, as well as the different degraders used and their thickness.

**Table 1.** Beam energy properties for each degrader settings used in the simulation. Ticks indicate the degraders necessary to obtain the corresponding energies.

| $E_{\text{nominal}}$ MeV | $\overline{E}_{\text{simulated}}$ MeV | $\sigma$ MeV | Plate 1 0.5 mm | Plate 2 1 mm | Plate 3 2 mm | Plate 4 4 mm | Plate 5 8 mm | Plate 6 16 mm | Plate 7 38 mm |
|---|---|---|---|---|---|---|---|---|---|
| 70 | 78.91 | 9.06 | ✓ | ✓ | ✓ | | | ✓ | ✓ |
| 80 | 86.53 | 8.53 | | | ✓ | | | ✓ | ✓ |
| 90 | 95.32 | 7.86 | | | | | | ✓ | ✓ |
| 100 | 106.13 | 7.44 | ✓ | ✓ | | ✓ | ✓ | | ✓ |
| 150 | 152.7 | 6.4 | ✓ | | | | | | ✓ |
| 200 | 203.31 | 4.87 | | ✓ | | ✓ | | ✓ | |
| 250 | 251.75 | 3.68 | ✓ | ✓ | | | | | |

## 4. Low energy cut-off of the RM shield

The shield design is based on compromise between weight, size, and material. A good trade-off was found to be copper material. A cubic box with curved corners was the chosen shape —see figure 4. Such box is intended to stop soft-energy charged nuclei before they reach the PIN diodes in the RM. The performance of the shield will be assessed as a function of the energy of the proton beam as well as its angle of incidence on the shield.

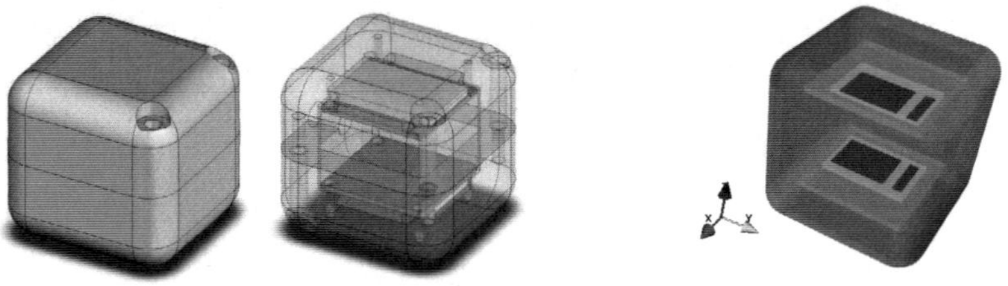

**Figure 4.** Design of the RM shield containing the diodes, their substrates and the PCBs. Left side: CAD model. Right side: GEANT4 model.

Simulations show that the energy cut-off below which a proton can not penetrate the RM shield is approximately 75 MeV. Normalised coincident count rates measured by the simulation as a function of the proton energy for angles between 0 and 30 degrees are shown in figure 5. Normalised count rate is obtained dividing the detection rate by the beam flux measured outside the RM shield. The energy cut-off is measured from the 50 % detection rate.

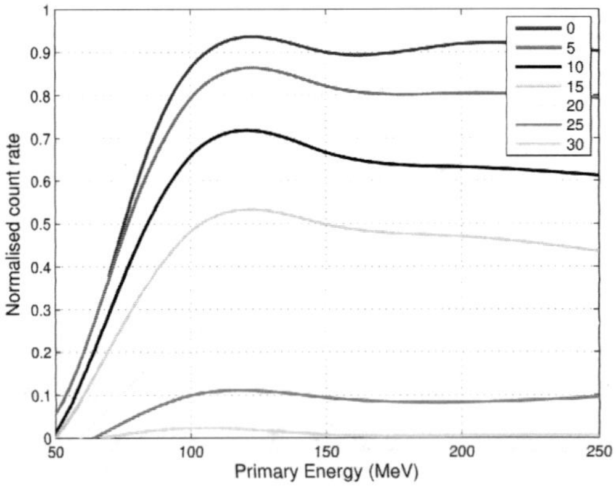


**Figure 5.** Low energy cut-off of the FM RM shield for different angles of incidence (from 0 to 30 degrees).

## 5. Angular acceptance of the FM RM

Although the Galactic Cosmic Ray (GCR) flux is isotropic, Solar Energetic Particle (SEP) events may be directional, therefore it is important to know the angular acceptance of the RM, particularly for coincidence detection, which will be used for spectral measurements. The geometrical angular acceptance of the FM RM was calculated by simulating a directional flux of protons. By

Journal of Physics: Conference Series **228** (2010) 012039 doi:10.1088/1742-6596/228/1/012039

plotting the normalised rate against RM angle we can determine the efficiency of the telescopic configuration of the diodes.

The normalised count rates as functions of the FM RM orientation with respect to the beam (Z axis) is shown in figure 6. Differences between X and Y axis rotation of the FM RM in figure 7 are due to the non-symmetric active area of the detector (wider along the Y axis –see figure 4).

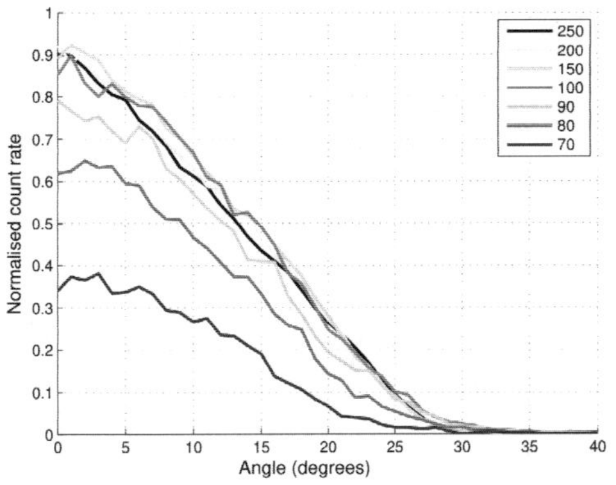


**Figure 6.** Acceptance angle for coincident particle detection as a function of FM RM orientation (Y axis rotation).

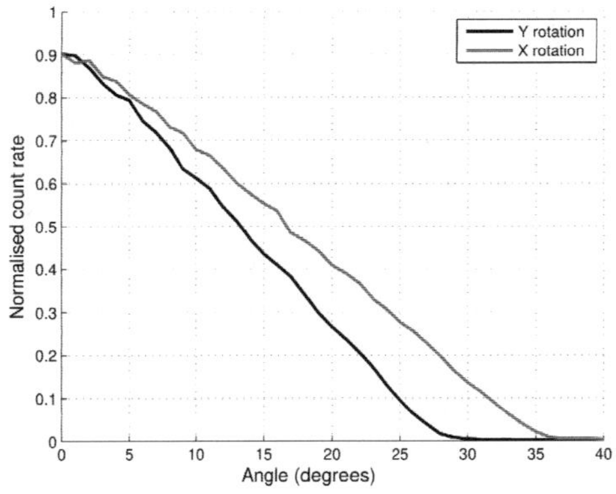


**Figure 7.** Comparison between X (Red) and Y (black) axis rotation of the FM RM for 250 MeV energy protons.

## 6. Measured energy deposited in PIN diodes

The energy deposited in the PIN diodes was simulated for orientations between 0 and 40 degrees (Y axis rotation) with nominal beam energies between 70 MeV and 250 MeV.

Figure 8 shows the energy deposited by the proton beam in the PIN diodes: low energy protons deposit higher energies than high energy protons, while the latter have a smaller spread in deposited energy. At 60 MeV almost no protons penetrate the copper shield and hence do not reach the diodes. Additionally (not shown in the figure), for angles of 30 degrees and above, the number of coincident events is largely suppressed by the geometric acceptance of the diode arrangement. Also, oblique beam incidence on the FM RM causes the distribution peaks to shift towards higher energies, as this increases the path-length of the protons inside the PIN diodes.

Figure 9 gives a comparison between experimental data from the prototype RM beam test and GEANT4 simulations for the prototype [3] and for the FM RM, with 100 MeV protons, showing good agreement between them. The main discrepancies between the FM and the prototype RM design are shown below:

- A silver layer surrounds the copper shielding in the FM RM, needed to avoid copper oxidation.
- Structural differences in the aluminium box housing of the FM RM.
- Data management by means of a Field Programmable Gate Array (FPGA) is added in the electronic circuitry for the FM RM.

All these details have been taken into consideration in the GEANT4 model of the FM RM.

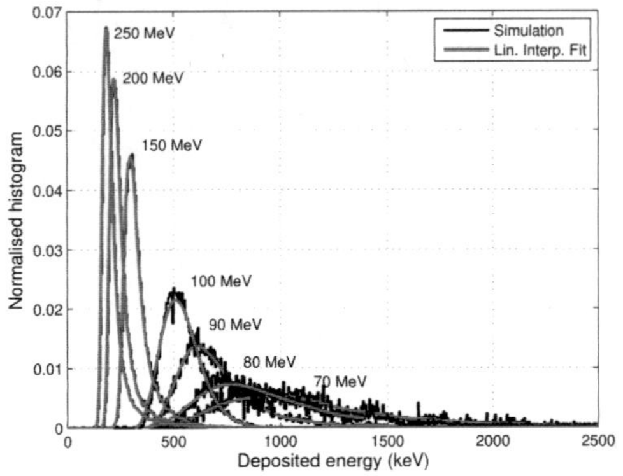


**Figure 8.** Simulations of the energy distribution deposited in the diode at 0 degree incidence. Labels on curves indicate the mean energy of the incident beam.

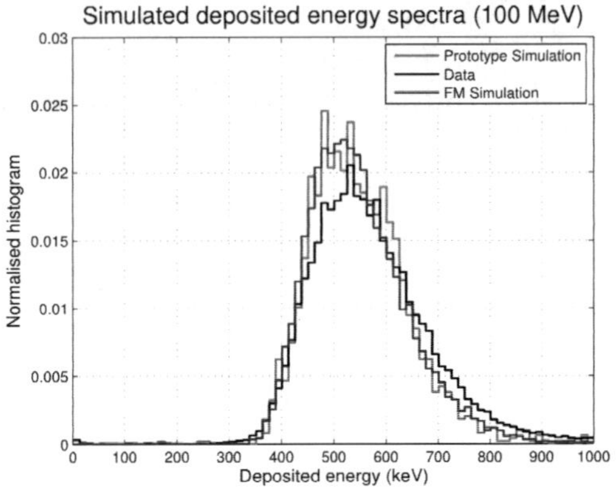


**Figure 9.** Comparison between real data in the prototype RM and GEANT4 simulations results for the prototype and for the FM RM.

## 7. Conclusions

The results of the simulation of the RM can be summarised as follows:

- The shield surrounding the PIN diodes prevents the detection of particles with energy less than $\sim 75$ MeV.
- The acceptance angle of the coincident mode of the FM RM is $\sim 25$ degree (worst case: Y rotation).
- The ADC dynamic range of the deposited energy measurements (0-5 MeV) is sufficient to record the energy deposited by all protons at 70 MeV.
- Simulations and experimental data from the prototype RM are in good agreement.
- The FM RM test should confirm that all the above is actually verified in the real device.

There are the basic facts which the FM RM test will have to validate.

## Acknowledgments

We acknowledge support from Project No. ESP2004-01647 of Plan Nacional del Espacio of the Spanish Ministry of Science and Innovation (MICINN).

## References

[1] Anza S *et al* 2005 *Class. Quantum Grav.* **22** 125
[2] Araújo H M, Wass P, Shaul D, Rochester G and Summer T J 2005 *Astropart. Phys.* **22** 451-69
[3] Wass P J, Araújo H, Boatella C, Chmeissani M, Hajdas W, Lobo A, Puigdengoles C and Summer T 2006 *Proc. 6th Int. LISA Symp.(Greenbelt)* Vol 873 (AIP Conf. Proc.) pp 225-9
[4] Cañizares P *et al* 2009 *Class. Quantum Grav.* **26** 11

Journal of Physics: Conference Series **228** (2010) 012040 doi:10.1088/1742-6596/228/1/012040

# Short-term forecasting of solar energetic ions on board LISA

**C. Grimani[1], H.M. Araújo[2], M. Fabi[3], N. Finetti[4], A. Lobo[5], I. Mateos[5], D.N.A. Shaul[2], T.J. Sumner[2]**

[1] Istituto di Fisica Università degli Studi di Urbino "Carlo Bo", Urbino (PU) and Istituto Nazionale di Fisica Nucleare, Florence, Italy
[2] Imperial College, London, UK
[3] Istituto di Fisica Università degli studi di Urbino "Carlo Bo", Italy
[4] Dipartimento di Fisica, Università degli Studi dell'Aquila and Istituto Nazionale di Fisica Nucleare, Laboratori Nazionali del Gran Sasso, Gruppo collegato dell'Aquila, L'Aquila, Italy
[5] Institut d'Estudis Espacials de Catalunya (IEEC), Barcelona, Spain

E-mail: `catia.grimani@uniurb.it`

**Abstract.** LISA (Laser Interferometer Space Antenna) and LISA Pathfinder (LISA-PF) free-falling test-masses are charged by galactic and solar energetic particles. This process generates spurious forces on the test masses which appear as noise in the experiments. It was shown that relativistic solar electron detection can be used for up-to-one-hour forecasting of incoming energetic ions at 1 $AU$. Warning of incoming solar energetic particle events could allow us to optimize the test-mass discharging. The current LISA-PF radiation monitor design needs to be upgraded if solar electron detection is to be implemented in LISA.

## 1. Solar relativistic electron and non-relativistic proton onset at 1 $AU$

Protons and helium nuclei constitute approximately 98% in composition of both galactic cosmic rays and solar particles. Electrons amount to about 1%. Electrons and ions with kinetic energies larger than a few MeV and 100 $MeV(/n)$, respectively, can penetrate the spacecraft material charging the LISA and LISA-PF test masses [1, 2]. Unfortunately, very little is known about solar particle energy spectrum evolution during strong solar events above 100 $MeV(/n)$.

Particle detectors will be placed onboard both LISA-PF and LISA missions to monitor the galactic and solar protons and helium nuclei incident on the spacecraft. The radiation monitors designed for LISA-PF consist of two silicon wafers of $1.4 \times 1.05$ $cm^2$ area placed in a telescopic arrangement at a distance of 2 $cm$ from each ones (figure 1) [3]. The geometrical factor of each silicon layer for an isotropic incidence is 9 $cm^2$ $sr$ and for coincident events is about one tenth of that. The silicon telescopes are located inside a copper box of 6.4 $mm$ thickness in order to limit the energy of protons and helium nuclei traversing these detectors to a few tens of MeV(/n) (figure 2). This energy cutoff is similar to the minimum energy of the most abundant components of cosmic rays penetrating the test masses (100 $MeV(/n)$) [4]. No electron monitoring will be carried out on LISA-PF. We have found that both radiation monitor countrate and charge deposited in the test masses vary by several orders of magnitude within a few tens of minutes at the occurrence of strong solar events [5]. The radiation monitor delivers data accumulated over periods of 614.4 $s$ [3]. During this same period of time, all solar events

© 2010 IOP Publishing Ltd

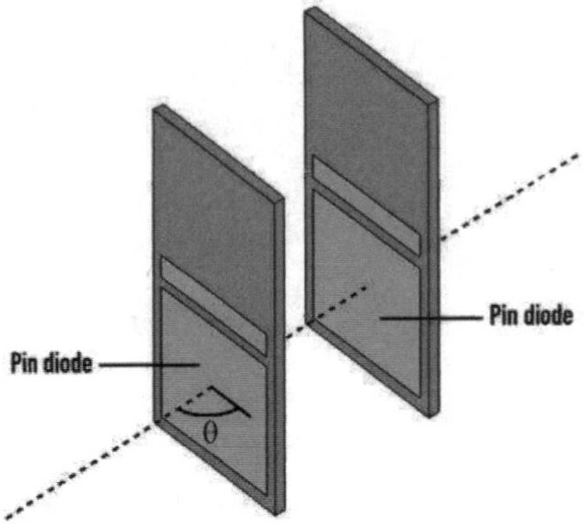


**Figure 1.** Sketch of silicon detectors of the LISA-PF radiation monitors.

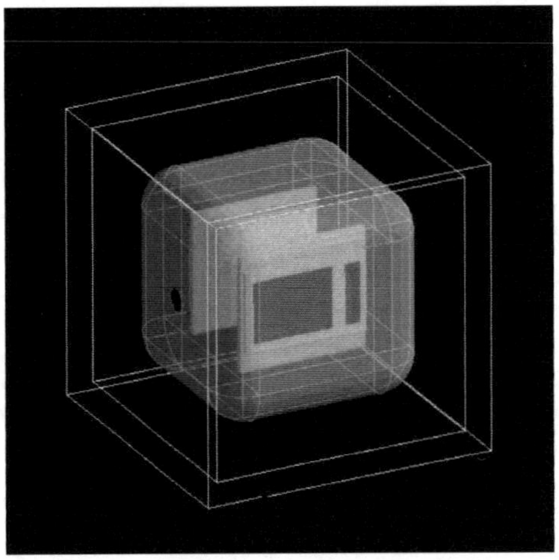

**Figure 2.** LISA-PF shielding copper box surrounding silicon wafers.

generating a proton flux larger than the statistical fluctuations of galactic cosmic rays will be detected. Optimization of test-mass discharging can be considered accordingly [6]. Work by Posner [7] has shown that during strong SEP events, relativistic electrons always reach 1 $AU$ in advance of non-relativistic ions. It was found that the intensity increase of both electron and ion fluxes is similar and depends mainly on the magnetic longitude distance (magnetic connection) between spacecraft and flare. Correlations are found also between the early time profiles of electron intensity and increase with upcoming proton intensities. Electrons in the energy range 0.3-1.2 $MeV$ and 31-50 $MeV$ proton data from COSTEP [8] on SOHO [9] and GOES 8 [10] were studied in the Posner work. Events exceeding 10 $pfu$ (proton flux units) above 10 $MeV$ (fluxes above 10 $MeV$ greater than 10 $protons/(cm^2\ sr\ s)$) were considered for analysis.

Following the Posner results, we estimated the minimum, average and maximum delay of non-relativistic protons with respect to relativistic electrons of solar origin in reaching 1 $AU$. Scatter-free particle propagation along the interplanetary magnetic field lines for a magnetically well connected event (pitch angle 0°; path=1.2 $AU$) was assumed in figure 3 for minimum time delay determination. Maximum time delays of approximately 1 $h$ were estimated. Posner points

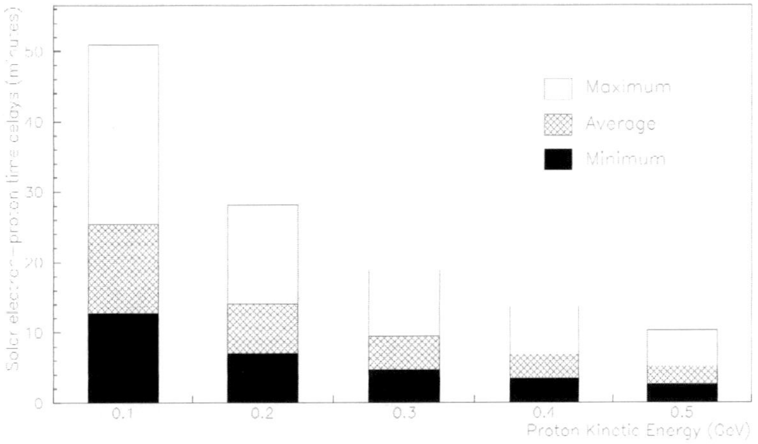


**Figure 3.** Minimum, average and maximum time delays between solar relativistic electrons and non-relativistic protons of different kinetic energies reaching 1 $AU$.

out that only rarely is a major proton intensity increase observed after 2 or 3 h following an electron intensity increase. As an example, the proton spectrum of a typical medium-strong event such as that dated 7 May 1978 (fluence between $10^6$ and $10^7$ $protons/cm^2$ above 30 $MeV$) is peaked at about 300 $MeV$ at onset [5]. Therefore, according to figure 3, the time interval to be expected between solar relativistic electrons and non-relativistic protons reaching 1 $AU$ ranges between 4 and 20 min. In other words, for a May 7th-like flare it will be possible to predict, within the above mentioned interval of time, a variation of the test-mass charging of 3 orders of magnitude from the moment solar protons reach the spacecraft and the peak of the event occurring in approximately 25 min [2]. Stronger events would result in higher test mass charging rates.

In figure 4 we have compared the expected electron and proton intensity increase versus time at the onset of a well connected event. Particle propagation along a path of 2 $AU$ in the interplanetary magnetic field was assumed in figure 5. We point out that the increase of electron and proton intensities, I(t), reported in arbitrary units in figures 4 and 5, at the beginning of a solar event, as a function of time (t) appears linear on a lin-log plot and therefore, it can be represented by the following expression:

$$I(t) = Ae^{\gamma t} \tag{1}$$

In figures 4 and 5 it can be seen that both electron and proton fluxes show a similar intensity increase. Sensitivity of an onboard monitor to large $\gamma$ variations in electron intensity within a few minutes would allow us to issue a warning of incoming SEPs on LISA. In addition to the main connection distance effect influencing SEP onset at 1 $AU$, particle transport might play an important role. The interplanetary magnetic field sector structure affecting the onset of a few hundreds of keV electrons could reduce the accuracy of proton onset forecast. According to the Posner work (see figure 2 in [7]), this occurred in one case out of a sample of 65 well connected events.

In the rigidity (particle momentum per unit charge) range 0.3-300 $MV$, the mean free paths of electrons and protons are found to change by a factor of 20-50 from event to event [11], [12]. This corresponds to a maximum proton kinetic energy of 50 $MeV$.

Journal of Physics: Conference Series **228** (2010) 012040 doi:10.1088/1742-6596/228/1/012040

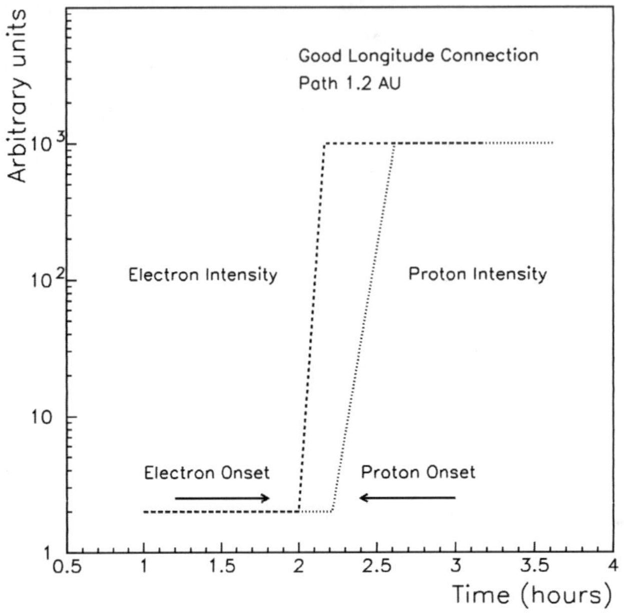


**Figure 4.** Expected time variation of the intensity of electron and proton fluxes at the onset of a solar event well connected to the spacecraft.

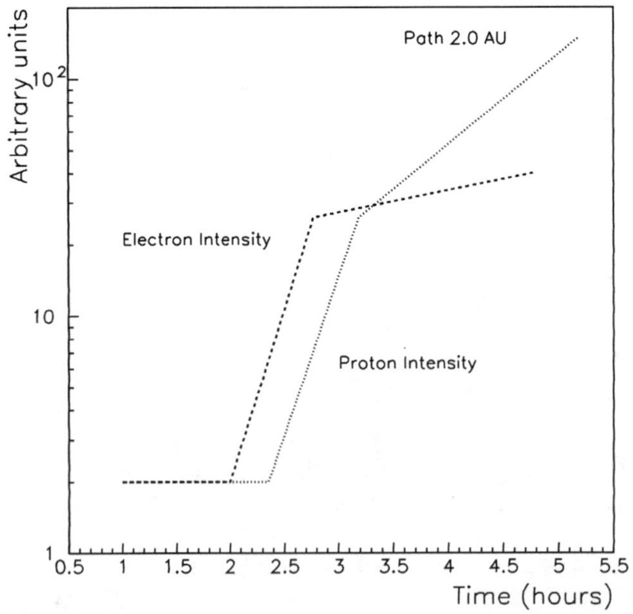


**Figure 5.** As figure 4, assuming particle paths of 2.0 $AU$ instead of 1.2 $AU$.

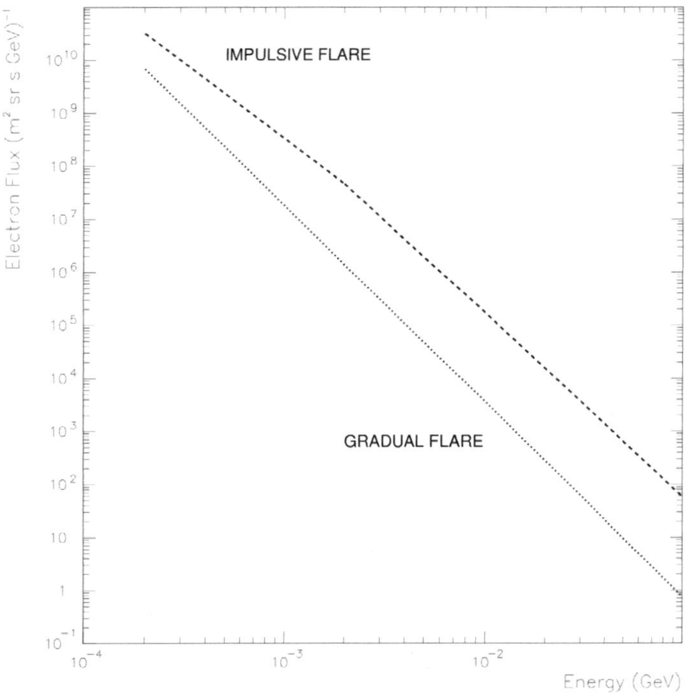


**Figure 6.** Interpolated solar electron fluxes associated with the November 3rd (dotted line) and September 7th 1973 (dashed line) solar flares. The measurement trend was extrapolated to 100 $MeV$.

The order of magnitude of the mean free path variations is close to the scatter of rise parameters of 1 $MV$ electrons and 100 $MV$ protons about the regression curve with magnetic connection distance [7].

The onset times of relativistic electrons are related to the coronal mass ejection (CME) speed for events occurring outside the fast propagation region (25° - 90° range of the angular connection) corresponding to approximately 57% of the events [13]. In particular, the delay time between electron onset and flare is close to the time of CME propagation to the observer's magnetic field line [14]. Interplanetary shocks accelerate only small amounts of electrons, therefore, solar electron detection cannot be used to forecast interplanetary shock transit. However, interplanetary shocks do not accelerate ions above 30 $MeV/n$ and therefore LISA-PF and LISA are not affected by interplanetary acceleration.

## 2. Solar electron detection requirements on LISA

It is well known that only a subset of solar events generate particles above 100 $MeV(/n)$. SEP forecasting on LISA should be addressed to these events only. To this purpose, we suggest monitoring electron intensity variations for particle energies above a few hundreds of keV on board LISA. This choice is due to the fact that only during the strongest events a relevant electron flux is generated above these energies. Unfortunately, we will not be able to discriminate between strong pure impulsive events which accelerate protons and nuclei below a few tens of

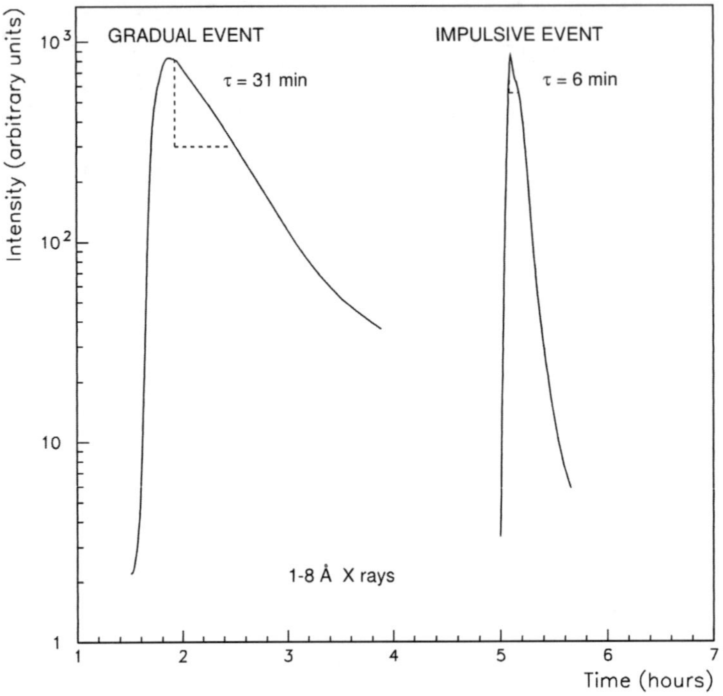


**Figure 7.** X-ray intensity near peak for typical gradual and impulsive events.

MeV, and gradual events. Relevant electron fluxes up to tens of MeV are associated with both kind of events. The main difference is that electron spectra observed for impulsive events present a double power-law slope, breaking at a few MeV, while gradual events present electron spectra modeled by a single power law, up to tens of MeV [15]. Examples o the two types of spectra are presented in figure 6.

It is not feasible to measure differential particle fluxes on LISA, therefore this separation will not be possible. However, simultaneous observations of solar events on other experiments devoted to solar physics, monitoring X-ray and $\gamma$-ray fluxes, might help in discriminating between pure impulsive and gradual events within 10 min from occurrence (see [13] and references therein). It was shown that X-ray events can be classified into two classes: gradual and impulsive, by considering the duration of 1-8 $\AA$ emission. If the time interval ($\tau$) when the peak intensity decreases by a factor 1/e is >10 min ($\leq$10 min) the event is gradual (impulsive). In figure 7 the trend of X-ray intensity associated with gradual and impulsive events are shown.

Solar events reaching Earth and near Earth detectors, such as LISA, are mostly generated in the western hemisphere because of the interplanetary magnetic field topology. Impulsive events with heliographic longitude $<-30°$ never produce SEPs reaching Earth. Electrons in the MeV range can be detected more than 80° from the flare longitude. However, these events are not associated with SEPs and they can be identified since the electron peak intensities are small and decrease with increasing connection angle [16]. Conversely, SEPs associated with gradual events presenting heliographic longitudes ranging between $-120°$ and $+180°$ are observed. False warnings would be issued on LISA only for those strong impulsive events generating electrons

Journal of Physics: Conference Series **228** (2010) 012040 doi:10.1088/1742-6596/228/1/012040

detected on each spacecraft within the 10 min needed to discriminate between impulsive and gradual events through X-ray observations. In the work by Laurenza et al. [13], SEP events analyzed were those exceeding 10 $pfu$ above 10 $MeV$, as in the work by Posner. These authors find that 7% is the fraction of impulsive events of the whole sample of impulsive and gradual events for which it was possible to determine the heliographic longitude.

Electron intensity read-out would be necessary every minute at most on board LISA. The best connected events (up to 20 degrees) are expected to present very sharp electron onsets, of the order of 4 min. A dead-time in $e^-$ intensity read-out larger than 4-5 min would prevent us from detecting in advance the most intense solar events. In case a SEP event occurs after other intense events, we expect a maximum uncertainty of 10 min in proton onset prediction. We have shown that [5] at the onset of a medium-strong solar event the incident proton flux increases of approximately one order of magnitude in about ten minutes. In the same interval of time, this variation can be detected above the statistical fluctuations associated with a previous event.

## 3. Detector characteristics for solar electrons on board LISA

Multi-layer solid-state detectors a few hundreds $\mu m$ thick can be used for solar electron, proton and helium nucleus identification. Active anticoincidence shielding would be required around solid state detectors to identify charged particles. Fast-pulse-height analysis would be necessary for electron and ion separation [6]. X-rays would be absorbed in the shielding material, the anticoincidence would in any case avoid contamination of Compton electrons produced inside the detector. Total weight and power consumption can be limited to below a few kilograms and 2 $W$, respectively (similar to the EPHIN [17] sensor on SOHO). A geometrical factor of a few $cm^2 sr$ is required.

The detector aperture should point in the direction of the nominal interplanetary magnetic field at 1 $AU$, 45° west of the spacecraft-Sun line. Spacecraft rotation would probably require pointing the particle detector in the spacecraft-Sun line in order to maintain, on average, a good particle detection.

## 4. LISA dead-time reduction

SEP forecasting could help in better matching test-mass charging [1] and discharging rates on LISA [6]. Consequently, an electron monitor would allow for mission noise and, possibly, dead time reduction [2].

As an example, in Vocca et al. [18] it was shown that the 7 May 1978 medium-strong solar event would have increased the test-mass charging of 1 order of magnitude every 15 min during the first hour after the onset, limiting the LISA sensitivity in the low frequency range soon after the onset. During solar events of similar or larger intensity, the test-mass discharging is expected to increase with respect to solar quiet time, in order to allow for reliable data analysis. We point out that during solar maximum periods the occurrence of strong solar events might amount to up to a fraction of 20% of the mission time. Solar events of fluence larger than $10^9$ protons/cm$^2$ might damage the radiation monitors, preventing us from further solar event prediction. The probability of occurrence of solar events as a function of intensity is reported in [5].

## 5. Conclusions

We have shown that solar electron detection on board LISA would allow us for short-term forecasting of incoming energetic ions of solar origin. Basically all events magnetically well connected to the spacecraft will be detected. A maximum issue of false warnings of 7% was estimated. Optimization of test mass discharging and experiment dead-time reduction might be consequently achieved. Minor modifications with respect to the LISA-PF radiation monitor design would be required in order to obtain a major experiment performance improvement.

## Acknowledgements
Two of us (CG and MF) are greatly indebted to Dr. Vincenzo Acconcia as a member of SIA (IT Center, Urbino University) for providing his valuable technical help. This work was supported by the Italian Space Agency within the program "Study Exploration of the Solar System".

## References
[1] Araújo H M *et al* 2005 *Astropart. Phys.* **22** 451 469
[2] Grimani C *et al* 2009 *Class. Quantum Grav.* **26** 215004
[3] Cañizares P *et al* 2009 *Class. Quantum Grav.* **26** 094005
[4] Wass P 2007 *PHD Thesis* Imperial College of London University of London
[5] Grimani C *et al* 2009 *Class. Quantum Grav.* **26** 094018
[6] Shaul D N A *et al* 2006 *AIP Conf. Proc.* vol 873 p 172
[7] Posner A 2007 *Space Weather* **5** S05001
[8] Müller-Mellin R *et al* 1995 *Sol. Phys.* **162** 483 504
[9] http://www2.physik.uni-kiel.de/SOHO/phpeph/ephin.htm
[10] http://spidr.ngdc.noaa.gov/spidr/index.jsp
[11] Dröge W 2000 *Ap. J.* **537** 1073 1079
[12] Dröge W Ruffolo D and Khumlumlert T 1997 *Proc. 25th Int. Cosmic-Ray Conf. (Durban)* vol 1 p 137
[13] Laurenza M *et al* 2006 *Proc. 20th Europ. Cosmic-Ray Symp. (Lisbon)*
[14] Stolpovskii V G *et al* 1997 *Proc. 25th Int. Cosmic-Ray Conf. (Durban)* vol 1 p 189
[15] Dröge W 1995 *Proc. 24th Int. Cosmic-Ray Conf. (Rome)* vol 4 p 187
[16] Wibberenz G and Cane H V 2006 *Ap. J.* **650** 1199 1207
[17] http://www.ieap.uni-kiel.de/et/ag-heber/costep/ephin.php
[18] Vocca H *et al* 2005 *Class. Quantum Grav.* **22** S319 S325

# ADC non-linear error corrections for low-noise temperature measurements in the LISA band

**J Sanjuán[1,2], A Lobo[1,2], J Ramos-Castro[3], N Mateos[1] and M Díaz-Aguiló[4]**

[1]Institut de Ciències de l'Espai, *CSIC*, Fac. de Ciències, Torre C5, 08193 Bellaterra, Spain
[2]Institut d'Estudis Espacials de Catalunya (*IEEC*), Gran Capità 2-4, 08034 Barcelona, Spain
[3]Dep. Eng. Electrònica, *UPC*, Campus Nord, Ed. C4, J Girona 1-3, 08034 Barcelona, Spain
[4]Dep. Física Aplicada, *UPC*, Campus Nord, Ed. B4/B5, J Girona 1-3, 08034 Barcelona, Spain

E-mail: `sanjuan@ieec.fcr.es`

**Abstract.** Temperature fluctuations degrade the performance of different subsystems in the *LISA* mission. For instance, they can exert stray forces on the test masses and thus hamper the required drag-free accuracy. Also, the interferometric system performance depends on the stability of the temperature in the optical elements. Therefore, monitoring the temperature in specific points of the *LISA* subsystems is required. These measurements will be useful to identify the sources of excess noise caused by temperature fluctuations. The required temperature stability is still to be defined, but a figure around $10\,\mu\mathrm{K\,Hz^{-1/2}}$ from 0.1 mHz to 0.1 Hz can be a good rough guess. The temperature measurement subsystem on board the *LISA* Pathfinder mission exhibits noise levels of $10\,\mu\mathrm{K\,Hz^{-1/2}}$ for $f > 0.1$ mHz. For *LISA*, based on the above hypothesis, the measurement system should overcome limitations related to the analog-to-digital conversion stage which degrades the performance of the measurement when temperature drifts. Investigations on the mitigation of such noise will be here presented.

## 1. Introduction

*LISA* is a space-based gravitational wave detector with the primary scientific goal of observing *GW*s from astronomical sources in the frequency range from 0.1 mHz to 0.1 Hz [1]. However, the detection of *GW*s requires differential measurements of distance at the pico-meter level between two bodies separated by 5 million kilometres. This fact implies challenging requirements in the test masses' (*TM*s) residual acceleration noise in order to ensure *GW* detection with high signal-to-noise ratio. Otherwise *GW* signals will be buried in the random *TM*s motion associated to the forces acting on them. The *LISA* top level requirement, in terms of *TM* acceleration noise, is [2]

$$S^{1/2}_{\delta a,LISA}(\omega) \leq 3\times 10^{-15}\cdot\left\{\left[1+\left(\frac{\omega/2\pi}{8\ \mathrm{mHz}}\right)^4\right]\left[1+\left(\frac{0.1\ \mathrm{mHz}}{\omega/2\pi}\right)\right]\right\}^{\frac{1}{2}}\ \mathrm{m\,s^{-2}\,Hz^{-1/2}} \tag{1}$$

from 0.1 mHz to 0.1 Hz. One of the noise sources that can jeopardise the *LISA* performance is temperature fluctuations affecting key subsystems of the mission. More specifically, thermal stability is necessary to, first, ensure that the acceleration noise caused by temperature

© 2010 IOP Publishing Ltd

fluctuations[1] in the *TMs* is kept below the top level requirement —see Eq. (1); and second, to ensure the stability of the optical elements properties of the interferometric systems[2] [3].

The very demanding requirements for the *LISA* mission led to recognise the need of a technology demonstration mission: the so-called *LISA* Pathfinder mission [4]. The main goal of *LPF* is to check if the residual acceleration in Eq. (1) can be achieved, i.e., it must verify the ability to set a *TM* in purely gravitational free fall to the level required by Eq. (1), although in *LPF* the requirements have been relaxed one order of magnitude below the *LISA* requirements both in amplitude and in frequency.

The *LISA* Technology Package[3] (*LTP*) includes a temperature measurement subsystem (*TMS*) in order to assess whether or not the required thermal stability is met, and to provide information of the actual coupling between temperature fluctuations and acceleration noise. At this moment the *LTP TMS* flight model version has been already designed and tested, and is compliant with the *LTP* requirements [5] —see Fig. 1. The *TMS* must exhibit noise levels of $10^{-5}\,\mathrm{K\,Hz^{-1/2}}$ from 1 mHz to 30 mHz in order to measure the expected temperature fluctuations in the *LTP*[4] with a good signal-to-noise ratio. However, such system exhibits *extra* noise in the submilli-Hertz range when the measured temperature drifts ($\sim \mu\mathrm{K\,s^{-1}}$). Although this is not a problem for the *LTP*, investigations are needed to overcome such limitation in view of *LISA*: the measurement bandwidth (*MBW*) stretches down to 0.1 mHz, the floor noise level requirement is one order of magnitude more demanding and, therefore, weaker temperature drifts can deteriorate the performance of the *TMS*.

The source of the excess noise has been seen to be due to non-idealities of the transfer curve of the analog-to-digital converter (*ADC*). A 16-bit successive approximation register (*SAR*) *ADC* is the only suitable option due to space qualification constraints. As shown in Fig. 1, the noise in the temperature measurement increases noticeably for $f < 0.1$ mHz when the measured temperature drifts are higher than $0.5\,\mu\mathrm{K\,s^{-1}}$. This paper focuses on the identification and mitigation of this problem. In Sec. 2 we review the theoretical basis of the noise introduced by a non-ideal *ADC*. In Sec. 3 we show how the effect of the *ADC* degrades the performance of *TMS* and how the theoretically expected noise and the experimental results agree. In Sec. 4 we describe a method to mitigate the noise coming from the *ADC*, the so-called *dithering techniques*. The test set-up and the experimental results using the dithering technique are shown in Sec. 5. Finally, Sec. 6 highlights the most relevant results.

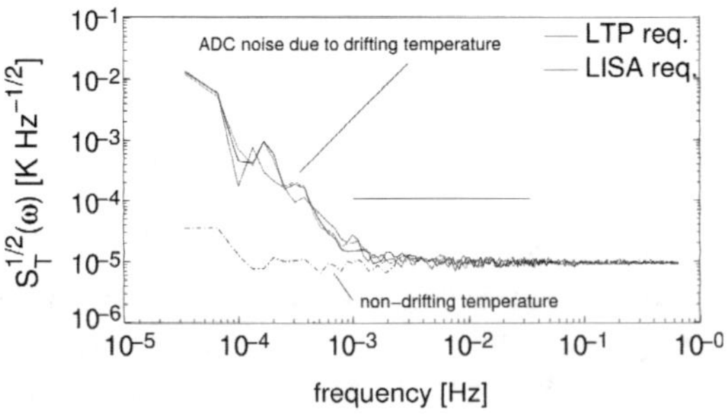


**Figure 1.** *LTP TMS* performance. The excess noise at low frequency ($f$ <1 mHz) appears when the measured temperature drifts ("ADC noise due to drifting temperature" trace). It vanishes when the signal is kept constant ("non-drifting temperature" trace).

[1] Radiometer effect, radiation pressure or outgassing
[2] Changes of the index of refraction or dilatation (and contraction) cause an error in the acceleration measurement.
[3] The science module of the European Space Agency (*ESA*) in LPF.
[4] They are $10^{-4}\,\mathrm{K\,Hz^{-1/2}}$ from 1 mHz to 30 mHz.

## 2. Quantisation noise in *SAR ADCs*

In real *SAR ADCs* the quantisation steps are not uniform due to mismatches in their internal topology (tolerances in the capacitors, $\delta C_k$, of the *SAR* array). Such non-uniformity is specified with two parameters: differential non-linearity (*DNL*) and integral non-linearity (*INL*) [6]. The latter represents the difference between the real and the ideal *ADC* transfer curves. The noise introduced by this error is the one limiting the temperature measurement performance when drifting input signals are present. The voltage error $\epsilon_k$[5] induced by an error in the $k$-th bit is:

$$\epsilon_k(v) \simeq \frac{b_k(v)\,\delta C_k}{(2^N-1)\,C}\,V_{\rm FS} = b_k(v)\,\frac{\delta C_k}{C}\,\Delta \qquad (k=0 \text{ is the } LSB) \tag{2}$$

where $b_k$ is the binary digit (it is set to 0 or 1 depending on the sampled voltage), $\delta C_k$ is $C_{k,\rm real} - 2^k C$, $\Delta$ ($\simeq V_{\rm FS}/2^N$) is the *ADC* quantisation step and $N$ is the number of bits of the *ADC*. *LSB* stands for least significant bit. Equation (2) can be expressed as[6] [7]:

$$|\mathcal{Q}_k(\xi)| = \Delta \sum_{n=-\infty}^{\infty} \frac{\sin\frac{n\pi\epsilon_k}{2^{k+1}\Delta}}{n}\,\frac{\sin\frac{n\pi}{2}}{\sin\frac{n\pi}{2^{k+1}}}\,\delta\!\left(\xi - \frac{\pi n}{2^k\Delta}\right) \tag{3}$$

Equation (3) is graphically represented in Fig. 2 (center and bottom panels) where a superimposed periodic pattern in the quantisation error of the *ADC* with 1 *LSB* of amplitude and periodicity of $2^{k+1}\Delta$ (caused by the *INL* errors) is clearly seen.

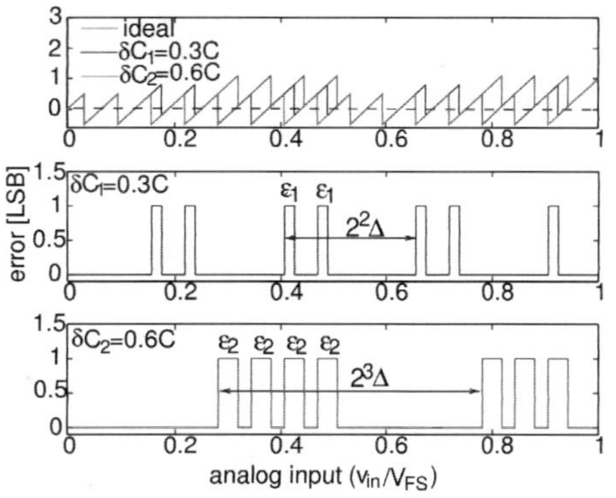


**Figure 2.** Top: quantisation error functions for an ideal *ADC* (red) and for two non-ideal *ADCs* with different error bits (black and blue). Centre and bottom: error differences between the non-ideal and ideal *ADCs* [$\epsilon_k(v)$].

## 3. Effect of *ADC* non-idealities on the *TMS*

To estimate *INL* induced errors in the temperature measurements let us assume the input signal is a straight line for a certain time interval:

$$v(t) = a + bt \;\text{ if }\; 0 \le t \le t_{\rm max}, \;\text{ otherwise } v(t) = 0 \tag{4}$$

[5] The difference between a real *ADC* (non-uniform quantisation steps) transfer curve and an ideal *ADC* (uniform quantisation steps) transfer curve.

[6] We have defined the Fourier transform in the voltage domain by $\mathcal{Q}(\xi) \equiv \int_{-\infty}^{\infty} q(v)e^{-i\xi v}dv$. Not to be mistaken by the usual time/frequency Fourier transforms. We use special symbols for this to avoid confusion. $\xi$ is the Fourier-conjugate variable of $v$, and is accordingly measured in $2\pi$ volts$^{-1}$.

In this case, Eq. (3) becomes (in the frequency, $2\pi$ Hz, domain)[7] [8]:

$$|\tilde{q}_k(\omega)| = \Delta \sum_{n=-\infty}^{\infty} \frac{\sin \frac{n\pi\epsilon_k}{2^{k+1}\Delta}}{n} \frac{\sin \frac{n\pi}{2}}{\sin \frac{n\pi}{2^{k+1}}} \, \delta\left(\omega - \frac{\pi n|b|}{2^k \Delta}\right) \tag{5}$$

and the main frequency component introduced by an error in the $k$-th bit is located at[8]:

$$\omega_{1_k} = \frac{\pi 2^N}{2^k V_{\mathrm{FS}}} |\dot{v}(t)| \ , \quad \dot{\omega}_{1_k} = \frac{\pi 2^N}{2^k V_{\mathrm{FS}}} |\ddot{v}(t)| \tag{6}$$

From Eqs. (6) some conclusion can be drawn:

- high-slope signals at the *ADC* input translate into high-frequency components at the output of the *ADC*,
- errors in the higher bits show up as noise at lower frequency than errors in the lower bits,
- consequently, high-slope inputs combined with errors in high bits may show up as excess noise in the bandwidth of interest,
- the fundamental frequency associated to a bit error varies with variations of $\dot{v}(t)$, hence the error in the bit spreads across the frequency band when $\ddot{v}(t) \neq 0$.

Figure 3 shows experimental data and the theoretical expected evolution of the power spectrum density with time —cf. Eq. (6). They are in excellent agreement.

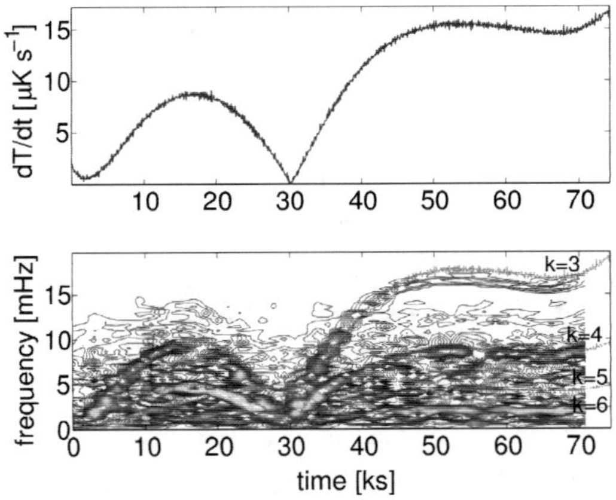


**Figure 3.** Top: absolute value of the input signal slope, $|\dot{v}(t)|$, of a temperature measurement run. The sensitivity at the input of the *ADC* is $\simeq 1.35\,\mathrm{V\,K^{-1}}$, hence, temperature drifts vary between 0 and $9.3\,\mu\mathrm{K\,s^{-1}}$. Bottom: spectrogram of the top panel measurement. The energy of the signal is concentrated in specific frequencies which change with time precisely following $|\dot{v}(t)|$, as predicted by Eq. (6), and as shown in red dashed traces labelled by the order of the corresponding bit error (in this case from $k = 3$ to $k = 6$).

[7] Dirac $\delta$-functions are the result of infinite length integration intervals in Fourier transform calculations. In any practical situation the $\delta$-functions remain sinc-functions.

[8] These expressions are valid when the temperature signals can be conveniently approximated by a series of concatenated ramps with adequate slopes and lengths.

## 4. Mitigation of non-ideal *ADC* errors: dithering techniques

Two different methods have been used to mitigate the effect of the non-idealities of the *ADCs* on the temperature measurements. Both methods are based on the same technique: *dithering signals*. This technique is based on the injection of suitable analog signals to the signal of interest before quantisation in order to *linearise* the *ADC* transfer curve. In the following we briefly review the theoretical basis of this technique. If a dither voltage, $d$, with a certain probability density function (*PDF*), $p(d)$, is added (before quantisation) to the signal of interest, $v$, the average quantisation error at the output of the *ADC* is [9]:

$$\langle q(v)\rangle = \int_{-\infty}^{\infty} q(v+z)\,p(z)\,dz \qquad (7)$$

$$\langle \mathcal{Q}(\xi)\rangle = \mathcal{Q}(\xi)\,\mathcal{P}^*(\xi) \qquad (8)$$

where $q(v)$ [and $Q(\xi)$] is the quantization error of a real *ADC* —see Sec. 2—, $p(d)$ is the *PDF* of the dither signal [and $P(\xi)$ is the Fourier transform as defined in footnote 6)] and $\langle q(v)\rangle$ [and $\langle Q(\xi)\rangle$] is the quantisation error function at the output of the *ADC* considering we inject a dither signal. In our experiments we have tested two different dither signals: (i) Gaussian noise dither added out of the *MBW* and (ii) triangular wave dither of frequency 50 Hz. These signals are characterised by the following equations:

$$\text{Gaussian noise}: \quad p(d) = \frac{1}{(2\pi)^{1/2}\sigma}\,e^{-\frac{d^2}{2\sigma^2}} \rightarrow \mathcal{P}(\xi) = e^{-\sigma^2\xi^2/2} \qquad (9)$$

$$\text{Triangular wave}: \quad p(d) = \frac{1}{D} \rightarrow \mathcal{P}(\xi) = \mathrm{sinc}(D/2)\xi \qquad (10)$$

where $\sigma$ and $D$ are the standard deviation of the Gaussian noise and the amplitude of the triangular wave, respectively; and $\xi = \frac{2\pi}{2^{k+1}\Delta}$. Figure 4 shows the error components introduced by error bits ($k$=0 and $k$=3) and the effect of the dither signals on them (red traces).

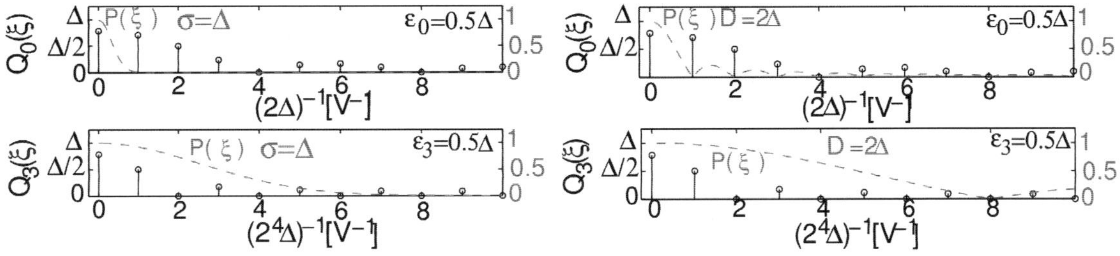


**Figure 4.** *ADC* quantisation error as given in Eq. (3). They are indicated by the circles on top of vertical strokes. Left: the red dashed line represents the Gaussian dither attenuation profile with $\sigma = \Delta$ which suffices to mitigate the error in the bit $k = 0$ but not the errors in $k = 3$. To do so higher values of $\sigma$ are required. Right: *idem* for the triangular wave dither of amplitude $D = 2\Delta$. Note the different scales in the $x$-axis. The conversion to Hertz is easily done by using $\xi = \omega/|\dot{v}(t)|$.

## 5. Test set-up and results

The test set-up in order to check the performance of the dithering methods implemented consisted in attaching a set of thermistors inside a thermal insulator designed to screen out temperature fluctuations to the required level in the milli-Hertz range, i.e., $< 10^{-5}\,\mathrm{K\,Hz^{-1/2}}$ at the milli-Hertz [3]. A temperature control was implemented to generate temperature ramps of different slopes (from $0.5\,\mu\mathrm{K\,s^{-1}}$ to $16\,\mu\mathrm{K\,s^{-1}}$) inside the insulator. In this manner we could observe the effect of the dither signals in the *TMS* under different temperature slopes.

Four configurations were tested under identical conditions. First, a 16-bit *SAR ADC* (AD977 of Analog Devices[9]) without any dither signal was tested. Second, we tested the same *ADC* but injecting Gaussian noise dither before quantisation. The third configuration consisted in using the triangular wave dither signal. Finally, a higher resolution 24-bit Delta-Sigma *ADC*[10] was also used for comparison. Results are shown in Fig. 5 for different slopes and configurations.

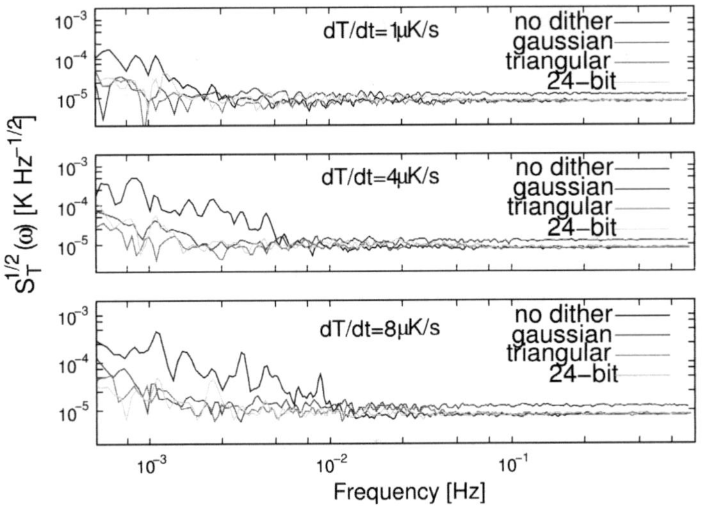


**Figure 5.** Linear Spectral Density for different configurations. As can be seen in absence of dither, the noise of the *ADC* increases (in amplitude and bandwidth) with the slope of the input signal. Gaussian noise dither attenuates this effect although the floor noise is slightly higher than the typical one. Triangular wave dither works satisfactorily and the floor noise remains untouched. The 24-bit Delta-Sigma *ADC* works also well for the tested slopes.

## 6. Conclusions

The analysis and results obtained lead to the following conclusions:

- Extra noise caused by *INL* errors of the *ADC* was seen at the lower end of the *LTP MBW* when drifting temperatures were measured with the *LTP* temperature measurement subsystem.
- For temperature slopes $>0.5\,\mu\mathrm{K\,s^{-1}}$ the extra noise challenges the performance of the *LTP* measuring system[11], the problem growing more severe at lower frequencies (*LISA* band).
- The use of dithering techniques mitigates the *INL* errors introduced by the *ADC*. However the injection of Gaussian noise dither increases the floor noise of the system in the *MBW*. Alternatively, the triangular wave dither does not add noise in the *MBW* and, therefore, results in a more efficient attenuation of *INL* errors. Dither technique is not foreseen to fly in *LPF*, but the results obtained herein are important in view of designing a more robust

[9] The *ADC* used for the space model is the Texas Instruments ADS7809 which is based in the same structure.
[10] Such *ADC* should exhibit less non-linearity problems. Space qualified version of these *ADC*s are not available.
[11] The expected slopes in the *LTP* (in nominal conditions) are lower than $0.5\,\mu\mathrm{K\,s^{-1}}$.

*TMS* for the *LISA* mission. The dither signal to be injected in the system will depend on the expected temperature drifts in *LISA* and, apparently, the triangular dither signal should be able to cope with the requirements of the mission. This would avoid the use of 24-bit Delta-Sigma *ADCs* which they are not suitable for multiplexed systems as in the case of the *TMS* and, furthermore, at present they are not space qualified.

## Acknowledgments
Support for this work came from Project ESP2004-01647 of Plan Nacional del Espacio of the Spanish Ministry of Science and Innovation (MICINN). JS acknowledges a grant from MICINN.

## References
[1] Bender P *et al.* 2000 ESA Report No ESA-SCI(2000)11
[2] The LISA International Science Team 2008 ESA-NASA Report No LISA-ScRD-Iss5-Rev1
[3] Lobo A, Nofrarias M, Ramos-Castro J and Sanjuán J 2006 *Class Quantum Grav* **23** 5177
[4] Vitale S *et al* 2005 Science Requirements and Top-level Architecture Definition for the LISA Technology Package (*LTP*) on Board *LISA* Pathfinder (SMART-2), report no. LTPA-UTN-ScRD-Iss003-Rev1
[5] Sanjuán J, Lobo A, Nofrarias M, Ramos-Castro J and Riu P 2007 *Rev Sci Instr* **78** 104904
[6] Hoeschele DF 1994 *Analog-to-digital and digital-to-analog conversion techniques* (New York, NY: Wiley)
[7] Wagdy MF 1996 *IEEE Trans Instrum Meas* **45** 610
[8] Sanjuán J 2009 *Development and validation of the thermal diagnostics instrumentation in LISA Pathfinder* PhD dissertation
[9] Wagdy MF 1989 *IEEE Trans Instrum Meas* **38** 850

8th Edoardo Amaldi Conference on Gravitational Waves IOP Publishing
Journal of Physics: Conference Series **228** (2010) 012042 doi:10.1088/1742-6596/228/1/012042

# Fiber modulators and fiber amplifiers for LISA

**M Tröbs[1], S Barke[1], J Möbius[2,3], M Engelbrecht[2,4], Th Theeg[2], D Kracht[2], B Sheard[1], G Heinzel[1] and K Danzmann[1]**

[1] AEI Hannover, (MPI für Gravitationsphysik und Leibniz Universität Hannover), Callinstr. 38, 30167 Hannover, Germany

[2] Laser Zentrum Hannover, Hollerithallee 8, 30419 Hannover, Germany

[3] present address: Viscom AG, Carl-Buderus-Straße 9-15, 30455 Hannover, Germany

[4] present address: Menlo Systems GmbH, Am Klopferspitz 19, 82152 Martinsried (Munich), Germany

E-mail: `michael.troebs@aei.mpg.de`

**Abstract.** We present the sideband phase characteristics of a fiber-coupled integrated electro-optical modulator (EOM) at a modulation frequency of 2 GHz for Fourier frequencies from 0.1 mHz to 1 Hz. The upper phase noise limit was almost an order of magnitude better than required for LISA. The EOM's phase dependencies on temperature and transmitted optical power were measured and found to be uncritical. Additionally we have investigated three optical amplifiers emitting 1 W. Their differential phase noise and optical pathlength noise as one contribution to differential phase noise were measured. The measured differential phase noise was within the requirement. The dependencies of differential phase noise on pump power were measured and requirements for the operation of the amplifier on the LISA satellite derived.

## 1. Introduction

In the LISA mission [1] the phase of the beat note between pairs of laser beams will be measured to detect pathlength variations caused by gravitational waves. Each satellite requires a stable frequency standard (ultra-stable oscillator, USO) for this phase measurement. There are no suitable space-qualified frequency standards known to the authors providing the required accuracy. Due to the Doppler shift resulting from the relative velocities of the spacecraft (S/C), the beat notes lie between 2 and 20 MHz. To reach the required sensitivity the clock noise must be removed in post-processing [2]. Electro-optical modulators (EOMs) will be used to imprint the clock noise of the local USO as phase modulation sidebands onto the laser beam travelling to the remote spacecraft where this clock noise will be compared to the local clock noise. The differential clock noise will be recorded and sent to Earth for post-processing. No component in the clock noise transfer chain may degrade the phase fidelity of the phase modulation sidebands with respect to the carrier.

We have investigated the phase fidelity of two components in the clock noise transfer chain, namely an EOM and three fiber amplifiers (FAs). The requirements for EOM and fiber amplifier differential phase noise used here are based on $1\,\mathrm{pm}/\sqrt{\mathrm{Hz}}$ equivalent displacement noise at the heterodyne frequency. By using a high modulation frequency $f_{\mathrm{SB}}$ for the EOM sidebands to be transmitted through the amplifier, the phase noise requirement between carrier and sideband is relaxed by the factor $f_{\mathrm{SB}}/f_{\mathrm{het}}$ where $f_{\mathrm{het}} = 2\ldots20\,\mathrm{MHz}$ is the heterodyne frequency onboard

© 2010 IOP Publishing Ltd

the LISA spacecraft. In our case with $f_{SB} = 2\,\text{GHz}$ this results in a differential phase noise requirement $\phi_{RQ}$ of

$$\phi_{RQ}(f) = 0.6\,\frac{\text{mrad}}{\sqrt{\text{Hz}}} \cdot \sqrt{1 + \left(\frac{2.8\,\text{mHz}}{f}\right)^4} \tag{1}$$

where the commonly used frequency shape factor has been used [3, 4, 5].

## 2. EOM single sideband phase characteristics

We have investigated an integrated optical phase modulator based on LiNbO3 / MgO:LiNbO3 crystals for wavelengths of 1064 nm ± 60 nm [6]. The phase of the light could be modulated with frequencies ranging from 100 Hz to 2 GHz. At 2 GHz modulation frequency the half-wave voltage measured 8.6 V. The modulator could handle up to 300 mW of optical power with a 4 dB insertion loss and was equipped with polarisation maintaining single-mode fibers.

### *2.1. EOM single sideband phase noise*

In contrast to Klipstein et al. [3] who measured the relative phase noise beween the sidebands of two EOMs, we have chosen a setup which allowed us to investigate the phase noise of a single EOM. The setup is shown in Fig. 1. Two laser beams were superimposed at a beam splitter with one beam first sent through an EOM driven by a 2 GHz + 1.6 kHz signal. Both then were passed through a mode cleaning fiber (not shown in the schematic setup) and sent to two photodiodes. The beatnote signal of one photodiode was mixed with a 2 GHz reference signal, low-pass filtered and used to offset phase lock both lasers to the reference frequency. At the other photodiode, we measured the carrier-carrier beat note at 2 GHz, mixed it down using the EOM driver signal and low-pass filtered the output. In the end the phase of the resulting 1.6 kHz signal was measured by a software phasemeter operating with the same principle as the phasemeter of the LISA Technology Package (LTP) [7]. The phase of the sideband-carrier beat note at 1.6 kHz which contained phase shifts induced by the EOM was also measured. To remove setup noise such as residual phase lock error and phase noise of the EOM driver signal, we subtracted the phase of the mixed-down carrier-carrier signal from the sideband-carrier beat.

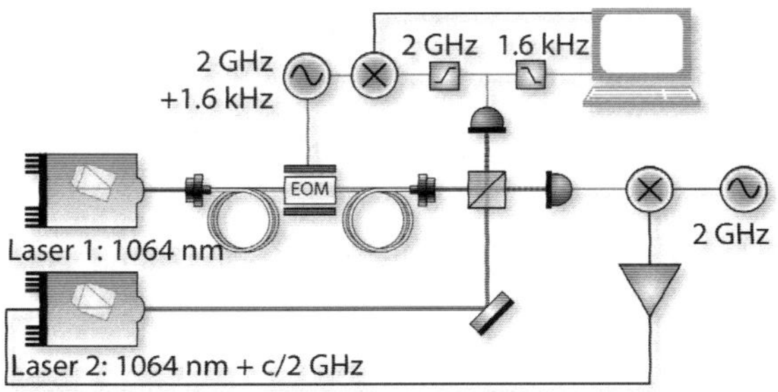


**Figure 1.** Schematic setup for EOM phase noise measurements.

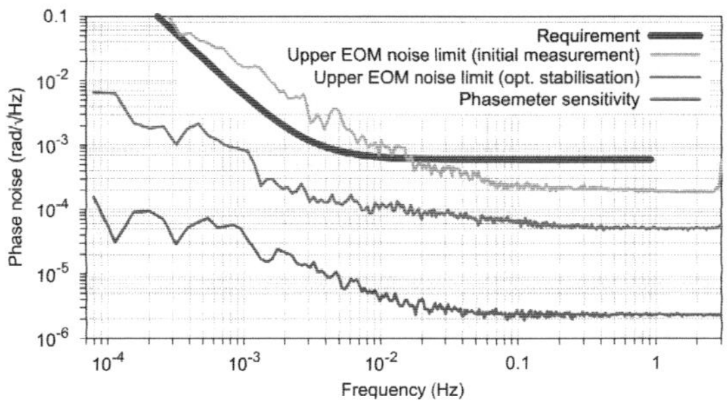


**Figure 2.** Differential phasenoise measurement.

Figure 2 shows the measurement results. We showed that the upper phase noise limit of the EOM (red trace) was almost an order of magnitude better than the LISA mission requirements (grey trace). For this measurement the 2 GHz beat-note power was stabilized to a constant reference by tuning the laser power (pump diode current). This technique removed power-level related phase shifts in the electronic mixers that were visible in the inital measurement that was performed without this stabilization (yellow trace). The program LPSD [8, 9] was used to generate the traces shown in Fig. 2 and in all following graphs showing spectral densities.

*2.2. Dependencies of EOM phaseshift on temperature and light power*

We measured the dependency of the relative phase between the carrier and one sideband generated inside the EOM on the temperature. For doing so we placed the EOM in an aluminium housing with thermo-electric coolers (TEC) attached to it. Platinum resistors in the aluminium housing were used to measure the EOM temperature. While the EOM temperature was varied, the resulting phase shift was recorded using the setup shown in Fig. 1. Figure 3 shows the measurement result of the phase shift induced by the EOM versus its temperature (red for heating and orange for cooling periods). A linear fit is shown in blue which represents a dependency of $2.1 \pm 0.2\,\mathrm{mrad/K}$.

The temperature of the output fiber of the EOM was also altered and the resulting phase shift between carrier and sideband was measured to be $1.5 \pm 0.2\,\mathrm{mrad/K}$ for the 1.5 m long output fiber. Combining both coefficients we obtained a total phase dependency on the EOM's temperature of about $3.6\,\mathrm{mrad/K}$.

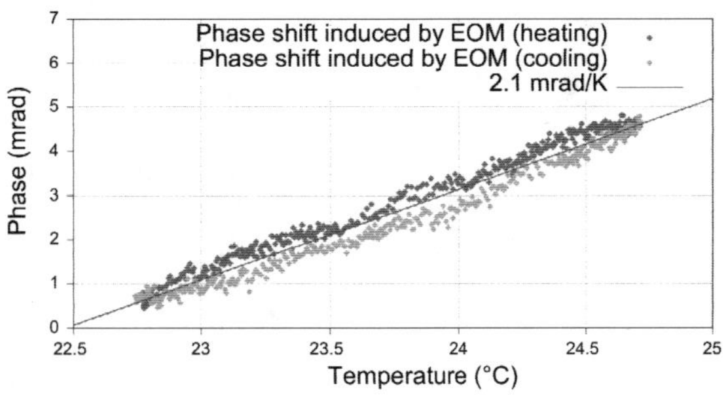


**Figure 3.** Measured dependency of EOM phasenoise on temperature.

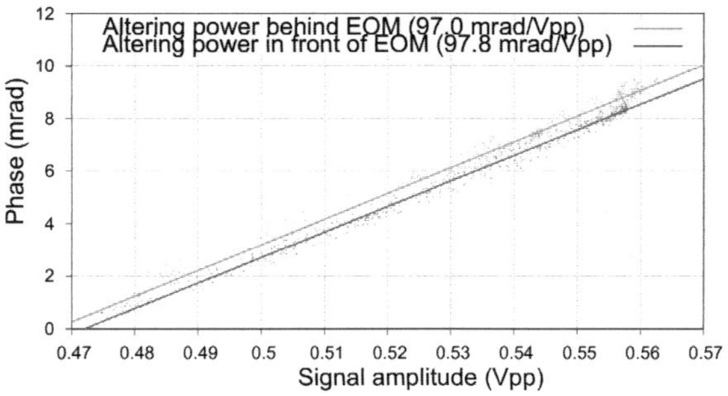


**Figure 4.** Measured dependency of EOM phasenoise on optical power.

We have also determined the photorefractive effect (phase dependence on transmitted optical power) by altering the laser power going through the EOM. Figure 4 shows the differential phase plotted against the beat signal amplitude. The optical input power was varied by detuning the fiber coupling to the EOM and the resulting phase shift between carrier and sibeband was measured (red trace). Since altering the laser power causes phase shifts not only in the EOM but also in e.g. the mixer, we have made a second measurement where the optical power behind the EOM was varied by detuning the coupling to the mode cleaning fiber and the resulting phaseshift was measured (yellow trace). The influence of the EOM—and hence the photorefractive effect—is the difference between the slope coefficients of the two traces; this was computed to be $0.0026 \pm 0.0022\,\mathrm{mrad/mW}$.

Both, the phase dependency of the EOM on temperature and on optical power is compatible with the expected noise levels onboard the LISA satellites.

## 3. Fiber amplifier phase stability

In previous experiments [10, 11] the length noise of a fiber amplifier was measured. First measurements of phase noise between carrier and sideband on fiber amplifiers have been reported in [5].

Journal of Physics: Conference Series **228** (2010) 012042 doi:10.1088/1742-6596/228/1/012042

### *3.1. Amplifiers under test*

We have investigated three Ytterbium-doped fiber amplifiers emitting 1 W of output power that were pumped by diode lasers at 976 nm in the direction of the seed beam.

The first amplifier consisted of two identical stages using 0.4 m polarization-maintaining (PM) Yb-doped single-mode fiber with 490 dB/m absorption at the pump wavelength and 5.7 μm core diameter by INO. It was not fully polarization-maintaining since non-PM wavelength division multiplexers (WDMs) were used to couple the pump light in the active fiber. This amplifier will be called core-pumped amplifier (CPA) in the following. It was seeded with 30 mW and pumped by approximately 2 W.

The second amplifier consisted of a single stage using 3 m Yb-doped double-clad fiber (Yb1200-10/125DC-PM by Liekki) with a 10 μm core diameter, a 125 μm cladding diameter, and numerical apertures of 0.08 and 0.46 for the core and pump cladding, respectively. It was fully polarization-maintaining and the pump diode was cooled by a fan. This amplifier will be called double-clad amplifier 1 (DCA1). It was seeded by 45 mW and pumped by 3.5 W

The third amplifier also consisted of a single stage using 7.5 m Yb-doped double-clad fiber (PM-YDF-5/130 by Nufern) with a 5 μm core diameter, a 130 μm pump cladding diameter, and numerical apertures of 0.13 and 0.46 for the core and pump cladding, respectively. The amplifier was fully polarization-maintaining. The pump diode was placed on a heat sink with thermo-electric cooling without fans. This amplifier will be called double-clad amplifier 2 (DCA2). It was seeded by 40 mW and pumped by 2.3 W. The seedpowers that were used are in the output power range of the LISA Pathfinder laser, that emits 45 mW and is forseen as seed source for a fiber amplifier within LISA.

### *3.2. Optical pathlength noise measurements*

Optical pathlength changes in an amplifier, $\Delta s$, due to e.g. changes in ambient temperature, pump power, or seed power lead to differential phase changes, $\Delta\phi$, of sidebands at $f_{\rm SB}$ with respect to the carrier of

$$\Delta\phi = \frac{2\pi f_{\rm SB}}{c}\Delta s, \tag{2}$$

where $c$ is the speed of light in vacuum. We have measured the length noise of the three amplifiers with the setup shown in Fig. 5 and used Eq. (2) to calculate the effect of amplifier length changes on phase noise between carrier and a 2 GHz sideband. The amplifier under test

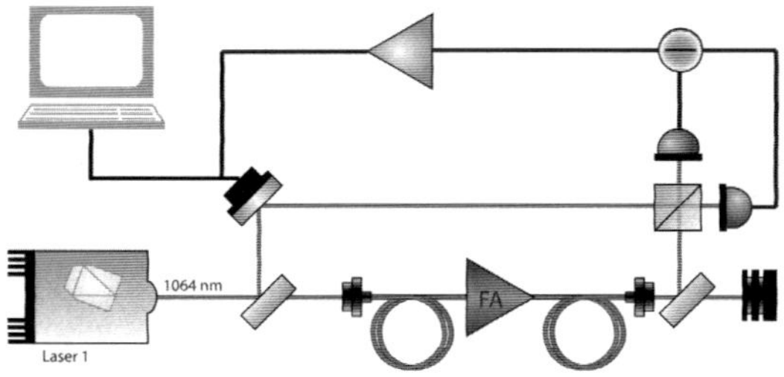


**Figure 5.** Simplified setup to measure fiber amplifier length noise.

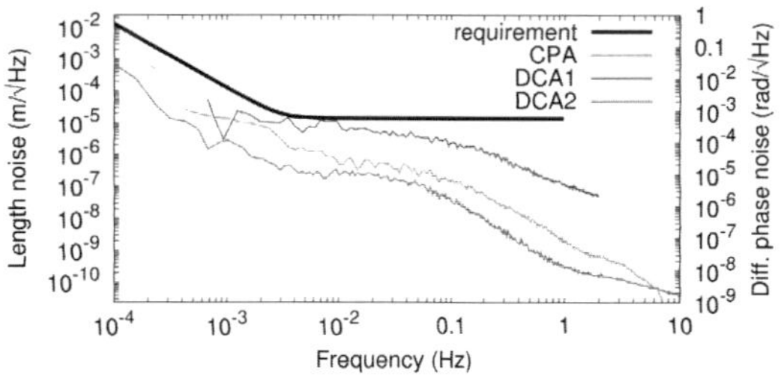


**Figure 6.** Measured length noise of the fiber amplifiers.

was placed in one arm of a homodyne Mach-Zehnder interferometer. The output phase was detected with a balanced detector and stabilized on mid-fringe by acting on a PZT-mounted mirror in the other arm. The calibrated actuator signal was used as measure for the amplifier length noise. The setup was enclosed in a thermal shield made of extruded polystyrene. The amplifiers were covered with an additional thermal isolation layer.

Figure 6 shows the measured optical pathlength noise of the fiber amplifiers on the left axis. The right axis shows the projected differential phase noise between carrier and a 2 GHz sideband caused by the optical pathlength noise as given by Eq. (2). All three amplifiers showed lower optical pathlength noise for higher Fourier frequencies. Double-clad amplifier 1 showed the highest optical pathlength noise of the three amplifiers. It touched the requirement between 3 mHz and 10 mHz. The core-pumped amplifier and double-clad amplifier 2 showed lower optical pathlength noise than double-clad amplifier 1, at 10 mHz by factors of 10 and 20, respectively. The optical pathlength noise depends on fiber length, the type of fiber and its jacketing [12], pump power noise, seed power noise, and ambient temperature noise. The three amplifiers consisted of different lengths and types of fibers and used different pump diodes. The fan used within DCA1 but not within CPA and DCA2 might be a cause for the higher length noise.

### *3.3. Differential phase noise measurements*

Figure 7 shows the simplified experimental setup used for measuring differential phase noise. Light from two lasers was interfered at a beam splitter and detected by a (fiber-coupled) photodiode. A control loop was used to offset-phase lock the lasers with a 2 GHz frequency difference. The same photodiode signal and that of a (fiber-coupled) photodiode behind the amplifier were mixed with a 2 GHz + 1.6 kHz local oscillator. The phases of the resulting 1.6 kHz signals were measured using a LISA Pathfinder-style phasemeter [7], that was implemented using a PC with DAQ card and self-written software. Since the mixers showed a signal amplitude dependent phase shift, the amplitudes of the 1.6 kHz signals were stabilized by acting on the light power incident on the photodiodes by PZT-mounted tilting mirrors that altered the coupling efficiency to the fiber-coupled photodidoes. The setup was again enclosed in a thermal shield made of extruded polystyrene and the amplifier under test was covered with an additional isolation layer.

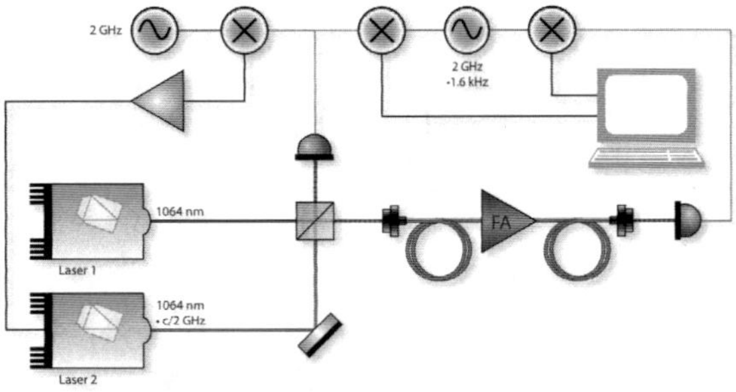


**Figure 7.** Simplified setup to measure fiber amplifier differential phase noise.

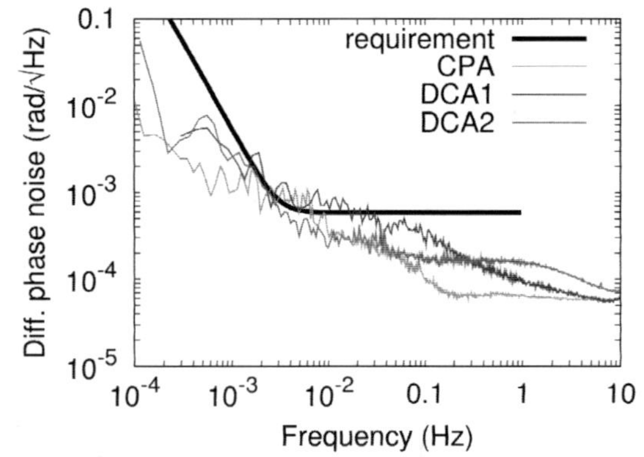


**Figure 8.** Measured differential phase noise of the fiber amplifiers.

Figure 8 shows the measured differential phase noise for the three amplifiers. All amplifiers showed differential phase noise within the LISA requirement. Between 3 mHz and 10 mHz double-clad amplifier 1 was limited by optical pathlength noise, the other two differential phase noise measurements were limited by setup noise (traces not shown).

### *3.4. Dependencies of differential phase on pump power*

Dependencies of differential phase on pump power were measured. The strongest measured dependencies and resulting noise requirements are shown in Table 1. While the two double-clad

**Table 1.** Measured strongest dependencies of differential phase noise on pump current and resulting noise requirements.

| Parameter | Unit | CPA | DCA1 | DCA2 |
|---|---|---|---|---|
| Diff. phase / pump power | (mrad/W) | 32 | 58 | 161 |
| Noise requirement | $(\mathrm{mW}/\sqrt{\mathrm{Hz}})$ | 18 | 10 | 4 |

amplifier used one pump diode each, the core-pumped amplifier used four pump diodes. For the measurement, three pump diodes were operated at constant power, while one pump diode power was changed and the differential phase was measured. With increasing length of the active fiber, the coefficient from pump power to differential phase increased. The noise requirements shown in Table 1 each result in differential phase noise at the requirement. When multiple operating parameters are taken into account and requirements for them are made, a suitable noise budget has to be allocated to each noise source such that the combined effect results in differential phase noise within the overall requirement.

## 4. Conclusions

In conclusion, we have measured an upper limit to the phase noise introduced by an EOM between carrier and a 2 GHz sideband and found this phase noise to be almost one order of magnitude below the requirement. The dependency of EOM phase shift on temperature of 3.6 mrad/K and on optical power of $0.0026 \pm 0.0022$ mrad/mW were measured.

Three fiber amplifiers emitting 1 W were investigated. For all fiber amplifiers the phase noise between carrier and 2 GHz sideband was within the requirement. Optical pathlength noise as one source of differential phase noise was measured for the three amplifiers. Only the differential phase noise of double-clad amplifier 1 was limited by its optical pathlength noise for Fourier frequencies between 3 mHz and 10 mHz. Maximum dependencies of differential phase noise on pump power were measured. None required the use of an active pump power stabilization.

## Acknowledgments

We acknowledge support by Deutsches Zentrum für Luft- und Raumfahrt (DLR) (reference 50 OQ 0601) and thank the German Research Foundation for funding the Cluster of Excellence QUEST – Centre for Quantum Engineering and Space-Time Research.

## References

[1] Danzmann K and Rüdiger A 2003 *Class. Quantum Grav.* **20** S1–S9
[2] Hellings R W 2001 *Phys. Rev. D* **64** 022002
[3] Klipstein W, Halverson P G, Peters R, Cruz R and Shaddock D 2006 *LASER INTERFEROMETER SPACE ANTENNA: 6th International LISA Symposium* vol 873 ed Merkowitz S M (American Insitute of Physics) pp 312–318
[4] Barke S, Tröbs M, Sheard B, Heinzel G and Danzmann K 2010 *Appl. Phys. B* **98** 33–39
[5] Tröbs M, Barke S, Möbius J, Engelbrecht M, Kracht D, d'Arcio L, Heinzel G and Danzmann K 2009 *J. Phys.: Conf. Ser.* **154** 012016
[6] JENOPTIC Laser, Optik, Systeme GmbH, Jena, Germany Integrated Optical Phase Modulator / Fiber-coupled electro-optical light-modulator http://www.jenoptik-los.de/cms.php?pageid=788&lang=0
[7] Heinzel G, Wand V, Garcia A, Jennrich O, Braxmaier C, Robertson D, Middleton K, Hoyland D, Rüdiger A, Schilling R, Johann U and Danzmann K 2004 *Class. Quantum Grav.* **21** S581–S587
[8] Tröbs M and Heinzel G 2006 *Measurement* **39** 120–129
[9] Tröbs M and Heinzel G 2009 *Measurement* **42** 170
[10] Tröbs M, Weßels P and Fallnich C 2005 *Opt. Lett.* **30** 789–791
[11] Tröbs M, Weßels P and Fallnich C 2005 *Opt. Express.* **13** 2224–2235
[12] Tateda M, Tanaka S and Sugawara Y 1980 *Appl. Opt.* **19** 770–773 ISSN 0003-6935

# Fiber laser development for LISA

**Kenji Numata[1,2,*], Jeffrey R. Chen[3], Jordan Camp[2]**

[1]Department of Astronomy, University of Maryland, College Park, Maryland, 20742, USA
[2]NASA Goddard Space Flight Center, Gravitational astrophysics branch, Code 663, Greenbelt, Maryland, 20771, USA
[3]NASA Goddard Space Flight Center, Laser and electro-optics branch, Code 554, Greenbelt, Maryland, 20771, USA

E-mail: *kenji.numata@nasa.gov

**Abstract.** We have developed a linearly-polarized Ytterbium-doped fiber ring laser with single longitudinal-mode output at 1064 nm for LISA and other space applications. Single longitudinal-mode selection was achieved by using a fiber Bragg grating (FBG) and a fiber Fabry-Perot (FFP). The FFP also serves as a frequency-reference within our ring laser. Our laser exhibits comparable low frequency and intensity noise to Non-Planar Ring Oscillator (NPRO). By using a fiber-coupled phase modulator as a frequency actuator, the laser frequency can be electro-optically tuned at a rate of 100 kHz. It appears that our fiber ring laser is promising for space applications where robustness of fiber optics is desirable.

## 1. Introduction

Single frequency fiber laser technology has made great advances over the last ten years and is overcoming limitations of traditional bulk-optics based lasers such as the Non-Planar Ring Oscillator (NPRO). The NPRO exhibits low frequency fluctuations due to small deformations of bulk crystal that forms the laser cavity [1], and has been widely used in low-noise, single-frequency applications.

On the other hand, there is a great interest to develop single frequency fiber lasers. Compared to a NPRO, a fiber laser offers significant advantages: 1) A fiber laser is virtually alignment free due to the wave-guided laser cavity and pump laser path, and thus more robust against mechanical disturbances; 2) The fiber waveguide maintains single mode and linear polarized laser beam that can be readily coupled into fiber amplifiers; 3) A strong magnet is not needed; 4) A fiber laser is also contamination free due to the closed cavity; 5) It is easier to implement component redundancies in a fiber laser.

The high robustness and efficiency of fiber lasers are particularly attractive for space applications. It has been proposed for LISA mission [2] that a fiber-coupled waveguide phase modulator and a Ytterbium(Yb)-doped fiber amplifier be incorporated in the laser transmitter in order to modulate and amplify the laser. This makes a fiber oscillator more attractive for its inherent fiber coupled output.

We have developed a Yb-doped fiber ring laser that emits linearly-polarized, single longitudinal-mode, and continuous-wave light at 1064 nm for space applications such as LISA and GRACE follow-on [3]. This laser was built solely with commercially available components. Single longitudinal-mode was selected by two filters in series, a fiber Bragg grating (FBG) and a

© 2010 IOP Publishing Ltd

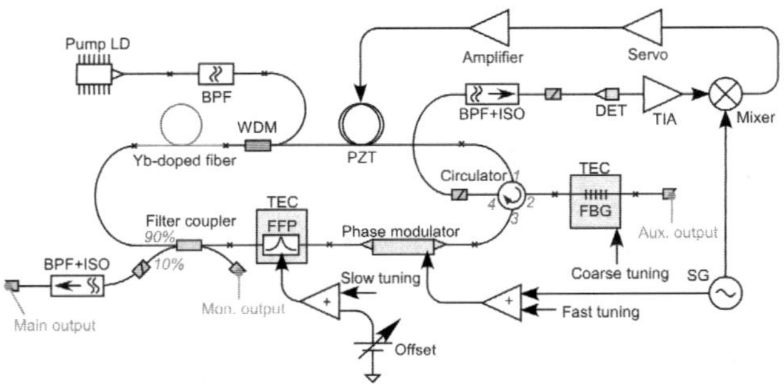


**Figure 1.** Ring laser configuration. WDM: wavelength division multiplexing coupler, FBG: fiber Bragg grating, FFP: fiber Fabry-Perot, BPF: band-pass filter, ISO: isolator, DET: detector, TIA: transimpedance amplifier, TEC: thermo electric cooler, SG: signal generator.

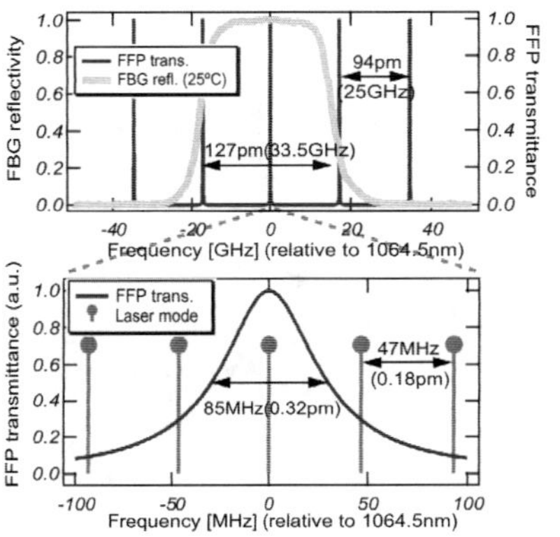


**Figure 2.** Mode selection. Bottom figure magnifies the central region of top figure.

fiber Fabry-Perot (FFP). We achieved mode-hop free operation and low frequency and intensity noise performances comparable to a commercial NPRO. The optical frequency of the laser can be varied through slow and fast actuators to facilitate frequency stabilization using external references. Coarse but slow frequency tuning was achieved by changing FFP spacing (and FBG temperature for large change), and fast tuning was enabled through an intra-cavity phase modulator. It appears that our fiber ring laser is promising for space applications. The details of this laser are described in the following sections.

## 2. Experimental setup

Figure 1 shows our ring laser configuration. The Yb-doped gain fiber in the ring cavity was core-pumped by a laser diode (LD) through a wavelength division multiplexing (WDM) coupler. Single longitudinal mode selection was achieved by cascading a FBG and a FFP. This configuration is similar to earlier work done at 1.5 $\mu$m range using Erbium-doped, non-polarization maintaining (PM) fibers [5, 6]. By using PM Yb fiber and PM components, our ring laser produces stable and linearly-polarized output at 1064 nm. We used different control schemes and introduced fast frequency tuning to minimize noise.

### *2.1. Filters and mode selection*

Figure 2 illustrates how single longitudinal mode was selected. We used a FBG as a coarse filter to select lasing wavelength out of the Yb's wide gain bandwidth that spans over 100 nm around 1030 nm. The FBG was written on the slow axis of a PM980 fiber. The FBG's peak reflectivity and reflection bandwidth were 99 % and 0.127 nm (33.5 GHz), respectively. The center wavelength of its reflection was 1064.5 nm at room temperature. The FBG was inserted into a temperature-controlled copper block for coarse wavelength tuning. The FBG was spliced to port 2 of a 4-port circulator, so that the FBG was used in its reflection mode. The light going into port 1, 2, and 3 comes out from port 2, 3, and 4, respectively. The unidirectional operation of this fiber ring laser was achieved by this 4-port circulator.

The FFP was used as the second filter to select one of the longitudinal modes within the FBG bandwidth. In FFP, a Fabry-Perot (FP) cavity is formed between two PM980 fiber ends. Free spectral range (FSR) and finesse of the FP cavity was 25 GHz and 290, respectively, and corresponding bandwidth was ~85 MHz. The FFP's 25-GHz FSR restricted the lasing to the

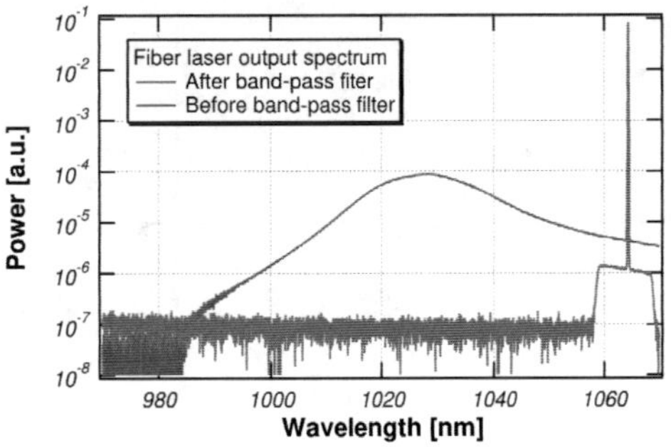


**Figure 3.** Output optical spectra. The ASE is filtered out by a band-pass filter.

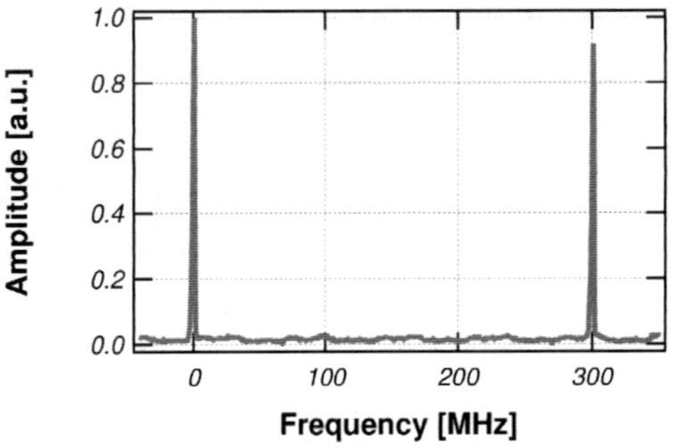


**Figure 4.** Single longitudinal laser mode measured by a scanning Fabry-Perot. An identical mode showed up twice as two peaks due to 300-MHz FSR of the scanning FP.

center of the FBG bandwidth. The 85-MHz FFP's BW then selected one longitudinal mode of the ~4.4-m laser cavity, whose FSR was about 47 MHz.

### *2.2. Control systems*

In order to keep single-mode oscillation, the lasing longitudinal mode and the FFP resonance must be aligned. Pound-Drever-Hall (PDH) technique [4] was used to lock the cavity to the FFP. The reflected light from FFP is passed though a intra-cavity lithium-niobate phase modulator that phase modulates the light at 80 MHz, and is directed through port 4 of the circulator to a fiber-coupled detector. The detected signal is demodulated using a mixer. An isolator with an integral bandpass filter is placed in front of the detector to remove amplified spontaneous emissions (ASEs) from the laser.

Once lasing is achieved, the demodulated signal at the mixer represents the difference between the laser frequency and the FFP resonance. The signal was filtered and fed back to a piezo actuator (PZT) around which a section of the ring cavity fiber was coiled, forcing laser frequency to follow the FFP resonance. Thus, the FFP serves as a frequency reference in this control scheme. Control bandwidth of this loop was about 1 kHz. Temperature of the FFP was actively stabilized by a thermo electric cooler. The phase modulator was used also for tuning the laser frequency by changing its optical length with an applied voltage. The output intensity of the laser was actively stabilized by monitoring the main output and by controlling the pump current.

### *2.3. Pump source and gain media*

The pump LD was single mode, PM, fiber-coupled, and single-longitudinal-mode laser. The output wavelength was internally stabilized to 976 nm, where our gain fiber had maximum absorption. The pump light was coupled into the cavity through a PM WDM coupler after passing through a narrow-band filter at 976 nm. The filter prevented the ASE and 1064 nm laser from reaching the pump LD.

The gain fiber was a double-cladding, single mode, PM, Yb-doped fiber. We used it as a single-clad fiber, pumping its 6-$\mu$m core. The small signal absorption of the core was 1200 dB/m at 976 nm. The length of the gain fiber was about 40 cm.

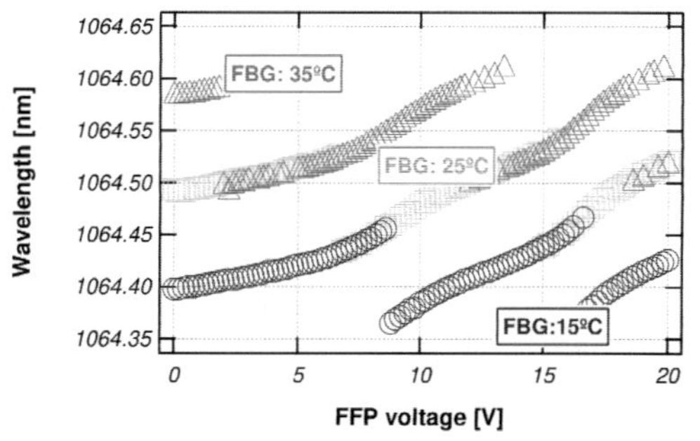


**Figure 5.** Wavelength tuning by FBG temperature and FFP spacing. Results with three different FBG temperatures are shown.

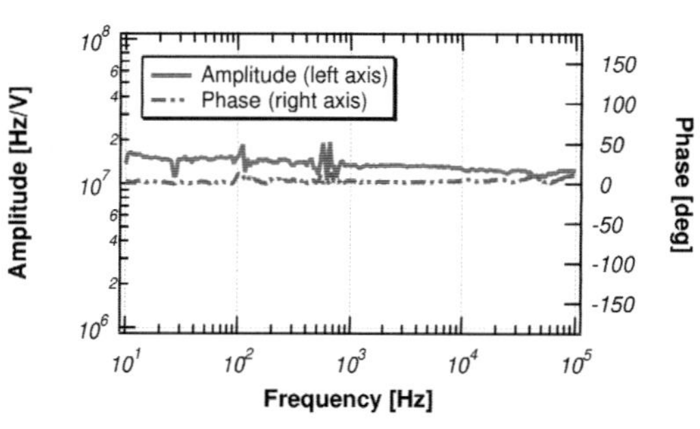


**Figure 6.** Frequency tuning transfer function of the phase modulator used in our ring laser.

### *2.4. Other laser components*

A PM filter coupler was used as an output coupler of the laser. 10 % of the light was extracted from the laser cavity, and then was filtered by an isolator with integrated ASE filter. The location of this output coupler and the coupling ratio were not optimized.

Polarization parallel to fast-axis was blocked in the circulator and isolators. This prevented lasing along fast axis and improved the polarization extinction ratio.

## 3. Experimental results

### *3.1. Output optical power and spectrum*

Limited by commercially available components, our experiment was intended to prove the design concept and was by no means optimized. The excessive insertion loss in the FFP (4.3 dB), the phase modulator (2.6 dB), and the circulator (4.9 dB) resulted in high pump threshold of ∼400 mW and an output power of ∼0.2 mW under maximum available pump power of ∼600 mW. Placing the output coupler after lossy components and the low coupling ratio (10 %) also contributed to the low efficiency. The output polarization extinction ratio was better than 20 dB.

Figure 3 shows the output optical spectrum. The ASE component centered around 1030 nm was filtered out by the ASE filter integrated with the output isolator.

Figure 4 shows the detailed optical spectrum measured by a scanning Fabry-Perot cavity with 300-MHz FSR. It can be seen that the fiber laser oscillates in a single longitudinal mode and the spectrum linewidth was below ∼1 MHz (resolution limited). The control system prevented mode-hopping.

### *3.2. Frequency tuning*

Coarse wavelength (frequency) tuning was achieved by changing the FFP spacing and the FBG temperature for long range tuning. Figure 5 shows the result of such coarse tuning. The temperature was tuned between 15°C and 35°C, resulting in a center wavelength shifting of 0.25 nm (66 GHz). The FFP spacing was changed by varying the voltage applied to the PZT stage of the FFP. The wavelength can be fast tuned through the FFP PZT in a 1 kHz bandwidth.

The optical length of the cavity can also be fast tuned by varying the voltage applied to the phase modulator. We merged the modulation signal and the fast tuning signal with a wide-bandwidth operational amplifier. As shown in Fig.6, the transfer function of such frequency tuning remains flat within the 100 kHz measurement range. The phase modulator enables

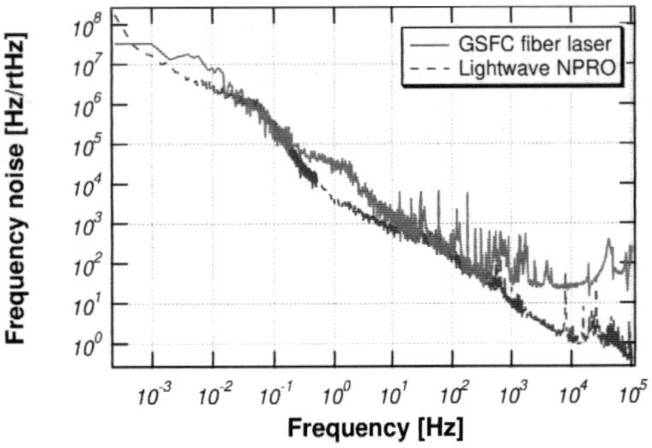


**Figure 7.** Frequency noise of fiber laser and NPRO. Below 10 Hz, the measurements were done by taking beatnotes against another NPRO.

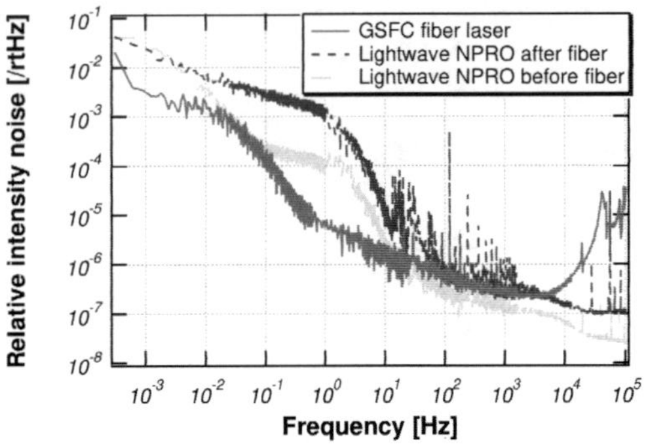


**Figure 8.** Relative intensity noise of fiber laser and NPRO. All were measured with internal intensity stabilization turned on.

tuning much faster than commercial NPROs and fiber lasers, in which mechanical deformation is used as a method to change cavity lengths [7].

### *3.3. Frequency and intensity noise*

Figure 7 shows the frequency noise spectrum of our fiber laser in comparison with that of commercial NPRO laser from Lightwave [8]. Below 10 Hz, the frequency noise was measured by taking beatnotes between a NPRO and the fiber laser and between two NPROs, respectively. Above 10 Hz, the frequency noise was measured with a fiber Mach-Zehnder interferometer with asymmetric arms.

Below 1 kHz, our fiber laser had comparable frequency noise to the NPRO. The frequency noise of our fiber laser was lower around 0.1 Hz, and the measured beatnote noise was limited by the NPRO. Our fiber laser exhibited frequency noise peaks around 1 kHz due to acoustic and electronic noise. Above 1 kHz, our fiber laser had larger frequency noise than the NPRO, due to the relaxation oscillation in our fiber laser.

Figure 8 shows the relative intensity noise of the fiber laser and the NPRO. The NPRO had larger intensity noise once its output was fiber-coupled, due to beam pointing fluctuations. Below 10 Hz, our fiber laser had about 10 times lower intensity noise than the fiber-coupled NPRO. Above 10 kHz, our fiber laser had larger intensity noise due to the relaxation oscillation at $\sim$40 kHz. Similar relaxation oscillation frequency has been observed in Yb fiber lasers [9].

We also connected the output of our fiber laser to a dual stage, core pumping, PM, Yb-doped fiber amplifier, in order to stabilize its output intensity by controlling pump current of the amplifier. We confirmed that the relative intensity noise can be stabilized down to $\sim 10^{-4}/\sqrt{\text{Hz}}$ level below 0.1 Hz, which satisfies the LISA's low-frequency intensity noise requirement.

## 4. Discussion

In order to improve efficiency of the fiber laser, we are experimenting with simpler optical configurations that minimize optical losses. One way to reduce loss is to remove the output filter coupler, and use the leakage through the FBG as the laser output. A low-reflectivity FBG will increase the output coupling ratio. The optical loss in the FFP was one of the main causes of low efficiency. In our next steps, the FFP will be replaced by a phase-shifted FBG or a fiber-coupled solid etalon, which should have smaller insertion loss and higher dimensional stability. We also expect to improve the noise performance by these modifications. We can also produce more output power by adding another pump laser diode to boost the pump power. Frequency stabilization of our fiber ring laser to optical cavity or iodine is also planned.

For space applications, it is important to have internal redundancy, especially for the pump LDs. In the case of single-mode, PM, core-pumping LD that we used, additional pump LDs can be easily added by polarization combining and by pumping in both directions without introducing large insertion losses.

The reliability of our fiber laser optical components is planned to be tested in 2010 in collaboration with Lucent Government Solutions (LGS), including vibration, thermal cycling, and radiation. At LGS, environmental tests of 2-W Yb fiber amplifier components for use in LISA have started. We expect to complete these fiber laser and amplifier tests in a year, to help identify the final laser configuration.

## 5. Summary

We developed a fiber ring laser for space applications including LISA. Our fiber laser offers comparable frequency and intensity noise to an NPRO, but also faster frequency tuning, higher polarization extinction ratio, inherently fiber-coupled output, and open architecture in which all optical components are commercially standard and testable. Future work will include solving problems associated with high insertion losses by using simpler optical configuration and different narrow-band filters. Space qualification has been started to verify the robustness of fiber optic components. We believe the evolving technologies of fiber lasers and amplifiers will become the lasers of choice for space interferometer applications within the time frame of LISA, and this work is an important initial step for it.

## References

[1] T.J. Kane and R.L. Byer, Opt. Lett., 10 (1985) 65.
[2] P. Bender, K. Danzmann, and the LISA Study Team, "Laser interferometer space antenna for the detection of gravitational waves, pre-Phase A report," Tech. Rep. MPQ233, Max-Planck-Institut für Quantenoptik, Gärching (1998). 2nd ed.
[3] M. Stephens, et al., "Demonstration of an Interferometric Laser Ranging System for a Follow-On Gravity Mission to GRACE," in *Proceedings of IEEE International Conference on Geoscience and Remote Sensing Symposium*, (IEEE, 2006), pp. 1115–1118.
[4] J.L. Hall et al., Appl. Phys. Lett., 39 (1981) 680.
[5] H. Inaba et al., IEEE J. Quantum. Electron., 38 (2002) 1325.
[6] X.P. Cheng et al., IEEE Photon. Tech. Lett., 20 (2008) 976.
[7] M. Tröbs et al., J. Opt. Soc. Am. B, 26 (2009) 1137.
[8] Now owned by JDS Uniphase Corporation (Milpitas, CA, USA).
[9] S. Huang et al., Laser Phys. Lett., 2 (2005) 498.

# Arm locking with Doppler estimation errors

**Yinan Yu, Vinzenz Wand, Shawn Mitryk, Guido Mueller**
Department of Physics, University of Florida, Gainesville, FL 32611, USA
E-mail: yinan@phys.ufl.edu

**Abstract.** At the University of Florida we developed the University of Florida LISA Interferometer Simulator (UFLIS) in order to study LISA interferometry with hardware in the loop at a system level. One of the proposed laser frequency stabilization techniques in LISA is arm locking. Arm locking uses an adequately filtered linear combination of the LISA arm signals as a frequency reference. We will report about experiments in which we demonstrated arm locking using UFLIS. During these experiments we also discovered a problem associated with the Doppler shift of the return beam. The initial arm locking publications assumed that this Doppler shift can perfectly be subtracted inside the phasemeter or adds an insignificant offset to the sensor signal. However, the remaining Doppler knowledge error will cause a constant change in the laser frequency if unaccounted for. Several ways to circumvent this problem have been identified. We performed detailed simulations and started preliminary experiments to verify the performance of the proposed new controller designs.

## 1. Introduction

The Laser Interferometer Space Antenna (LISA) is a joint NASA-ESA mission to detect gravitational waves in the frequency region of $3\times10^{-5}$ Hz to 1 Hz by means of laser interferometry [1]. LISA synthesizes three spacecraft arranged in an equilateral triangle constellation with each arm length of $5 \times 10^9$ m to reach a strain sensitivity of $10^{-21}/\sqrt{\text{Hz}}$ at 3 mHz. A variety of laser frequency stabilization techniques need to be introduced to suppress the laser frequency/phase noise to achieve this sensitivity. The first step is to stabilize the laser frequency to a resonant optical cavity or other local frequency reference. The second step is arm locking, which uses the LISA arms as a stable reference [2, 3]. The last step is known as time delay interferometry (TDI). This is a data post-processing technique to synthesize individual interferometer outputs in order to suppress the laser frequency/phase noise and clock noise to reach the LISA requirement.

The University of Florida LISA Interferometer Simulator (UFLIS) [4] is a benchtop hardware simulator used to simulate LISA interferometry and gravitational wave detections under LISA-like conditions. It incorporates the light travel time as well as variable Doppler shifts in each interferometer arm using electronic phase delays (EPD) [5]. The EPD unit includes an IQ phasemeter which extracts the frequency fluctuations of the incoming signal and sends the frequency information to a memory buffer. At the memory buffer the frequency information is stored for a certain amount of time and then is sent to a numerically control oscillator (NCO) to regenerate a delayed copy of the original signal. During this routine, a Doppler frequency can be added to the nominal frequency of the original signal.

This feature allows us to experimentally study the Doppler induced frequency pulling effect in arm locking as well as proposed solutions to solve this problem. In this paper we will report

© 2010 IOP Publishing Ltd

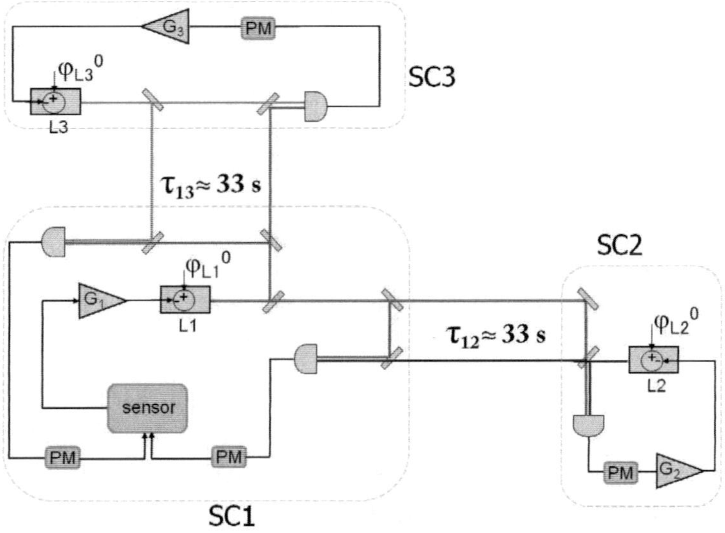


**Figure 1.** The baseline design of arm locking. The field from Laser 1 (L1) on the local SC1 is split and send to both far SC. The phase of the beat signals between L1 and the lasers L2 and L3 on SC2 and SC3 is measured with local phasemeter (PM) and used to phase lock L2 and L3 to L1. Similar beat signals are also measured on SC1 and used to generate the arm locking sensor signal.

on the current status of our arm locking experiments regarding the Doppler impact and on time-domain simulations of dual arm locking.

## 2. Arm Locking Architecture

### *2.1. Arm locking concepts*

Figure 1 displays the baseline design for the discussed arm locking schemes. Among the three spacecraft (SC) of LISA, SC1 is used as the local or master spacecraft. The lasers on SC2 and SC3 are phase-locked to the incoming laser field of the laser from SC1. Therefore, once the laser fields from SC2 and SC3 are transmitted to SC1 and interfere with the local laser field, the phase of the interferometry signals is given by

$$S_{1i}(t) = \phi_1^0(t) - \phi_1^0(t - \tau_i), \quad i = 2, 3 \tag{1}$$

where $\phi_1^0(t)$ is the instantaneous phase of the pre-stabilized laser on SC1 and $\tau_i = \tau_{1i} + \tau_{i1} \approx$ 33s is the round trip time between SC1 and SC2/3. This relation assumes that all other noise sources such as the phase-locked loop error, clock noise and shot noise as well as the phase modulations from gravitational waves are negligible. These noise sources will limit the performance of arm locking but are not essential for the here discussed effects.

In the Laplace domain, we can derive the transfer function of the two LISA arms from the interferometer outputs described above:

$$P_{1i}(s) = \frac{\widetilde{S}_{1i}(s)}{\widetilde{\phi}_1^0(s)} = 1 - e^{-s\tau_i}, \quad i = 2, 3 \tag{2}$$

where $s = \sigma + i\omega$ is the Laplace variable. Then the two interferometer outputs are manipulated to construct an arm locking error signal. This process is described by a $2 \times 1$ mapping vector

$S_k$ [6, 7]. Then the sensor transfer function will take the form

$$H_k(s) = P(s) \cdot S_k, \tag{3}$$

where

$$P(s) = \begin{bmatrix} P_{12}(s) \\ P_{13}(s) \end{bmatrix} \tag{4}$$

The specific form of the mapping vector $S_k$ determines what arm locking scheme is utilized.

*2.1.1. Single arm locking* The simplest form of arm locking uses only one arm, $S_1 = [1, 0]$, to generate the error signal. For single arm locking the sensor transfer function simply takes the form

$$H_1(s) = P_{12} = 1 - e^{-s\tau_2}. \tag{5}$$

The delay term in the sensor transfer function places additional constraints on the controller design. The sensor is insensitive at the multiple frequencies of $1/\tau_2 \approx 30$ mHz (nulls). Also, the significant phase loss from the delay time causes noise enhancement at nulls as well as limits the gain above 30 mHz. Furthermore, some of the nulls are well located in the LISA band. This would significantly jeopardize the sensitivity of the LISA at these frequencies.

*2.1.2. Dual arm locking* In dual arm locking both of the two interferometry signals are used to construct the arm locking sensor. It has been demonstrated that the information on the second arm could effectively push the first null in the sensor response beyond the LISA band. The mapping vector of dual arm locking takes the form of $[1 - \frac{E(s)}{s\Delta\tau}, 1 + \frac{E(s)}{s\Delta\tau}]$, where $\Delta\tau = (\tau_2 - \tau_3)/2$ and $E(s)$ is an additional low-pass filter to maintain the stability of the sensor for frequencies above $1/\Delta\tau$. Consequently, the sensor transfer function of dual arm locking is given by

$$\begin{aligned} H_D(s) &= \begin{bmatrix} P_{12}(s) \\ P_{13}(s) \end{bmatrix} \cdot [1 - \frac{E(s)}{s\Delta\tau}, 1 + \frac{E(s)}{s\Delta\tau}] \\ &= P_{12}(s) + P_{13}(s) - \frac{E(s)}{s\Delta\tau}(P_{12}(s) - P_{13}(s)) \\ &= P_+(s) - \frac{E(s)}{s\Delta\tau} P_-(s) \end{aligned} \tag{6}$$

where $P_+(s) = P_{12}(s) + P_{13}(s)$ and $P_-(s) = P_{12}(s) - P_{13}(s)$ are known as the common arm sensor and the differential arm sensor, respectively. The $1/s$ factor ensures that $P_-(s)$ dominates at low frequencies. The magnitude of the dual arm locking sensor is almost flat for all frequencies below $1/\Delta\tau$ and this feature allows a controller with higher gain within the LISA band.

*2.2. Doppler knowledge error and the frequency pulling*

The frequency of the laser beat signal depends on the known offset used at the phase-locked loop at the far spacecraft and the not so well known Doppler shift caused by the relative S/C motion. Ideally, the phasemeter demodulates the beat signal with the difference frequency of the two lasers. Consequently, the preset frequency at the phasemeter needs to be updated in real-time due to the time-variable Doppler shift, which is unfortunately difficult to estimate at a very accurate level. It appears to be possible to limit the Doppler estimation error to $\sim 20$ Hz per 200 s averaging time [6] by introducing additional functionality in on-board data processing.

The remaining offset will be integrated up in the arm locking controller, which causes a frequency pulling in the stabilized laser frequency. Generally a linear drift in frequency is mostly harmless in the data post-processing. However, a large frequency drift introduces additional requirements on the tunability of the arm locked laser and consequently on that of all six lasers in the LISA constellation. In the following sections we will quantify the frequency pulling induced by Doppler errors by investigating the frequency response of a generic arm locking loop.

*2.2.1. Single arm locking* Figure 2 shows a generic single arm locking loop with a Doppler error input in the Laplace domain. The transfer functions of the single arm locking sensor and controller are $G(s)$ and $H(s)$, respectively. For single arm locking, the Doppler error is introduced in the demodulation at the phasemeter right before the controller. In this diagram, the closed-loop transfer function from the Doppler error input to the stabilized laser noise output is given by

$$TF_{\rm D}(s) = \frac{G(s)}{1+H(s)G(s)}. \tag{7}$$

In a standard control loop when the controller yields the high gain limit $G(s) \to \infty$, the transfer function is reduced to

$$TF_{\rm D}(s) \approx \frac{1}{H_1(s)} \approx \frac{1}{\tau s} \qquad \text{(at low frequencies)}. \tag{8}$$

The frequency response of a DC-coupled single arm locking loop is equivalent to an integrator, which accumulates the Doppler error over time and causes a drift in the output frequency. In the time domain, the instantaneous output frequency is given by the convolution between the instantaneous Doppler error input and the transfer function described above. Over short time intervals the Doppler error is constant and the laser frequency changes by

$$\begin{aligned} \Delta\omega_{\rm L}(t) &= \frac{1}{\tau}\int \Delta\omega_{\rm D}{\rm dt} \\ &= \frac{\Delta\omega_{\rm D}}{\tau}t \end{aligned} \tag{9}$$

One potential solution is to reduce the controller gain gradually at low frequencies. This is known as AC-coupling. In the low gain limit, we then have

$$TF_{\rm D}(s) \approx G(s) \ll 1 \qquad \text{(at low frequencies)}. \tag{10}$$

In the low gain limit, the laser frequency changes simply by the Doppler error multiplied with the low controller gain. Thus, in the presence of an AC-coupled controller the Doppler error will not be accumulated by the controller and consequently the residual effect is no more than an insignificant offset added to the output laser frequency.

*2.2.2. Dual arm locking* In dual arm locking the two arms generate two independent Doppler errors. These can be combined to construct a common and a differential Doppler error which will show up in the common and differential arm sensors and will be amplified by the arm locking gain:

$$TF_{\rm D+}(s) = \frac{G(s)}{1+H_{\rm D}(s)G(s)} \tag{11}$$

$$TF_{\rm D-}(s) = \frac{G(s)E(s)}{\Delta\tau s[1+H_{\rm D}(s)G(s)]}, \tag{12}$$

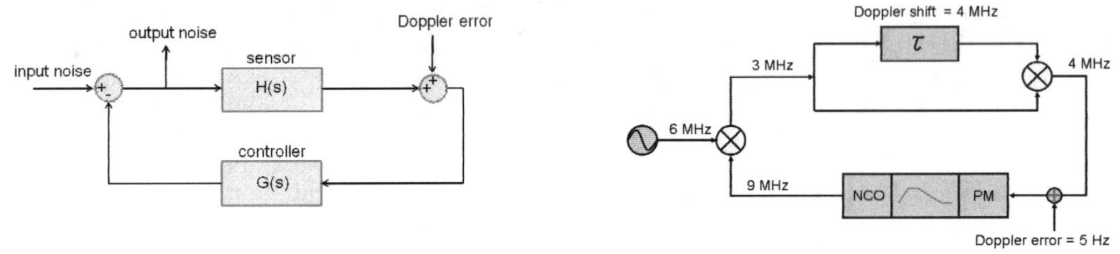


**Figure 2.** Left: A generic single arm locking loop with a Doppler error input in the Laplace domain. The Doppler error is introduced in the demodulation at the phasemeter. We define the (effective) sensor $H(s)$ to be the components before the Doppler error point and the (effective) controller $G(s)$ to be the components after the Doppler error point. Right: The experimental configuration of single arm locking in the presence of a Doppler error of 5 Hz. An input free-running VCO signal is demodulated by a tracking NCO signal. The output of the mixer is sent to a single arm locking sensor and shifted by a Doppler frequency. The controller filter (represented by the red curve) is AC-coupled in order to remove the frequency pulling effect from the Doppler error.

In the high gain limit Equation (10) is approximately equal to $1/H_{\mathrm{D}}(s)$. Since the magnitude of the dual arm locking sensor is flat at low frequencies, a common Doppler error will only cause a frequency offset in dual arm locking.[1]

In the differential arm, we have $E(s) \approx 1$ and $H_{\mathrm{D}}(s) \approx 2$ at low frequencies and in the high gain limit the equation approaches $1/2\Delta\tau s$. This frequency response corresponds to a frequency drift rate of $\Delta\omega_{\mathrm{D}-}/2\Delta\tau$. Several kinds of approaches have been proposed to reduce this frequency pulling. Some of them will be discussed in more detail in Section 4.

## 3. Single arm locking experiments in the presence of a Doppler error

The feasibility of single arm locking has been studied experimentally by several highly idealized verification experiments. All of these benchtop experiments simulated an ideal situation without any Doppler shift [8, 9] or when the Doppler frequency was perfectly accounted for at the phasemeter [3, 10]. This is unrealistic and the difference between the real and the estimated Doppler frequency will generate a Doppler error which will cause the above discussed laser frequency pulling effect. The Doppler frequency can be estimated using various techniques. For example laser ranging will also measure the relative velocity between the spacecraft or the averaged laser beat frequency allows to estimate the Doppler frequency directly. The performance of these techniques depends on many factors such as the initial laser frequency noise, integration time, differential clock noise, or the ability to exchange data between the S/C. Expected Doppler errors also depend on the functionality installed in the phasemeter and range from a few Hz to a few kHz [6]. In this section we will discuss a pioneering experiment in which a constant Doppler error is introduced into our single arm locking loop and a designed AC-coupled controller is dedicated to eliminate the frequency pulling.

In our preliminary experiment we use a single arm locking loop featuring the EPD unit to stabilize the frequency of a numerical control oscillator (NCO) to a noisy voltage controlled oscillator (VCO). Figure 2 illustrates the experimental configuration, in which a 6 MHz free

[1] This characteristics is to be distinguished from the frequency response of a common arm locking loop to a common Doppler error. In that case the arm locking sensor is replaced with $H_+(s) = P_+(s)$. Consequently, in the high gain limit a common Doppler error $\Delta\omega_{\mathrm{D}+} = \Delta\omega_{\mathrm{D}1} + \Delta\omega_{\mathrm{D}2}$ will cause the frequency pulling with a drift rate of $\Delta\omega_{\mathrm{D}+}/2\bar{\tau}$.

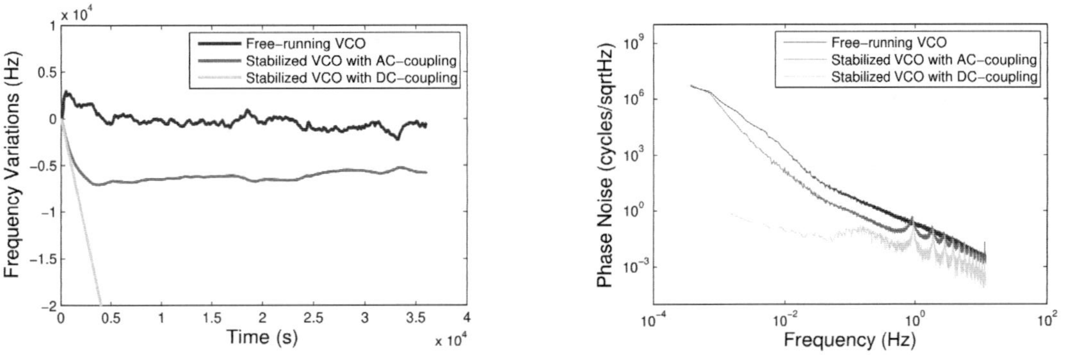


**Figure 3.** The measurement results of the experiment shown in Figure 2. Left: time series of a free-running VCO signal, a stabilized VCO signal with the AC-coupled controller and a stabilized VCO signal with a standard DC-coupled controller. The AC-coupled controller is capable of removing the frequency pulling that shows up in the situation with the DC-coupled controller. Right: linear spectral densities of the phase noise of the three time series. Although the Doppler error does not cause frequency pulling with the AC-coupled controller, the controller gain is significantly limited and the noise suppression performance is compromised.

running VCO signal is demodulated with a 9 MHz NCO signal. The beat signal of 3 MHz is then split into a prompt signal and a signal which is delayed by 1 s as well as shifted by 4 MHz. The delayed signal is mixed with the prompt signal. The output of this mixer is the Doppler shifted single arm locking error signal. The phasemeter will demodulate the error signal with a pre-set frequency.

Previous experiments have shown the noise suppression when the preset frequency matches the Doppler frequency perfectly. Figure 3 shows for the first time the performance of arm locking when the pre-set frequency does not match the Doppler frequency; here we used an offset of 5 Hz. The left graph shows the 10 h time series of the free running VCO-NCO beat (blue), the time series of the beat when stabilized by an AC-coupled controller (red), and the time series when stabilized by a standard DC-coupled controller (green). The right graph shows the linear spectral densities of the phase fluctuations of the three beat signals. The DC-coupled controller has the best in-band performance but the frequency is ramping down with the expected rate of 5 Hz/s. Note that we use a fairly high low unity gain frequency (UGF) compared to LISA to perform the measurement in a reasonable time. This limits the noise suppression in our experiments compared to the noise suppression expected in LISA.

## 4. Dual arm locking simulations

We are currently in the process of setting up a dual arm locking experiment. Here we report about time-domain simulations of dual arm locking with Doppler errors in LISA. Our time-domain simulation is running on a Matlab Simulink model with floating point arithmetic. For our simulated noise source we use the phase noise integrated from 30 Hz/$\sqrt{\text{Hz}}$ white frequency noise.

### *4.1. Dual arm locking with AC-coupled controller*

We designed an AC-coupled controller based on the zeros and poles listed in Table 1. The other parameters used in our simulation are listed in Table 2.

The black curve in Figure 4 is the open-loop transfer function which is given by the dual arm locking sensor (the blue curve) multiplied by the AC-coupled controller (the red curve) with a

Journal of Physics: Conference Series **228** (2010) 012044 doi:10.1088/1742-6596/228/1/012044

**Table 1.** Parameters of the AC-coupled filter used in simulations

| zeros (Hz) | poles (Hz) |
|---|---|
| $10^{-7}$ | $5 \times 10^{-6}$ |
| $10^{-7}$ | $5 \times 10^{-6}$ |
| $10^{-6}$ | $3 \times 10^{-3}$ |
| $5 \times 10^{-5}$ | $3 \times 10^{-3}$ |
| $5 \times 10^{-5}$ | $3 \times 10^{-3}$ |
| 0.1 | $3 \times 10^{-3}$ |
| 0.3 | $3 \times 10^{-3}$ |

**Table 2.** Parameters of the time-domain simulation of AC-coupled dual arm locking.

| $\bar{\tau}$ | $\Delta\tau$ | $\Delta\omega_{D1}$ | $\Delta\omega_{D2}$ | $G_0$ |
|---|---|---|---|---|
| 33 s | 0.255 s | 2 kHz | 3 kHz | 100 |

loop gain factor $G_0$. The AC-coupled controller consists of a $f^{-1/2}$ slope low-pass filter for the high frequency suppression and a high pass filter to reduce the frequency pulling. The open-loop transfer function provides a bandwidth of around 42 kHz and a unity gain frequency of 15 $\mu$Hz at low frequencies. The yellow curve and the cyan curve demonstrate the effective controller $G_{\text{eff}}$ and the effective sensor $H_{\text{eff}}$ in the differential arm respectively.

The closed-loop frequency response to the differential Doppler error is illustrated by the green curve in the figure. Depending on the gain level of the open-loop transfer function $H_{\text{eff}}(s)G_{\text{eff}}(s)$, the closed-loop frequency response consists of three segments in different frequency regions:

$$TF_{D-}(s) = \begin{cases} \frac{1}{H_{\text{eff}}(s)} \approx \frac{1}{2\Delta\tau s}, & \text{when } f \gg f_{UG} \\ \frac{G_{\text{eff}(s)}}{1+H_{\text{eff}}(s)G_{\text{eff}(s)}}, & \text{when } f \sim f_{UG} \\ G_{\text{eff}(s)}, & \text{when } f \ll f_{UG} \end{cases} \tag{13}$$

where $f_{UG}$ is the lower UGF of the open loop transfer function (black curve in right graph of Fig. 4).

We are more interested in the region where the closed-loop frequency response is approximately equal to the effective controller, i.e., $TF_{D-}(s) = G_{\text{eff}(s)} = G_0 G(s)/(\Delta\tau s)$. The $1/s$ slope in the frequency response indicates a Doppler error induced pulling still occurs in the output frequency. However, compared with the large gain factor of $1/(2\Delta\tau)$ in the DC-coupling situation, the gain factor now is significantly reduced due to the low gain of the AC-coupled controller. If we recall that the gain factor of $1/s$ is directly proportional to the drift rate of the frequency pulling, it is straightforward to derive the reduced drift rate given by

$$\left(\frac{\mathrm{d}\omega_L}{\mathrm{d}t}\right)_{\text{AC-coupled}} = 2G_0 G(s) \left(\frac{\mathrm{d}\omega_L}{\mathrm{d}t}\right)_{\text{DC-coupled}}. \tag{14}$$

This relation indicates that the reduced drift rate is also directly proportional to the loop gain. Our time-domain simulation is based on a long arm length mismatch of $2\Delta\tau = 0.51$ s and an inaccurate differential Doppler estimation of $\Delta\omega_{D-} = 1$ kHz. Without the AC-coupling scheme

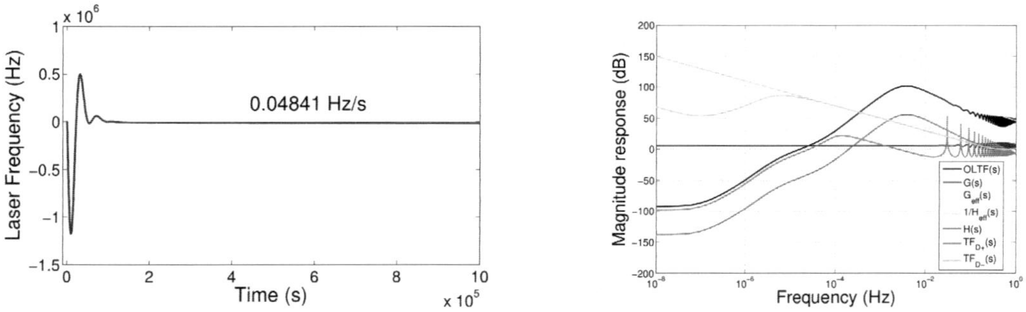


**Figure 4.** Left: Simulated time series of the output frequency change due to a 1 kHz differential Doppler error in a dual arm locking loop with an AC-coupled controller. The initial transients during locking acquisition also appear in the output frequency. Without the AC-coupling the drift rate would be 1961 Hz/s. Right: Modelled frequency responses of a dual arm locking loop to common and differential Doppler errors in the presence of an AC-coupled controller. Without the AC-coupled controller the frequency response to a differential Doppler error (the green curve) would follow $1/H_{\text{eff}}(s)$ (the cyan curve). In the presence of the AC-coupled controller it starts to deviate when passing the unity gain frequency and the gain is consequently scaled down by the controller gain.

the drift rate would be 1 kHz/0.51 s $\approx$ 1961 Hz/s for a high gain DC-coupled controller. Figure 4 shows the time series of the variation of the laser frequency using our AC-coupled controller. The drift rate has been reduced to 0.04841 Hz/s which is exactly what we expect based on the frequency domain analysis presented above. Note that we reduced the laser frequency noise in this simulation to zero to be able to determine the drift rate with this accuracy.

### *4.2. Dual arm locking with modified sensor*

A second option to reduce the frequency pulling effect has also been proposed by Kirk McKenzie [6]. The basic idea of the modified dual arm locking sensor is to combine the benefits of common arm locking and dual arm locking by amplifying the common arm sensor below $1/\bar{\tau}$ and keeping the dual arm sensor above $1/\bar{\tau}$. This has the additional advantages that many other technical noise sources are reduced as well.

We modified our original dual arm locking sensor based on the same principle but used a simpler approach. The common arm signal is integrated by $1/s$ and scaled by $1/\bar{\tau}$ such that the frequency response of the common arm signal dominates at frequencies below $1/\bar{\tau}$. On the other hand, the original dual arm locking sensor signal is filtered by a high-pass filter $F(s)$ designed in a manner that the combined frequency response stays well below the common arm at frequencies below $1/\bar{\tau}$ and smoothly returns to the original level. Therefore, the sensor transfer function of modified dual arm locking is given by

$$H_{\text{MD}}(s) = \frac{1}{\bar{\tau}s}P_{+}(s) + F(s)H_{\text{D}}(s) \tag{15}$$

The parameters of our simulation model are listed in Table 3 and the corresponding modified sensor transfer function is shown in Figure 5.

To analyse the frequency response of modified dual arm locking to Doppler errors, we follow the same routine by considering the common arm and differential arm independently. First, the expression of $H_{\text{MD}}(s)$ allows us to rewrite the sensor transfer function in the form of a combination of common arm and differential arm, i.e.,

**Table 3.** Parameters of the time-domain simulation of modified dual arm locking.

| $\bar{\tau}$ | $\Delta\tau$ | $\Delta\omega_{\mathrm{D1}}$ | $\Delta\omega_{\mathrm{D2}}$ | $G_0$ |
|---|---|---|---|---|
| 33 s | 0.165 s | 4 kHz | 2.5 kHz | 10 |

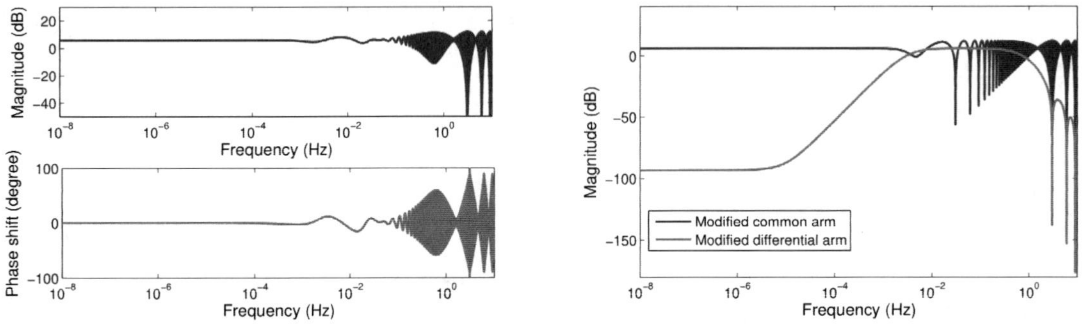


**Figure 5.** Left: The frequency response and phase response of our modified dual arm locking sensor. Right: Frequency responses in the common arm and differential arm of our modified sensor. Comparing to the original frequency responses, the common arm is integrated and high-pass filtered such that it dominates at frequencies below $1/\bar{\tau}$ and above $1/\Delta\tau$, while the differential arm is high-pass filtered such that the gain at frequencies below $1/\bar{\tau}$ is efficiently suppressed.

$$\begin{aligned} H_{\mathrm{MD}}(s) &= \frac{1}{\bar{\tau}s}P_+(s) + F(s)H_{\mathrm{D}}(s) \\ &= \frac{1}{\bar{\tau}s}P_+(s) + F(s)(P_+(s) - \frac{E(s)}{\Delta\tau s}P_-(s)) \\ &= (\frac{1}{\bar{\tau}s} + F(s))P_+(s) - \frac{E(s)F(s)}{\Delta\tau s}P_-(s) \end{aligned} \tag{16}$$

Consequently, the effective common arm sensor and effective common arm controller are given by

$$H_{+\mathrm{eff}}(s) = P_+(s) \tag{17}$$

$$G_{+\mathrm{eff}}(s) = G_0(\frac{1}{\bar{\tau}s} + F(s))G(s). \tag{18}$$

The effective differential arm sensor and effective differential arm controller are given by

$$H_{-\mathrm{eff}}(s) = P_-(s) \tag{19}$$

$$G_{-\mathrm{eff}}(s) = \frac{G_0E(s)F(s)G(s)}{\Delta\tau s}. \tag{20}$$

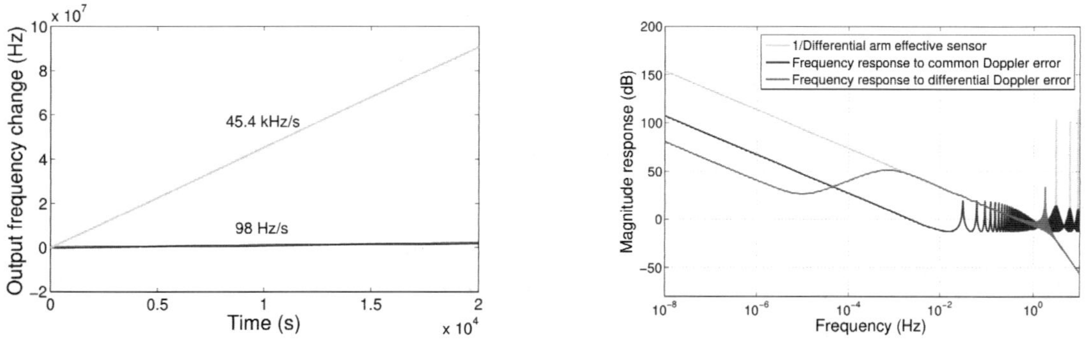


**Figure 6.** Left: Simulated time series of the output frequency change due to a 6.5 kHz common Doppler error and a 1.5 kHz differential Doppler error in a dual arm locking loop with a modified dual arm locking sensor. The modified sensor is capable of reducing the frequency drift rate to the level of common arm locking (6.5 kHz/(2 × 33 s) ≈ 98 Hz/s). Without the modified sensor the drift rate would be 45.4 kHz/s due to the differential Doppler error. Right: Modelled frequency responses to common and differential Doppler errors in the modified dual arm locking loop. Without the modified sensor the frequency response to a differential Doppler error would follow $1/H_{-\text{eff}}(s)$ (the green curve). Due to the high-pass filter implemented in the modified sensor the gain of the differential arm is gradually suppressed lower than the gain of the common arm for frequencies below $1/\bar{\tau}$.

Substitute the relations into Equation (6) and we are able to derive the closed-loop frequency response for both paths. If we only consider the high gain limit, the frequency response to the common arm Doppler error is now

$$TF_{\mathrm{D}+}(s) \approx \frac{1}{P_+(s)} \approx \frac{1}{2\bar{\tau}s} \qquad \text{(at low frequencies).} \tag{21}$$

This frequency response is the same as the case for common arm locking, corresponding to a frequency drift rate of $\Delta\omega_{\mathrm{D}+}/2\bar{\tau}$. The frequency response to the differential arm Doppler error takes the form

$$TF_{\mathrm{D}-}(s) = \frac{\frac{G_0 E(s)F(s)G(s)}{\Delta\tau s}}{1 + P_-(s)\cdot\frac{G_0 E(s)F(s)G(s)}{\Delta\tau s}}. \tag{22}$$

At low frequencies we have $E(s) \approx 1$ and $P_-(s) \approx 2\Delta\tau s$, therefore the above equation is simplified into

$$TF_{\mathrm{D}-}(s) = \frac{1}{\Delta\tau s}\cdot\frac{G_0 F(s)G(s)}{1 + 2G_0 F(s)G(s)} \tag{23}$$

The transfer function will be reduced significantly if the magnitude of $F(s)$ falls fast enough towards lower frequencies to overcome the gains of $G_0$ and $G(s)$. In this case the drift rate of the lasers scales with $1/(2\bar{\tau})$ rather than $1/(2\Delta\tau)$.

The frequency responses of both common and differential arms are plotted in Figure 6. In this modified dual arm locking model the common arm Doppler error dominates over the differential Doppler error. For our case the Doppler error in each arm was 4 kHz and 2.5 kHz, respectively. Based on the calculation above the reduced frequency drift rate should be 6.5 kHz/(2 · 33 s) = 98.5 Hz/s. The simulated time series is shown in Figure 6, featuring a reduced drift rate of 98 Hz/s while the original drift rate due to the differential Doppler error would be 45.4 kHz/s.

## 5. Conclusions

We discussed the impact of the relative SC motion and the induced Doppler shifts on arm locking. We discussed how the Doppler offset produces a ramp in the laser frequency in a standard 'high gain at low frequencies' controller. We also showed how this ramp depends on the final sensor design. This paper also presents the first ever experimental proof of arm locking in a realistic, non-static LISA configuration. We developed an AC-coupled controller which significantly reduces the induced ramp in the laser frequency. In addition to the single arm locking experiments, we also present detailed simulation results for the dual arm locking and modified dual arm locking sensor with Doppler offsets. The numerical results confirm the theoretically expected changes in the laser frequency and show that the modified dual arm locking sensor will also limit the changes in the laser frequency. This work was supported by NASA grant 07-ATFP07-0116.

## References

[1] LISA science requirement document version 3.0 2005 http://www.srl.caltech.edu/lisa/documents/SRD_v3_0_sb.pdf 1

[2] Sheard B, Gray M, McClelland D and Shaddock D 2003 Laser frequency stabilization by locking to a LISA arm *Phys. Lett.* A **320** 9-21 1

[3] Wand V, Yu Y, Mitryk S, Sweeney D, Preston A, Tanner D, Mueller G, Thorpe J I and Livas J 2009 Implementation of armlocking with a delay of 1 second in the presence of Doppler shifts *J. Phys.: Conf. Series* **154** 012024 1, 5

[4] Thorpe J I, Cruz R J, Sankar S R, Mueller G and McNamara P 2004 First step toward a benchtop model of the Laser Interferometer Space Antenna *Opt. Lett.* **29** 2843-45 1

[5] Thorpe J I, Cruz R J, Sankar S and Mueller G 2005 Electronic phase delay: A first step towards a bench-top model of LISA *Class. Quantum Grav.* **22** S227-34 1

[6] McKenzie K, Spero R E, Shaddock D A 2009 The performance of arm locking in LISA arXiv:0908.0290v2 [gr-qc] 3, 5, 8

[7] Sutton A and Shaddock D A 2008 Laser frequency stabilization by dual arm locking for LISA *Phys. Rev.* D **78** 082001 3

[8] Marin A *et al* 2005 Phase locking to a LISA arm: First results on a hardware model *Class. Quantum Grav.* **22** S235-42 5

[9] Thorpe J I and Mueller G 2005 Experimental verification of arm-locking for LISA using electronic phase delay *Phys. Lett.* A **342** 199-204 5

[10] Livas J C, Thorpe J I, Numata K, Mitryk S, Mueller G and Wand V 2009 Frequency-tunable pre-stabilized lasers for LISA via sideband locking *Class. Quantum Grav.* **26** 094016 5

Journal of Physics: Conference Series **228** (2010) 012045 doi:10.1088/1742-6596/228/1/012045

# Ranging and phase measurement for LISA

**Juan José Esteban[1], Antonio F. García[1], Johannes Eichholz[1], Antonio M. Peinado[2], Iouri Bykov[1], Gerhard Heinzel[1], Karsten Danzmann[1].**

[1]Max-Planck-Institute for Gravitational Physics (Albert-Einstein-Institute) and Leibniz University of Hannover, Germany

[2]Department of signal theory, networking and communications, University of Granada, Spain

E-mail: juan.jose.esteban@aei.mpg.de

**Abstract.**
The LISA phase measurement system (PMS) will provide interferometric phase readout of the primary heterodyne signal at microcycle sensitivity, ranging measurements at sub-meter accuracy and data communication at rates of several kilobits per seconds. Our investigations are focused on inter-spacecraft laser ranging and data transfer for LISA using Direct Sequence Spread Spectrum (DS/SS) modulation onto the laser links. We present the setup of an optical experimental to test the levels of performance achievable with a single laser link as well as a new hardware prototype based on FPGA (Field Programmable Gate Array) processing to perform high-accuracy phase readout of the optical signal, ranging measurements, data communication and is suitable for clock noise demodulation and digital laser-phase locking.

## 1. Introduction

The Laser Interferometer Space Antenna, LISA, will be a huge unequal armlength interferometer in space for gravitational wave detection [1]. LISA will consist of three spacecraft separated by 5 million kilometers forming a triangular formation and communicating via three bidirectional laser links. As consequence of variations in the spacecraft-to-spacecraft distance, the LISA arms are unequal and time-varying (change by approximately $\pm 1\%$ over one year orbit, $\simeq 50.000\,\mathrm{km}$), resulting in a limitation of the interferometric sensitivity due to laser frequency noise. In order to suppress this effect and make LISA virtually work like an equal arm interferometer, one of the techniques planned is a post-processing algorithm on ground, known as time-delay interferometry (TDI) [2]. It combines onboard ranging measurements with phase information of the laser beams to suppress the effect of laser frequency noise by up to seven orders of magnitude. This technique relies on interspacecraft distance tracking with resolution of the order of a few meters and therefore any error in the ranging measurements will turn into an error in the outcome of the post-processing algorithm [3]. In order to achieve inter-spacecraft laser ranging and data communication, the main carrier of the laser link is phase modulated with pseudo-random noise (PRN) and the travel time can be measured via the correlation of the local and incoming PRN codes. The phase measurement system (PMS) processes the interferometric signal coming from the photodetectors on the optical bench to measure its phase with microcycle accuracy, performs ranging at sub-meter resolutions and enables data transfer at several kilobits per seconds.

© 2010 IOP Publishing Ltd

In order to test the modulation scheme in a single LISA arm, an optical experiment has been built in our laboratories. Figure 1 shows the experimental setup: two offset phase-locked lasers are modulated using fibre-coupled electro-optic modulators (EOM) and injected onto a monolithic optical bench for interference. The beatnote is digitised and processed in a custom-designed PMS breadboard which implements the phase readout via a phase-lock loop (PLL) architecture and the ranging capability via a delay-lock loop (DLL) architecture.

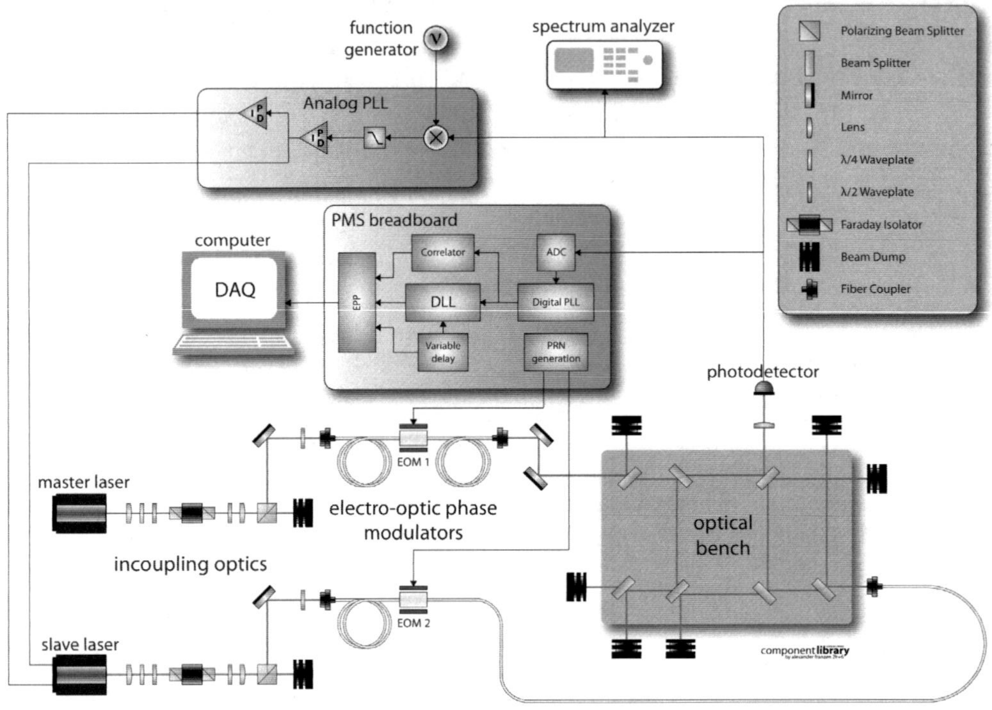


**Figure 1.** General schematic of the experiment setup to test the laser modulation scheme and PMS performance: two phase-locked lasers are modulated using an EOM and injected onto a Mach-Zehnder interferometer. The beatnote is detected and processed in the PMS breadboard in order to obtain phase readout, ranging measurement and data transfer.

## 2. Ranging and data communication

Ranging is based on the correlation properties of the pseudo-random noise (PRN) codes. There are a total of six laser beams exchanged between the LISA satellites, and therefore a PRN code has been designed for each one of them. In the LISA topology, each laser is used simultaneously in different interferometric measurements producing a beatnote modulated with more than one code. The main design driver for the codes is that after interference between any given two laser, a single PRN can be tracked separately from each other and without incurring in significant mutual interference. The set of six PRN sequences was designed by numerical optimisation and with an even length of 1024 chips[1]. The chipping rate is about 1.5 Mbps, limited by the photodectector bandwidth, and in which the data sequences are encoded at 97 Kbps. This produces

[1] The size of the code family is a compromise between efficient use in a digital processor and the correlation properties achieved at the running chipping rate

a periodicity of the code every 200 km over the 5 million kilometer armlength, and therefore, an initial positioning system is required. The deep-space network (DNS), combined with the star tracker onboard each satellite, will provide a positioning uncertainty of about 25 km [4]. After this initial positioning, a more accurate distance determination will be achieved using the DLL architecture proposed. Note that only when the cross-correlation is made by the same PRN code with the same delay, does a peak appear in the correlation (see Figure 2). This way, the correlation peak serves as a timestamp if the start of the PRN is synchronised with the clock of the remote spacecraft.

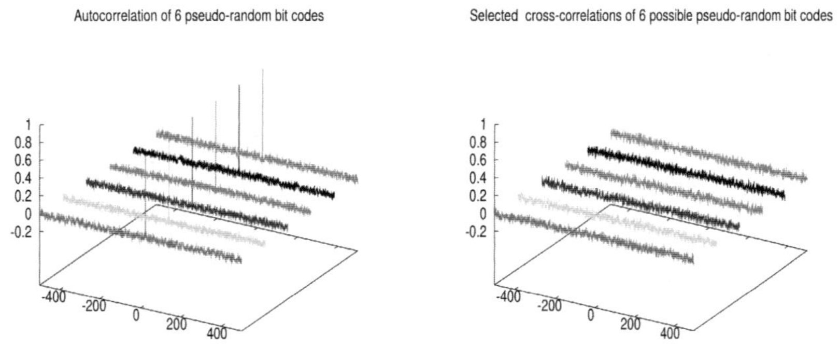


**Figure 2.** Autocorrelation of each pseudocode (left) and cross-correlation between a possible set of two different pseudocodes (right).

The presence of sudden phase changes modulated to the laser beam could compromise the phase stability of the LISA fringe signal. However, recent experiments pointed out that medium phase modulation could be applied without degradation of the phase fidelity of the science signal [5]. Consequently, a low-index phase modulation has been chosen using about 1% of the carrier power for ranging and data communication. Figure 3 shows the spectrum of the beatnote measured on the optical laboratory where an equivalent modulation index of m=0.16 rad was phase modulated using a EOMs.

## 3. Phase measurement system

The main task to be implemented by the LISA phasemeter is the phase readout of the beatnote at a required sensitivity of $2\pi \times 10^{-6}\,\mathrm{rad}/\sqrt{\mathrm{Hz}}$ on the frequency range from 0.1 mHz to 100 mHz. This task corresponds with the core processing of the PMS and is referred as the main science measurement, since the gravitational wave information is contained here. The most suitable architecture to implement this measurement is based on a digital phase-locked loop (DPLL)[6, 7] scheme and the general block diagram is represented in Figure 4: the beatnote is fed into an in-phase/quadrature (I/Q) demodulator to acquire its phase. A control loop locks the phase of a Numerically Controlled Oscillator (NCO) to the incoming beatnote. The phase measurement is formed in a floating-point unit as the sum of a raw phase estimation from the NCO and the arctangent of I and Q component. Once the phase is being processed, the PRN modulated onto the phase of the main carrier can be tracked in back-end processing using the fast residual phase error as input signal to the delay-locked loop (DLL)[8] architecture.

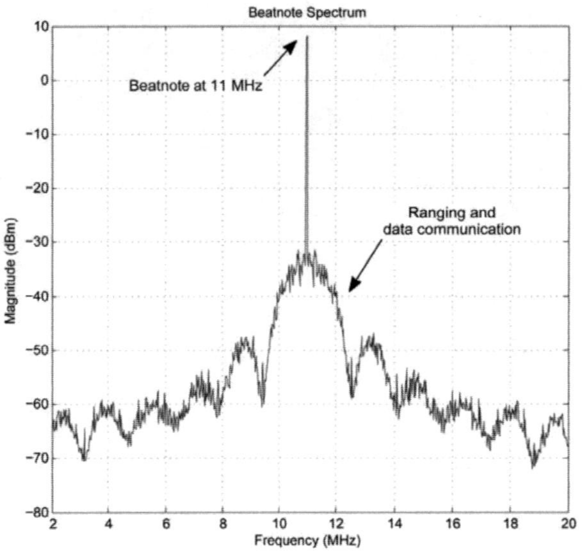


**Figure 3.** Spectrum of the electrical beat signal of two offset-phase locked lasers modulated by two PRN sequence with a modulation index of m=0.16 rad.

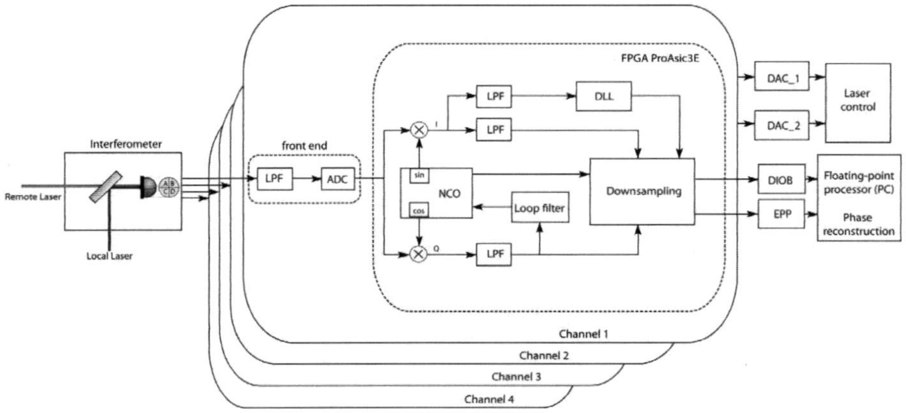


**Figure 4.** General PMS block diagram. FPGA: Field Programmable Gate array, ADC: Analog Digital Converter, DAC: Digital Analog Converter, LPF: Low Pass Filter, NCO: Numerically Controlled Oscillator, DLL: Delay-locked loop, EPP: Enhanced Parallel port, DIOB: Digital Input/Output board.

The DLL correlates the incoming phase signal with three versions of the same reference PRN: a punctual, an early and a late one (see block diagram in Figure 5). The early and late versions of the reference code[2] are delayed in the current implementation by plus and minus half a chip respectively, and the punctual version is not delayed with respect to the transmitted PRN. The punctual correlator is responsible for data recovery and peak detection, whereas the difference between early and late correlators is used as the error signal in a control loop to update the delay of the code generator to the input signal, thus providing tracking between the incoming and the local PRN. The control logic switches between two different modes of working:

[2] "early" (shifted by $+\Delta/2$) and "late" (shifted by $-\Delta/2$) where $\Delta \approx 0.1\,T_c \ldots T_c$

- Acquisition mode: determines delay between the local and incoming PRN sequences at $\mu$s accuracy (one chip length). The local PRN is shifted with a coarse resolution of one code period until a peak of correlation is detected at the correct delay. Lower code period could be used, but it would increase the acquisition time. The acquition time for the current implementation is about 0.67 s.
- Tracking mode: once the acquisition is finished, it determines the timing delay with higher resolution (ns accuracy) and enables data transfer. The estimated delay at measurements rate of 1.5 kHz updates the pseudocode generator to produce the three copies of the local PRN at a resolution of 20 ns.

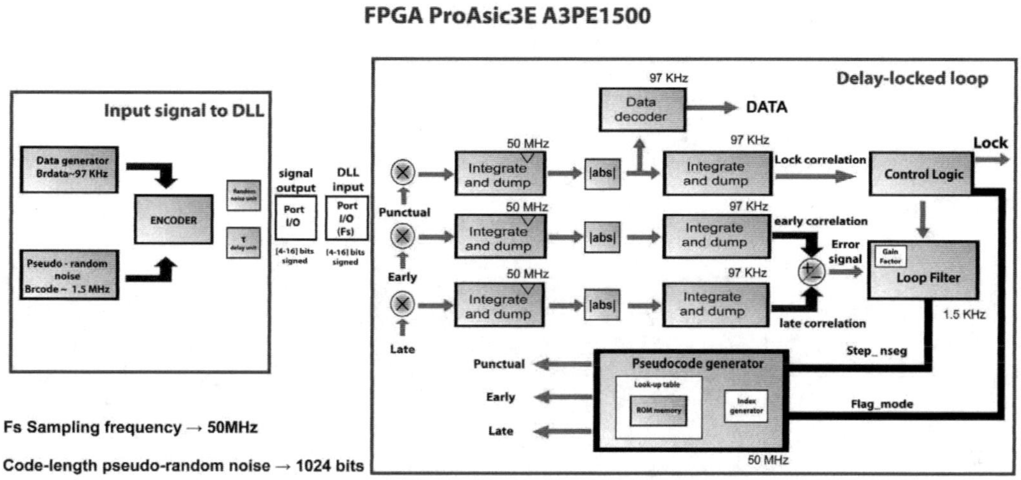


**Figure 5.** General block diagram of the DLL implemented in the same board as the phasemeter. The phase signal coming from the DPLL was internally simulated to provide data encoding, known shot noise and variable delay as input to the DLL.

## 4. Hardware

The hardware implementation of the PMS breadboard (see Picture 6) is based on a space-compatible FPGA processor running at 50 MHz, in which the DPLL and DLL architectures have been programmed and integrated. The breadboard has been designed with four independent A/D channels being able to measure in parallel the phase of each quadrant photodiodes. Additionally, the breadboard implements two independent D/A channels for digital offset phase-locking purposes and two EOM drivers for PRN modulation.

The EOMs and Mach-Zehnder interferometer used in the optical experiment are shown in Figure 7. Two Fibre-coupled EOM with frequencies up to 3 GHz have been tested in our Institute for clock noise sideband modulation[9] and are currently used for low-phase PRN modulation. The Mach-Zehnder interferometer bonded on a Zerodur-baseplate provides a stable optical bench environment with beam extraction at five points and is suitable for future testing of laser ranging with weak-light optical powers.

**Figure 6.** PMS breadboard. Main processor unit: ProAsic3E Actel FPGA with 3 Million system gates. Four input channels (AD9446-100), two output channels (AD9744-210), output interfaces: Digital Input/Output at 2 GB/s and parallel port at 1 MB/s.

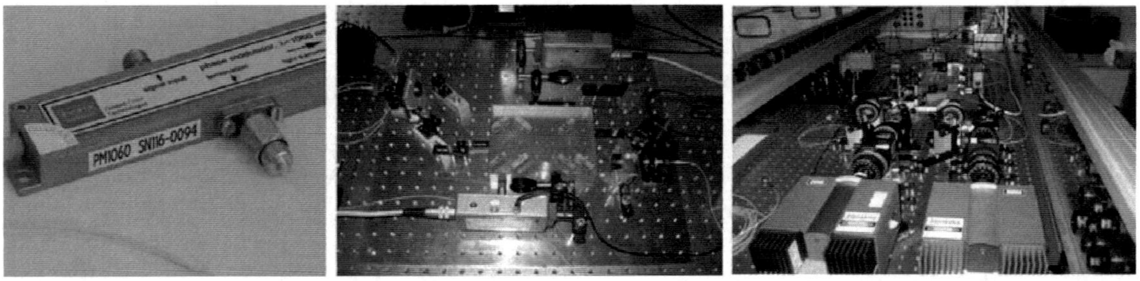

**Figure 7.** Fibre coupled electro-optic modulator (left), Mach-Zehnder interferometer (centre), lasers and optics (right)..

The functionality of our PMS breadboard has been verified and its performance is being tested in the laboratory[7]. The functionality of the implemented DLL architecture proposed in section 3 has been also verified achieving an ideal ranging resolution of 20 ns at a measurement rate of 1.5 kHz[10]. The system is still under active development, however simulated results show an expected distance determination of sub-meter precision with lower measurement rates and in the presence of realistic LISA-like noise sources, i.e., interference with a second PRN, data encoded and an equivalent shot noise of about $60\,\mu\mathrm{rad}/\sqrt{\mathrm{Hz}}$. The effective ranging accuracy can be improved by post-processing, taking into account optical Doppler shifts measurements in the PMS and orbit integration with Kalman filters[3].

## 5. Conclusions

This paper presents the optical experiments and relevant hardware components involved in the design of a delay-locked loop scheme for tracking the absolute armlength in LISA. The custom-designed Phase Measurement System breadboard can fit the architecture proposed. It functionality with optics signal is being confirmed with respect to the previous electrical testing. We thus conclude that the current experimental setup can be expected to be used to verify

[3] This technique is similar to the applied in the Global Positioning System (GPS)

in a more realistic environment how the different modulation schemes affect the phase fidelity of the main science measurement. Besides, it can also be used to verify the phase redout of performance in a single LISA arm.

## 6. References

[1] The LISA Study Team 1998 *Laser Interferometer Space Antenna for the Detection and Observation of Gravitational Waves: Pre-Phase A Report* 2nd edn **MPG233**. Max-Planck-Institute for Quantum Optics.
[2] Massimo T, Shaddock D, Sylvestre J, Armstrong J 2003 Implementation of time-delay interferometry for LISA *Phys. Rev. D***67**.
[3] The LISA Frequency Control Team. *LISA Frequency Control White Paper.* Not published yet. July 2009
[4] Folkner W. M 2009 LISA orbit selection and stability 2001 *Class. Quamtum Grav.* **18** 4053-4057.
[5] Pollack S E, Stebbins R T 2006 A demonstration of LISA laser communication *Class. Quantum Grav.* **23** 4201-4213.
[6] Shaddock D, Ware B, Halverson P, Spero R, Klipstein B 2006 Overview of the LISA Phasemeter *AIP Conf Proc.* **873** 654-660.
[7] Bykov I, Esteban J.J, García Marín A.F, Heinzel G, Danzmann K 2008 LISA phasemeter development: Advanced prototyping J. Phys.: Conf. Series. **154**.
[8] Esteban J.J, Bykov I, García Marín A.F, Heinzel G, Danzmann K 2008 Optical ranging and data transfer development for LISA J. Phys.: Conf. Series. **154**.
[9] Simon B, Troebs M, Sheard B, Heinzel G, Danzmann K 2009 Phase noise contribution of EOMs and HF cables J. Phys.: Conf. Series. **154**.
[10] Esteban J.J, García Marín A.F, Bykov I, Heinzel G, Danzmann K 2009 Free-space laser ranging and data communication *IEEE Conf Proc.* 275-281.

Journal of Physics: Conference Series **228** (2010) 012046 doi:10.1088/1742-6596/228/1/012046

# Test-mass module for DECIGO Pathfinder

**Shuichi Sato[1], Yasuo Torii[2] , Yaka Wakabayashi[3], Yumiko Ejiri[3], Rieko Suzuki[3], Akitoshi Ueda[2], Seiji Kawamura[2], Akito Araya[4], Masaki Ando[5], Yoshiyuki Obuchi[2], Norio Okada[2]**

[1]Hosei University, 3-7-2, kajinocho, Koganei, Tokyo, 184-8584, Japan
[2]National Astronomical Observatory of Japan, 2-21-1 Osawa, Mitaka, Tokyo, 181-8588, Japan
[3]Ochanomizu University, 2-1-1 Ohtsuka, Bunkyo-ku, Tokyo, 112-8610, Japan
[4]ERI, The University of Tokyo, 1-1-1, Yayoi, Bunkyo-ku, Tokyo, 113-0032, Japan
[5]Kyoto University, Kitashirakawa, Sakyo, Kyoto, 606-8502, Japan

E-mail: `sato.shuichi@k.hosei.ac.jp`

**Abstract.**
DECIGO Path nder (DPF) is the rst precursory satellite mission for DECIGO (DECi-hertz Interferometer Gravitational wave Observatory), which is a future space gravitational wave antenna. The key instrument of DPF is a Fabry-Perot laser interferometer with freely floating test-masses, demonstrating precision laser interferometry, stabilized laser system and drag-free control system in orbit. A test-mass module is one of the sub-components of the DPF instrument, realizing non-contact suspension of the test mass in space, which provides capabilities of local sensing/actuation, test-mass lock mechanism and charge management system. In this article, the test-mass module system is reviewed.

## 1. DECIGO and DECIGO Pathfinder (DPF)

DECIGO (DECi-hertz Interferometer Gravitational wave Observatory) is the planned Japanese space gravitational wave antenna mission, which is targeting the observation of gravitational waves from astrophysically and cosmologically signi cant sources mainly between 0.1 Hz and 10 Hz, to open a new window of observation for gravitational wave astronomy, and thus for the universe [1–4]. The scope of DECIGO is to bridge the frequency gap between the LISA band and the band of terrestrial detectors such as Advanced LIGO and LCGT [5–6]. The major advantage of DECIGO specializing in this frequency band is that the expected limit due to confusion noise caused by irresolvable gravitational wave signals from many compact objects, such as white dwarf binaries in our Galaxy, is quite low above 0.1 Hz [7]. Therefore there is a potentially extremely deep window in this band. Full success of DECIGO is expected to yield fruitful sciences: an observation of stochastic GWs from inflation of the universe, a characterization of dark energy with precise observation of distant neutron star merger, a dark-matter identi cation by observing cosmological black holes, and an understanding of the formation mechanism of supermassive black holes.

The pre-conceptual design of DECIGO consists of three drag-free spacecraft which keep their triangular con guration using a formation flying technique. The separation between the spacecraft (proof masses) is designed to be 1,000 km, and their relative displacements are measured by a di erential Fabry-Perot (FP) interferometer. DECIGO is expected to be launched in the middle of the 2020s. Before that, we plan to launch two precursor satellites:

© 2010 IOP Publishing Ltd

Journal of Physics: Conference Series **228** (2010) 012046 doi:10.1088/1742-6596/228/1/012046

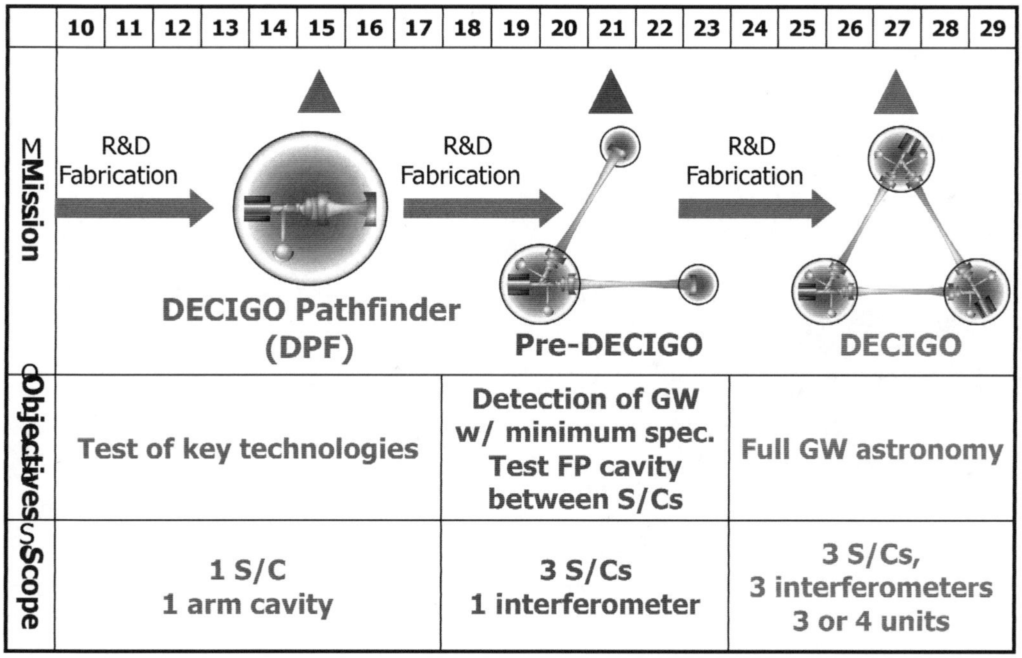


**Figure 1.** Roadmap of DECIGO, together with precursor satellites DECIGO Path nder and Pre-DECIGO.

DECIGO Path nder (DPF) [8–9] and Pre-DECIGO as shown in Fig.1. Because DECIGO will be an extremely large mission, it is important to increase the technical readiness level and demonstrate feasibility before its launch. The major objective of these milestone missions is a demonstration of key technologies needed for DECIGO just as the LISA Path nder [10] does for LISA. In addition, we also hope to extract some scienti c achievements with the limited equipment allowed for these satellites of the precursor phases. DPF tests the key technologies for DECIGO such as drag-free control of the spacecraft, stabilized laser system in space and precision laser metrology in space. At the same time, as DPF will have gentle sensitivity to gravitational waves, it is expected that DPF will put some upper limit to the gravitational waves from sources around the center of our galaxy.

DPF (see Fig.2) will employ a small-sized drag-free spacecraft that contains two freely falling test masses whose relative displacement is measured with a Fabry-Perot interferometer. A short Fabry-Perot cavity of 30 cm with nesse of 100 is illuminated by the frequency-stabilized Yb:YAG laser light yielding an output power of 100 mW. The test masses are clamped tightly during launch and released gently in orbit. DPF is supposed to be delivered into a geocentric sun-synchronous orbit with an altitude of 500 km. DPF will have a strain sensitivity of $2\ \ 10^{\ 15}\,\mathrm{m}/\sqrt{\mathrm{Hz}}$ around the frequency band of 0.1–1 Hz. The primary objective of DPF is to demonstrate key technologies for DECIGO such as the drag-free control system, the FP cavity precision metrology system in orbit, a frequency-stabilized laser in orbit, and the clamp release mechanism. In addition, the scienti c objective of DPF is to detect rather unlikely inspiral events of intermediate-mass black holes ($10^3\ \ 10^4$ M ) in our galaxy; it is possible to detect such events with the targeted sensitivity of DPF. Another prospective achievement is earth gravity monitoring. The precision sensing of the test mass using a laser sensor, and feeding all the error signals back to the test mass, works as a sensitive accelerometer in space.

From the combination of this accelerometer and GPS information, one can derive the earth's

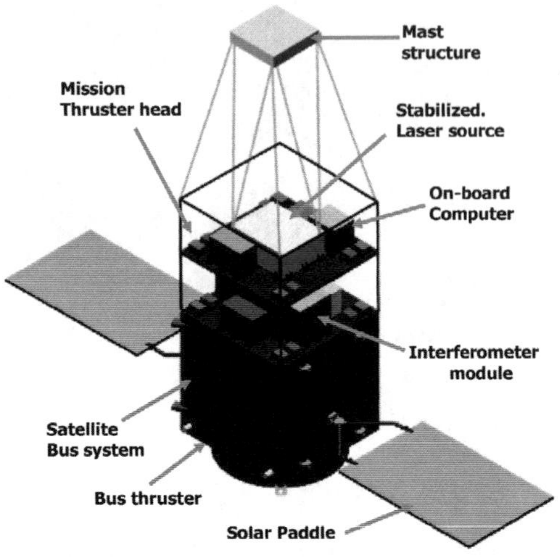


**Figure 2.** Schematic of DPF satellite.

## 2. Test-mass module (TMM)

The spacecraft of DECIGO Path nder will be a technical demonstration of Fabry-Perot laser interferometry. It is a shrunk version of DECIGO instruments, and also a laser interferometer accelerometer in orbit. Laser module, interferometer module and thruster actuation system serve as key components of DPF to demonstrate a stabilized laser system, precision laser interferometry and drag-free control in space. Thus, DPF will have two freely floating test masses in orbit, which are references in space-time for GWs and also references for earth's gravity. A caging system, called test-mass module (TMM) for DPF, will house the test mass, serving as contact-free suspension including sensing/actuation, test-mass locking/positioning capabilities and charge management system (see Fig.3). The test-mass modules will house one test mass each, so DFP will have two TMMs to compose the FP interferometer inside the interferometer module. TMM is, thus, a complex of functionalities composed of several sub-components. The baseline of this caging system is similar to that of LISA Path nder, but the design of the bread-board model for DPF is a simpli ed version to check the functionalities of each sub-component.

- Test masses (TMs): 70 mm cubes made of aluminum for the bread-board model (BBM). A dielectric multi-layer coated mirror is attached on one side to work as part of the FP interferometer mirror, together with three pairs of corner cube reflectors for the laser sensor (LS).
- Electrostatic Sensor and Actuator (ES/EA): works as local sensor for relative displacements between test mass and spacecraft, which is as sensitive as $10\ ^{10}\,\mathrm{m}/\sqrt{\mathrm{Hz}}$, and as actuator to apply a force to the test mass. In order to sense/actuate six degrees of freedom of the test mass, a pair of electrodes are facing the test mass on all surfaces, together with additional four injection electrodes. The potential of the test mass is modulated with an injection signal, then the displacement information of the test mass appears as modulated signal after a di erential transformer. Therefore, the demodulated signal gives a nearly linear error signal around the operating point. As for actuation, appropriate combinations of modulated feedback signals are applied to four electrodes, facing +x and x for example, to apply forces to the test mass, for the translation and for the rotational degrees of freedom.

Journal of Physics: Conference Series **228** (2010) 012046 doi:10.1088/1742-6596/228/1/012046

**Figure 3.** Assembly diagram of the test-mass module for the bread-board model of DPF. The test mass is housed inside a caging frame, then each surface is enclosed by electrode plates and laser sensor modules.

- Laser sensor (LS): Michelson interferometer displacement sensor. This is a redundant local sensor for test-mass motion, but more sensitive than the electrostatic sensor by a factor of 1000, corresponding $10\ ^{13}\,\mathrm{m}/\sqrt{\mathrm{Hz}}$ displacement sensitivity. Frequency stable 1.5  m light from a distributed feedback (DFB) laser illuminates the Michelson interferometer composed of a  xed reference corner cube and one on the test mass. In order to sense all degrees of freedom of the test mass, including translation and rotation, three pairs of laser sensor units are facing three orthogonal faces. By feeding back all the error signals to EA, the the control system works as a high-sensitivity, all-degree accelerometer, which is supposed to be used mainly for earth gravity monitoring.
- Test-mass lock mechanism: Motorized equipment to hold the test mass during launch and also in orbit. Some kinds of piezo motors are now under test for the clamp-release mechanism, which provides capabilities of precision positioning and soft release of the test mass in orbit. Another device is foreseen for the launch-lock mechanism, which is to lock the test mass for protection during launch, however the design is still in review.
- Charge-control system: UV-LED based charge management system. The photo-electric electrons produced by UV light on metal surfaces can be transferred in both direction, from test mass to the electrodes and vice versa, by applying appropriate potentials to the electrodes. The basic system had been developed at the Stanford SSGD group and redesigned to  t to the TMM of DPF.

## 3. Current status

DPF aims to be one of the "small satellite mission series" that had been initiated by the Japanese space agency, ISAS/JAXA. This program is to launch at least 3 small satellites in the upcoming 5 years using standard bus systems, whose scope is to reduce the cost of missions signi cantly compared with the conventional missions, and thus to increase the chance to go to space for a variety of  elds. The  rst mission of this series will be a planet observing satellite (Sprint-A EXCEED), and the second will be a plasma observing mission (Sprint-B ERG), then DPF aspires for Sprint-C, which is expected to be launched around 2015. All components of DPF, including TMM, are under development to establish their technical feasibility and to advance their technical readiness to such a level as to be recognized as a phase-A mission.

## References

[1] Seto N, Kawamura S and Nakamura T 2001 Possibility of direct measurement of the acceleration of the universe using 0.1 Hz band laser interferometer gravitational wave antenna in space *Phys. Rev. Lett.* **87** 221103

[2] Kawamura S et al 2006 The Japanese Space Gravitational Wave Antenna - DECIGO *Class. Quantum Grav.* **23** S125

[3] Kawamura S et al 2008 The Japanese Space Gravitational Wave Antenna - DECIGO *Journ. of Phys.: Conf. Ser.* **120** 032004

[4] Kawamura S et al 2008 The Japanese Space Gravitational Wave Antenna; DECIGO *Journ. of Phys.: Conf. Ser.* **122** 012006

[5] LISA: System and Technology Study Report, ESA document ESA-SCI (2000)

[6] Kuroda K et al 2002 Japanese large-scale interferometers *Class. Quantum Grav.* **19** 1237

[7] Farmer A J and Phinney E S 2003 The gravitational wave background from cosmological compact binaries *Mon. Not. R. Astron. Soc.* **346** 1197

[8] Ando M et al 2008 DECIGO Path nder *Journ. of Phys.: Conf. Ser.* **120** 032005

[9] Ando M et al 2008 DECIGO Path nder *in this volume*

[10] Anza S et al 2005 The LTP experiment on the LISA Path nder mission *Class. Quantum Grav.* **22** S125-S138

# Structural design and analysis of test mass module for DECIGO Pathfinder

**Y Wakabayashi[1], Y Obuchi[2], N Okada[2] , Y Torii[2], Y Ejiri[1], R Suzuki[1], A Ueda[2], S Kawamura[2], A Araya[3], M Ando[4], S Sato[5] and A Sugamoto[1]**

[1] Department of Physics, Ochanomizu University, 2-1-1 Ohtsuka, Bunkyo-ku, Tokyo, 112-8610, Japan
[2] National Astronomical Observatory of Japan, 2-21-1 Osawa, Mitaka, Tokyo, 181-8588, Japan
[3] Earthquake Research Institute, University of Tokyo, 1-1-1, Yayoi, Bunkyo-ku, Tokyo, 113-0032, Japan
[4] Division of Physics and Astronomy, Graduate School of Science, Kyoto University, Kitashirakawa, Sakyo, Kyoto, 606-8502, Japan
[5] Faculty of Engineering, Hosei University, 3-7-2, kajinocho, Koganeishi, Tokyo, 184-8584, Japan

E-mail: `yaka.wakabayashi@nao.ac.jp`

**Abstract.** Deci-hertz Interferometer Gravitational-Wave Observatory: DECIGO is a project aimed at future detection of deci-hertz gravitational waves in space. DECIGO Pathfinder: DPF is a precursor mission to test the key technologies with one spacecraft. Our work in this article was to examine the strength of the DPF test mass module to ensure that it is sufficiently robust for launch with a launch vehicle. We designed the test mass module, and examined the structural strength of this model by structural analysis, Quasi-static acceleration analysis and Modal analysis using FEA (Finite Element Analysis). We found that the results of each analysis fulfilled all requirements. We are confident that the DPF test mass module will withstand Quasi-static acceleration or coupling with vibration of launch vehicle during launch, if the design matches the current design. For more detail, further analysis including Response analysis and Thermal analysis are recommended. In addition, it will be necessary to lighten the model in the next step.

## 1. Introduction

DECIGO (Deci-hertz Interferometer Gravitational-Wave Observatory)[1][2] is the future Japanese space gravitational wave antenna. The purpose of DECIGO is to observe gravitational waves from astrophysically and cosmologically significant sources at the frequency band mainly between 0.1-10Hz, thus, to open a new window of observation for gravitational wave astronomy. DECIGO will be formed by three drag-free spacecraft which are separated from one another by 1000km. We will detect gravitational waves by measuring relative displacements between each spacecraft as a Fabry-Perot Michelson interferometer. For this magnificent mission to succeed, we will need to make use of some advanced techniques, for example drag-free control, precise measurement, etc. At present, we have a roadmap to launch two milestone missions before DECIGO, DECIGO Pathfinder (DPF) and Pre-DECIGO. DPF[3] is the first milestone satellite currently being developed. The purposes of DPF are to verify each technique that will be used

© 2010 IOP Publishing Ltd

in DECIGO, to detect gravitational waves in our galaxy, and to monitor the Earth's gravity. We will also use a Fabry-Perot Interferometer as a sensor for gravitation waves. The differences between DECIGO and DPF are that the arm length will be 30cm and we use only one Fabry-Perot cavity within one spacecraft. The Fabry-Perot Interferometer is formed by two mirrors adhering to the test masses, which act as free proof masses. Each test mass is held by the test mass module, whose functions are to hold the test mass inside floating on the geodesics in orbit, and to guard the test mass from the effects of launch.

This article describes the structural design and analysis of the test mass module. We designed the test mass module to satisfy certain requirements. As a first step in the design, we worked on the prototype (Bread-Board Model). In addition, we checked the rigidity of the model using FEA methods (Finite Element Analysis). Our work is very important in examining how the module functions without breaking down during launch and in orbit.

## 2. DPF / Test mass module

### *2.1. DPF*

DPF will be the first precursory satellite for DECIGO. One of the purposes of DPF is the demonstration of each technique that will be used in DECIGO (Drag-free control, Precise measurement, Stabilized Laser, etc.). The other purposes are detection of gravitational waves in our galaxy at 0.1-1Hz frequency band and Earth gravity monitoring. DPF will be a small satellite weighing about 350kg. We assume that DPF will be launched by a next-generation solid propellant rocket from JAXA (Japan Aerospace Exploration Agency) as one of the "small scientific satellite series." The DPF orbit will be sun-synchronous (Altitude: 500km). The mission payload part of DPF (Figure 1) will be the size of $950 \times 950 \times 1100$ mm. Included in the DPF are main units, Interferometer Module, Stabilized Laser, Mission Thrusters for drag-free control, and some Control Units. The Interferometer Module contains a Fabry-Perot cavity for detection of gravitational waves (Figure 2). The arm length will be 30cm. We can detect the gravitational wave signals by measuring the relative displacement of two mirrors in the Fabry-Perot Interferometer. The interferometer is formed by two mirrors bonded on the test mass which act as free proof mass. To act as free proof mass, each test mass will be drag-free controlled by S/C. In the next section, we will provide further details of the test mass module.

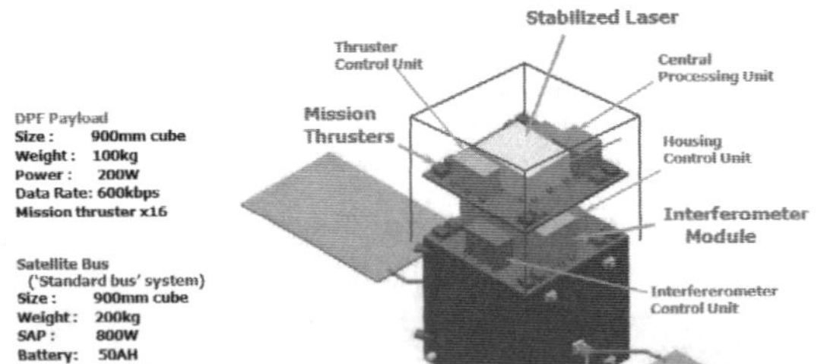


**Figure 1.** DPF mission payload

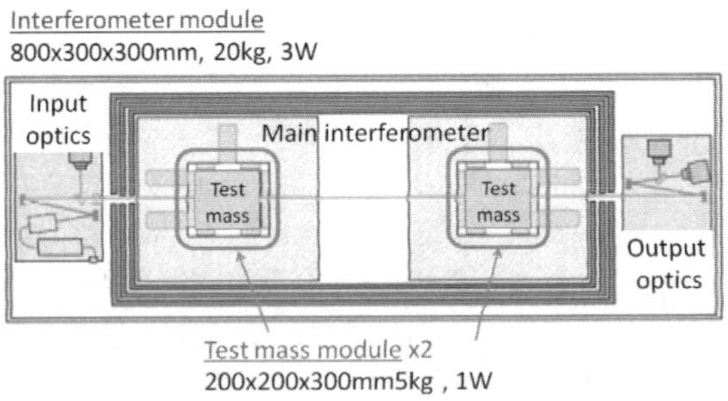


**Figure 2.** Interferometer module

### *2.2. Test mass module*

We will use a Fabry-Perot Interferometer as a sensor for gravitational wave (Figure 2). We will measure the relative displacement between two proof masses of Fabry-Perot Interferometer. The test masses will act as free proof masses. The proof mass has to freely float in space, so we will use drag-free control techniques to achieve free mass. This means that the test masses will be supported without contact in orbit. Therefore, the test masses must be stabilized with the satellite during launch. The test mass module controls the test mass without contact in orbit, and holds the test mass tight during launch. To achieve this, we will use certain mechanisms.

**(1)** Test mass (acts as proof mass including mirror for interferometer, metallic cube)
**(2)** Electrostatic sensor and actuator (to sense and actuate the position of test mass, some electrodes as capacitor with metallic test mass)
**(3)** Laser sensor unit (for Earth gravity monitor, sense the position of test mass with Michelson Interferometer, with arm length of about 23mm)
**(4)** Launch lock mechanism (to hold test mass during launch using motors)
**(5)** Clamp and release mechanism (for precise positioning and clamp-release of test mass in orbit using motors)
**(6)** Caging frame (to hold other components, metallic frame)
**(7)** UV-LED (discharge of test mass)

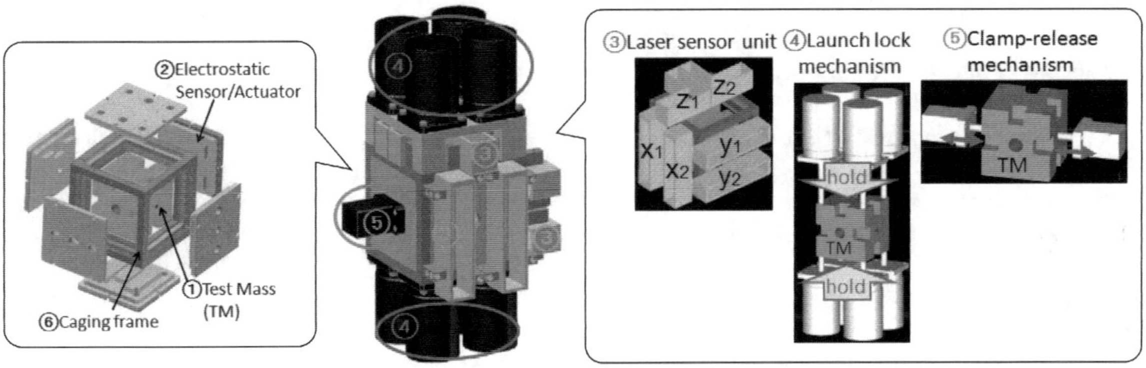


**Figure 3.** Design of the test mass module (BBM)

## 3. Structural design of DPF test mass module

We will have three steps of development for test mass module, Bread-Board Model (BBM), Engineering Model (EM) and Flight Model (FM). As a first step, we designed test mass module as BBM.

The module was designed to contain the components in the above section except UV-LED as seen in Figure 3. The required size is 200 × 200 × 300mm and weight is 5kg. The size of the test mass is 70mm cube (blue cube in Figure 3). The test mass has a hole at 15mm in diameter, which is the path for the laser of Fabry-Perot Interferometer, and there is a mirror for the Interferometer on one side of the hole. The test mass also has three pairs of corner cube mirrors for Laser Sensor Units. The test mass is inside the caging frame, which acts as a holder for the test mass and base of other components. The caging frame is a 110mm cube ((6) in Figure 3). On each surface, there are three electrodes ((2) in Figure 3), face to face with each surface of the test mass. The center electrode is an injection electrode for sensor (area: 18×40mm), and there are two electrodes for the sensor and actuator (area: 18×62mm) on the right and left sides of the injection electrode. The gap between the test mass and the electrode is 1mm. Three pairs of the laser sensor units are on the caging frame. The units surround three surfaces of the test mass ((3) in Figure 3). The units for the X and Y surfaces are linear (size: $26 \times 140 \times 32$mm), and those for the Z surface are L-shaped (size: $26 \times 85 \times 32 + 24 \times 62 \times 32$mm). Each unit has optics for the Michelson Interferometer, which measure the displacement of the test mass with the corner cube mirrors on the test mass. We will use some kind of motor as a launch lock mechanism. Eight motors will push top and bottom of the test mass during launch ((4) in Figure 3). We will also employ motors for the clamp and release mechanism. Motors for clamp and release are on the right and left sides ((5) in Figure 3). They can move the test mass to the center of the caging frame and release it at very slow velocity. We have not yet installed a UV-LED unit in this model but we will include it in the next step.

## 4. Structural analysis of DPF test mass module

Every subsystem including the test mass module will experience extraordinary disturbance, mainly during launch, so we have to test the strength of each component to prepare for the disturbance. We have to perform some structural analyses that correspond to mechanical environmental tests and thermal vacuum tests. Mechanical environmental tests are run on quasi-static acceleration (Axis direction and Cross axle direction), modal analysis, random vibration, vibroacoustic, shock. The thermal vacuum test is on radiant heat. We performed the quasi-static acceleration analysis and modal analysis in this paper.

## 5. Quasi-static acceleration analysis

### *5.1. Method of analysis*

The first procedure is modeling the test mass module using 3D CAD software (Autodesk Inventor Professional 2009) for optimized quasi-static acceleration analysis. All main components, except the optics for the laser sensor, are contained and each part is attached with a screw. The key points for analysis are that the mass, shape and position of each part match those of the design, and the physical properties of each part are the same as the FM (physical properties: test mass; Au-Pt, casing frame; aluminum, electrostatic sensor/actuator; sapphire, motor rod; steel). The next procedure is to build a Finite Elements Model (FE Model) and to apply certain conditions to the FE Model using CAE software (Autodesk Inventor Professional 2009 Advanced Simulation Technology Preview). The number of elements is about 50,000. We clamped the bottom. We applied quasi-static acceleration (Axis direction) 24G (235.4m/s$^2$) as Figure 4. As a final procedure, we used software to solve the model as linear statics analysis under the above conditions.

### *5.2. Results of analysis*

Figure 5 shows the stress of the test mass module and electrode plate as an example of tested parts. Left of Figure 5 is the stress of the test mass module. The color bar shows stress value (red: high stress, blue: low stress). Because we used different materials for each part, the physical properties of each differ; so we verified each one. Right of Figure 5 is the stress of one part: electrostatic sensor and actuator. The maximum and minimum value of the maximum principal stress is +4.2MPa (tensile) and -0.2MPa (compressive). The minimum of Safety Factor of the part is 535.7 (tensile), because of each strength of sapphire is 2,250MPa (tensile) and 2,950MPa (compressive). Therefore, this part will not break. Results concerning the other parts also demonstrated them to be robust and unbreakable.

## 6. Modal analysis

Modal analysis yields resonant frequencies and mode shapes at the frequency. The requirement of the frequency of fundamental mode for axis direction is above 50Hz, and that for cross axle direction is above 30Hz. We used two methods of modal analysis for validation; (1)Spring - Mass Model, (2)Finite Element Analysis (FEA).

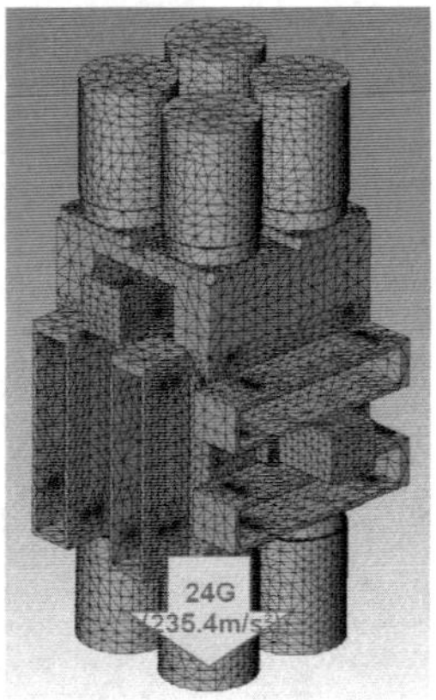


**Figure 4.** FE Model: Quasi-static acceleration (Axis direction) 24G (235.4m/s$^2$)

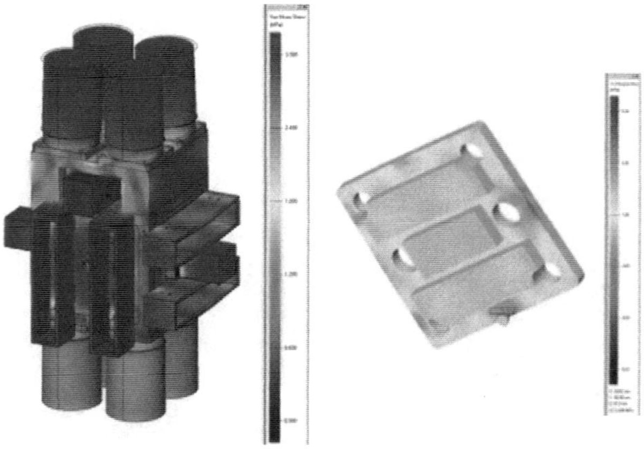

**Figure 5.** Results of analysis: Stress (left: module, right: Electrode plate for sensor and actuator )

### *6.1. Spring-mass model*

The spring-mass model was used to obtain a rough estimate of resonant frequency. We regard the test mass module as two mass systems for the spring-mass model. The first system is caging frame, electrode plate and launch lock motors. The total mass of the system is $m_1$, and the spring constant is $k_1$ (the upper right of Figure 6), which is given by the estimation by considering the first system to be a beam clamped on one side. The second system is the test mass and the rods of launch lock motors. The total mass of the system is $m_2$, and the spring constant is $k_2$ (the lower right of Figure 6), which is given by the estimation by considering the second system to also be a beam clamped on one side. We calculated each spring constant by considering each system to be a leaf spring. The equations of motion of this spring-mass system is given as follows.

$$m_1\frac{d^2x_1}{dt^2} = -k_1x_1 - (x_1 - x_2)k_2 \tag{1}$$

$$m_2\frac{d^2x_2}{dt^2} = -(x_2 - x_1)k_2 \tag{2}$$

$x_1$ and $x_2$ are the displacements of the mass of each system. The analytic solution of these equations give the resonant frequency (angular frequency) of the system as follows.

$$\omega_{\mp}^2 = \frac{1}{2}\{\omega_{11}^2 + (1+\mu)\omega_{22}^2 \mp \sqrt{[\omega_{11}^2 + (1+\mu)\omega_{22}^2]^2 - 4\omega_{11}^2\omega_{22}^2}\} \tag{3}$$

$$\omega_{11}^2 = \frac{k_1}{m_1}; \omega_{22}^2 = \frac{k_2}{m_2}; \mu = \frac{m_2}{m_1}$$

$$- : first\ mode, \quad + : second\ mode$$

$\omega$ is the angular frequency of the system.

According to the expression, the first mode resonant frequency of the system is 434Hz, which is high enough to fulfill requirements.

### *6.2. Finite element analysis*

We used CAE software (NX I-DEAS 5) for FEA.

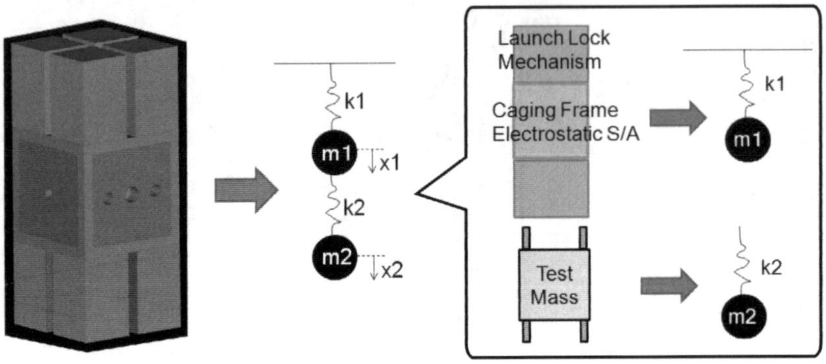


**Figure 6.** spring-mass system

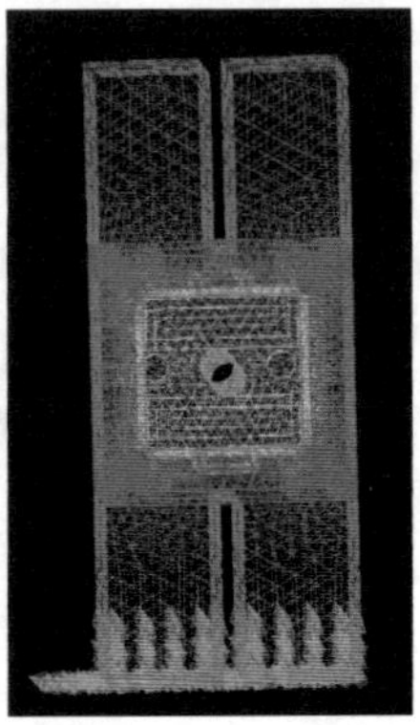

**Figure 7.** FE Model for model analysis: clamp the bottom

The model was optimized for modal analysis. We want to obtain the resonant frequency of the main structure, so there are no laser sensors or clamp and release mechanisms, which are outside of the caging frame and very light compared with the test mass and launch lock mechanism. Each part was bonded with the others, and the physical properties of each part match those of the flight model. The key feature is that the mass and stiffness of each part are correct. We based the FE Model on the model. The number of elements is about 20,000, which is optimum for the model. We clamped the bottom of the model to estimate most low frequency (Figure 7). We used software to solve the model using modal analysis under the above conditions. As a result, the resonant frequency of first mode is 764Hz (cross axle direction), and the lowest resonant frequency of the axis direction is 2.56kHz, which are high enough to satisfy requirements.

### *6.3. Conclusion of modal analysis*

The resonant frequency of the test mass module of DPF is high enough according to the results of analysis; spring-mass model and FEA. Therefore, the module will not resonate with the launch vehicle.

## 7. Conclusions

The DPF test mass module was designed as BBM including most necessary components. According to the results of the structural analysis, the module will be strong enough for launch with the launch vehicle. It can be concluded that the structural design is so far successful. Further analysis, including Response analysis and Thermal analysis is recommended for more detail. We will lighten the model in the next step, EM, because the current design is still too heavy to satisfy requirements. In addition, we are now making a BBM test mass module, and we will demonstrate that the system and each component work as required.

## References

[1] Kawamura S, *et al* 2008 *J. phys.: Conf. Series* **120** 032004
[2] Sato S, *et al* 2009 *J. phys.: Conf. Series***154** 012040
[3] Ando M *et al* 2008 *J. phys.: Conf. Series***120** 032005

# Geophysical constraint on a relic background of the dilatons

**Sachie Shiomi**[1]

Space Geodesy Laboratory, Department of Civil Engineering, National Chiao Tung University, Hsinchu, Taiwan 300, Republic of China

**Abstract.** According to a scenario in string cosmology, a relic background of light dilatons can be a significant component of the dark matter in the Universe. A new approach of searching for such a dilatonic background by observing Earth's surface gravity was proposed in my previous work. In this paper, the concept of the geophysical search is briefly reviewed, and the geophysical constraint on the dilaton background is presented as a function of the strength of the dilaton coupling, $q_b^2$. For simplicity, I focus on massless dilatons and assume a simple Earth model. With the current upper limit on $q_b^2$, we obtain the upper limit on the dimensionless energy density of the massless background, $\Omega_{DW} h_{100}^2 \leq 6 \times 10^{-7}$, which is about one-order of magnitude more stringent than the one from astrophysical observations, at the frequency of $\sim 7 \times 10^{-5}$ Hz. If the magnitude of $q_b^2$ is experimentally found to be smaller than the current upper limit by one order of magnitude, the geophysical upper limit on $\Omega_{DW} h_{100}^2$ becomes less stringent and comparable to the one obtained from the astrophysical observations.

## 1. Introduction

The scalar gravitational fields (or the dilatons) have appeared in scalar-tensor theories of gravitation and also in theories towards the unification of the fundamental forces in nature. These theories predict the existence of scalar waves. Their sources fall into two categories. One is cosmological and the other is astronomical. The former is predicted as a relic background of the dilatons in string cosmology [1]. The latter is as emission from a spherical symmetric gravitational collapse of a star [2, 3]. The detection or experimental constraints on the scalar waves would play an important role for the development of the unified theories and the verification of gravitational theories. In this paper, I focus on the search for scalar waves from the cosmological origin.

Scalar waves can interact with detectors in two ways [4]: (1) indirectly, through the geodesic coupling of the detectors to the scalar component of the metric fluctuations (e.g. [5]) and (2) directly, through the effective dilatonic charges of the detectors, which depend on the internal chemical composition of the detectors. In the direct coupling, the response of the detectors is nongeodesic in general [4]. A new approach of searching for the direct-coupling scalar waves using the geophysical tool of superconducting gravimeters has been proposed [6]. Here, I briefly review the concept of the geophysical search (Section 2). Also, I give a further discussion on the preliminary upper limits on the dilaton background, obtained in my previous work [6], and present a geophysical constraint on a relic background of massless dilatons (Section 3).

[1] Present address: Institute for Cosmic Ray Research, The University of Tokyo, Kashiwa, Chiba 277-8582, Japan.

© 2010 IOP Publishing Ltd

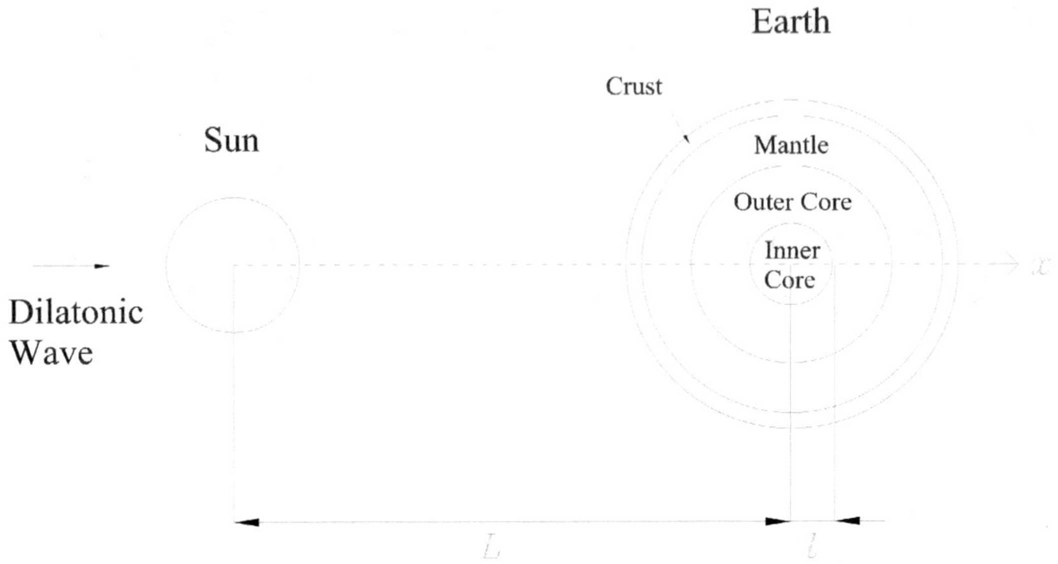


**Figure 1.** The concept of the geophysical search for dilatonic waves (not drawn to scale), quoted from [6]. As dilatonic waves propagate along the Sun-Earth line, the Earth's inner core oscillates with amplitude $l$. Such oscillations would cause changes in surface gravity, which can be searched for with superconducting gravimeters installed at surface of the Earth.

## 2. Geophysical search for composition-dependent dilatonic waves

### *2.1. The concept*

The Earth's interior can mainly be classified into four layers: the crust, the mantle, the liquid outer core, and the solid inner core (see Figure 1). The solid inner core is held at the center of the Earth mainly by the gravitational pull from the rest part of the Earth.

Unified theories predict the existence of dilatonic charges that depend on the internal chemical structure of matter. It should be noted that the dilatonic charge of the inner core is different from the one of the rest part of the Earth, because of the difference in their chemical compositions (see Equation (4) below for the definition of dilatonic charges). The inner core is made from dense elements, such as Iron and Nickel and its average density is 13,000 kg $m^{-3}$, while the rest part of the Earth is mainly made from less dense materials, such as silicon oxides, and its average density is about 5,400 kg $m^{-3}$.

Because of the difference in the dilatonic charges, impinging dilatonic waves on the Earth would couple differently to the inner core and the rest part of the Earth. As a result, there would be relative translational motions between the inner core and the rest part of the Earth, or oscillations of the inner core relative to the rest part of the Earth. Such relative motions cause surface gravity changes that can be searched for with superconducting gravimeters, the most sensitive instruments to measure surface gravity at the frequencies of our interest.

Unlike the tensor gravitational waves predicted by general relativity, which are transverse, dilatonic waves are longitudinal waves. When dilatonic waves propagate along the Sun-Earth line, we would observe strains due to the dilatonic waves along the line (Figure 1). The maximum amplitude of the translational motion would be observed when the dilatonic waves have the same frequencies as the natural oscillation frequencies of the inner core.

### *2.2. Equation of motion*

The equation of motion of the inner core can be given by (Equation (1) of [6]):

$$\ddot{l}^i \approx -\gamma \dot{l}^i - \omega_0^2 l^i - \Delta q (L^k \partial_k + 1)\partial^i \phi, \tag{1}$$

where $i$ and $k$ indicate the space-time components of the parameters, $\Delta q$ is the difference in dilatonic charge between the inner core and the rest part of the Earth (for the definition, see Equation (6) below.), $L$ is the rest separation between the Sun and the Earth (see Figure 1), and $\phi$ is the dilaton field. The gravitational stiffness and the friction coefficient are respectively given by

$$\omega_0^2 \approx \frac{4}{3}\pi G \frac{\rho_{ic} - \rho_{oc}}{\rho_{ic}} \rho_{oc} \approx 1.8 \times 10^{-7} \mathrm{s}^{-2} \approx \{2\pi(4.1\ \mathrm{h})^{-1}\}^2, \tag{2}$$

$$\gamma \equiv \frac{6\pi\eta r_{ic}}{m_{ic}} \approx 2.3 \times 10^{-16} \eta\ \mathrm{s}^{-1}, \tag{3}$$

where $G = 6.67 \times 10^{-11}$ N m$^2$ kg$^{-2}$ is the gravitational constant, $m_{ic} \approx 9.8 \times 10^{22}$ kg is the mass of the inner core, and $\eta$ is the effective viscosity of the outer core.

Dilatonic charges can be defined as a relative strength of scalar to gravitational forces, by summing over all the components $n$ of the test body (Equation (22) of [4]):

$$q = \frac{\sum_n m_n q_n}{\sum_n m_n}. \tag{4}$$

For a test body $A$ made from an ordinary matter with mass $M$, the dilatonic charge is given by

$$q_A \simeq \frac{B m_b q_b}{M} \equiv \frac{B}{\mu} q_b, \tag{5}$$

where $B$, $m_b$, and $q_b$ are the baryon number, baryon mass, and dimensionless-fundamental barionic charge, respectively. The difference in dilatonic charge between the inner core ($ic$) and the rest part of the Earth ($rpe$) can be given by

$$\Delta q \simeq \left\{ \left(\frac{B}{\mu}\right)_{ic} - \left(\frac{B}{\mu}\right)_{rpe} \right\} q_b \equiv \Delta \left(\frac{B}{\mu}\right) q_b. \tag{6}$$

The magnitude of $q_b$ is constrained by tests of the equivalence principle [7].

### *2.3. Estimation of the upper limits*

The spectrum of the displacement of the inner core ($l$) at resonance can be given by (Equation (6) of [6]):

$$S_l(f_0) = \frac{(\Delta q)^2 (L^2\kappa^2 + 1)\kappa^2 c^4}{4\omega_0^2 \gamma^2} S_h(f_0) \tag{7}$$

where $c$ is the speed of light and $\kappa = \omega_0/c$ is the wave length. The resonant frequency is $f_0 = \omega_0/2\pi \simeq 6.8 \times 10^{-5}$ Hz. $S_h(f)$ is the power spectrum of the strain due to massless dilatonic waves: $h \equiv 2\phi/c^2$.

From residual gravity data of superconducting gravimeters, an upper limit on the spectrum of the displacement is given to be $S_l(f_0) < 1.1 \times 10^{-4}$ mHz$^{-1/2}$. The dimensionless energy density for a stochastic background of massless dilaton is related to the strain spectrum by

$$\Omega_{DW}(f) = \frac{\pi^2 f^3}{3H_0^2} S_h(f) \tag{8}$$

With the equation (7) and the upper bound on the displacement from gravity data, we obtain the upper limit on the dimensionless energy density [6]:

$$\Omega_{DW}(f_0)h_{100}^2 < 2.2 \times 10^{-17}\eta^2. \tag{9}$$

This upper limit is proportional to $\eta^2$. However, the magnitude of $\eta$ is not well determined. A handbook of physical constants [8] shows different values of $\eta$, ranging from $10^{-3}$ Pa s to $10^{12}$ Pa s, depending on different methods to estimate the magnitude. A more resent estimate of an upper bound on $\eta$ given by nutation data is $\eta < 1.7 \times 10^5$ Pa s [9].

Using this resent upper bound from nutation data, we obtain $\Omega_{DW}h_{100}^2 \leq 6 \times 10^{-7}$ at the resonant frequency. This is below the astrophysical constraints on a massless background obtained from Nucleosynthesis and measurements of the cosmic microwave background [10]: $\Omega h_{100}^2 \leq 10^{-5}$. Expected upper limits on the massless dilaton background for different values of the effective viscosity is given in Figure 2 of [6].

## 3. Geophysical constraint on the massless dilaton background

From the equation (7), one can see that the upper limit on the dimensionless energy density depends on the magnitude of the strength of the dilaton coupling $q_b^2$. In the above estimation of the upper limit on the dimensionless energy density (the condition (9)), we have used the upper bound of $q_b^2 \sim 1.5 \times 10^{-10}$ at the Yukawa range of our interest, obtained by tests of the equivalence principle [7]. In Figure 2, the upper limit on the massless dilaton background at the resonant frequency of $\sim 7 \times 10^{-5}$ Hz is drawn against the strength of the dilaton coupling $q_b^2$, using the recent estimate of the upper bound on the effective viscosity of $\eta \sim 1.7 \times 10^5$ Pa s. At the frequency of our interest, there are the astrophysical constraints on the massless background (the dotted line in Figure 2), obtained from Nucleosynthesis and measurements of cosmic microwave background [10]. The astrophysical constraints are independent of the strength of the dilaton coupling. The region below the dotted line and above the solid curve is constrained by the geophysical search.

## 4. Discussions and Summary

In my previous work [6], I have proposed the geophysical search for dilatonic waves and presented preliminary upper limits on a massless dilatonic background by assuming a simple Earth model. In this work, I have discussed the upper limit further and presented the constraint on the dimensionless energy density of the massless dilaton background as a function of the strength of the dilaton coupling, $q_b^2$ (Figure 2).

The magnitude of $q_b^2$ is constrained by tests of the equivalence principle [7]. By using the most stringent constraint on $q_b^2$, I have obtained the upper limit on the dimensionless energy density of a massless dilaton background: $\Omega_{DW}h_{100}^2 \leq 6 \times 10^{-7}$. This upper limit is below the one obtained from the astrophysical observations: $\Omega h_{100}^2 \leq 10^{-5}$. If the sensitivity of testing the equivalence principle increases and the upper limit on $q_b^2$ is improved by one order of magnitude, the upper limit on $\Omega_{DW}$ that can be placed by the geophysical search becomes less stringent and comparable to the one obtained by the astrophysical observations (Figure 2).

I have used the recent upper bound on the effective viscosity of the liquid outer core for the estimation of the constraint. The determination of the effective viscosity is crucial for the accurate estimate of the constraint.

For simplicity, I have assumed massless dilatons. However, according to a scenario in string cosmology, a relic background of light dilatons could be a significant component of dark matter in the Universe [1]. Unlike the massless background, the intensity of light mass-dilaton background is not constrained by the astrophysical observations and its dimensionless energy density could be as large as unity. This indicates that the signals could be larger by about five orders of

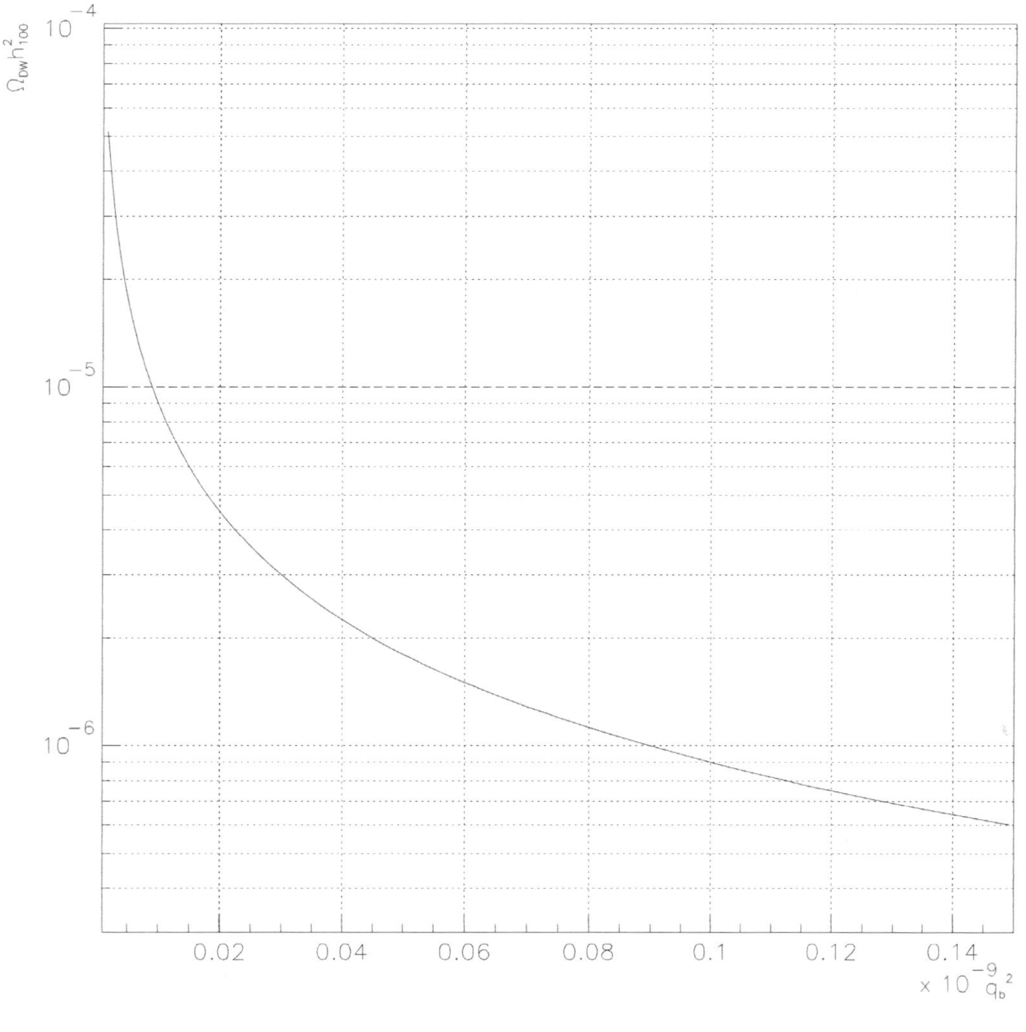


**Figure 2.** Constraint on the dimensionless energy density of the massless dilaton background (the solid curve) at the resonant frequency of $\sim 7 \times 10^{-5}$ Hz, obtained using the recent upper bound on the effective viscosity: $\eta \simeq 1.7{\times}10^5$ Pa s (estimated from nutation data [9]). The dotted line indicates the astrophysical upper limit on the massless background [10]. The region below the dotted line and above the solid curve is constrained by this geophysical search. The strength of the dilaton coupling is constrained to be $q_b^2 \leq 1.5 \times 10^{-10}$ by the tests of the equivalence principle [7].

magnitude than the one of the massless background; in spite of the weakness of the dilaton coupling, it might be within reach of experiments [1]. The possibility of testing the cosmological scenario by the geophysical approach has to be studied further.

## 5. Conclusions

In my previous work, I have proposed the geophysical search for dilatonic waves and presented preliminary upper limits on a massless dilatonic background by assuming a simple Earth model. In this work, I have discussed the upper limit further and presented the constraint on the dimensionless energy density of the massless dilaton background as a function of the strength of the dilaton coupling, $q_b^2$. The geophysical method is sensitive to the direct coupling of the dilaton fields to matter and the constraint obtained by this method depends on the strength of the dilaton coupling.

Using the most stringent upper limit on $q_b^2$ by tests of the equivalence principle, we have obtained the upper limit on the dimensionless energy density of the massless dilaton background, $\Omega_{DW}h_{100}^2 \leq 6 \times 10^{-7}$, at the frequency of $\sim 7 \times 10^{-5}$ Hz. This upper limit is about one-order of magnitude more stringent than the one from the astrophysical observations. If the magnitude of $q_b^2$ is experimentally found to be smaller than the current upper limit by one order of magnitude, the geophysical upper limit on $\Omega_{DW}h_{100}^2$ becomes less stringent and comparable to the one obtained from the astrophysical observations.

I have discussed the massless dilaton background for simplicity. The constraint on the relic background of light dilatons, predicted in string cosmology, has to be studied in the future.

## Acknowledgments

This work was funded by the Ministry of the Interior and National Science Council of Taiwan.

## References

[1] Gasperini M and Veneziano G 2003 *Phys. Rept.* **373** 1; Gasperini M 2007 *Elements of String Cosmology* (Cambridge University Press)
[2] Shibata M, Nakao K and Nakamura T 1994 *Phys. Rev. D* **50** 7304
[3] Harada T, Chiba T, Nakao K and Nakamura T 1997 *Phys. Rev. D* **55** 2024
[4] Gasperini M 1999 *Phys. Lett. B* **470** 67
[5] Maggiore M and Nicolis A 2000 *Phys. Rev. Lett.* **62** 024004
[6] Shiomi S 2008 *Phys. Rev. D* **78** 042001
[7] Schlamminger S, Choi K-Y, Wagner T A, Gundlach J H and Adelberger E G 2008 *Phys. Rev. Lett.* **100** 041101
[8] Secco R A 1995 *Mineral Physics and Crystallography: A Handbook of Physical Constants*, edited by T. J. Ahrens (American Geophysical Union, Washington) p 218
[9] Mathews P M and Guo J Y 2005 *J. Geophys. Res.* **110** B02402
[10] Cyburt R H, Fields B D, Olive K A and Skillman E 2005 *Astropart. Phys.* **23** 311; Smith T L, Pierpaoli E and Kamionkowski M 2006 *Phys. Rev. Lett.* **97** 021301

Journal of Physics: Conference Series **228** (2010) 012049 doi:10.1088/1742-6596/228/1/012049

# Effect of spin precession on bounding the mass of the graviton using gravitational waves from massive black hole binaries

**A. Stavridis and C.M. Will**

McDonnell Center for the Space Sciences, Department of Physics, Washington University, St. Louis, Missouri 63130, USA

E-mail: astavrid@physics.wustl.edu,cmw@physics.wustl.edu

**Abstract.** Observations of gravitational waves from massive binary black hole systems at cosmological distances can be used to search for a dependence of the speed of propagation of the waves on wavelength, and thereby to bound the mass of a hypothetical graviton. We study the effects of precession of the spins of the black holes and of the orbital angular momentum on the process of parameter estimation based on the method of matched filtering of gravitational-wave signals vs. theoretical template waveforms. For the proposed space interferometer *LISA*, we show that precession, and the accompanying modulations of the gravitational waveforms, are effective in breaking degeneracies among the parameters being estimated, and effectively restore the achievable graviton-mass bounds to levels obtainable from binary inspirals *without* spin. For spinning, precessing binary black hole systems of equal masses $10^6\,M_\odot$ at 3 Gpc, the lower bounds on the graviton Compton wavelength achievable are of the order of $5 \times 10^{16}$ km.

## Summary of the results

The anticipated launch of the Laser Interferometer Space Antenna (*LISA*) in the 2020 time frame will provide a promising new tool for doing astrophysics with massive binary black hole systems. The inspiral and merger of massive black holes (MBH) with masses of the order of $10^5$ - $10^7$ $M_\odot$ will be detectable to large distances in *LISA*'s sensitive frequency band between $10^{-5}$ and 1 Hz. The detection of gravitational waves (GW) from MBH systems will allow us to infer important astrophysical and astronomical information, such as the masses and spins of the black holes, the location of the system on the sky and its distance from the solar system.

Another important aspect of MBH binaries is the possibility of testing general relativity itself. In previous papers we have studied the bounds that could be placed on alternative theories of gravity such as scalar-tensor theories of the Brans-Dicke type, and theories in which gravitational waves propagate with a wavelength-dependent speed, as if the "graviton" were massive [1, 2, 3, 4, 5, 6]. Specifically, in [5] we showed that the inclusion of aligned, non-precessing spins weakens the bounds obtainable on the graviton mass by almost an order of magnitude. This is because the parameters that characterize the inpiraling binary are highly correlated, so that the addition of parameters (the spins) into the estimation process effectively dilutes the available information, leading to weakened bounds or estimates on most parameters.

However, Vecchio [7] pointed out that when the effects of precession of spins are incorporated into the gravitational waveforms, i.e. when the spins are not aligned with the orbital angular

© 2010 IOP Publishing Ltd

momentum, the accuracy of parameter estimation can be improved. He studied the so called "simple precession" case where either one of the bodies has zero spin, or the black hole masses are equal, and only spin-orbit interactions are included. The modulations of the amplitude and phase of the gravitational waveform induced by the precession of the spin(s) and by the precession of the orbital plane effectively *adds* information to the estimation process, partially decouples some of the parameters, and thus leads to restored accuracy. Lang and Hughes [8] extended Vecchio's work to include arbitrary spins and masses, and also spin-spin interactions, and found significant improvements in the accuracy of mass measurements as well as sky localization. In addition, they showed that the magnitudes of the spins of the binary members, especially for low redshift systems at $z \simeq 1$, could be measured with accuracies of the order of $10^{-2}$.

In this contribution we describe the results of an independent code for analysing binary inspiral with precessing spin and for carrying out parameter estimation based on the method of matched filtering, but extended to include the effects of a massive graviton. In addition to confirming the central conclusions of Lang and Hughes [8], we show that spin precession significantly improve the bounds that can be placed on the mass of the graviton. In parallel work, we have shown that including higher signal harmonics in the PN waveform (but without spins) also leads to improved bounds on the graviton mass [6].

We will not present here all the details of the analysis, but we rather refer the interested reader to the published article [9] for all the computational and method details. Our main conclusion, shown in Figs. 1 and 2 is that the inclusion of spin precession effects increases the lower bound on the graviton Compton wavelength $\lambda_{\rm g}$ by almost an order of magnitude, on average, with respect to the one calculated for the same non precessing system. Recall that $\lambda_g$ is related to the mass of the graviton by $\lambda_g = h/m_g c$, where $h$ is Planck's constant and $c$ is the speed of light, so that a lower bound on $\lambda_g$ represents an upper bound on $m_g$.

The total number of parameters for the studied system are the following 16: The individual masses of the binary system $\ln(m_i)$, luminocity distance $\ln(L_D)$, the two dimensionless spin parameters $\chi_i = S_i/m_i^2$ $(0 < \chi_i < 1)$, the time and phase of coalescence $t_c$, $\phi_c$, two angles for the binary position at the sky, $\phi_S$, $\cos\theta_S$, two angles for the initial angular momentum vectors $\phi_L$, $\cos\theta_L$, four angles for the initial orientations of the spins of the two bodies, $\phi_{S1}$, $\cos\theta_{S1}$, $\phi_{S2}$, $\cos\theta_{S2}$ and finally the massive graviton parameter $\beta_g = \frac{\pi^2 D\mathcal{M}}{\lambda_g^2(1+z)}$, where $D$ is a distance factor, $\mathcal{M}$ is the chirp mass and $z$ the redshift. For the Monte Carlo simulations we fix the pair of black hole masses, the redshift (consequently the luminocity distance) and we take random values for the 8 angle parameters, the 2 dimensionless spin parameters and the phase and time of coalescence.

Indeed, the new bounds, labeled MG+SO+PREC and MG+SO+SS+PREC in Fig. 1, which incorporate spin-orbit (SO) effects only and spin-orbit and spin-spin (SS) effects, respectively, along with the effect of a massive graviton (MG), are comparable to those inferred from an identical system without spin effects at all, labeled MG. This improvement is independent of mass, as seen in Fig. 2, which plots median lower bounds on $\lambda_g$ for systems without spin (MG), with non precessing spins (MG+SO+SS), and with precessing spins (MG+SO+SS+PREC) for various pairs of masses spanning two orders of magnitude in total mass.

In Figures 3,4 we plot the distribution of the correlation coefficients between the massive graviton parameter $\beta_{\rm g}$, and the two dimensionless spin parameters $\chi_i$ (left panel) and the two masses $m_i$ (right panel) for the $10^6 + 10^6 M_\odot$ black hole case. The correlations are quite mild, with most of the values ranging between 0 and 0.8, in contrast to the nonprecessing case [5], where correlation coefficients larger than 0.9 were routine. This illustrates clearly the strong decorrelating effect of the spin precession.

The effect of choosing arbitrary coalescence times $t_c$ is illustrated in Fig. 5 where the distributions of lower bounds on the graviton Compton wavelength $\lambda_g$ are shown for fixed and random values of $t_c$ . It is clear from the graph that randomizing $t_c$, leads to somewhat smaller lower bounds, with a tail at low values of the bounds, depicting the effect of signal loss in some of the cases. Fig. 5 also shows that using two *LISA* arm combinations generally leads to improved bounds.

## Acknowledgments

This paper is based in large part on sections of [9] This work was supported in part by the National Science Foundation Grant No. PHY 06-52448, the National Aeronautics and Space Administration, Grant No. NNG-06GI60G, and the Centre National de la Recherche Scientifique, Programme Internationale de la Coopération Scientifique (CNRS-PICS), Grant No. 4396. We are grateful for the hospitality of the Institut d'Astrophysique de Paris, where parts of this work were carried out. All the computer runs for this paper were performed at the HPC center of the Physics Department of Washington University.

## References

[1] Will C M 1994 *Phys. Rev. D* **50** 6058–6067
[2] Will C M 1998 *Phys. Rev. D* **57** 2061–2068
[3] Scharre P D and Will C M 2002 *Phys. Rev. D* 042002
[4] Will C M and Yunes N 2004 *Classical and Quantum Gravity* **21** 4367–4381
[5] Berti E, Buonanno A and Will C M 2005 *Phys. Rev. D* **71** 084025
[6] Arun K G and Will C M 2009 *Class. Quant. Grav.* **26** 155002 (*Preprint* 0904.1190)
[7] Vecchio A 2004 *Phys. Rev. D* **70** 042001
[8] Lang R N and Hughes S A 2006 *Phys. Rev. D* **74** 122001
[9] Stavridis A and Will C M 2009 *Phys. Rev. D* **80** 044002

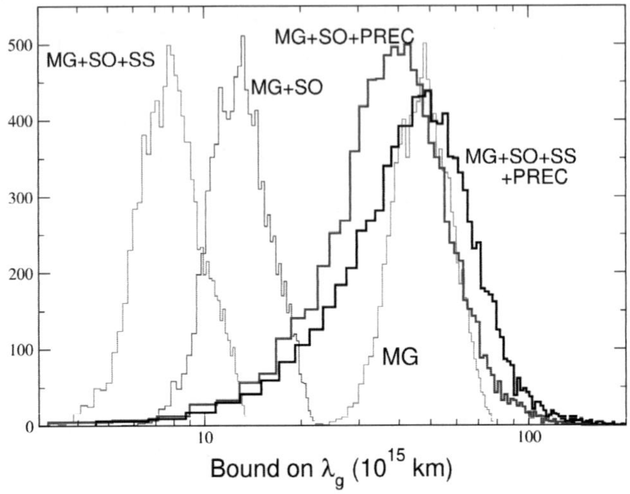


**Figure 1.** Distribution of lower bounds on the graviton Compton wavelength $\lambda_g$ (in units of $10^{15}$ km) for $10^4$ equal-mass ($10^6\,M_\odot$) black-hole binaries at redshift $z = 0.55$, or a luminosity distance 3 Gpc, randomly located on the sky. Number of bins is set to 50. First three histograms (narrow lines; red, blue, green in the color version) assume either no spins or aligned spins with spin-orbit (SO) and/or spin-spin (SS) coupling. Final two histograms (thick lines; violet and black in the color version) include precession induced by non-aligned spins

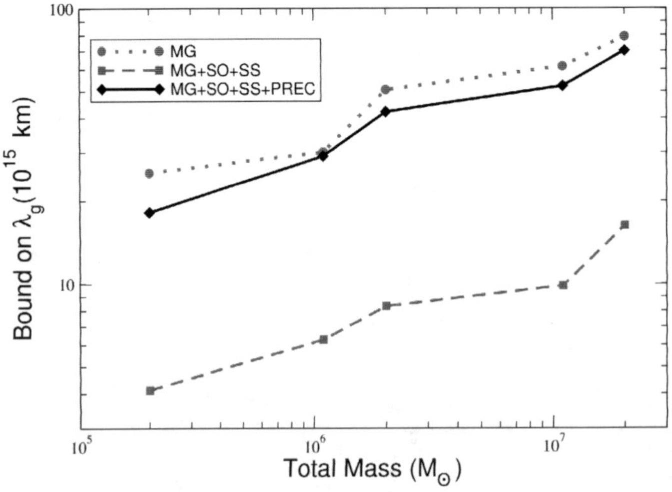


**Figure 2.** Median lower bounds on the graviton Compton wavelength $\lambda_g$ (in units of $10^{15}$ km) for $10^4$ black-hole binaries at redshift $z = 0.55$, or a luminosity distance 3 Gpc, randomly located on the sky. Systems contain black holes of mass $(1,1) \times 10^5$, $(1,10) \times 10^5$, $(1,1) \times 10^6$, $(1,10) \times 10^6$ and $(1,1) \times 10^7\,M_\odot$, from left to right, respectively.

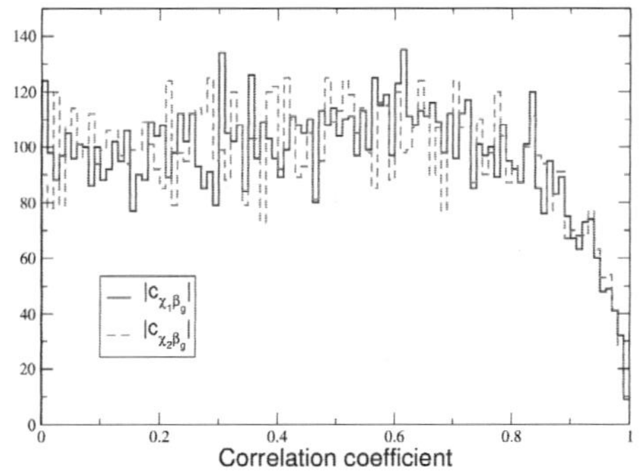


**Figure 3.** Distribution of the correlation coefficients between spin parameters $\chi_i$ and the massive graviton parameter $\beta_g$

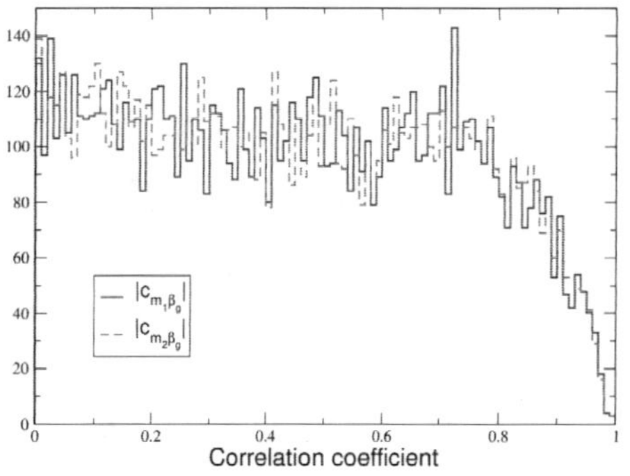


**Figure 4.** Distribution of the correlation coefficients between individual masses $m_i$ and the massive graviton parameter $\beta_g$

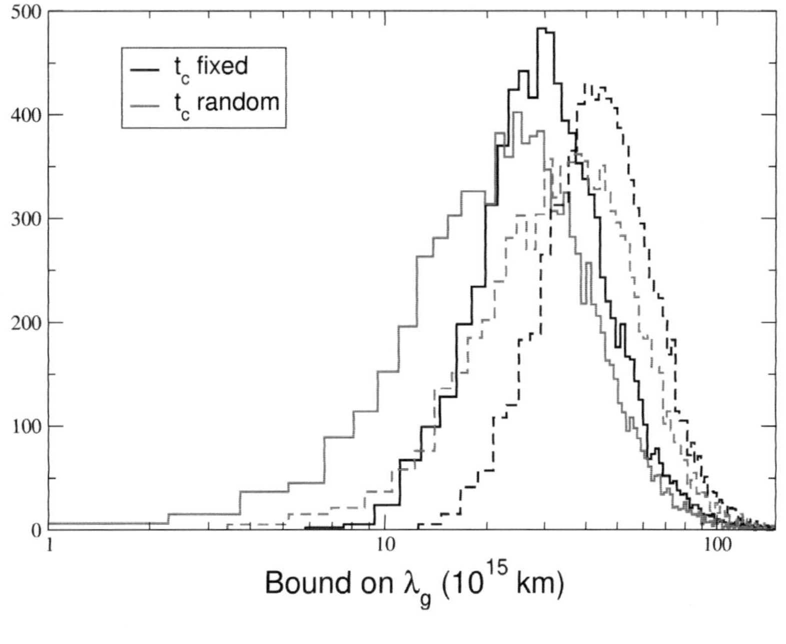


**Figure 5.** Distribution of lower bounds on $\lambda_g$ (in units of $10^{15}$ km) for $10^4$ binaries including spin precession. The system is a $10^6 + 10^6$ $M_\odot$ BBH at $z = 0.55$ (3 Gpc). Red curve is for $t_c$ fixed to one year; black curve is for random values of $t_c$ in the one year interval of the *LISA* mission. Solid (dashed) lines refer to one (two) *LISA* detectors respectively.

# The importance of precession in modelling the direction of the final spin from a black-hole merger

**Enrico Barausse**

Center for Fundamental Physics, University of Maryland, College Park, MD 20742-4111, USA

**Abstract.** The prediction of the spin of the black hole resulting from the merger of a generic black-hole binary system is of great importance to study the cosmological evolution of supermassive black holes. Several attempts have been recently made to model the spin via simple expressions exploiting the results of numerical-relativity simulations. Here, I first review the derivation of a formula, proposed in Ref. [1], which accurately predicts the final spin magnitude and direction when applied to binaries with separations of hundred or thousands of gravitational radii. This makes my formula particularly suitable for cosmological merger-trees and N-body simulations, which provide the spins and angular momentum of the two black holes when their separation is of thousands of gravitational radii. More importantly, I investigate the physical reason behind the good agreement between my formula and numerical relativity simulations, and nail it down to the fact that my formula takes into account the post-Newtonian precession of the spins and angular momentum in a consistent manner.

The dynamics of black-hole (BH) binaries is a very complex problem which has been solved only very recently through time-expensive numerical-relativity (NR) calculations. However, it has been shown that the dimensionless spin of the remnant from a BH binary merger, $\boldsymbol{a}_{\rm fin} = \boldsymbol{S}_{\rm fin}/M_{\rm fin}^2$, can be described with simple prescriptions based on point particles [3, 4, 5], on fits to the NR data [6, 7, 8, 9, 10], or on a combination of the two approaches [11]. These formulas are useful because they provide information over the entire 7-dimensional space of parameters for BH binaries in quasi-circular orbits, namely: the mass ratio $q \equiv M_2/M_1$ and the six components of the initial dimensionless spin vectors $\boldsymbol{a}_{1,2} = \boldsymbol{S}_{1,2}/M_{1,2}^2$. Such parameter space could in principle be investigated entirely via NR calculations; in practice, however, the simulations are very expensive and restricted to $q = 0.1$–$1$. Also, these formulas have applications in astrophysics, where they could provide information on the properties of massive-star binary systems; in cosmology, where supermassive BHs (SMBHs) are believed to assemble through accretion and mergers; in gravitational-wave astronomy, where the *a priori* knowledge of the final spin can help the detection.

While the different expressions for the spin norm, $|\boldsymbol{a}_{\rm fin}|$, are in good agreement with the results of NR simulations, the predictions for the final spin direction, $\hat{\boldsymbol{a}}_{\rm fin} \equiv \boldsymbol{a}_{\rm fin}/|\boldsymbol{a}_{\rm fin}|$, do not agree well with one another and are all essentially *imprecise* when the binaries are widely separated. This is because all expressions are built from and model the typical NR binaries and hence the dynamics of the last few orbits before the merger. Because it does not account systematically for the precession of the orbital angular momentum $\boldsymbol{L}$, the prediction for $\hat{\boldsymbol{a}}_{\rm fin}$ depends on the separation of the binary and is therefore of little use for applications, such as cosmological merger-trees or N-body simulations, that provide the spins of the two BHs at separations of thousands of gravitational radii. Although one could use the PN equations to

© 2010 IOP Publishing Ltd

evolve a widely-separated binary to a separation of few gravitational radii and then apply the formulas, this makes the formulas difficult to use and implement. In Ref. [1] we followed instead a different approach and showed that it is possible to derive a formula for the final spin that takes into account the precession of the spins and that is, therefore, applicable to binaries with arbitrary separations.

The derivation of the formula, for more details about which we refer the reader to Ref. [1], is based on the following assumptions:

*(i) When the spins are parallel to* $\boldsymbol{L}$ *(either aligned or antialigned), the formula must reduce to the fit of the NR results presented in Refs. [1, 11], namely*

$$a_{\rm fin} = \tilde{a} + \tilde{a}\nu(s_4\tilde{a} + s_5\nu + t_0) + \nu(2\sqrt{3} + t_2\nu + t_3\nu^2)\,, \qquad (1)$$

where $\nu \equiv M_1M_2/(M_1+M_2)^2$ is the symmetric mass ratio, $\tilde{a} \equiv (a_1 + a_2q^2)/(1+q^2)$, and[1]

$$\begin{aligned} s_4 &= -0.1229 \pm 0.0075\,, \quad s_5 = 0.4537 \pm 0.1463\,, \\ t_0 &= -2.8904 \pm 0.0359\,, \quad t_2 = -3.5171 \pm 0.1208\,, \quad t_3 = 2.5763 \pm 0.4833\,. \end{aligned} \qquad (2)$$

*(ii) The mass* $M_{\rm rad}$ *radiated to gravitational waves can be neglected i.e.,* $M_{\rm fin} = M \equiv M_1 + M_2$. The reason why assumption *(i)* is reasonable here is that $M_{\rm rad}$ is largest for aligned binaries but these are also the ones fitted by expression (1). In this way, the mass losses are automatically accounted for by the values of the coefficients $t_0$, $t_2$, $t_3$, $s_4$ and $s_5$.

*(iii) The norms* $|\boldsymbol{S}_1|$, $|\boldsymbol{S}_2|$, $|\tilde{\boldsymbol{\ell}}|$ *do not depend on the separation of the binary* $r$, with $\tilde{\boldsymbol{\ell}}$ being

$$\tilde{\boldsymbol{\ell}}(r) \equiv \boldsymbol{S}_{\rm fin} - [\boldsymbol{S}_1(r) + \boldsymbol{S}_2(r)] = \boldsymbol{L}(r) - \boldsymbol{J}_{\rm rad}(r)\,, \qquad (3)$$

where $\boldsymbol{S}_1(r)$, $\boldsymbol{S}_2(r)$ and $\boldsymbol{L}(r)$ are the spins and the orbital angular momentum at the separation $r$ and $\boldsymbol{J}_{\rm rad}(r)$ is the angular momentum radiated from $r$ to the merger. While the constancy of $|\boldsymbol{S}_1|$ and $|\boldsymbol{S}_2|$ is a very good assumption for BHs, the constancy of $|\tilde{\boldsymbol{\ell}}|$ is heuristic and based on the idea that the merger takes place at an "effective" innermost stable circular orbit (ISCO), so that $|\tilde{\boldsymbol{\ell}}|$ can be interpreted as the residual orbital angular momentum contributing to $\boldsymbol{S}_{\rm fin}$.

*(iv) The final spin* $\boldsymbol{S}_{\rm fin}$ *is parallel to the initial total angular momentum* $\boldsymbol{J}(r_{\rm in}) \equiv \boldsymbol{S}_1(r_{\rm in}) + \boldsymbol{S}_2(r_{\rm in}) + \boldsymbol{L}(r_{\rm in})$.

*(v) The angle between* $\boldsymbol{L}$ *and* $\boldsymbol{S} \equiv \boldsymbol{S}_1 + \boldsymbol{S}_2$ *and the angle between* $\boldsymbol{S}_1$ *and* $\boldsymbol{S}_1$ *are constant during the inspiral, although* $\boldsymbol{L}$ *and* $\boldsymbol{S}$ *precess around* $\boldsymbol{J}$.

Assumptions *(iv)* and *(v)* are motivated by the PN approximation. It has been in fact shown by Ref. [13] that within the adiabatic approximation the secular angular-momentum losses via gravitational radiation are along $\boldsymbol{J}$. This is because as $\boldsymbol{L}$ rotates around $\boldsymbol{J}$, the emission orthogonal to $\boldsymbol{J}$ averages out. Therefore, $\boldsymbol{S}$ and $\boldsymbol{L}$ essentially precess around the direction $\hat{\boldsymbol{J}}$, which remains nearly constant (*cf.* the detailed discussion in Ref. [13]), and the angle between $\boldsymbol{L}$ and $\boldsymbol{S}$ and that between the two spins remain constant as well.

*(vi) When the initial spin vectors are equal and opposite and the masses are equal, the spin of the final BH is the same as for nonspinning binaries.* Besides being physically reasonable – if the spins are equal and opposite, their contributions are expected to cancel out – this assumption is confirmed by NR simulations and by the leading-order PN spin-spin and spin-orbit couplings.

[1] The number of free parameters of the fit is actually *four*, since $t_0$, $t_2$, $t_3$, $s_4$ and $s_5$ must satisfy the constraint

$$a_{\rm fin} = \frac{\sqrt{3}}{2} + \frac{t_2}{16} + \frac{t_3}{64} = 0.68646 \pm 0.00004\,.$$

which follows from the results obtained by Ref. [12] for equal-mass non-spinning BHs.

Journal of Physics: Conference Series **228** (2010) 012050 doi:10.1088/1742-6596/228/1/012050

These assumptions are sufficient to derive an expression for both the magnitude and the direction of the final spin [1]. In particular, the final spin norm is given by

$$|\boldsymbol{a}_{\rm fin}| = \tfrac{1}{(1+q)^2}\Big[|\boldsymbol{a}_1|^2 + |\boldsymbol{a}_2|^2 q^4 + 2|\boldsymbol{a}_2||\boldsymbol{a}_1| q^2 \cos\alpha + 2\left(|\boldsymbol{a}_1|\cos\beta + |\boldsymbol{a}_2| q^2 \cos\gamma\right)|\boldsymbol{\ell}| q + |\boldsymbol{\ell}|^2 q^2\Big]^{1/2}, \tag{4}$$

where

$$|\boldsymbol{\ell}| = 2\sqrt{3} + t_2\nu + t_3\nu^2 + \frac{s_4}{(1+q^2)^2}\left(|\boldsymbol{a}_1|^2 + |\boldsymbol{a}_2|^2 q^4 + 2|\boldsymbol{a}_1||\boldsymbol{a}_2| q^2\cos\alpha\right) + \left(\frac{s_5\nu + t_0 + 2}{1+q^2}\right)\left(|\boldsymbol{a}_1|\cos\beta + |\boldsymbol{a}_2| q^2\cos\gamma\right),$$

and the angles $\alpha, \beta$ and $\gamma$ are defined by

$$\cos\alpha \equiv \hat{\boldsymbol{a}}_1(r_{\rm in})\cdot\hat{\boldsymbol{a}}_2(r_{\rm in}), \quad \cos\beta \equiv \hat{\boldsymbol{a}}_1(r_{\rm in})\cdot\hat{\boldsymbol{L}}(r_{\rm in}), \quad \cos\gamma \equiv \hat{\boldsymbol{a}}_2(r_{\rm in})\cdot\hat{\boldsymbol{L}}(r_{\rm in}). \tag{5}$$

The angle $\theta_{\rm fin}$ between the final spin and the initial orbital angular momentum $\boldsymbol{L}(r_{\rm in})$ is instead

$$\cos\theta_{\rm fin} = \hat{\boldsymbol{L}}(r_{\rm in})\cdot\hat{\boldsymbol{J}}(r_{\rm in}). \tag{6}$$

We will now test our expressions (4) and (5) for $|\boldsymbol{a}_{\rm fin}|$ and our expression (6) for $\theta_{\rm fin}$ against the NR simulations for *generic* binaries (*i.e.*, with spins *not* parallel to $\boldsymbol{L}$) published so far. Also, we will compare our predictions (AEI) with those of similar formulas suggested by Refs. [4] (BKL), [11] (AEI old) and [8] (FAU). The comparison consists of two steps. First, we compare the different predictions using as input the initial data of the NR simulations, in which the binaries have *small separations* ($r_{\rm in} \lesssim 10\,M$). Second, using binaries at *large separations* ($r_{\rm in} \le 2\times 10^4\,M$), for which the dynamics starts being dominated by gravitational-wave emission and thus of direct relevance for cosmological investigations. More precisely, we evolve the NR initial configurations back in time up to a separation of $2\times 10^4\,M$ using the PN equations in the quasi-circular limit[2], calculating the predictions of the different formulas at each step and considering the maximum error for each formula. (The results that we present, however, do not change significantly if we integrate only up to a separation of $\sim 200M$.)

The upper left panel of Fig. 1 shows the predictions of the various formulas for $|\boldsymbol{a}_{\rm fin}|$, when applied to the small-separation configurations corresponding to the initial data of the NR simulations (see caption for details). In particular, it reports the error $||\boldsymbol{a}_{\rm fin,NR}| - |\boldsymbol{a}_{\rm fin,*}||$, where "*" stands either for "AEI" (which, as mentioned in Ref. [1], gives the same predictions for $|\boldsymbol{a}_{\rm fin}|$ as "AEI old"), "FAU" or "BKL". The lower left panel shows instead the *maximum* error when the configurations are evolved back in time up to $r_{\rm in} = 2\times 10^4 M$. Although the AEI expression is slightly better, all the formulas give accurate predictions for the final spin *norm*, both for small and large separations. Note that the larger errors for the indices 1–10, which correspond to the simulations of Ref. [14] with small mass ratios ($q = 0.13$–$0.17$), are most likely due to the larger truncation errors affecting those simulations (see Ref. [14], sec. IIIA).

The situation is very different when considering the final spin *direction*. In particular, the right panels of Fig. 1 report the inclination-angle error, $|\theta_{\rm fin,*} - \arccos[\hat{\boldsymbol{L}}(r_{\rm in})\cdot\hat{\boldsymbol{a}}_{\rm fin,NR}]|$, for the data in the left panels, but for those of Ref. [17], for which the final spin direction was not published. Clearly, when considering small-separation binaries (upper right panel), our new formula performs slightly better than the "BKL" and "AEI old" formulas, but it is not better

[2] In particular, following Ref. [2], we use the precession equations for the spins and the angular momentum at 2 PN order and the rate of change of the frequency at 3.5 PN order (with spin terms included up to 2 PN order).

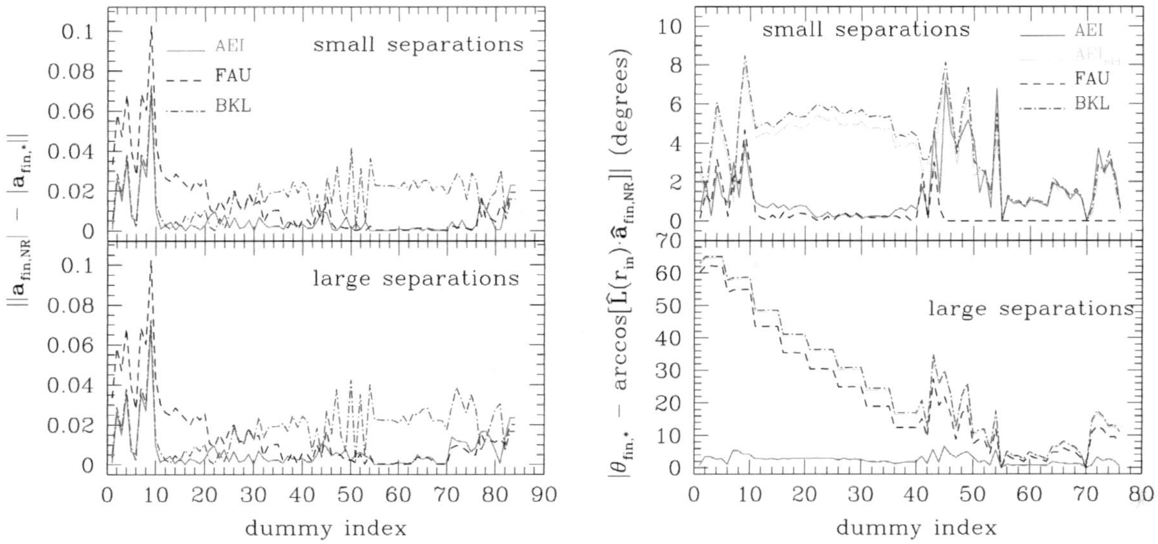


**Figure 1.** *Left:* The upper panel shows the error $||\boldsymbol{a}_{\rm fin,NR}| - |\boldsymbol{a}_{\rm fin,*}||$ ("*" being either "AEI", "FAU" or "BKL") of the various formulas for the final spin norm, when applied to the *small-separation* configurations corresponding to the initial data of the simulations of Refs. [14] (indices 1-40), [15] (indices 41-42), [16] (index 43), [8] (indices 44-76) and [17] (indices 77-84). The lower panel shows instead the *maximum* error $||\boldsymbol{a}_{\rm fin,NR}|-|\boldsymbol{a}_{\rm fin,*}||$ when the configurations are evolved back in time up to *large separations* of $r = 2\times10^4 M$. Although the new AEI expression is slightly better, all formulas give accurate predictions for $|\boldsymbol{a}_{\rm fin}|$, both for small and large separations. Note that the larger errors for the indices 1–10 are due to the larger truncation errors affecting those simulations (see text for details). *Right:* The same as in the left panel but for the inclination angle error $|\theta_{\rm fin,*} - \arccos[\hat{\boldsymbol{L}}(r_{\rm in})\cdot\hat{\boldsymbol{a}}_{\rm fin,NR}]|$ and without the data of Ref. [17], for which the final spin direction has not been published. The new AEI expression is accurate both for small and large separations, while the other ones become imprecise for large separations.

than the "FAU" formula. Indeed, the latter is exact by construction for the indices 34–66. This is because for such data the final spin direction has not been published and it has been here reconstructed using the FAU formula applied to the configurations of Table II in Ref. [8]. However, when considering large-separation binaries (lower right panel), our new formula clearly performs much better than all the other ones. For instance, our error for $\theta_{\rm fin}$ is below 7 degrees, for *any* separation, while it can be as large as 70 degrees with the older formulas. (The "steps" in the lower right panel reflect the different sequences of Table I of Ref. [14].)

To highlight the role played by the precession of the spins in the prediction of the final spin direction, in the left panel of Fig. 2 we focus on the binary "SP3" of Ref. [15] (index 41 in Fig. 1). In this configuration the spins precess strongly and the final spin flips relative to $\boldsymbol{S}(r_{\rm in})$. In particular, $\boldsymbol{S}_1$ and $\boldsymbol{S}_2$ are initially parallel, orthogonal to $\boldsymbol{L}$ and have: $a_1 = a_2 \approx 0.5$, $q = 1$. The filled hexagon in Fig. 2 is the numerical result as obtained with an initial separation $r_{\rm in} \simeq 6.6\,M$, while different lines show the angle $\theta_{\rm fin}$ obtained with the various formulas as a function of the binary separation. More specifically, using the initial conditions of the NR simulation we have evolved the "SP3" binary back in time using the PN equations up to a separation of $2\times10^4 M$, applying the various formulas at each step to compute $\theta_{\rm fin,*} = \arccos[\hat{\boldsymbol{L}}(r_{\rm in})\cdot\hat{\boldsymbol{a}}_{\rm fin,*}]$. The line "PN+NR" reports instead the angle $\arccos[\hat{\boldsymbol{L}}(r_{\rm in})\cdot\hat{\boldsymbol{a}}_{\rm fin,NR}]$. In the right panel, we consider the configuration "Q13TH00" of Ref. [14] (index 1 in Fig. 1). This configuration features a smaller, non-spinning BH and a larger, spinning BH with spin $a \approx 0.81$ lying on the equatorial plane and

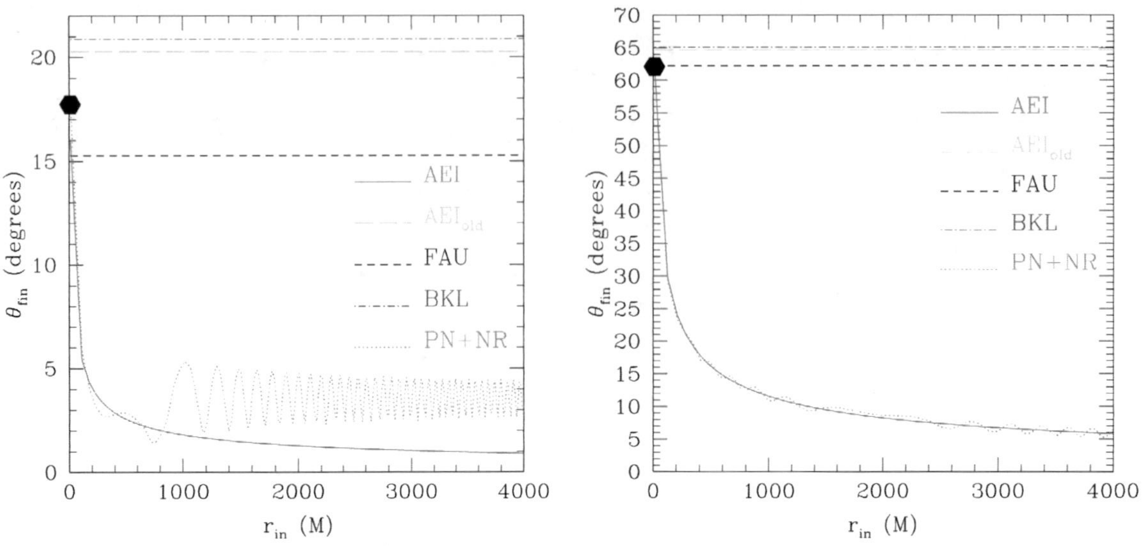


**Figure 2.** *Left:* Predictions of the different formulas for $\theta_{\rm fin}$ for the binary "SP3" of Ref. [15] as a function of the initial separation $r_{\rm in}$. Shown with the "NR+PN" curve is the angle $\arccos\,[\hat{\boldsymbol{L}}(r_{\rm in})\cdot\hat{\boldsymbol{a}}_{\rm fin,NR}]$, where $\hat{\boldsymbol{L}}(r_{\rm in})$ is obtained by integrating the PN equations starting from the initial conditions of the NR simulation, while the filled hexagon shows $\theta_{\rm fin}$ as computed by the NR simulation. *Right:* The same as in the left panel, but for the configuration "Q13TH00" of Ref. [14]. Clearly, only our formula reproduces the "NR+PN" results to within 2-3 degrees for any $r_{\rm in}$, while the other formulas have errors that grow very large for separations $r_{\rm in}\gtrsim 200M$.

orthogonal to the separation ($r_{\rm in}\approx 6.4M$). The mass ratio is $q\approx 0.13$, which makes this binary interesting as it allows to test the behavior of the various formulas in a region of parameter space which so far has been sampled only sparsely by NR simulations, which usually deal with $q\approx 1$. (This is because the cost of a NR simulation grows as $1/q^2$, a factor $1/q$ coming from the slow inspiral of a small mass ratio binary, and another factor $1/q$ appearing because if the mass ratio is small, one needs much resolution to resolve the two very different scales of the problem.)

Three results are clear from these comparisons. First: all previous formulas provide a reasonable estimate of $\theta_{\rm fin}$ but only when the binary has a separation $r_{\rm in}\ll 200M$. Second: Only our formula provides a reasonable estimate at all separations; indeed, the "PN+NR" results are reproduced to within the accuracy of the NR simulation, *i.e.* 2-3 degrees. All the other formulas, instead, predict the same value for $\theta_{\rm fin}$ for all $r_{\rm in}$, because they take as an input the angles $\arccos(\hat{\boldsymbol{S}}_{1,2}\cdot\hat{\boldsymbol{L}})$ and $\arccos(\hat{\boldsymbol{S}}_1\cdot\hat{\boldsymbol{S}}_2)$, which are constant under the quasi-circular PN evolution [2]. Therefore, the other formulas become rapidly become imprecise when the binary's separation increases, reaching errors as large as 60 degrees for $r_{\rm in}\gtrsim 200M$. Third: the error of the old formulas when applied to large separation binaries is larger for small mass ratio systems (*e.g.* the maximum error is $\sim 13$ degree for the comparable-mass configuration SP3, while it is $\sim 60$ degrees for the small mass-ratio binary Q13TH00). This had to be expected, again on the grounds that the old formulas predict the same $\theta_{\rm fin}$ for all $r_{\rm in}$. As a result of this, in fact, the maximum error of the old formulas is roughly given by their prediction for $\theta_{\rm fin}$ at small separations, because the correct $\theta_{\rm fin}$ becomes small for large $r_{\rm in}$ (*cf.* the "PN+NR" curves in Fig. 2). This prediction can be very large for small $q$ if the angle between $\hat{\boldsymbol{S}}$ and $\hat{\boldsymbol{L}}$ at small separations is large. This can be easily understood by noting that at small separations the old formulas give roughly the same results as our formula: Therefore, for $q\approx 0$ their prediction is

$\cos(\theta_{\rm fin}) \approx \hat{\boldsymbol{L}}(r_{\rm in}) \cdot \hat{\boldsymbol{S}}(r_{\rm in})$ [because for $q \approx 0$, $\hat{\boldsymbol{J}}(r_{\rm in}) \approx \hat{\boldsymbol{S}}(r_{\rm in})$].

We stress that such a good agreement between our formula and the data for the final spin, irrespective of the separation of the binary to which the formula is applied, emerges because we have consistently taken into account the effect of precession through assumptions *(iv)* and *(v)*, while that effect is not accounted for in the other formulas. For instance, in our earlier formula of Ref. [11] ("AEIold"), assumptions *(iv)* and *(v)* were replaced by the assumption that the angular momentum emitted during the inspiral is parallel to $\boldsymbol{L}(r_{\rm in})$ [*cf.* assumption *(iii)* of Ref. [11]]. This is not true unless one neglects spin-orbit precession (as was indeed stressed explicitly in Ref. [11]). A similar assumption was also made in Ref. [4], while Ref. [17] admittedly recognized that their formula might not be valid at arbitrary separations, as it is based on a fit to NR simulations, which, as already mentioned, have initial separations $r_{\rm in} \sim 10M$.

In conclusion: I have reviewed the assumptions needed to derive a new formula predicting the spin of the BH resulting from the merger of two BHs in quasi-circular orbits and having arbitrary initial masses and spins [1]. This formula includes the effect of the precession of the spins through assumptions *(iv)* and *(v)*, and can therefore be applied to widely separated binaries, such as those relevant for cosmological applications, for which the other available formulas become imprecise. I stress that requiring that the formulas for the final spin work also when applied to large separation binaries is necessary to ensure that these formulas can be readily used in cosmological applications. Cosmological simulations typically provide the initial spins and angular momentum of SMBH binaries when their separation is still of about 0.1 pc (see Ref. [18] for a recent example), which corresponds to $2 \times 10^5 M$ if $M = 10^8 M_{\odot}$. This separation roughly corresponds to the point at which the evolution of the binary starts being driven by gravitational-wave emission. While it would be in principle possible to use the PN equations and evolve the widely-separated binaries provided by cosmological simulations to small separations, read-off the relevant information and apply the formulas for the final spin, this procedure would be impractical and potentially very time-consuming (*cf.* again Ref. [18], which simulates tens of thousands of SMBH binaries in order to get significant statistics).

I am grateful to my coauthor L. Rezzolla for countless discussions on the issues examined in this contribution. I also acknowledge support from NSF Grant PHY-0603762.

## References

[1] Barausse E and Rezzolla L 2009, Astrophys. J.**704** L40
[2] Buonanno A, Chen Y and Vallisneri M 2006, Phys. Rev. D**74** 029904
[3] Hughes S A and Blandford R D 2003, Astrophys. J. Lett.**585** L101
[4] Buonanno A, Kidder L E and Lehner L 2008, Phys. Rev. D**77** 026004
[5] Kesden M 2008, Phys. Rev. D**78** 084030
[6] Rezzolla L, Dorband E N, Reisswig C, Diener P, Pollney D, Schnetter E and Szilágyi B 2008, Astrophys. J.**679** 1422
[7] Rezzolla L, Diener P, Dorband E N, Pollney D, Reisswig C, Schnetter E and Seiler J 2008, Astrophys. J. Lett.**674** L29
[8] Tichy W and Marronetti P 2008, Phys. Rev. D**78** 081501
[9] Boyle L, Kesden M and Nissanke S 2008, Phys. Rev. Lett., **100** 151101
[10] Boyle L and Kesden M 2008, Phys. Rev. D**78** 024017
[11] Rezzolla L, Barausse E, Dorband E N, Pollney D, Reisswig C, Seiler J and Husa S 2008, Phys. Rev. D**78** 044002
[12] Scheel M A, Boyle M, Chu T, Kidder L E, Matthews K D and Pfeiffer H P 2009, Phys. Rev. D**79** 024003
[13] Apostolatos T A, Cutler C, Sussman G J and Thorne K S 1994, Phys. Rev. D**49** 6274
[14] Lousto C O and Zlochower Y 2009, Phys. Rev. D**79** 064018
[15] Campanelli M, Lousto C O, Zlochower Y, Krishnan B and Merritt D 2007, Phys. Rev. D**75** 064030
[16] Campanelli M, Lousto C, Zlochower Y and Merritt D 2007, Astrophys. J. Lett.**659** L5
[17] Tichy W and Marronetti P 2007, Phys. Rev. D**76** 061502
[18] Dotti M, Volonteri M, Perego A, Colpi M, Ruszkowski M and Haardt F 2010, Mon. Not. Roy. Astron. Soc.**402** 682

# Testing effective quantum gravity with gravitational waves from extreme mass ratio inspirals

**N. Yunes**[1] **and C. F. Sopuerta**[2]

[1]Institute for Gravitational Physics and Geometry, Center for Gravitational Wave Physics, Department of Physics, The Pennsylvania State University, University Park, PA 16802-6300
[2]Institut de Ciències de l'Espai (CSIC-IEEC), Facultat de Ciències, Campus UAB, Torre C5 parells, Bellaterra, 08193 Barcelona, Spain.

E-mail: [1]nyunes@princeton.edu

**Abstract.**
Testing deviation of GR is one of the main goals of the proposed *Laser Interferometer Space Antenna*. For the first time, we consistently compute the generation of gravitational waves from extreme-mass ratio inspirals (stellar compact objects into supermassive black holes) in a well-motivated alternative theory of gravity, that to date remains weakly constrained by double binary pulsar observations. The theory we concentrate on is Chern-Simons (CS) modified gravity, a 4-D, effective theory that is motivated both from string theory and loop-quantum gravity, and which enhances the Einstein-Hilbert action through the addition of a dynamical scalar field and the parity-violating Pontryagin density. We show that although point particles continue to follow geodesics in the modified theory, the background about which they inspiral is a modification to the Kerr metric, which imprints a CS correction on the gravitational waves emitted. CS modified gravitational waves are sufficiently different from the General Relativistic expectation that they lead to significant dephasing after 3 weeks of evolution, but such dephasing will probably not prevent detection of these signals, but instead lead to a systematic error in the determination of parameters. We end with a study of radiation-reaction in the modified theory and show that, to leading-order, energy-momentum emission is not CS modified, except possibly for the subdominant effect of scalar-field emission. The inclusion of radiation-reaction will allow for tests of CS modified gravity with space-borne detectors that might be two orders of magnitude larger than current binary pulsar bounds.

## 1. Motivation

One of the primary goals of the *Laser Interferometer Space Antenna* (LISA) is to study possible deviations from General Relativity (GR) in the strong field, where gravity is strong, speeds are large and binary pulsar or Solar System experiments are inapplicable. One route to such tests has been to study deviations from the Kerr metric *within GR*, through the determination of the multipolar structure of the gravitational field. In GR, however, the Kerr metric is known to be the only physically-reasonable representation of the exterior gravitational field of spinning compact objects, provided cosmic censorship and causality holds.

Another route to test GR is to consider alternative theories of gravity and study the imprint these leave on gravitational waves (GWs) emitted by LISA sources. Here we report on the first consistent study of GW generation in a well-motivated alternative theory of gravity by the

© 2010 IOP Publishing Ltd

inspiral of a small compact object into a spinning, supermassive BHs (a so-called extreme-mass ratio inspiral or EMRI) [1].

Due to the plethora of alternative theories available, it becomes difficult to justify the choice of one theory over another. For this purpose, it is convenient to propose certain criteria for a theory to be *a reasonable candidate* to test GR with LISA [2]:

- *Metric theories of gravity*: the gravitational field is described by a symmetric metric tensor that satisfies the weak-equivalence principle [3].
- *Weak-field consistency*: reduction to GR for weak gravitational fields and small velocities, such that experimental and observational tests are passed.
- *Strong-field inconsistency*: deviations from GR in the dynamical strong-field, where gravity is strong and speeds are large.

The weak-field consistency criterion also implies the existence and stability of physical solutions, such as the Newtonian limit of the Schwarzschild metric to describe physics in the Solar System. Additional criteria can also be proposed, such as well-posedness of the initial-value formulation, the existence of a well-defined and complete relativistic action, and some type of motivation from fundamental physics. These additional criteria, however, might be too stringent since only GR has been sufficiently studied to determine whether they are satisfied.

A modification to GR that passes all the criteria in the itemized list above is Chern-Simons (CS) modified gravity [4]. In this 4D theory, the Einstein-Hilbert action is modified through the addition of the product of a dynamical scalar field and the Pontryagin density. Since the scalar field couples only to the quadratic curvature correction to the action, the weak-equivalence principle remains satisfied and the theory is indeed metric. Due to its quadratic curvature nature, this theory is also weak-field consistent but strong-field inconsistent. Stable solutions, like the Schwarzschild metric and a modified Kerr metric [5], have also been found, although there are only qualitative arguments that suggest Cauchy well-posedness [6]. As for motivation, the CS modification arises generically and unavoidingly in the low-energy effective limit of string theory, both in the perturbative sector due to the Green-Schwarz anomaly-canceling mechanism [7], and in the non-perturbative sector due to D-instanton charges and duality symmetries [8]. The CS correction can also be shown to arise in loop quantum gravity, when the Barbero-Immirzi parameter is promoted to a field in the presence of fermions [9, 10]. In fact, one can show generically (eg. in cosmological inflation) that the CS correction is one of a few terms that generically arise in effective field theories when one expands the action in the curvature tensor [11].

Two different version of CS modified gravity exist, a non-dynamical and a dynamical one, where the former has been strongly constrained by theoretical and experimental considerations, while the latter has only bee weakly constrained by binary pulsar observations. The non-dynamical version assumes the scalar field is *a priori* prescribed and non-evolving, which forces what would have been an evolution equation for the scalar field into an additional constraint on the solution space of the theory. This additional constraint has been shown to be too restrictive, essentially disallowing spinning black hole (BH) solutions [12] and forbidding perturbations of non-spinning BHs [13]. From an observational standpoint, the non-dynamical version has been strongly constrained in the neighbourhood of neutron stars through binary pulsar observations [14]; GWs would allow a constrain in regions of spacetime between source and observer due to corrections to GW propagation [15, 16].

Dynamical CS gravity has been only weakly constrained by binary pulsar observations [5]. Recently, [17] studied perturbations of non-spinning, Schwarzschild BHs in the dynamical theory and found certain instabilities, which they used to argue a strong constraint could be placed due to the observationally confirmed existence of rotating BHs. This result, however, is strongly dependent on the assumption that there is no background scalar field to which the metric perturbations could couple to. Such an assumption is not valid for rotating BHs, even in the limit of small rotation, as found in [5]. Then, the work of [17] can be considered as a good motivation to explore in more detail oscillations of BHs in the dynamical theory. As a first step, one would study the behaviour of *linear* oscillations for rotating BHs (which has not been done yet as it can be prohibitively difficult). Then, in the case that instabilities are still present, one could ask whether *non-linear* effects can suppress the linear instabilities. If the answer is in the negative, one can then try to put constraints on the theory based on astrophysical observations.

We here study EMRIs in dynamical CS modified gravity. We first show that point-particles follow geodesics in this theory as in GR, although the supermassive BH background is modified. Such a modification to geodesic motion leads to corrections in the generation of GWs that lead to a dephasing between CS waves and those expected in GR. This dephasing will not prevent detection of GWs with LISA, but instead it will bias the estimation of parameters, leading to an uncontrolled systematic error. We conclude with a study of radiation-reaction in dynamical CS modified gravity and we proof that to leading order this is the same as in GR, except perhaps for subdominant energy-momentum emission by the scalar field. The inclusion of radiation-reaction will lead to stronger GW modifications that will break the degeneracy between the CS coupling parameter and the system parameters, thus allowing for a more stringent test of the dynamical theory relative to the binary pulsar constraint.

The remainder of this Proceedings describes the above results in more detail: Section 2 presents the basics of the theory; Section 3 describes the generation of GWs; Section 4 shows the difference between GR and CS GWs and discusses the inclusion of radiation-reaction; Section 5 concludes and points to future work. We employ the conventions in Misner, Thorne and Wheeler [3], described in more detail in [1].

## 2. Chern-Simons Basics

We here summarize the basics of CS modified gravity and we refer the reader to the review paper [4] for further details.

### *2.1. Field Equations*

Consider then the Lagrangian density

$$\mathcal{L} = \kappa R + \frac{\alpha}{4}\vartheta \, {}^*R\,R - \frac{\beta}{2}\left[g^{\mu\nu}\left(\nabla_\mu\vartheta\right)\left(\nabla_\nu\vartheta\right) + 2V(\vartheta)\right] + \mathcal{L}_{\rm mat}, \tag{1}$$

where the gravitational constant is $\kappa = (16\pi G)^{-1}$; the coupling constant associated with the gravitational CS correction is $\alpha$, while that associated with the strength of the CS scalar field $\vartheta$ is $\beta$. The parity-violating Pontryagin density ${}^*R\,R := \epsilon^{\alpha\beta\mu\nu} R_{\alpha\beta\gamma\delta} R^{\gamma\delta}{}_{\mu\nu}/2$, where the asterisk denotes the dual tensor, constructed using the antisymmetric Levi-Civita tensor $\epsilon^{\alpha\beta\mu\nu}$. The potential term $V(\vartheta)$ will be assumed to vanish, as it does in string theory due to shift symmetry. The last term in (1) represents the Lagrangian density for external matter degrees of freedom.

Variation of the action with respect to all degrees of freedom leads to the CS field equations:

$$G_{\mu\nu} + \frac{\alpha}{\kappa} C_{\mu\nu} = \frac{1}{2\kappa}\left(T^{\rm mat}_{\mu\nu} + T^{(\vartheta)}_{\mu\nu}\right), \tag{2}$$

$$\beta \Box \vartheta = \beta \frac{dV}{d\vartheta} - \frac{\alpha}{4} {}^*R\,R, \tag{3}$$

where $T^{\rm mat}_{\mu\nu}$ is a matter stress-energy tensor, $T^{(\vartheta)}_{\mu\nu}$ is the stress-energy of the CS scalar field and $C^{\mu\nu}$ is the so-called C-tensor, both of which are given by

$$T^{(\vartheta)}_{\mu\nu} = \beta\left[(\nabla_\mu\vartheta)(\nabla_\nu\vartheta) - \frac{1}{2}g_{\mu\nu}(\nabla^\sigma\vartheta)(\nabla_\sigma\vartheta) - g_{\mu\nu}V(\vartheta)\right].$$

$$C^{\alpha\beta} = (\nabla_\sigma\vartheta)\,\epsilon^{\sigma\delta\nu(\alpha}\nabla_\nu R^{\beta)}{}_\delta + (\nabla_\sigma\nabla_\delta\vartheta)\;{}^*R^{\delta(\alpha\beta)\sigma}. \tag{4}$$

As we are here interested in the dynamical version of dynamical CS modified gravity, we restrict attention to coupling constants $\beta \neq 0$. If $\beta = 0$, one would recover non-dynamical CS theory and the equation of motion for $\vartheta$ would reduce to the Pontryagin constraint $^{*}R\,R = 0$, which over-constrains the field equations [13]. We consider the dynamical theory, where we define $\xi := \alpha^2/(\kappa\beta)$ as an expansion parameter for the so-called small-coupling approximation (see [5] for details), in which a modified Kerr solution is obtained.

### 2.2. Gravitational Waves in CS theory

Let us decompose the metric tensor as $g_{\mu\nu} = \bar{g}_{\mu\nu} + \epsilon\, h_{\mu\nu} + \mathcal{O}(\epsilon^2)$, where $\bar{g}_{\mu\nu}$ is some background metric, $h_{\mu\nu}$ is a metric perturbation, which in the radiation-zone (many GW wavelengths away from the center of mass) can be associated with GWs, and $\epsilon$ is a book-keeping perturbation parameter associated with the GW strength.

The CS field equations can now be expanded in $\epsilon$. In the radiation zone, however, we can further simplify the analysis by focusing on a flat background, $\bar{g}_{\mu\nu} = \eta_{\mu\nu}$. We then find

$$
\begin{aligned}
-\frac{1}{\kappa}\bar{T}_{\mu\nu} &= \epsilon\Box_\eta h_{\mu\nu} + \frac{\alpha}{\kappa}(\partial_\sigma\vartheta)\epsilon\; \bar{\epsilon}^{\sigma\delta\xi}{}_{(\mu}\Box_\eta h_{\nu)\delta,\xi} \\
&+\frac{\alpha}{\kappa}(\partial_\sigma{}^\gamma\vartheta)\epsilon\; \bar{\epsilon}^{\sigma\delta\xi}{}_{(\mu}\left[h_{\xi\gamma,|\nu)\delta} - h_{\nu)\xi,\gamma\delta}\right] + \mathcal{O}(\epsilon^2) \qquad (5)\\
\beta\Box\vartheta &= -\frac{\alpha}{2}\epsilon^2\; \bar{\epsilon}^{\alpha\beta\mu\nu} h_{\alpha\delta,\gamma\beta} h_\nu{}^{[\gamma,\delta]}{}_\mu + \mathcal{O}(\epsilon^3)\,, \qquad (6)
\end{aligned}
$$

where $\bar{T}_{\mu\nu} = T_{\mu\nu} - (1/2)\eta_{\mu\nu}T$ is the trace-reversed version of the total stress-energy tensor $T_{\mu\nu} = T^{\rm mat}_{\mu\nu} + T^{(\vartheta)}_{\mu\nu}$, $\Box_\eta = \eta^{\mu\nu}\partial_\mu\partial_\nu$ is the flat-space D'Alembertian and $\bar{\epsilon}^{\mu\nu\alpha\beta}$ is the flat-space Levi-Civita tensor density (not to be confused with $\epsilon$). We work in Lorenz gauge, $\left(h^{\mu\nu} - \frac{1}{2}h\bar{g}^{\mu\nu}\right)_{|\nu} = 0$, where $h = \bar{g}^{\mu\nu}h_{\mu\nu}$ is the trace of $h_{\mu\nu}$ with respect to $\bar{g}_{\mu\nu}$, and we restrict the gauge further to a transverse-traceless one by $h = 0$.

The GW generation formula are then identical to the GR ones to leading order in $\epsilon$, as one can see in (5). This is because the CS correction is proportional both to the scalar field and the metric perturbation, where the former is sourced by the Pontryagin density [see (6)], which itself is of $\mathcal{O}(\epsilon^2)$ in the radiation zone. One can make this argument more formally accurate by decomposing the scalar field itself into a background plus a perturbation and then performing a multiple-scale perturbative expansion, but we relegate such details to [1]. Furthermore, one can show that there are only two independent GW polarizations in dynamical CS gravity, as is the case in GR, in spite of the extra scalar field present in the theory. This can be done by studying the geodesic deviation equation or plane wave solutions, as found in [1, 6].

## 3. EMRIs in CS Gravity

We now consider the generation of GWs by EMRIs in dynamical CS gravity. We shall approximate these waves in the so-called semi-relativistic approximation [18], where the motion is assumed geodesic and GWs are assumed to propagate in flat spacetime. This simplifying assumptions have recently been shown to be accurate relative to more precise black-hole perturbation theory calculations [19]. We begin with a description of the background upon which geodesics are evolved. We then show that point-particle orbits can truly be described with geodesics and explicitly derive the latter. We end by presenting the GW generation formula in the semi-relativistic approximation.

### 3.1. Background Geometry

Via the small-coupling approximation, complimented by a slow-rotation approximation, one can solve for the exterior gravitational field of a rotating BH in dynamical CS modified gravity.

The line element is given by $ds^2 = ds^2_{\rm Kerr} + 5\xi a/(4r^4)\left[1 + 12M/(7r) + 27M^2/(10r^2)\right]\sin^2\theta dt d\phi$, where $ds^2_{\rm Kerr}$ is the line element for the Kerr metric and we have used Boyer-Lindquist type coordinates [5].

The CS modified metric remains stationary and axisymmetric, allowing for the definition of multipole moments appropriately. The multipolar structure of the modified metric remains completely determined by only two moments (no-hair or two-hair theorem): the mass monopole and the current dipole. The relation, however, between these two moments and higher order ones is modified from the GR expectation at multipole $\ell > 4$. In spite of the CS scalar field, the dynamical theory satisfies the no-hair theorem, as the scalar field is fully determined by the geometry.

### 3.2. Motion of Massive Bodies

One can show that test-bodies follow geodesics in CS modified gravity [1]. This is because the divergence of the field equations (2) leads to (3) for the non-matter degrees of freedom, forcing the divergence of the matter stress energy tensor to vanish. This then proves dynamical CS modified gravity satisfies the weak-equivalence principle, thus rendering it a metric theory as we anticipated in the Introduction.

The motion of small compact object with an extreme-mass ratio can then be modeled, to zeroth order, by geodesics in the modified Kerr background of the previous section. As the background remains stationary and axisymmetric, there still exist timelike and azimuthal, commuting Killing vectors that define a conserved energy $E$ and angular momentum $L$ per unit mass. Moreover, the modified background also possesses an additional constant of the motion (the Carter constant, $Q$) associated with a Killing tensor. Using these constants and the orthonormality condition for timelike geodesics, we can derive the geodesic equations $\dot{x}^\mu = \dot{x}^\mu_{\rm Kerr} + \delta x^\mu_{\rm CS}$, where an overhead dot stands for partial time derivative and $x^\mu = [t, r, \theta, \phi]$ are Boyer-Lindquist coordinates (where the polar angle $\theta$ is not to be confused with the CS scalar field $\vartheta$). The correction factor $\delta x^\mu_{\rm CS} = [L, 2EL, 0, -E]\delta g^{\rm CS}_\phi$, where $\delta g^{\rm CS}_\phi = \xi a/(112 r^8 f)(70r^2 + 120r\,M + 189M^2)$ and $f := 1 - 2M/r$.

The geodesic equations allow one to compute some physical observables that waveforms are sensitive to. One such observable is the innermost-stable circular orbit (ISCO) location, which is CS shifted by [5]: $R_{\rm ISCO} = 6M \mp 4\sqrt{6}a/3 - 7a^2/(18M) \pm 77\sqrt{6}a\xi/(5184M^4)$, where the upper (lower) signs correspond to co-rotating (counter-rotating) geodesics. Notice that the CS correction works *against* the spin effects. Other observables of interest are the fundamental orbital frequencies associated with orbital motion [20, 21]. In CS theory, the temporal, radial and azimuthal frequencies are modified in an unilluminating way that we avoid here [1], if one employs the time coordinate introduced in [22]. Therefore, CS geodesics correspond to GR geodesics with different fundamental frequencies, which maps to different system parameters.

### 3.3. GWs for EMRIs

The first simplification of the semi-relativistic approximation (that of modeling orbital motion as geodesic) has already been discussed in the previous section, and so we here concentrate on the second semi-relativistic simplification: GWs propagating on flat spacetime. We implement this approximation by modeling the solution of the linearized GW generation formula in 5 via a multipolar expansion (see eg. [23]). In particular, we employ a quadrupole-octopole approximation, where the plus- and cross-polarizations of the waveforms are given by

$$h_{+,\times} = \frac{1}{r}\varepsilon^{ij}_{+,\times}\left[\ddot{I}_{ij} - 2n^\mu \ddot{S}_{\mu ij} + n^\mu \dddot{I}_{\mu ij}\right], \tag{7}$$

where $r$ is the flat-space distance from the source to the observer, $n^\mu = [1, x^i/r]$, $\epsilon^{ij}_{+,\times}$ are polarization tensors, and $I_{ij}$, $S_{ijk}$ and $I_{ijk}$ are the symmetric and trace-free projections of the

Journal of Physics: Conference Series **228** (2010) 012051 doi:10.1088/1742-6596/228/1/012051

mass quadrupole, current octopole and mass octopole respectively, defined as integrals of the product of point-particle stress-energy tensor with the point-particle trajectories (see eg. [24] for explicit expressions).

An important detail in the above implementation is the relation between the coordinates used in the geodesic evolution (Boyer-Lindquist) and those employed in the metric reconstruction (Cartesian). We here choose to identify Boyer-Lindquist coordinates with flat-space spherical coordinates, such that $(x, y, z) = r(\sin\theta\cos\phi, \sin\theta\sin\phi, \cos\theta)$. Another important detail is that the above approximation is truly a slow-motion one, which implies waveforms are accurate provided the pericenter orbital velocity is much smaller than the speed of light. In spite of this, the semi-relativistic approximation does capture the correct *qualitative* behaviour of EMRI orbits and waveforms.

## 4. Testing CS theory with LISA and Radiation-Reaction

We here describe some of the differences between a GR and a CS EMRI waveform. For this, we shall concentrate on the following test orbits: (A) $(a, \xi, r_{\rm peri}, e, \theta_{\rm inc}) = (0.1\,M, 0.1M^4, 12\,M, 0.2, 0.1)$, (B) $(a, \xi, r_{\rm peri}, e, \theta_{\rm inc}) = (0.2\,M, 0.2M^4, 8\,M, 0.4, 0.2)$, (C) $(a, \xi, r_{\rm peri}, e, \theta_{\rm inc}) = (0.4\,M, 0.4M^4, 6\,M, 0.6, 0.3)$, where $r_{\rm peri}$ stands for pericenter distance, $e$ for orbital eccentricity and $\theta_{\rm inc}$ is related to the inclination angle. Waveforms are measured by $z$-axis observers at a distance of 8 kpc and test orbits have been ordered from least to most relativistic. All orbits assume the central BH has $M = M_\bullet$, where $M_\bullet = 4.5 \times 10^6 M_\odot$ is approximately the mass of the BH at the center of the Milky Way, while the SCO has $m = 35M_\odot$, which leads to a mass ratio $\mu \sim 7.8 \times 10^{-6}$. We conclude with a discussion of radiation-reaction in CS theory.

### *4.1. Trajectory and Waveform Dephasing*

The trajectories are obtained by solving the geodesic equations discussed above for a total time of $T = 5 \times 10^5 M$, obtaining on the order of $10^3$-$10^4$ cycles. (see eg. [1] for details). Geodesics are initialized with the same orbital data when performing GR and CS evolutions. That is, given some set of conserved quantities $(E, L, Q)$, one can derive three initial orbital parameters $(p, e, \theta_{\rm inc})$, but these will differ in GR and CS due to the latter's geodesic corrections. We here consider orbits with the *same* orbital parameters, instead of the same constants of motion.

The orbits chosen present rather generic behaviour with a stage of zoom-whirl, where the particle whirls violently for several cycles close to pericenter and then zooms out to large radius only to return again close to pericenter, during all of which there is both in-plane and out-of-the-plane precession. The left panel of 1 shows the projection of the orbital trajectory onto the $x$-$y$ plane (orbit C) assuming a CS modified Kerr background BH (black line) and a Kerr background (light gray line). We present only the last $17500M$ of geodesic evolution, during which we can clearly observe that the trajectories have dephased significantly . We find that orbit $C$ presents the biggest dephasing, followed by $B$ and then $A$.

The right-panel of 1 shows the difference in the plus-polarized waveforms in CS gravity and GR. As a reference, the maximum magnitude of the GR GW polarization $h_+$ is approximately $|h_+| < 8 \times 10^{-17}$ (orbit $A$), $|h_+| < 1.25 \times 10^{-16}$ (orbit $B$), and $|h_+| < 2.25 \times 10^{-16}$ (orbit $C$). Observe that by the end of the simulation ($\sim$ 128 days of data using $M = M_\bullet$), orbit $A$ waveforms have dephased by 0.3% , while orbit $B$ waveforms have dephased by 16%, and orbit $C$ waveforms by 90% relative to the maximum amplitude of the respective GR waveforms. Similar behaviour is observed for the other polarization.

### *4.2. Radiation-Reaction in Dynamical CS Gravity*

Radiation-reaction can be incorporated via a radiative, adiabatic approximation [22, 25, 26], where the constants of the motion are evolved using balance laws that relate orbital energy loss

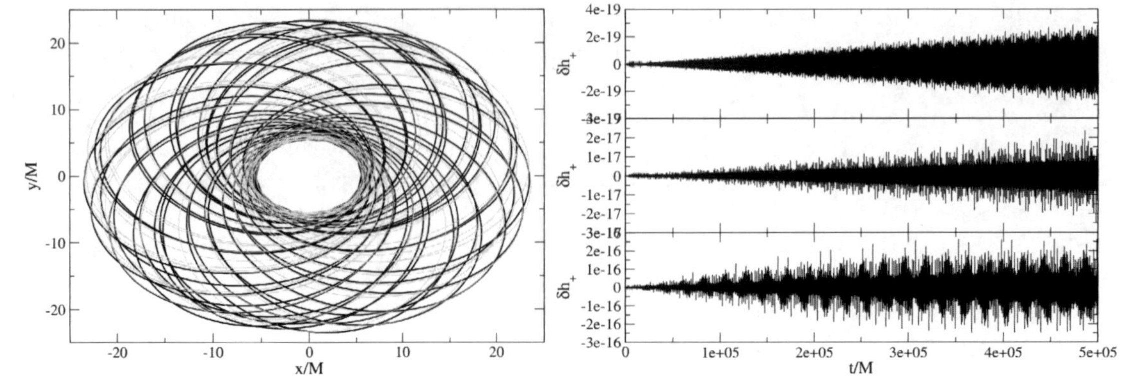


**Figure 1.** Left: Two-dimensional projection onto the $x$-$y$ plane of orbit $C$ in a Kerr background (light gray) and in the CS corrected Kerr metric (black). Observe that a large dephasing has built up by the end of the evolution. Right: Plus-polarized GW dephasing as a function of time. The top panel shows results for orbit $A$, the middle one for $B$ and the bottom one for $C$. Observe that after 2/5 of the evolution, the waveform has dephased significantly.

to GW energy emission. Such an approximation is valid provided the change in the constant of the motion occurs on a much larger timescale relative to the orbital one.

The balance laws can be obtained by studying the effective stress-energy tensor associated with GWs in the framework of the short-wavelength approximation [27, 28]. We thus decompose both the geometry and the CS scalar field into a slowly- or non-oscillating background and a rapidly-oscillating perturbation. The linear part of the modified field equations describe the evolution of the perturbation, while the averaged second-order part describes how the background is modified by GWs. This later part serves as an effective GW stress-energy tensor, $T_{\mu\nu}^{\rm GW}$ (the Isaacson tensor),

$$T_{\mu\nu}^{\rm GW} = -2\kappa \Big\{ < {}^{(2)}R_{\mu\nu}[h] > - \frac{1}{2}\bar{g}_{\mu\nu} < {}^{(2)}R[h] > \\ + \frac{\alpha}{\kappa}\left( < {}^{(2)}C_{\mu\nu}[\bar{\vartheta}, h] > + < {}^{(1)}C_{\mu\nu}[\tilde{\vartheta}, h] > \right) \Big\} \tag{8}$$

where the superscript preceding any given quantity stands for the perturbative order of that quantity with respect to metric perturbations. Restricting attention to the transverse-traceless gauge and using properties of the averaging procedure, one can show that all terms associated with the CS correction in 8 identically vanish, yielding the same expressions for the Isaacson tensor as in GR.

The above result proofs that orbital backreaction in dynamical CS modified gravity is *similar* but not identical to that expected in GR. First, in addition to GW emission, orbital energy is also lost via scalar field emission, although the latter is a subleading effect. Second, even if one neglects such scalar field emission, GW emission depends on time-derivatives of the waveforms, which we showed in the previous subsection are different in CS gravity.

### *4.3. A possible LISA Implementation of a CS Test*

What is the effect of the CS correction in LISA data analysis? As shown through the semi-relativistic waveforms, GR and CS waveforms dephase given sufficient time, but such dephasing will not prevent detection. Instead such dephasing will only affect parameter estimation. That is, the CS correction modifies the fundamental geodesic frequencies, and thus, a CS geodesic is equivalent to a GR geodesic with different orbital parameters.

How big should the CS coupling parameter be to induce a sufficiently large dephasing that would contaminate parameter estimation? 2 shows the average of a normalized dephasing measure as a function of time for orbit $C$ with different $\xi$ (see caption). This dephasing measure

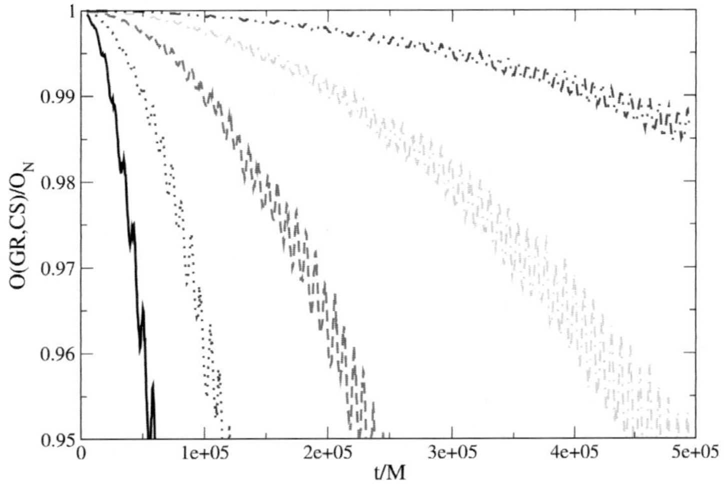


**Figure 2.** Normalized dephasing measure $\mathcal{O}(\textsc{gr},\textsc{cs})$ as a function of time for orbit $C$ with: $\xi = 0.4M^4$ (solid black), $\xi = 0.2M^4$ (dotted blue), $\xi = 0.1M^4$ (dashed red), $\xi = 0.05M^4$ (dot-dashed green), and $\xi = 0.025M^4$ (dot-dot-dashed violet). Observe that a significant dephasing is found even for $\xi = 10^{-2}$.

is nothing but the integrand of the overlap with averaged beam-pattern functions in the time-domain: $\bigcirc(\textsc{gr},\textsc{cs}) := [h_+^{\rm GR} h_+^{\rm CS} + h_\times^{\rm GR} h_\times^{\rm CS}]$ normalized by $\bigcirc_N(\textsc{gr},\textsc{cs}) := \sqrt{\bigcirc(\textsc{gr},\textsc{gr})\ \bigcirc\ (\textsc{cs},\textsc{cs})}$. This figures shows that the CS correction could lead to a significant dephasing, and thus bias in parameter estimation, even for a CS parameter of $\mathcal{O}(10^{-2})$ after only 4 months of data.

The inclusion of radiation-reaction is crucial for breaking the degeneracy between the CS correction $\xi$ and the system parameters, *i.e.* to distinguish between a GR and a CS geodesic. One expects that radiation-reaction will not only break this degeneracy, but also lead to larger deviation from the GR expectation. A rough estimate of the accuracy to which CS gravity could be constrained via a LISA observation is $\xi^{1/4} \lesssim 10^5\,{\rm km}\,\left(\delta/10^{-6}\right)^{1/4}(M/M_\bullet)$, where $\delta$ is the accuracy to which $\xi$ can be measured, which depends on the integration time, the signal-to-noise ratio, the type of orbit considered and how much radiation-reaction affects the orbit. Notice that intermediate-mass ratio inspirals (IMRIs) are favored over EMRIs. This result is to be compared with the binary pulsar constrained $\xi^{1/4} \lesssim 10^4$ km [5]. We then see that an IMRI with $M = 10^3 M_\odot$ could place a constraint two-orders of magnitude more stringent than the binary pulsar one. Moreover, a GW test can constrain the dynamical behaviour of the theory in the neighbourhood of BHs, which is simply not possible with binary pulsar observations.

## 5. Conclusions

Dynamical CS gravity is a viable theory that can only be constrained dynamically via GW observations. We have constructed geodesics in the modified theory and derived radiation-reaction formula that should allow for the construction of CS-modified EMRI waveforms. Such waveforms could be used in LISA data analysis pipelines to test the theory at least two-orders of magnitude better than current binary pulsar observations.

## Acknowledgments

We thank Leor Barack and Jonathan Gair for comments and suggestions. We acknowledge support from NSF grant PHY-0745779, the Ramón y Cajal Programme (MEC, Spain), by a Marie Curie International Reintegration Grant (MIRG-CT-2007-205005/PHY) within the 7th European Community Framework Programme, and from the contract ESP2007-61712 of MEC, Spain.

## References

[1] Sopuerta C F and Yunes N 2009 *Phys. Rev.* **D80** 064006 (*Preprint* 0904.4501)
[2] Yunes N and Pretorius F 2009 *to be submitted*
[3] Misner C W, Thorne K and Wheeler J A 1973 *Gravitation* (San Francisco: W. H. Freeman & Co.)
[4] Alexander S and Yunes N 2009 *Phys. Rept.* **480** 1–55 (*Preprint* 0907.2562)
[5] Yunes N and Pretorius F 2009 *Phys. Rev.* **D79** 084043 (*Preprint* 0902.4669)
[6] Grumiller D, Mann R B and McNees R 2008 *Phys. Rev.* **D78** 081502 (*Preprint* 0803.1485)
[7] Polchinski J 1998 *String theory. Vol. 2: Superstring theory and beyond* (Cambridge, UK: Cambridge University Press)
[8] Alexander S H S and Gates S James J 2006 *JCAP* **0606** 018 (*Preprint* hep-th/0409014)
[9] Taveras V and Yunes N 2008 *Phys. Rev.* **D78** 064070 (*Preprint* 0807.2652)
[10] Mercuri S and Taveras V 2009 (*Preprint* 0903.4407)
[11] Weinberg S 2008 *Phys. Rev.* **D77** 123541 (*Preprint* 0804.4291)
[12] Grumiller D and Yunes N 2008 *Phys. Rev.* **D77** 044015 (*Preprint* 0711.1868)
[13] Yunes N and Sopuerta C F 2008 *Phys. Rev.* **D77** 064007 (*Preprint* 0712.1028)
[14] Yunes N and Spergel D N 2009 *Phys. Rev.* **D80** 042004 (*Preprint* 0810.5541)
[15] Alexander S, Finn L S and Yunes N 2008 *Phys. Rev.* **D78** 066005 (*Preprint* 0712.2542)
[16] Yunes N and Finn L S 2009 *J. Phys. Conf. Ser.* **154** 012041 (*Preprint* 0811.0181)
[17] Cardoso V and Gualtieri L 2009 (*Preprint* 0907.5008)
[18] Ruffini R and Sasaki M 1981 *Prog. Theor. Phys.* **66** 1627–1638
[19] Babak S, Fang H, Gair J R, Glampedakis K and Hughes S A 2007 *Phys. Rev.* **D75** 024005 (*Preprint* gr-qc/0607007)
[20] Schmidt W 2002 *Class. Quant. Grav.* **19** 2743 (*Preprint* gr-qc/0202090)
[21] Drasco S and Hughes S A 2004 *Phys. Rev.* **D69** 044015 (*Preprint* astro-ph/0308479)
[22] Mino Y 2003 *Phys. Rev.* **D67** 084027 (*Preprint* gr-qc/0302075)
[23] Thorne K S 1980 *Rev. Mod. Phys.* **52** 299–339
[24] Yunes N, Sopuerta C F, Rubbo L J and Holley-Bockelmann K 2008 *Astrophys. J.* **675** 604–613 (*Preprint* 0704.2612)
[25] Sago N, Tanaka T, Hikida W and Nakano H 2005 *Prog. Theor. Phys.* **114** 509–514 (*Preprint* gr-qc/0506092)
[26] Hughes S A, Drasco S, Flanagan E E and Franklin J 2005 *Phys. Rev. Lett.* **94** 221101 (*Preprint* gr-qc/0504015)
[27] Isaacson R A 1968 *Phys. Rev.* **166** 1263–1271
[28] Isaacson R A 1968 *Phys. Rev.* **166** 1272–1279

# Relativistic orbits and Gravitational Waves from gravitomagnetic corrections

**F Garufi , S Capozziello, M De Laurentis, L Forte and L Milano**

Dipartimento di Scienze Fisiche, Università di Napoli Federico II and INFN Sez. di Napoli, Complesso Universitario di Monte S. Angelo, via Cintia, I-80126 Napoli, Italy

E-mail: Fabio.Garufi@na.infn.it

**Abstract.** Corrections to the relativistic theory of orbits are discussed considering higher order approximations induced by gravitomagnetic effects. Beside the standard periastron effect of General Relativity (GR), a new nutation effect was found due to the $c^{-3}$ orbital correction. According to the presence of that new nutation effect we studied, via the quadrupole approximation, the gravitational waveforms emitted by a compact object (neutron star (NS) or black hole (BH)) orbiting around a massive black hole (MBH). To obtain the emitted gravitational wave (GW) amplitudes, a numerical solution of the equations is given in different mass ratios and initial conditions. We conclude that the effects we studied could be of interest for the future space laser interferometric GW antenna LISA.

## 1. Introduction

The magnetic field is produced by the motion of electric-charge, i.e. the electric current. The analogy with gravity consists in the fact that a mass-current can produce a "gravitomagnetic" (GM) field. The formal analogy between electromagnetic and gravitational fields was explored in the framework of GR [1], it was also shown that a rotating mass generates a gravitomagnetic field. We want to study how the relativistic theory of orbits and the production of GW is affected by GM corrections. The corrections, which are off-diagonal terms in the metric, can be seen as further powers in the expansion in $c^{-1}$ (up to $c^{-3}$). Nevertheless, the effects on the orbit behaviour involves not only the precession at peri-astron but also nutation corrections. Our approach suggests that, in the weak field approximation, when considering higher order corrections in the equations of motion, the GM effects can be particularly significant. In systems approaching to strong field regimes, these corrections give rise to chaotic behaviors in the transients dividing stable from unstable orbits [2]. In general, such contributions are discarded since they are assumed to be too small but they have to be taken into account as soon as the $v/c$ ratio is significant. It is important to stress that we are not investigating simply frame-dragging effects in the Post-Newtonian (PN) approach [3] but the GM effects emerging from the weak-field approximation. In this sense, as soon as higher order terms in $v/c$ come out, their effect could give rise to, in principle, observable signatures both in orbital and in GW emission (see [2]).

© 2010 IOP Publishing Ltd

## 2. Gravitomagnetic effects: the weak field limit

In the framework of the linearized weak-field and slow motion approximation of GR, the ensemble of the so-called GM effects are induced by the off-diagonal components of the space-time metric tensor that are proportional to the components of the matter current density of the source. It is possible to take into account two types of mass currents in gravity. The former is induced by the matter source rotation around its center of mass: it generates the intrinsic GM field which is closely related to the angular momentum (spin) of the rotating body. The latter is due to the translational motion of the source: it is responsible of the extrinsic GM field. Starting from the Einstein field equations in the weak field approximation, one obtains the GM equations and then the corrections in the metric [4].

Starting from the weak field approximation of the gravitational field:

$$g_{\mu\nu}(x) = \eta_{\mu\nu} + h_{\mu\nu}(x) \quad |h_{\mu\nu}(x)| \ll 1$$

where $\eta_{\mu\nu}$ is the Minkowski tensor and $h_{\mu\nu}$ a small perturbation[1], in the approximation $|\mathbf{v}| \ll c$ and by using the Einstein equations and the harmonic gauge condition, one finds that, up to leading order in $v/c$, the metric is determined by the usual gravitational potential $\Phi$ and by a vector potential $V^l$,

$$V^l = -G\int \frac{\rho v^l}{|\mathbf{x}-\mathbf{x}'|}d^3x' \,. \tag{1}$$

depending on the orbital velocities $v^l$ of the bodies. The metric results to be:

$$ds^2 = \left(1+\frac{2\Phi}{c^2}\right)c^2dt^2 - \frac{8\delta_{lj}V^l}{c^3}cdtdx^j - \left(1-\frac{2\Phi}{c^2}\right)\delta_{lj}dx^ldx^j \tag{2}$$

The difference with respect to the standard PN approximation is that the time-spatial terms are generally discarded, while here, as we will see, they give rise to observable effects. This fact could be immediatly seen on the geodesic motion. By calculating the affine connection related to the metric (2), the geodesic equations are obtained:

$$\ddot{x}^\alpha + \Gamma^\alpha_{\mu\nu}\dot{x}^\mu\dot{x}^\nu = 0 \tag{3}$$

where the dot indicate differentiation with respect to the affine parameter. In order to put in evidence the GM contributions, let us explicitly calculate the Christoffel symbols at lower orders. By some straightforward calculations, one gets

$$\Gamma^0_{00} = 0 \qquad \Gamma^0_{0j} = \frac{1}{c^2}\frac{\partial\Phi}{\partial x^j} \qquad \Gamma^0_{ij} = -\frac{2}{c^3}\left(\frac{\partial V^i}{\partial x^j} + \frac{\partial V^j}{\partial x^i}\right)$$
$$\Gamma^k_{00} = \frac{1}{c^2}\frac{\partial\Phi}{\partial x^k} \quad \Gamma^k_{0j} = \frac{2}{c^3}\left(\frac{\partial V^k}{\partial x^j} - \frac{\partial V^j}{\partial x^k}\right) \quad \Gamma^k_{ij} = -\frac{1}{c^2}\left(\frac{\partial\Phi}{\partial x^j}\delta^k_i + \frac{\partial\Phi}{\partial x^i}\delta^k_j - \frac{\partial\Phi}{\partial x^k}\delta_{ij}\right) \tag{4}$$

Let us stress that we are considering the GM effects on the orbital motion, so the Christoffel symbols result accordingly corrected. Usually, in the standard weak field approximation, vector potentials are not considered. In our approximation, we retain terms up to order $\Phi/c^2$ and $V^j/c^3$. Our aim is to show that, in several cases like in tight binary stars, it is not correct to discard higher order terms in $v/c$ since physically interesting effects could come out.

Considering only the spatial components of the geodesic equations:

$$\frac{d^2x^k}{d\sigma^2} + \frac{1}{c^2}\frac{\partial\Phi}{\partial x^k}\left(c\frac{dt}{d\sigma}\right)^2 + \frac{1}{c^2}\frac{\partial\Phi}{\partial x^k}\delta_{ij}\frac{dx^i}{d\sigma}\frac{dx^j}{d\sigma} - \frac{2}{c^2}\frac{\partial\Phi}{\partial x^l}\frac{dx^l}{d\sigma}\frac{dx^k}{d\sigma} +$$
$$+\frac{4}{c^3}\left(\frac{\partial V^k}{\partial x^j} - \delta_{jm}\frac{\partial V^m}{\partial x^k}\right)c\frac{dt}{d\sigma}\frac{dx^j}{d\sigma} = 0 \tag{5}$$

[1] Greek indices running from 0 to 3 and latin from 1 to 3

we obtain the orbit equations. Calling $dl_{euclid} = \delta_{ij}dx^i dx^j$, and $e^k = \frac{dx^k}{dl_{euclid}}$ we have, in vector form,

$$\frac{d\mathbf{e}}{dl_{euclid}} = -\frac{2}{c^2}\left[\nabla\Phi - \mathbf{e}(\mathbf{e}\cdot\nabla\Phi)\right] + \frac{4}{c^3}\left[\mathbf{e}\wedge(\nabla\wedge\mathbf{V})\right] \tag{6}$$

The GM term is the second one in Eq.(6) and it is usually discarded since considered not relevant. This is not true if $v/c$ is non negligible as in the cases of tight binary systems or point masses approaching to BH.

We can then write the Lagrangian and derive the orbital equations of motion by means of the Euler-Lagrange equations (see [2]):

$$\ddot{r} = \frac{1}{cr\left(rc^2+2GM\right)}\left\{c\left(rc^2+GM\right)\left(\dot{\theta}^2+\sin^2\theta\dot{\phi}^2\right)r^2+ \right.$$
$$\left. -4GM\dot{t}\left[\cos\theta(\cos\phi+\sin\phi)-\sin\theta)\dot{\theta}+\sin\theta(\cos\phi-\sin\phi)\dot{\phi}\right]r+cGM\dot{r}^2-cGM\dot{t}^2\right\} \tag{7}$$

$$\ddot{\phi} = -\frac{2}{r^2\left(rc^3+2GMc\right)}\left\{c\cot\theta\left(rc^2+2GM\right)\dot{\theta}\dot{\phi}r^2+ \right.$$
$$\left. +\dot{r}\left[2GM\csc\theta(\sin\phi-\cos\phi)\dot{t}+cr\left(rc^2+GM\right)\dot{\phi}\right]\right\} \tag{8}$$

$$\ddot{\theta} = \frac{1}{r^2\left(rc^3+2GMc\right)}\left\{c\cos\theta r^2\left(rc^2+2GM\right)\sin\theta\dot{\phi}^2+ \right.$$
$$\left. +\dot{r}\left[4GM(\cos\theta(\cos\phi+\sin\phi)-\sin\theta)\dot{t}-2cr\left(rc^2+GM\right)\dot{\theta}\right]\right\} \tag{9}$$

and, with the condition $\mathcal{L} = 1\,(m = 1, c = 1)$, find an expression for $\dot{r}$ that is a constrain equation related to the energy.

Beside the standard periastron precession of GR, if $\dot{r} \neq 0$, even with initial conditions $\dot{\theta} = 0$ and $\theta = \pi/2$, a nutation effect is induced by gravitomagnetism and stability depends on it. Our aim is to study how GM effects modify the orbital shapes and what are the relevant parameters to determine the stability of the problem.

## 3. Numerical approach to the solution

The solution of the system of differential equations (ODE) of motion (7-9) presents some difficulties since the equations are stiff. We have found solutions by using a Stiffness Switching Method to provide an automatic mean of switching between a non-stiff and a stiff solver coupled with a more conventional explicit Runge-Kutta method for the non-stiff part of differential equations. Time series of both $\dot{r}(t)$ and $\ddot{r}(t)$ together with the phase portrait $r(t), \dot{r}(t)$ are shown assuming as initial angular precession and nutation velocities values that are integer fractions of the initial radial velocity. In Fig. 1 we show as example the one with: $\dot{\varphi} = \frac{1}{10}\dot{r}$ and $\dot{\theta} = \frac{1}{10}\dot{\varphi}$

### *3.1. Orbits with gravitomagnetic corrections*

In this section we show some examples of orbits. In Fig. 2 we show the orbits with (red-line) and without GM correction (black-line). Finally in Fig. 3 the phase portrait with (red-line) and without (blue-line) GM orbital correction is shown. The example we are showing was obtained by solving the system with the parameters and initial conditions: $\mu \approx 1M_{\odot}$, $E = 0.95$, $\phi_0 = 0$, $\theta_0 = \frac{\pi}{2}$, $\dot{\theta}_0 = \frac{1}{10}\dot{\phi}_0$, $\dot{\phi}_0 = -\frac{1}{10}\dot{r}_0$ and $\dot{r}_0 = -\frac{1}{10}$ and $r_0 = 20\mu$.

## 4. Gravitational waves in the quadrupole approximation

Considering the orbital equations (see [2]), we know that direct signatures of gravitational radiation are its amplitude and waveform [4]. These observations give information on the nature

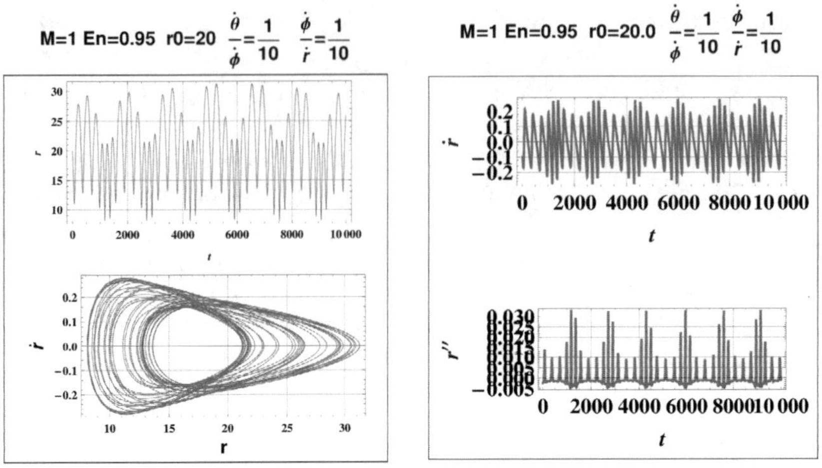


**Figure 1.** Plots of: $r(t)$ (upper left), phase portrait of $r(t)$ versus $\dot{r}(t)$ (bottom left), $\dot{r}(t)$ (upper right) and $\ddot{r}(t)$ (bottom right) system of reduced mass $1M_{\odot}$. The parameters and initial conditions are: $\mu \approx 1M_{\odot}$, $E = 0.95$,$\phi_0 = 0$, $\theta_0 = \frac{\pi}{2}$, $\dot{\theta}_0 = \frac{1}{10}\dot{\phi}_0$, $\dot{\phi}_0 = -\frac{1}{10}\dot{r}_0$ and $\dot{r}_0 = -\frac{1}{100}$ and $r_0 = 20\mu$. The stiffness is evident from the trend of $\dot{r}(t)$ and $\ddot{r}(t)$

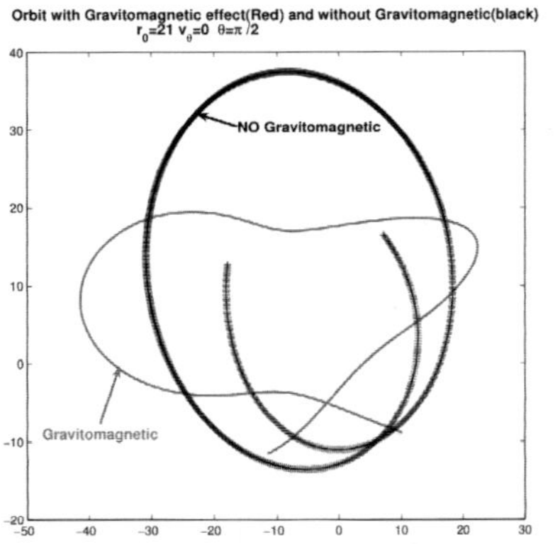


**Figure 2.** The orbit with GM correction (red-line) and without GM correction (black-line).

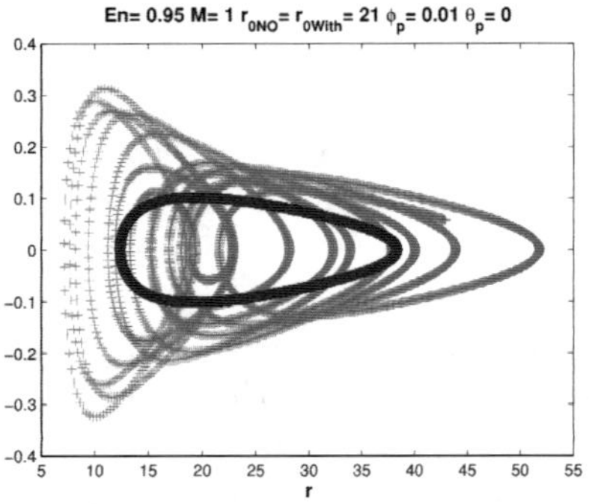


**Figure 3.** Phase portrait with (red-line) and without (blue-line) GM orbital correction for: $\mu \approx 1M_{\odot}$, $E = 0.95$, $\phi_0 = 0$, $\theta_0 = \pi/2$, $\dot{\theta}_0 = (1/10)\dot{\phi}_0$, $\dot{\phi}_0 = -(1/10)\dot{r}_0$ and $\dot{r}_0 = -1/10$ and $r_0 = 20\mu$.

of the GW source, on the propagating medium, and , in principle, on the gravitational theory producing such a radiation. The amplitude of GWs is, as usual, evaluated by

$$h^{jk}(t, R) = \frac{2G}{Rc^4}\ddot{Q}^{jk} \ , \tag{10}$$

being $R$ the distance between the source and the observer, $\{j, k\} = 1, 2$, and $Q_{ij}$ the quadrupole mass tensor

$$Q_{ij} = \sum_a m_a(3x_a^i x_a^j - \delta_{ij} r_a^2) \ . \tag{11}$$

being $G$ the Newton constant, $r_a$ the modulus of the vector radius of the $a - th$ particle and the sum running over all masses $m_a$ in the system. We computed the amplitude components with

GM corrections in geometrized units ([5]) and performed the numerical simulations in two cases: a NS of $1.4M_\odot$ orbiting around a Super-MBH (e.g. Sagittarius A* $10^6 M_\odot$) and a BH of $10M_\odot$ orbiting around a Super-MBH. Computations are performed with orbital radii measured in units of the system reduced mass $\mu$. Initial distances are sampled to show orbits from high eccentricity up to circularity ($e = \frac{r_{max}-r_{min}}{r_{max}+r_{min}}$). In Fig. 4-6 we show respectively the field velocities of the

**Table 1.** Data for GW for a NS of $1.4M_\odot$ orbiting around a Super-MBH

| $\frac{r_0}{\mu}$ | $e$ | $f(mHz)$ | $h$ | $h_+$ | $h_\times$ |
|---|---|---|---|---|---|
| 20 | 0.91 | $7.7 \cdot 10^{-2}$ | $2.0 \cdot 10^{-22}$ | $5.1 \cdot 10^{-23}$ | $5.1 \cdot 10^{-22}$ |
| 200 | 0.79 | $1.1 \cdot 10^{-1}$ | $1.2 \cdot 10^{-20}$ | $2.2 \cdot 10^{-21}$ | $3.1 \cdot 10^{-20}$ |
| 500 | 0.64 | $1.4 \cdot 10^{-1}$ | $6.9 \cdot 10^{-20}$ | $8.7 \cdot 10^{-21}$ | $1.7 \cdot 10^{-19}$ |
| 1000 | 0.44 | $1.9 \cdot 10^{-1}$ | $2.6 \cdot 10^{-19}$ | $6.4 \cdot 10^{-20}$ | $6.4 \cdot 10^{-19}$ |
| 1500 | 0.28 | $2.3 \cdot 10^{-1}$ | $4.8 \cdot 10^{-19}$ | $3.6 \cdot 10^{-20}$ | $1.2 \cdot 10^{-18}$ |
| 2000 | 0.14 | $2.7 \cdot 10^{-1}$ | $5.9 \cdot 10^{-19}$ | $4.9 \cdot 10^{-20}$ | $1.3 \cdot 10^{-18}$ |
| 2500 | 0.01 | $3.1 \cdot 10^{-1}$ | $5.9 \cdot 10^{-19}$ | $1.7 \cdot 10^{-20}$ | $9.2 \cdot 10^{-19}$ |

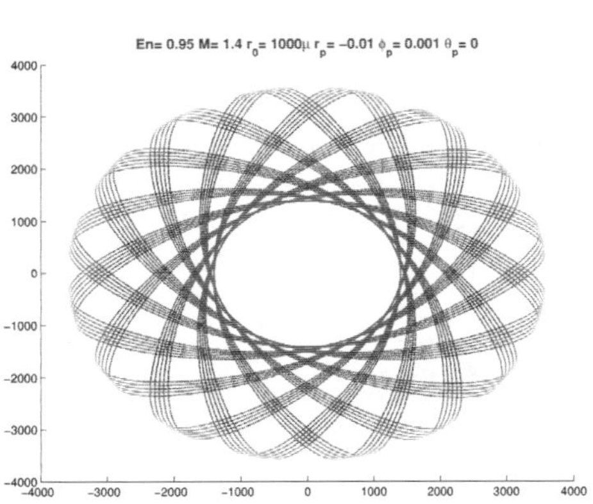


**Figure 4.** Plot of field velocities of the orbits.

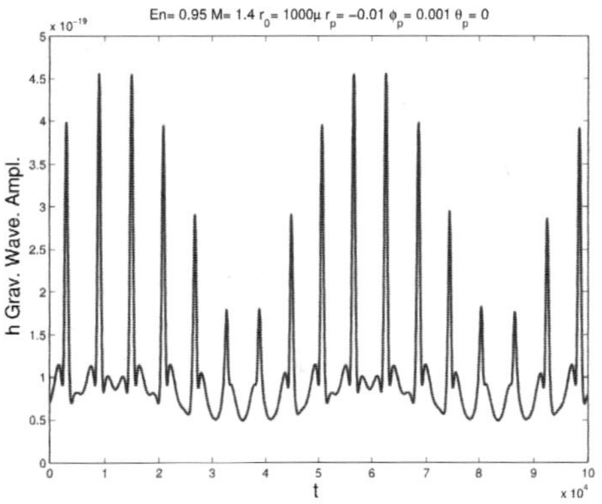


**Figure 5.** Total gravitational emission waveform $h$ for a neutron star of $1.4M_\odot$

orbits along the axes of maximum covariances, the total gravitational emission waveform $h$ and the gravitational waveform polarizations $h_+$ and $h_\times$ for a NS of $1.4M_\odot$. The waveform were computed for the Earth-distance from Sagittarius A (central Galactic Black Hole). We obtained the numerical examples solving the ODE system for the following parameters and initial conditions: $\mu \approx 1.4M_\odot$, $E = 0.95$,$\phi_0 = 0$, $\theta_0 = \frac{\pi}{2}$,$\dot{\theta}_0 = 0$, $\dot{\phi}_0 = -\frac{1}{10}\dot{r}_0$ and $\dot{r}_0 = -\frac{1}{100}$ and $r_0 = (20 \div 1000)\mu$. See also Fig. 4-6 and Table 1. Finally we show in Fig. 7 the plot of the estimated $h$ GW-strain-amplitudes for the considered binary sources at Galactic Center distance. The blue-line is the foreseen LISA sensitivity (one year integration + white dwarf background noise). The red diamonds ($1.4M\odot$) and the green circles ($10M\odot$) are the $h$ values for the systems we have studied.

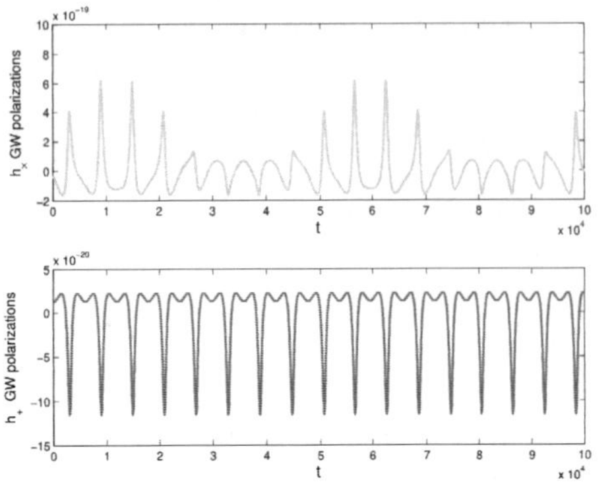


**Figure 6.** The gravitational waveform polarizations $h_+$ and $h_\times$ for a neutron star (NS) of $1.4M_\odot$

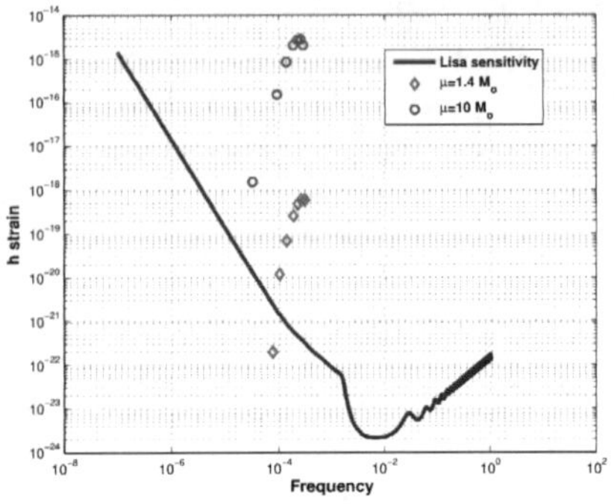


**Figure 7.** Plot of estimated mean values of gravitational emission in terms of strain $h$ for two binary sources from the galactic center with reduced mass ratio $\mu \approx 1.4M_\odot$ (red diamonds) and $\mu \approx 10M_\odot$ (green circles). The blue line is the foreseen LISA sensitivity curve. The waveforms were computed for the Earth-distance to Sagittarius A (central Galactic Black Hole).

## 5. Concluding Remarks

The GM effect could give rise to interesting phenomena in tight binding systems such as binaries of massive objects (neutron stars or black holes). The effects appears particularly interesting if $v/c$ is in the range $(10^{-1} \div 10^{-3})c$ as is, e.g. for objects captured by extremely massive black holes such as those at the Galactic Center. GM orbital corrections, after long integration time, induce precession and nutation effects capable of affecting the stability basin of the orbits. The global structure of such a basin is extremely sensitive to the initial radial velocities and angular velocities, the initial energy and masses which can determine possible transitions to chaotic behavior. In principle, GW emission could present signatures of GM corrections after suitable integration times in particular for the on going LISA space laser interferometric GW antenna.

## References

[1] H Thirring 1918 *Phys. Z.* **19** 204.
[2] S Capozziello M De Laurentis F Garufi L Milano 2009 *Physica Scripta* **79** 025901.
[3] L Blanchet 2006 Living Rev. Relativity 9 , 4
[4] C W Misner K S Thorne J A Wheeler 1973 *Gravitation* (New York: Freeman).
[5] S Capozziello M De Laurentis L Forte F Garufi L Milano 2009 *Preprint* arXiv:0909.0895 [gr-qc] submitted to Physica Scripta.

8th Edoardo Amaldi Conference on Gravitational Waves IOP Publishing
Journal of Physics: Conference Series **228** (2010) 012053 doi:10.1088/1742-6596/228/1/012053

# Secular momentum transport by gravitational waves from spinning compact binaries

**Zoltán Keresztes[1,2⋆], Balázs Mikóczi[3†], László Á. Gergely[1,2,4‡], Mátyás Vasúth[3¶]**

**Abstract.** We present a closed system of coupled first order differential equations governing the secular linear momentum loss of a compact binary due to emitted gravitational waves, with the leading order relativistic and spin-orbit perturbations included. In order to close the system, the secular evolution equations of the linear momentum derived from the dissipative dynamics are supplemented with the secular evolutions of the coupled angular variables, as derived from the conservative dynamics.

[1] Department of Theoretical Physics, University of Szeged, Tisza Lajos krt 84-86, Szeged 6720, Hungary
[2] Department of Experimental Physics, University of Szeged, Dóm tér 9, Szeged 6720, Hungary
[3] KFKI Research Institute for Particle and Nuclear Physics, Budapest 114, P.O.Box 49, H-1525 Hungary
[4] Institute for Advanced Study, Collegium Budapest, Szentháromság u 2, Budapest 1014, Hungary

E-mail: *zkeresztes@titan.physx.u-szeged.hu †mikoczi@rmki.kfki.hu
‡gergely@physx.u-szeged.hu ¶vasuth@rmki.kfki.hu

## 1. Introduction

The inspiral of compact objects in binary systems is driven by the emitted gravitational radiation which carries energy, angular and linear momenta away from the source. Global conservation of these quantities implies a radiative orbital evolution, including a possible recoil of the system. The energy and magnitude of orbital angular momentum of the binary determine the quasi-Keplerian orbit in the post-Newtonian (PN) regime. These quantities are known to high accuracy for binaries on eccentric orbits, with PN, spin-orbit (SO), spin-spin (SS), mass quadrupole - mass dipole (QM) and magnetic dipole - magnetic dipole (DD) coupling terms [1]. Moreover, the orientation of the (quasi-precessing) plan of motion, defined by the direction of the Newtonian orbital angular momentum, is known to high accuracy with all the above-mentioned contributions.

Due to asymmetries in the configuration of the source the radiation is often emitted anisotropically. The linear momentum loss of the binary leads to the recoil of the center of mass in the opposite direction, which in extreme situations results the kick-off of the binary from its host galaxy. This effect, however, vanishes for equal mass binaries due to the fact that the leading order relativistic contribution to the linear momentum loss scales with the mass difference $\delta m = m_1 - m_2$ of the orbiting bodies.

The gravitational recoil of a binary system and the classical linear momentum loss of the final black hole were discussed in [2, 3] with the inclusion of the lowest multipoles needed for

© 2010 IOP Publishing Ltd

the computation of momentum ejection. The first quasi-Keplerian analytic studies were given in [4], where the linear momentum flux of gravitational waves from a binary system of two point masses in Keplerian orbit were calculated. 1 PN corrections to the gravitational recoil were discussed in [5], while 2PN recoil effects for binaries on quasicircular orbits in [6]. Several numerical estimates for the kick velocity of non-spinning binaries were given, e.g. in [7, 8, 9], with the maximum recoil in this case found as $\sim$ 175 km/s [10]. Recently it has been shown [11], that the ringdown phase acts as an anti-kick as compared to the inspiral and plunge, the global result being consistent with the numerical estimates.

The rotation of the components adds however additional structure to the emitted radiation and recoil. Spin contributions to the linear momentum loss are analyzed in [12], and more recently in [13]. The SO contributions scale with the magnitude of spins which could be high for galactic black holes. Numerical analyses indicate that significant gravitational recoil can be obtained in spinning binaries [14, 15, 16, 17, 18] even with a high degree of symmetry in the configuration, i.e. for equal-mass binaries with antialigned initial spins in the orbital plane [19]. In a detailed review [20], Racine, Buonanno and Kidder have computed the instantaneous linear momentum flux emitted by spinning binaries at 2PN order with the inclusion of the next-to-leading order SO, SO tail and SS terms. Moreover, the recoil velocity as a function of the orbital frequency was given for quasicircular orbits.

In the present work we give the previously unknown *secular expressions* for the linear momentum loss of an inspiralling binary system, with the inclusion of leading order relativistic and SO contributions. The analytic approach presented here is suitable for characterize analytically the dependence of the recoil on binary parameters.

Section 2 contains elements of the conservative dynamics with the inclusion of first post-Newtonian and spin-orbit contributions. At the end of the section the secular conservative evolution of the relevant angle variables is given. In Section 3 we start from the expressions of the SO contributions to the instantaneous linear momentum loss given in [12], then we derive the dissipative secular linear momentum evolution. The two sets of secular evolutions couple to a closed system of first order differential equations.

The secular evolutions are derived as follows: first we rewrite the instantaneous evolutions in terms of the radial parametrization of quasi-Keplerian orbits [1]. As a result, all the radial functional dependencies are expressed in terms of a single variable $\chi$, the generalized true anomaly. Averaging these expressions over a radial period becomes particularly straightforward in terms of the complex version of $\chi$, which renders the problem to the computation of residues in the origin of the complex parameter plane [21, 22]. This procedure smears out short timescale effects and we obtain a simpler dynamics, suitable for monitoring the secular changes. The procedure is similar to our previous computations of the SO-induced secular changes of the energy, magnitude of orbital angular momentum and relative angles among the orbital angular momentum and spins [23]. An additional difficulty in the present computation arises from the vectorial character of the linear momentum, and the subsequent system of coupled differential equations.

Our description is generic, being valid for generic (non-circular, non-spherical) orbits.

*Notation.* For any vector $\mathbf{V}$ its magnitude is denoted as $V$ and its direction (a unit vector) as $\hat{\mathbf{V}}$.

## 2. Elements of the conservative dynamics

In this section we derive useful elements of the conservative dynamics, originating from the SO coupling. We derive the result in terms of:

(a) the physical parameters of the binary: total mass $m = m_1 + m_2$, mass ratio $\nu = m_2/m_1 \leq 1$, symmetric mass ratio $\eta = \mu/m$ (where $\mu = m_1 m_2/m$ is the reduced mass), dimensionless spin parameters $\chi_i = (G/c)^{-1} S_i/m_i^2$

(b) dynamical constants of motion: the energy $E$, the magnitude $L$ of the orbital angular momentum $\mathbf{L}$ and $A = \left(G^2m^2\mu^2 + 2EL^2/\mu\right)^{1/2}$ (this would be the length of the Laplace-Runge-Lenz vector of the Keplerian motion characterized by $E$ and $\mathbf{L}$)

(c) angular variables related to the orbital momenta: the relative angles of spins among themselves $\gamma$ and with the orbital angular momentum $\kappa_i$, the polar angles $\psi_i$ of the spins in the plane of motion (measured from the intersection $\hat{\mathbf{l}}$ of the plane of motion with the plane perpendicular to the total angular momentum $\mathbf{J} = \mathbf{L} + \mathbf{S}$, where $\mathbf{S} = \mathbf{S_1} + \mathbf{S_2}$), and

(d) angular variables characterizing the orbit: the inclination $\alpha$ of the orbital plane with respect to the plane perpendicular to $\mathbf{J}$, the angle $\phi_n$ between $\hat{\mathbf{l}}$ and an inertial axis $\hat{\mathbf{x}}$ taken in the plane perpendicular to $\mathbf{J}$, finally the angle $\psi_p$ span by the periastron $\hat{\mathbf{A}}_\mathbf{N}$ and $\hat{\mathbf{l}}$ (see Fig 1). These three angles will be referred occasionally as Euler angles, as three consecutive rotations with $\phi_n$, $\alpha$ and $\psi_p$ about the axes $z$, $x$ and again $z$ transform from an inertial system with $\hat{\mathbf{J}}$ as the $z$-axis to the system with the $x$-axis pointing towards the periastron and the $y$-axis in the plane of motion.

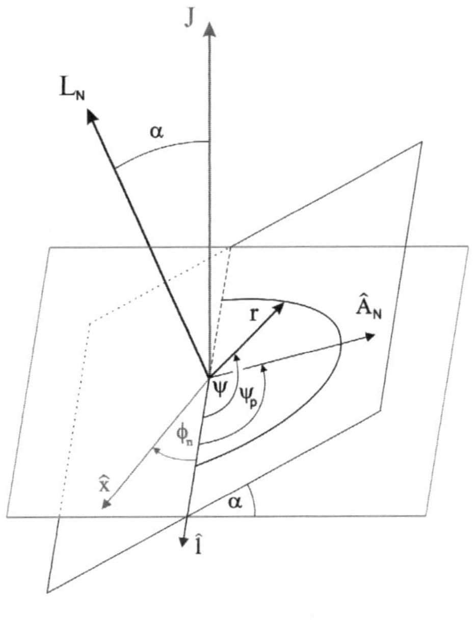


**Figure 1.** The angular variables.

### 2.1. *Lagrangian dynamics*

The Lagrangian characterizing the dynamics to 1.5 PN orders (tail terms are omitted) is

$$\mathcal{L} = \mathcal{L}_N + \mathcal{L}_{PN} + \mathcal{L}_{SO} \; , \tag{1}$$

with the contributions

$$\begin{aligned}\mathcal{L}_N &= \frac{\mu \mathbf{v}^2}{2} + \frac{Gm\mu}{r} \ , \\ \mathcal{L}_{PN} &= \frac{1}{8c^2}\left(1-3\eta\right)\mu v^4 + \frac{Gm\mu}{2rc^2}\left[\left(3+\eta\right)v^2 + \eta\dot{r}^2 - \frac{Gm}{r}\right] \ , \\ \mathcal{L}_{SO} &= \frac{G\mu}{2c^2 r^3}\mathbf{v}\cdot\left[\mathbf{r}\times\left(4\mathbf{S}+3\boldsymbol{\sigma}\right)\right] \ . \end{aligned} \tag{2}$$

Here $\boldsymbol{\sigma} = \nu \mathbf{S_1} + \nu^{-1}\mathbf{S_2}$ is the mass-weighted spin vector. The SO part of the Lagrangian applies to the Newton-Wigner spin supplementary condition and was given first in Ref. [1]. The Lagrangian gives

$$\mathbf{a} = \mathbf{a_N} + \mathbf{a_{PN}} + \mathbf{a_{SO}} \ , \tag{3}$$

with

$$\mathbf{a_N} = -\frac{Gm}{r^3}\mathbf{r} \ , \tag{4}$$

$$\mathbf{a_{PN}} = -\frac{Gm}{c^2 r^3}\left\{\mathbf{r}\left[\left(1+3\eta\right)\mathbf{v}^2 - 2\left(2+\eta\right)\frac{Gm}{r} - \frac{3}{2}\eta\dot{r}^2\right] - 2\left(2-\eta\right)r\dot{r}\mathbf{v}\right\} \ , \tag{5}$$

$$\mathbf{a_{SO}} = \frac{G}{c^2 r^3}\left[\frac{3}{2r^2}\mathbf{r}\left[\left(\mathbf{r}\times\mathbf{v}\right)\cdot\left(4\mathbf{S}+3\boldsymbol{\sigma}\right)\right] + \frac{3}{2}\frac{\dot{r}}{r}\mathbf{r}\times\left(4\mathbf{S}+3\boldsymbol{\sigma}\right) - \mathbf{v}\times\left(4\mathbf{S}+3\boldsymbol{\sigma}\right)\right] \ . \tag{6}$$

*2.2. Constants of motion*

The constants of motion are the energy $E$ and the total angular momentum $\mathbf{J}$. As the total orbital angular momentum undergoes a pure precessional motion [24, 25], its magnitude $L$ is also a conserved quantity.

The total orbital angular momentum $\mathbf{L} = \mathbf{r}\times\left(\partial\mathcal{L}/\partial\mathbf{v}\right)$ can be decomposed as

$$\mathbf{L} = \mathbf{L_N} + \mathbf{L_{PN}} + \mathbf{L_{SO}} \ , \tag{7}$$

with

$$\mathbf{L_N} = \mu\mathbf{r}\times\mathbf{v} \ , \tag{8}$$

$$\mathbf{L_{PN}} = \mathbf{L_N}\left[\frac{\left(1-3\eta\right)}{2}\frac{\mathbf{v}^2}{c^2} + \left(3+\eta\right)\frac{Gm}{c^2 r}\right] \ , \tag{9}$$

$$\mathbf{L_{SO}} = \frac{G\mu}{2c^2 r^3}\mathbf{r}\times\left[\mathbf{r}\times\left(4\mathbf{S}+3\boldsymbol{\sigma}\right)\right] \ . \tag{10}$$

We note here the useful approximate relation

$$L^2 = L_N^2 + \lambda_{PN} + \lambda_{SO} \ , \tag{11}$$

where

$$\lambda_{PN} \equiv 2\mathbf{L_N}\cdot\mathbf{L_{PN}} = 2L^2\left[\left(1-3\eta\right)\frac{E}{c^2} + 2\left(2-\eta\right)\frac{Gm\mu}{c^2 r}\right] \ , \tag{12}$$

$$\lambda_{SO} \equiv 2\mathbf{L_N}\cdot\mathbf{L_{SO}} = -\frac{G\mu L}{c^2 r}\sum_{i=1,j\neq i}^{2}\frac{4m_i+3m_j}{m_i}S_i\cos\kappa_i \ . \tag{13}$$

### 2.3. The evolutions of $\psi$ and $\phi_n$

The polar and azimuthal angles $\theta, \varphi$ of the reduced mass particle in the inertial system with $\hat{\mathbf{J}}$ as the $z$-axis can be related to the Euler angles ($\phi_n$, $\alpha$, $\psi$) characterizing a non-inertial reference system with the $x$-axis at the location of the reduced mass particle and the $y$-axis in the plane of motion [26]:

$$\sin\theta\cos\varphi = \cos\alpha\sin\phi_n\sin\psi + \cos\phi_n\cos\psi \ , \quad (14)$$

$$\sin\theta\sin\varphi = \cos\alpha\cos\phi_n\sin\psi - \sin\phi_n\cos\psi \ , \quad (15)$$

$$\cos\theta = \sin\alpha\sin\psi \ . \quad (16)$$

Tedious but straightforward algebra leads to

$$\dot{\psi} = \cos\alpha\ \dot{\varphi} - \frac{\sin\alpha\cos\psi}{\sin\theta}\dot{\theta} + \cos\alpha\ \dot{\phi}_n \ ,$$

$$\dot{\phi}_n = -\dot{\varphi} - \frac{\cot\alpha}{\sin\theta\cos\psi}\dot{\theta} + \frac{\tan\psi}{\sin^2\alpha}\frac{d}{dt}(\cos\alpha) \ . \quad (17)$$

The Newtonian orbital angular momentum, expressed in terms of $(\theta, \varphi)$ gives

$$L_N^2 = \mu^2 r^4(\dot{\theta}^2 + \sin^2\theta\ \dot{\varphi}^2) \ , \quad (\mathbf{L_N})_z = \mu r^2\sin^2\theta\ \dot{\varphi} \ . \quad (18)$$

By employing Eqs. (11), (16) and $(\mathbf{L_N})_z = L_N\cos\alpha$ we obtain the evolutions:

$$\sin\theta\ \dot{\theta} = -\frac{L}{\mu r^2}\sin\alpha\cos\psi\left(1 - \frac{\lambda_{PN} + \lambda_{SO}}{2L^2}\right) \ , \quad (19)$$

$$\sin^2\theta\ \dot{\varphi} = \frac{L}{\mu r^2}\cos\alpha\left(1 - \frac{\lambda_{PN} + \lambda_{SO}}{2L^2}\right) \ . \quad (20)$$

The minus sign in the first equation was chosen after taking the square root in such a way that when we insert these equations into Eqs. (17), we obtain

$$\dot{\psi} = \frac{L}{\mu r^2}\left(1 - \frac{\lambda_{PN} + \lambda_{SO}}{2L^2}\right) + \dot{\phi}_n\cos\alpha \ , \quad (21)$$

$$\dot{\phi}_n = \frac{\tan\psi}{\sin^2\alpha}\frac{d}{dt}(\cos\alpha) \ , \quad (22)$$

which reduce to the correct equations $\dot{\psi} = L/\mu r^2$ and $\phi_n$ =const at Newtonian order.

Provided that we can obtain $\dot{\alpha}$ by a complementary method, the evolutions of $\psi$ and $\phi_n$ can be given explicitly. This will be done in the following subsection.

### 2.4. The evolution of the orbital inclination

From $\cos\alpha = \hat{\mathbf{J}}\cdot\hat{\mathbf{L}}_\mathbf{N}$, as $\mathbf{J}$ is conserved, we get

$$\frac{d}{dt}(\cos\alpha) = \hat{\mathbf{J}}\cdot\frac{d}{dt}\left(\hat{\mathbf{L}}_\mathbf{N}\right) \ . \quad (23)$$

Employing Eq. (3) gives

$$\dot{\mathbf{L}}_\mathbf{N} = \mu\mathbf{r}\times\mathbf{a} = \frac{2Gm}{c^2r^2}(2-\eta)\,\dot{r}\mathbf{L_N} + \frac{G\mu}{c^2r^3}\left\{\left(\frac{3}{2}\frac{\dot{r}}{r}\mathbf{r} - \mathbf{v}\right)[\mathbf{r}\cdot(4\mathbf{S} + 3\boldsymbol{\sigma})] - \frac{r\dot{r}}{2}(4\mathbf{S} + 3\boldsymbol{\sigma})\right\} \ . \quad (24)$$

From here we can derive the time derivative of $\hat{\mathbf{L}}_{\mathbf{N}}$. We then find that the leading order evolution of $\alpha$ is of order $S^2/JL$:

$$\begin{aligned}\frac{d}{dt}(\cos\alpha) &= \frac{G\mu\dot{r}}{2c^2JLr^2}\left[(4+3\nu)S_1\cos\kappa_1+\left(4+3\nu^{-1}\right)S_2\cos\kappa_2\right](S_1\cos\kappa_1+S_2\cos\kappa_2)\\ &-\frac{G\mu\dot{r}}{2c^2JLr^2}\left\{(4+3\nu)S_1^2+\left[8+3\left(\nu+\nu^{-1}\right)\right]S_1S_2\cos\gamma+\left(4+3\nu^{-1}\right)S_2^2\right\}\\ &+\frac{G\mu}{c^2JLr^3}\left[(4+3\nu)\,\mathbf{r}\cdot\mathbf{S_1}+\left(4+3\nu^{-1}\right)\mathbf{r}\cdot\mathbf{S_2}\right]\left(\frac{3}{2}\frac{\dot{r}}{r}\,\mathbf{r}-\mathbf{v}\right)\cdot(\mathbf{S_1}+\mathbf{S_2})\ . \qquad (25)\end{aligned}$$

(We have employed $L_N = L$, valid to leading order accuracy.) The scalar products can be rewritten in terms of the angular variables as

$$\begin{aligned}\mathbf{r}\cdot\mathbf{S_i} &= rS_i\sin\kappa_i\cos(\psi_p+\chi-\psi_i)\ ,\\ \mathbf{v}\cdot\mathbf{S_i} &= \dot{r}S_i\sin\kappa_i\cos(\psi_p+\chi-\psi_i)-\frac{LS_i}{\mu r}\sin\kappa_i\sin(\psi_p+\chi-\psi_i)\ , \qquad (26)\end{aligned}$$

In Eq. (25), as all terms are of $S^2/J$ order, it is allowed to employ the Newtonian true anomaly parametrization of the orbit:

$$r=\frac{L^2}{\mu(Gm\mu+A\cos\chi)}\ ,\qquad \dot{r}=\frac{A}{L}\sin\chi\ . \qquad (27)$$

With these expressions, all the desired angular evolutions can be given explicitly as functions of $\chi$ (which, to Newtonian order is $\psi-\psi_p$, see Figure 1).

### *2.5. The secular evolution of the Euler angles*

We conclude this section by giving the secular evolutions of the Euler angles. The secular evolution of any function $f(\chi)$ is defined as $<\dot{f}>=T^{-1}\int_0^{2\pi}\dot{f}(\chi)\,\dot{\chi}^{-1}\,d\chi$. As the instantaneous angular evolutions (22) and (25) have no Newtonian contributions, we can employ the Newtonian period of the orbital motion $T = 2\pi Gm\left[\mu/(-2E)\right]^{3/2}$ in calculating their orbital average, obtaining:

$$\begin{aligned}\sin\alpha\left<\dot{\alpha}\right> &= \frac{3G\left(-2E\mu\right)^{3/2}}{2c^2JL^3}\left(\nu-\nu^{-1}\right)S_1S_2\sin\kappa_1\sin\kappa_2\sin\Delta\psi\ ,\\ \sin^2\alpha\left<\dot{\phi}_n\right> &= \frac{G\left(-2E\mu\right)^{3/2}}{2c^2JL^3}\left[\sum_{i=1}^{2}\left(4+3\nu^{3-2i}\right)S_i^2\sin\kappa_i^2\cos2\psi_i\right.\\ &\left.+S_1S_2\left[8+3\left(\nu+\nu^{-1}\right)\right]\sin\kappa_1\sin\kappa_2\cos2\overline{\psi}\right]\ . \qquad (28)\end{aligned}$$

We introduced here the shorthand notations $\Delta\psi=\psi_2-\psi_1$ and $\overline{\psi}=(\psi_1+\psi_2)/2$.

We also compute the change in the angle $\psi$ over a radial period by the above indicated method, as $\Delta\psi=\int_0^{2\pi}\dot{\psi}(\chi)\,\dot{\chi}^{-1}\,d\chi$. By employing Eq. (21) and the relevant terms[1] form Eq.

[1] The contributions $(dt/d\chi)_{PN}$ and $(dt/d\chi)_{SO}$ are corrected by a global sign and the factor $\mu/L$, respectively.

(18) of Ref. [1] we get:

$$\frac{dt}{d\chi} = \left(\frac{dt}{d\chi}\right)_N + \left(\frac{dt}{d\chi}\right)_{PN} + \left(\frac{dt}{d\chi}\right)_{SO} , \tag{29}$$

$$\left(\frac{dt}{d\chi}\right)_N = \frac{\mu r^2}{L} ,$$

$$\left(\frac{dt}{d\chi}\right)_{PN} = -\frac{\mu r^2}{2c^2L^3}\left[(\eta-13)\,G^2m^2\mu^2 + (3\eta-1)\,A^2 + (3\eta-8)Gm\mu A\cos\chi\right] ,$$

$$\left(\frac{dt}{d\chi}\right)_{SO} = -\frac{G\mu^3r^2}{2c^2L^4}\left(3Gm\mu + A\cos\chi\right)B_S , \tag{30}$$

with

$$B_S = (4+3\nu)\,S_1\cos\kappa_1 + \left(4+3\nu^{-1}\right)S_2\cos\kappa_2 . \tag{31}$$

The leading order term gives $\int_0^{2\pi}\left(L/\mu r^2\right)(dt/d\chi)_N\,d\chi = 2\pi$. Defining $\left\langle\dot\psi_p\right\rangle \equiv (\Delta\psi - 2\pi)/T$ we get the secular periastron precession rate as

$$\left\langle\dot\psi_p\right\rangle = \frac{Gm\,(-2E)^{3/2}\,\mu^{1/2}}{c^2L^2}\left(3-\eta\frac{B_S}{L}\right) + \cos\alpha\,\left\langle\dot\phi_n\right\rangle . \tag{32}$$

Eqs. (28) and (32) together with the evolution of the angles $\kappa_i$ given by the first two Eqs. (2.17) of [23] and the projections of $\mathbf{J}=\mathbf{L}+\mathbf{S_1}+\mathbf{S_2}$, determining $\psi_i$ in terms of the other variables form a closed system of differential equations for the conservative evolution of the angular variables.

## 3. Elements of the dissipative dynamics

The instantaneous loss in the linear momentum due to the gravitational radiation can be expressed as $\dot{\mathbf{P}}_i = -\{\frac{2}{63}\overset{(4)}{I}_{ijk}\overset{(3)}{I}_{jk} + \frac{16}{45}\epsilon_{ijk}\overset{(3)}{I}_{jl}\overset{(3)}{J}_{kl}\}$, where $I_{jk}$, $I_{ijk}$ and $J_{kl}$ are the mass quadrupole, mass octupole and current quadrupole moments [12]. The secular loss in the linear momentum is:

$$\left\langle\dot{\mathbf{P}}\right\rangle = \left\langle\dot{\mathbf{P}}\right\rangle_N + \left\langle\dot{\mathbf{P}}\right\rangle_{SO} , \tag{33}$$

with the leading order radiative (N) and leading order radiative SO contributions given by

$$\left\langle\dot{\mathbf{P}}\right\rangle_N = -\delta m\frac{G^2\mu^3A}{30c^7L^8}\left(\frac{-2E}{\mu}\right)^{3/2}\mathcal{A}_N\,\mathbf{U} , \tag{34}$$

$$\left\langle\dot{\mathbf{P}}\right\rangle_{SO} = \frac{G^2\mu^4A}{30c^7L^9}\left(\frac{-2E}{\mu}\right)^{3/2} \times\sum_{i=1}^{2}(-1)^{i-1}\left(1+\nu^{3-2i}\right)S_i\left[\mathcal{A}_{SO}\,\mathbf{U}\cos\kappa_i - \mathcal{B}_{SO}\mathbf{V}\sin\kappa_i\sin\left(\psi_i-\psi_p\right)\right] . \tag{35}$$

Here we have denoted

$$\mathbf{U} = \begin{pmatrix}\cos\alpha\sin\phi_n\cos\psi_p - \cos\phi_n\sin\psi_p\\ \cos\alpha\cos\phi_n\cos\psi_p + \sin\phi_n\sin\psi_p\\ \sin\alpha\cos\psi_p\end{pmatrix} , \quad \mathbf{V} = \begin{pmatrix}-\sin\alpha\sin\phi_n\\ -\sin\alpha\cos\phi_n\\ \cos\alpha\end{pmatrix} , \tag{36}$$

and

$$\mathcal{A}_N = 148E^2L^4 + 1060G^2m^2\mu^3EL^2 + 805G^4m^4\mu^6 , \tag{37}$$

$$\mathcal{A}_{SO} = 48E^2L^4 + 360G^2m^2\mu^3EL^2 + 280G^4m^4\mu^6 , \tag{38}$$

$$\mathcal{B}_{SO} = 108E^2L^4 + 780G^2m^2\mu^3EL^2 + 595G^4m^4\mu^6 . \tag{39}$$

## 4. Concluding Remarks

We have derived secular evolution equations, Eqs. (28) and (32) for the angular variables $\alpha, \phi_n$ and $\psi_p$ characterizing the orientation of the orbit, with the inclusion of the leading order relativistic and spin-orbit coupling contributions. Together with the evolution of the angles $\kappa_i$ given by the first two Eqs. (2.17) of [23] and the projections of $\mathbf{J} = \mathbf{L} + \mathbf{S_1} + \mathbf{S_2}$, determining $\psi_i$ in terms of the other variables, these form a closed system of differential equations for the conservative evolution of the angular variables.

We have complemented these with the equations expressing the respective secular losses of the linear momentum components, Eqs. (33). Then we have a closed system of coupled first order differential equations, giving the linear momentum loss during the inspiral in analytic form.

This system of equations is also suitable for numerical evolution, which in principle can lead to the value of the recoil velocity right before the plunge and also allows to determine the shift in the position of the binary during the inspiral. Our analytic approach is suitable for discussing the dependence of the recoil on various parameters characterizing the binary.

In order to gain higher accuracy, the results derived in this paper can be further generalized by the inclusion of higher order corrections: the spin-spin, mass quadrupole - mass monopole, and second order relativistic contributions.

## Acknowledgements

This work was supported by the Hungarian Scientific Research Fund (OTKA) grants no. 69036 and 68228, also by the Polányi and Sun Programs of the Hungarian National Office for Research and Technology (NKTH). LÁG is grateful to the organizers of the Amaldi8 meeting for support.

## References

[1] Keresztes Z, Mikóczi B and Gergely L Á 2005 *Phys. Rev.* D **72** 104022
[2] Peres A 1962 *Phys. Rev.* **128** 2471
[3] Bekenstein J D 1973 *Phys. Rev.* D **7** 949
[4] Fitchett M J 1983 *Mon. Not. R. Astron. Soc.* **203** 1049
[5] Wiseman A G 1992 *Phys. Rev.* D **46** 1517
[6] Blanchet L, Qusailah M S S and Will C M 2005 *Astrophys. J.* **635** 508
[7] Baker J G, Centrella J, Choi D I, Koppitz M, van Meter J and Miller M C 2006 *Astrophys. J.* **653** L93
[8] Herrmann F, Hinder I, Shoemaker D and Laguna P 2007 *Class. Quant. Grav.* **24** S33
[9] González J A, Sperhake U and Brügmann B 2009 *Phys. Rev.* D **79** 124006
[10] González J A, Sperhake U, Brügmann B, Hannam M and Husa S 2007 *Phys. Rev. Lett.* **98** 091101
[11] Le Tiec A, Blanchet L and Will C M 2010 *Class. Quant. Grav.* **27** 012001
[12] Kidder L 1995 *Phys. Rev.* D **52** 821
[13] Schnittman J D and Buonanno A 2007 *Astrophys. J.* **662** L63
[14] Herrmann F, Hinder I, Shoemaker D, Laguna P and Matzner R A 2007 *Astrophys. J.* **661** 430
[15] Koppitz M, Pollney D, Reisswig C, Rezzolla L, Thornburg J, Diener P and Schnetter E 2007 *Phys. Rev. Lett.* **99** 041102
[16] Campanelli M, Lousto C O, Zlochower Y and Merritt D 2007 *Astrophys. J.* **659** L5
[17] Schnittman J D et. al. 2008 *Phys. Rev.* D **77** 044031
[18] Lousto C O and Zlochower Y 2009 *Phys. Rev.* D **79** 064018
[19] Brügmann B, González J A, Hannam M, Husa S and Sperhake U 2008 *Phys. Rev.* D **77** 124047
[20] Racine É, Buonanno A and Kidder L 2009 *Phys. Rev.* D **80** 044010
[21] Gergely L Á, Perjés Z and Vasúth M 2000 *Astrophys. J. Suppl.* **126** 79
[22] Gergely L Á, Keresztes Z and Mikóczi B 2006 *Astrophys. J. Suppl.* **167** 286
[23] Gergely L Á, Perjés Z and Vasúth M 1998 *Phys. Rev.* D **58** 124001
[24] Barker B M and O'Connell R F 1975 *Phys. Rev.* D **12** 329
[25] Barker B M and O'Connell R F 1979 *Gen. Relativ. Gravit.* **2** 1428
[26] Gergely L Á, Perjés Z and Vasúth M 1998 *Phys. Rev.* D **57** 876
The angles $(\phi_n, \alpha, \psi_p, \psi)$ were denoted there as $(-\Phi, \iota, \psi_0, \Psi)$.

Journal of Physics: Conference Series **228** (2010) 012054 doi:10.1088/1742-6596/228/1/012054

# Gravitational waves and cosmological back reaction

**Claire Chevalier**[1]
**Université Pierre et Marie Curie-Paris6, UMR 7600, LPTMC, 4 place Jussieu, 75255 Paris, France.**

**Fabrice Debbasch**[2]
**Université Pierre et Marie Curie-Paris6, UMR 8112, ERGA-LERMA, 3 rue Galilée 94200 Ivry, France.**

**Yann Ollivier**[3]
**CNRS, UMPA, Ecole normale supérieure de Lyon, F-69364 Lyon Cedex 07, France.**

**Abstract.** Cosmological back reaction of gravitational waves and matter density fluctuations is revisited through a new general mean field theory of relativistic gravitation. Computations are done for all wave-lengths and Isaacson's results are confirmed for high-frequency waves. It is found that back reaction is typically much stronger for gravitational waves than for matter density fluctuations. In particular, a bath of gravitational waves of relative amplitude $10^{-5}$ and frequency $10^{-12}$ Hz (compatible with today's constraints) would generate an effective large-scale radiation of amplitude comparable to the unperturbed matter density of the universe.

## 1. Introduction

It is now well established [7] that the universe is, on large scale, expanding in an homogeneous and isotropic manner. Several authors [8, 9, 10, 11, 12, 13, 14, 15, 16, 17, 18] have recently argued that small-scale fluctuations around this large-scale expansion may, by non-linearity, contribute substantially to the large-scale energy repartition generating the expansion. This effect is termed cosmological back reaction.

This article focuses on the back reaction generated by gravitational waves (GWs) and evaluates it within the framework of the recently developed, general mean field theory of relativistic gravitation [8, 9, 5]. We first compute the back reaction due to a background of incoherent GWs of arbitrary frequency; in particular, we confirm, in the high frequency limit, the seminal result originally derived by Isaacson [23]. Our computation, combined with the current direct and indirect observational evidence, leaves only the so-called very long wave-length GWs ($10^{-12}$ Hz) as possible sources for non negligible back reaction. We also compute

[1] claire.chevalier26@gmail.com, corresponding author
[2] fabrice.debbasch@gmail.com
[3] yann.ollivier@umpa.ens-lyon.fr

© 2010 IOP Publishing Ltd

the back reaction due to matter density fluctuations and find it is typically much smaller than the back reaction produced by GWs.

## 2. Mean field theory

### 2.1. Notation

In this article, the metric has signature $(+,-,-,-)$. We shall use mixed components $T^{\nu}_{\mu}$ of the stress-energy tensor; with this signature, for a perfect fluid at rest with density $\rho$ and pressure $p$, we have $T^0_0 = \rho$ and $T^i_i = -p$.

### 2.2. General framework

Averaging classical gravitational fields necessitates a mean field theory of general relativity. Such a theory has been introduced in [8, 9]; we include a brief overview here for completeness. Some applications to black hole physics are presented in [19, 20, 21].

Let $\mathcal{M}$ be a fixed manifold and $g(\omega)$ be a Lorentzian metric on $\mathcal{M}$ indexed by a random parameter $\omega$. For example, $g(\omega)$ may represent a gravitational wave of random phase and wave vector around a given reference space-time. With each space-time $\mathcal{S}(\omega) = (\mathcal{M}, g(\omega))$ are associated the Einstein tensor $G(\omega)$ of the metric $g(\omega)$, and a stress-energy tensor $\mathcal{T}(\omega)$, satisfying the Einstein equation

$$G(\omega) = 8\pi\mathcal{T}(\omega). \tag{1}$$

Suppose one wants to extract from such a collection of space-times an average, mean gravitational field. There is a unique way to do so [8, 9] provided one imposes that (i) motions in the $\mathcal{S}(\omega)$'s average at least locally into motions in the mean field (ii) the mean gravitational field is described by General Relativity. The mean field has then to be represented by a mean metric $\bar{g} = \langle g(\omega)\rangle$ and the associated stress-energy tensor $\bar{\mathcal{T}}$ is linked by Einstein equation to the Einstein tensor $\bar{G}$ of $\bar{g}$: $\bar{\mathcal{T}} = (1/8\pi)\bar{G}$. Since the Einstein tensor depends non-linearly on the metric, $\bar{\mathcal{T}}$ does *not* coincide with $\langle\mathcal{T}(\omega)\rangle$ and the difference

$$\mathcal{T}^{\mathrm{app}\,\nu}_{\ \ \mu} = \bar{G}^{\nu}_{\mu}/8\pi - \left\langle \mathcal{T}^{\nu}_{\mu}(\omega)\right\rangle, \tag{2}$$

can be interpreted as the stress-energy of an 'apparent matter' which traces the net back reaction of the fluctuations on the average field. In particular, the vanishing of $\mathcal{T}(\omega)$ for all $\omega$ does not necessarily imply the vanishing of $\bar{\mathcal{T}}$. The mean stress-energy tensor $\bar{\mathcal{T}}$ can therefore be non-vanishing in regions where the unaveraged stress-energy tensor actually vanishes. We will see that this happens, for instance, with gravitational waves.

Note also that once the statistical ensemble of metrics has been chosen, there is no gauge choice involved above since the definition of apparent matter is manifestly covariant; however, the choice of the ensemble may, in some cases, reflect the point of view of some particular observer, akin to fixing a preferred referential for the observations.

### 2.3. Small amplitude fluctuations

We now investigate the case when the metrics $g_{\mu\nu}(\omega)$ are all close to a reference metric $g^{\mathrm{ref}}{}_{\mu\nu}$. More precisely, we assume that there is a small parameter $\varepsilon$ such that, for any value of the random parameter $\omega$,

$$g_{\mu\nu}(\omega) = g^{\mathrm{ref}}{}_{\mu\nu} + \varepsilon g^{(1)}{}_{\mu\nu}(\omega) + \varepsilon^2 g^{(2)}{}_{\mu\nu}(\omega) + O(\varepsilon^3) \tag{3}$$

and we will expand the theory above at second order in $\varepsilon$.

Consider 'centered' fluctuations *i.e.* fluctuations whose averages vanish at first order in $\varepsilon$: $\langle g^{(1)}(\omega)\rangle = 0$. A direct computation then delivers:

$$\mathcal{T}^{\mathrm{app}\,\nu}_{\ \ \mu} = -\frac{\varepsilon^2}{16\pi}\left\langle (\mathcal{D}^2 G^{\nu}_{\mu})(g^{(1)}(\omega), g^{(1)}(\omega))\right\rangle, \tag{4}$$

where $\mathcal{D}^2 G^\nu_\mu$ is the (functional) second derivative of the Einstein tensor with respect to the metric $g^{\text{ref}}$.

It is to be noted that the effect is at second order in $\varepsilon$, which was to be expected since at first order, gravitation is by definition linear. What is more interesting is that $g^{(2)}$ does not appear in the result. This reflects the fact that non-linearities acting on the second-order term $g^{(2)}$ will only produce higher-order terms.

Often, it makes more physical sense to define the fluctuations in terms of the sources rather than the metric, *i.e.* to first prescribe physically meaningful fluctuations $\mathcal{T}^{(1)}$ and $\mathcal{T}^{(2)}$ of the stress-energy tensor and then look for $g^{(1)}$ and $g^{(2)}$ solving the Einstein equation. Since $g^{(2)}$ vanishes from the above result, to compute the desired effect it is actually enough to solve the *linearized* Einstein equation around $g^{\text{ref}}$, for a given equation of state. In the sequel we will focus on choices of $g^{(1)}$ arising from physically interesting stress-energy tensors, such as gravitational waves or fluctuations of the density of matter.

## 3. Fluctuations around dust cosmologies

### 3.1. Basics

We now apply the above to the case of either gravitational waves or density fluctuations around a homogeneous and isotropic, spatially flat dust universe (flat Friedmann–Lemaï¿½tre–Robertson–Walker metric). The reference metric and stress-energy tensor of such a space-time are, in conformal coordinates [22]:

$$g^{\text{ref}} = a(\eta)^2(d\eta^2 - dx^2 - dy^2 - dz^2) \qquad T^0_0 = \rho(\eta) \quad T^0_i = T^j_i = 0 \tag{5}$$

where $a$ is the so-called expansion factor and $\rho$ is the energy density. The Einstein equation delivers $a(\eta) = C\eta^2$ and $8\pi\rho(\eta) = 3\dot{a}^2/a^4 = 12/C^2\eta^6$, where $C$ is an arbitrary (positive) constant. Proper time is $\tau = C\eta^3/3$ and the Hubble 'constant' is $H = \dfrac{1}{a}\dfrac{da}{d\tau} = \dfrac{\dot{a}}{a^2} = \dfrac{2C}{\eta^3}$.

The perturbations $g^{(1)}$ considered in this article will be of two types: gravitational waves and matter density fluctuations. They can be written as sums or integrals of spatial Fourier modes (this makes sense since $g^{\text{ref}}$ is spatially flat). Each term in such a series is of the form $F(\eta)\exp\left(i(\mathbf{q}.\mathbf{r} + \omega_{\mathbf{q}})\right)$ where $F(\eta)$ is some tensor, $\mathbf{q}$ is a three-dimensional wave vector, and $\omega_{\mathbf{q}}$ is a phase associated with mode $\mathbf{q}$.

Averaging a given mode $\mathbf{q}$ on spatial scales much larger than the wave-length $1/|\mathbf{q}|$ is equivalent to averaging this mode over the phase $\omega_{\mathbf{q}} \in [0; 2\pi]$. We therefore choose the set of all phases $(\omega_{\mathbf{q}})$ as our random parameter, and perform all averagings over these phases. By a simple superposition argument, which we omit, one can easily check that if several Fourier modes are present but statistically independent, then the averaging can be performed separately for each mode (at least at second order). Hence, in the sequel, we will use a single Fourier mode. Note finally that averaging over $\omega_{\mathbf{q}}$ can also be interpreted as statistically averaging over a superposition of interfering GWs of the same wave-number $\mathbf{q}$ but different phases.

### 3.2. Gravitational waves

Consider a single gravitational wave propagating along the above background [23]. This wave admits two polarizations [22]; since the background is isotropic, there is no loss of generality in assuming the wave propagates along, say the $x$-axis. The first-order metric perturbation then reads, for the first polarization:

$$g^{(1)}{}_{22}(\omega) = -a(\eta)^2 e^{i(q(x-\eta)+\omega)}(1 - i/q\eta)/\eta^2 \qquad g^{(1)}{}_{33} = -g^{(1)}{}_{22} \tag{6}$$

with the other components equal to 0. Here $q$ is the wave number in conformal coordinates, and $\omega \in [0; 2\pi]$.

Journal of Physics: Conference Series **228** (2010) 012054 doi:10.1088/1742-6596/228/1/012054

The statistical averaging corresponds to a uniform averaging over $\omega \in [0; 2\pi]$; this models situations in which the system is observed at a resolution much larger than the perturbation wave-length $1/q$.

The quantity $n_{\text{osc}} = q\eta$ measures the typical number of oscillations (periods) in that part of the universe accessible to an observer situated at time $\eta$. The relative amplitude of the perturbation $\varepsilon g^{(1)}$ at time $\eta$, compared to $g^{\text{ref}}$, is $\tilde{\varepsilon}(\eta) = \varepsilon/\eta^2$. We will express the results in terms of those quantities.

The apparent stress-energy corresponding to a superposition of statistically independent gravitational waves sharing a common frequency and amplitude, but propagating along random spatial directions reads :

$$\mathcal{T}^{\text{app}}{}^0_0 = \tilde{\varepsilon}(\eta)^2 n_{\text{osc}}^2 \, \frac{1 - 14/n_{\text{osc}}^2 - 39/2n_{\text{osc}}^4}{48} \, \rho(\eta) \tag{7}$$

$$\mathcal{T}^{\text{app}}{}^1_1 = \mathcal{T}^{\text{app}}{}^2_2 = \mathcal{T}^{\text{app}}{}^3_3 = -\tilde{\varepsilon}(\eta)^2 n_{\text{osc}}^2 \, \frac{1/3 - 4/3n_{\text{osc}}^2 - 45/6n_{\text{osc}}^4}{48} \, \rho(\eta). \tag{8}$$

In the particular case $n_{\text{osc}} \gg 1$ (high frequency), these expressions confirm Isaacson's result [23] that a background of gravitational waves of high frequency behaves like radiation, with positive pressure equal to a third of its energy density (at this order in $\tilde{\varepsilon}(\eta)^2 n_{\text{osc}}^2$). Since high frequency waves do not see the curvature of the background, this result is actually background independent and is also valid, *e.g.* in the presence of vacuum energy or standard electromagnetic radiation.

The first-order metric perturbation for the other polarization is given by

$$g^{(1)}{}_{23}(\omega) = (C^2\eta^4) e^{i(q(x-\eta)+\omega)}(1 - i/q\eta)/\eta^2. \tag{9}$$

The stress-energy tensor of the apparent matter associated with this polarization is identical to (7) and (8) and does not warrant separate discussion.

*Orders of magnitude.* The important factor in (7) and (8) is $\tilde{\varepsilon}(\eta)^2 n_{\text{osc}}^2$. The energy density and pressure of apparent matter are (at this order) quadratic, not only in the amplitude $\tilde{\varepsilon}(\eta)$ of the perturbation, but also in its 'frequency' $n_{\text{osc}}$. Thus, $\tilde{\varepsilon}(\eta) \ll 1$ does not necessarily translate into negligible energy density and pressure of apparent matter: the smallness of $\tilde{\varepsilon}(\eta)$ can be compensated by a sufficiently high frequency $n_{\text{osc}}$. Naturally, very stringent, frequency-dependent constraints exist on $\Omega_{\text{GW}}$ [24, 29]. Very low frequency waves ($10^{-17} \sim 10^{-15}$ Hz) are constrained by studying the power spectrum of the CMB B-polarizations; high frequency waves ($10^{-4} \sim 10^{4}$ Hz) are tested by laser interferometers, while the amplitude of intermediate frequency waves ($10^{-9} \sim 10^{-7}$ Hz) is indirectly bounded by studying millisecond pulsars. Current constraints definitely rule out any non negligible back reaction in all the above ranges. But a gravitational wave of frequency of order $10^{-12}$ Hz (corresponding to a wave-length of $1/n_{\text{osc}} = 10^{-5}$ times the radius of the observable Universe) and relative amplitude $\tilde{\varepsilon}(\eta) \approx 10^{-5}$ would, by (7) and (8), generate a $\mathcal{T}^{\text{app}}$ of order unity, which is quite compatible with the current constraints [26, 27] $\Omega_{\text{GW}}\, h_{100}^2 \leq 0.1$ and $\Omega_{\text{GW}}\, h_{100}^2 \leq 0.5$ around this frequency.

### *3.3. Fluctuations in the density of matter*

The first-order expressions for the metric and stress-energy tensors corresponding to a matter density fluctuation around a spatially flat, homogeneous and isotropic universe are well-known and given in [22]. These expressions can be used to compute the stress-energy of apparent matter from the formula (4).

We discuss here the simplest such perturbation, which reads:

$$g^{(1)}{}_{00} = 0 \qquad g^{(1)}{}_{11}(\omega) = a(\eta)^2\eta^2\cos(qx+\omega)$$
$$g^{(1)}{}_{22}(\omega) = g^{(1)}{}_{33}(\omega) = -a(\eta)^2\frac{10}{q^2}\cos(qx+\omega). \tag{10}$$

It corresponds to the following first-order stress-energy tensor perturbation

$$\mathcal{T}^{(1)}{}^0_0(\omega) = \rho\frac{\eta^2}{2}\cos(qx+\omega) \qquad \mathcal{T}^{(1)}{}^j_i = \mathcal{T}^{(1)}{}^j_0 = 0. \tag{11}$$

This stress-energy tensor describes a co-moving, shear-free spatial density fluctuation. Note that the relative amplitude of the perturbation increases with time, which traces the aggregating effect of gravitation.

The quantity $\tilde{\varepsilon}(\eta) = \varepsilon\eta^2$ measures the effective relative magnitude of the perturbation $\varepsilon g^{(1)}$ with respect to $g^{\text{ref}}$. The quantity $n_{\text{osc}} = q\eta$ represents the number of oscillations (periods) in that part of the universe accessible to an observer situated at time $\eta$; typically $n_{\text{osc}} \gg 1$.

As above, the averaging is over $\omega \in [0; 2\pi]$, and the stress-energy tensor of apparent matter can be obtained from (4) by direct computation. This gives

$$\mathcal{T}^{\text{app}}{}^0_0 = -\tilde{\varepsilon}(\eta)^2\frac{1-75/n_{\text{osc}}^2}{16\pi}\rho(\eta) \tag{12}$$

$$\mathcal{T}^{\text{app}}{}^1_1 = \tilde{\varepsilon}(\eta)^2\frac{25}{16\pi n_{\text{osc}}^2}\rho(\eta) \tag{13}$$

$$\mathcal{T}^{\text{app}}{}^2_2 = \mathcal{T}^{\text{app}}{}^3_3 = \tilde{\varepsilon}(\eta)^2\frac{7+50/n_{\text{osc}}^2}{32\pi}\rho(\eta) \tag{14}$$

and all the other terms are 0 at this order in $\varepsilon$.

The apparent matter associated with these fluctuations is thus characterized, at this order, by a negative energy density and a negative pressure. Loosely speaking, the negative energy could be interpreted in a semi-Newtonian setting as the gravitational energy of the fluctuations and the negative pressure represents the collapsing effects of gravitation.

There is an important difference with respect to the gravitational wave case above, namely that the effect simply scales like the square of the effective amplitude of the perturbation, with no $n_{\text{osc}}^2$ factor (compare (7–8)). Thus, the net large-scale effect of high-frequency gravitational waves is much more important than the net large-scale effect of matter density fluctuations of comparable wave-length, at least at this order in $\varepsilon$. Note however that matter density fluctuations 'may become quite large' [22] with time, and, thus, escape the perturbative regime in which our results are obtained.

## 4. Conclusion

We have investigated perturbatively how gravitational waves and matter density fluctuations influence the large-scale homogeneous an isotropic expansion of the universe. Our results indicate that the so-called back reaction effect is dominated by gravitational waves, rather than matter density fluctuations. The relative importance of the effective large-scale stress-energy generated by gravitational waves scales as the squared product of their amplitude by their frequency. Thus, even small amplitude waves can generate an important effect provided their frequencies are high enough. For example, it is found that waves of current amplitude $\sim 10^{-5}$ and current physical frequency $10^{-12}$ Hz (thus compatible with currently known constraints in this part of the spectrum) would generate a large-scale stress-energy comparable to the dust energy.

[1] Z. Burda, J. Jurkiewicz, and M.A. Nowak. Is Econophysics a solid Science? *Acta Phys. Pol. B*, 34 (1):87, 2003.
[2] J. Southern et al. Multi-scale computational modelling in biology and physiology. *Progr. Biophys. Mol. Bio.*, 96 (1–3):60–89, 2008.
[3] U. Frisch. *Turbulence.* Cambridge University Press, Cambridge, 1995.
[4] J. Cardy. *Scaling and renormalization in statistical physics.* Cambridge University Press, Cambridge, 1996.
[5] C. Chevalier, F. Debbasch and Y. Ollivier. Multiscale cosmological dynamics *Physica A*, 388 (24):5029, 2009.
[6] R.M. Wald. *General Relativity.* The University of Chicago Press, Chicago, 1984.
[7] S. Weinberg. *Cosmology.* Oxford University Press, 2008.
[8] F. Debbasch. What is a mean gravitational field? *Eur. Phys. J. B*, 37(2):257–270, 2004.
[9] F. Debbasch. Mean field theory and geodesics in general relativity. *Eur. Phys. J. B*, 43(1):143–154, 2005.
[10] T. Buchert. A cosmic equation of state for the inhomogeneous universe: can a global far-from-equilibrium state explain dark energy? *Class. Quantum Grav.*, 22(19):L113–L119, 2005.
[11] H. Alnes, M. Amarzguioui and ï¿½. Grï¿½n. An inhomogeneous alternative to dark energy? *Phys. Rev. D*, 73:083519, 2006.
[12] T. Buchert. On globally static and stationary cosmologies with or without a cosmological constant and the dark energy problem. *Class. Quantum Grav.*, 23:817–844, 2006.
[13] E.W. Kolb, S. Matarrese, and A. Riotto. On cosmic acceleration without dark energy. *New J. Phys.*, 8:322, 2006.
[14] A. Paranjape and T.P. Singh. The possibility of cosmic acceleration via spatial averaging in Lemaître–Tolman–Bondi models. *Class. Quant. Grav.*, 23:6955–6969, 2006.
[15] S.Räsänen. Constraints on backreaction in dust universes. *Class. Quantum Grav.*, 23(6):1823–1835, 2006.
[16] T. Buchert. Dark energy from structure: a status report. *Gen. Rel. Grav.*, 40:467–527, 2008.
[17] C. Chevalier, F. Debbasch, and Y. Ollivier. Large scale nonlinear effects of fluctuations in relativistic gravitation. *Nonlinear Analysis Series A: Theory, Methods & Applications, In Press.*
[18] J. Larena, J.M. Alimi, T. Buchert, M. Kuntz, and P.S. Corasaniti. Testing backreaction effects with observations. *arXiv:0808.1161*, 2008.
[19] F. Debbasch and Y. Ollivier. Observing a Schwarzschild black-hole with finite precision. *Astron. Astrophys.*, 433(2):397–404, 2005.
[20] C. Chevalier, M.D. Bustamante, and F. Debbasch. Thermal statistical ensembles of black holes. *Physica A*, 376:293–307, 2007.
[21] C. Chevalier and F. Debbasch. Thermal statistical ensembles of classical extreme black holes. *Physica A*, 388:628–638, 2009.
[22] L.D. Landau and E.M. Lifshitz. *The Classical Theory of Fields.* Pergamon Press, Oxford, 4th edition, 1975.
[23] R.A. Isaacson. Gravitational radiation in the limit of high frequency. II. Nonlinear terms and the effective stress tensor. *Phys. Rev.*, 166:1272, 1968.
[24] W. Zhao and Y. Zhang. Relic gravitational waves and their detection. *Phys. Rev. D*, 74:043503, 2006.
[25] S.E. Thorsett and R.J. Dewey. Pulsar timing limits on very low frequency stochastic gravitational radiation. *Phys. Rev. D*, 53:3468, 1996.
[26] C.R. Gwinn, T.M. Eubanks, T. Pyne, M. Birkinshaw and D.N. Matsakis. Quasar Proper Motions and Low-Frequency Gravitational Waves. *Astrophys. J.*, 485:87, 1997.
[27] GAIA project, ESA, 2000. http://www.rssd.esa.int/SA-general/Projects/GAIA_files/LATEX2HTML/node145.html
[28] D.R. Lorimer. Binary and Millisecond Pulsars. *Living Reviews in Relativity*, 2008. http://relativity.livingreviews.org/Articles/lrr-2008-8/
[29] B.S. Sathyaprakash and B.F. Schutz. Physics, Astrophysics and Cosmology with Gravitational Waves. *Living Reviews in Relativity*, 2009. http://relativity.livingreviews.org/Articles/lrr-2009-2/

# Dynamical spacetimes and gravitational radiation in a Fully Constrained Formulation

**Isabel Cordero-Carrión[1], Pablo Cerdá-Durán[2] and José María Ibáñez[1]**

[1] Departamento de Astronomía y Astrofísica, Universidad de Valencia, C/ Dr. Moliner 50, E-46100 Burjassot, Valencia, Spain

[2] Max-Planck-Institut für Astrophysik, Karl-Schwarzschild-Strasse 1, D-85741 Garching, Germany

E-mail: isabel.cordero@uv.es, cerda@mpa-garching.mpg.de, jose.m.ibanez@uv.es

**Abstract.** This contribution summarizes the recent work carried out to analyze the behavior of the hyperbolic sector of the Fully Constrained Formulation (FCF) derived in Bonazzola et al. 2004. The numerical experiments presented here allows one to be confident in the performances of the upgraded version of CoCoNuT's code by replacing the Conformally Flat Condition (CFC) approximation of the Einstein equations by the FCF.

## 1. Introduction

Most numerical codes used to solve the Einstein equations in order to obtain stationary or dynamical spacetimes generated by compact astrophysical objects are based on the 3+1 formalism (see, e.g., [1]). The Fully Constrained Formalism (FCF), recently proposed in [2, 3, 4], is a constrained evolution formulation of the full Einstein equations. It is a natural generalization of the Conformally Flat Condition (CFC) [5, 6], which is an approximation of the Einstein equations. FCF extends the CFC approximation and includes a hyperbolic system governing the gravitational radiation. Reader interested in details and motivation about FCF can address to the reference [2]. We present numerical simulations of this system in several cases.

## 2. Formalism

Given an asymptotically flat spacetime $(\mathcal{M}, g_{\mu\nu})$ we consider a $3+1$ splitting by spacelike hypersurfaces $\Sigma_t$, with $n^\mu$ being the timelike unit normal to $\Sigma_t$. Latin (greek) indices go from 1 (0) to 3. $\gamma_{\mu\nu} = g_{\mu\nu} + n_\mu n_\nu$ denotes the 3-metric on $\Sigma_t$ and $K_{\mu\nu} = -\frac{1}{2}\mathcal{L}_{\boldsymbol{n}}\gamma_{\mu\nu}$ the extrinsic curvature. With the lapse function $N$ and the shift vector $\beta^i$, the metric $g_{\mu\nu}$ is expressed in coordinates $(x^\mu)$ as $g_{\mu\nu}\, dx^\mu\, dx^\nu = -N^2\, dt^2 + \gamma_{ij}(dx^i + \beta^i\, dt)(dx^j + \beta^j\, dt)$. As in [2], we introduce a flat metric $f_{ij}$, which satisfies $\partial_t f_{ij} = 0$ and $f_{ij} \sim \gamma_{ij}$ at spatial infinity. We define $\gamma := \det \gamma_{ij}$ and $f := \det f_{ij}$. We introduce the following conformal decomposition: $\gamma_{ij} = \psi^4 \tilde{\gamma}_{ij}$. We define $h^{ij} := \tilde{\gamma}^{\,ij} - f^{ij}$. The gauge in [2] is *maximal slicing*, $K = 0$, and the so-called *generalized Dirac gauge*, $\mathcal{D}_k \tilde{\gamma}^{\,ki} = 0$, where $\mathcal{D}_k$ is the Levi–Civita connection associated with $f_{ij}$. The Einstein equations become a coupled elliptic-hyperbolic system: the elliptic sector acts on the variables $\psi$, $N$, and $\beta^i$, and the hyperbolic sector acts on $h^{ij}$ [2]. If $h^{ij} = 0$ is imposed, CFC is recovered.

© 2010 IOP Publishing Ltd

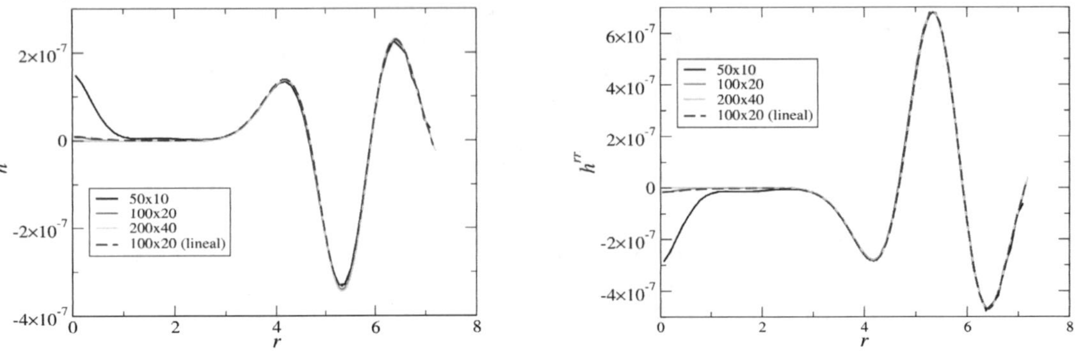


**Figure 1.** Radial profile of $h^{rr}$ at $t = 6$. Right figure corresponds to the equator and left figure corresponds to the pole. Continuous lines corresponds to $n_r = 50$ and $n_\theta = 100$ (radial and angular number of grid points), $n_r = 100$ and $n_\theta = 20$, and $n_r = 200$ and $n_\theta = 40$, in non-linear evolutions. Dashed line corresponds to $n_r = 100$ and $n_\theta = 20$, in a linear evolution.

We introduce the conformal decomposition $\hat{A}^{ij} = \psi^{10}\left(K^{ij} - \frac{1}{3}K\gamma^{ij}\right)$, where $K = \gamma^{ij}K_{ij}$. This decomposition is different from the one introduced in [2], but it is motivated by the local uniqueness properties of the elliptic equations shown in [4]. We rewrite the equation for $h^{ij}$ as a first order evolution system for the tensors $(h^{ij}, \hat{A}^{ij}, w_k^{ij})$, where $w_k^{ij} := \mathcal{D}_k\tilde{\gamma}^{ij}$ [7].

## 3. Numerical simulations

We evolve numerically the previous evolution system for $(h^{ij}, \hat{A}^{ij}, w_k^{ij})$. During the numerical simulations, we consider $N$, $\beta^i$, $\psi$ and the energy-momentum tensor as sources of the system. We perform the evolution of matter with the CoCoNuT code [8]. Some basic elements of this code are: i) fourth order finite differences scheme for the spatial derivatives and fourth order Runge-Kutta methods for the time derivative; ii) axisymmetry and symmetry with respect to the equatorial plane; iii) spherical orthonormal coordinates, $(r, \theta, \varphi)$; iv) a Sommerfeld condition at the outer boundary; and v) Kreiss-Oliger dissipative term in order to avoid the numerical noise of high frequency which appears during long-term simulations.

### *3.1. Teukolsky waves*

The first test is the evolution of a combination of ingoing and outgoing even-parity axisymmetric Teukolsky waves [9], in order to construct regular initial data at the center $r = 0$. The amplitude chosen is $10^{-5}$. These data are a solution of the linearized wave equation in a vacuum, they satisfy the Dirac gauge and are traceless (which is the linear approximation of unit determinant). The background is flat, i.e., $N = \psi = 1$, and $\beta^i = 0$. We display in Fig. 1 the radial profile of the component $h^{rr}$ at $t = 6$, at the equator and at the pole, with different values of the number of radial and angular points respectively. The analytical solution is recovered in the linear regime, as well as the velocity and the amplitude of the wave, and its decay with the radius. The absolute errors (L2 norm) in the numerical simulations are smaller than $8 \times 10^{-8}$ for all the non zero components. The Sommerfeld condition at the outer boundary produces ingoing reflections that do not grow in time and have an amplitude much smaller than the amplitude of the initial one. We obtain second order of convergence for all the non zero components of the tensor $h^{ij}$, due to the influence of the inner boundary conditions imposed at $r = 0$, $\theta = 0$ and $\theta = \pi/2$.

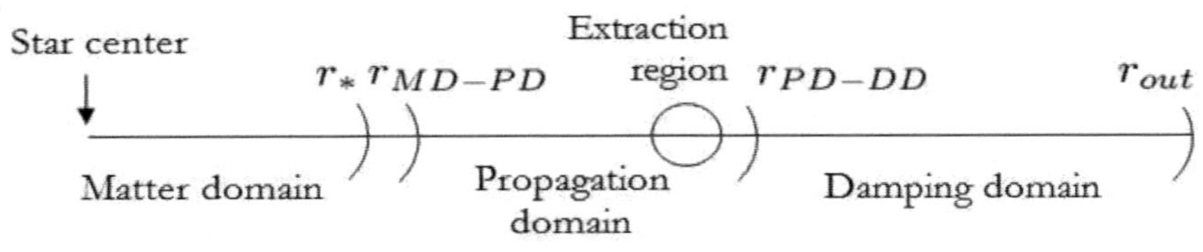


**Figure 2.** Scheme of the radial grid used in the code for rotating neutron star simulations.

*3.2. Equilibrium configuration of rotating neutron stars*

We consider an axisymmetric and uniformly rotating neutron star in equilibrium. The initial data have been obtained from LORENE [10], a library of spectral methods. These models, of different rotation parameters, have non-vanishing $\beta^i$, $N$, $\psi$ and matter fields. All these variables are kept fixed during the evolution of the tensor $h^{ij}$. The initial model is a neutron star with a 550 Hz rotation frequency, a 1.6 $M_\odot$ baryon mass and a 12.86 km coordinate equatorial radius.

Figure 2 shows a sketch of the grid, where the following domains have been defined: The matter domain (MD) contains the star. The propagation domain (PD) is the one where the waves are well-resolved. The region for extracting the gravitational radiation is placed at the outer boundary of the PD. The damping domain (DD) has a lower resolution in the radial direction. The wave travels far enough for imposing the Sommerfeld condition at the outer boundary before reaching it. Let $r_{\rm MD-PD}$ be the radius which separates the MD and the PD, and $r_{\rm PD-DD}$ be the radius which separates the PD and the DD. The star radius, $r_*$, is close to th radius $r_{\rm MD-PD}$ but still smaller in value. The logarithmic scale (in the radial direction) is a very useful tool for constructing this kind of grid with a reasonable number of points. We get the necessary accuracy in the MD and the PD, but the time stepping becomes more severe than the one required in an equally spaced radial grid.

As initial data we have used stationary values for the tensor $h^{ij}$ and some differences from stationarity for $\hat{A}^{ij}$. The aim at introducing these initial differences from stationarity is two-fold: on the one hand, the recovering of stationarity from a perturbed initial data, and, on the other hand, testing the outer boundary Sommerfeld condition imposed by the generation of the artificial wave. In this general background, those differences from stationarity introduced in the initial data, generate a perturbation that propagates to the outer boundary, leaving behind the stationary solution. In Fig. 3, we have plotted the absolute value of the component $h^{rr}$ in terms of the radius for different times. We can see the perturbation traveling towards the outer boundary, where the wave leaves the numerical domain. The differences from stationarity are interpreted as an initial perturbation. The reflections close to the outer boundary come from the numerical implementation of the Sommerfeld condition. Second order of convergence is obtained for all the non zero components of the tensor $h^{ij}$, as previously.

This test helps to understand how the Sommerfeld condition at the outer boundary works. This condition is imposed to treat the outgoing waves which reach the outer boundary, as the one caused by the initial data. If the background is not flat enough, the Sommerfeld condition can interpret the background as an outgoing wave. This behavior can be seen in Fig. 4 for the simulation whose outer boundary is placed closer, at $3 \times 10^7$. The Sommerfeld condition works properly for $r_{out} \gtrsim 3 \times 10^8$cm (around 300 stellar radii). In our calculations we have used 80

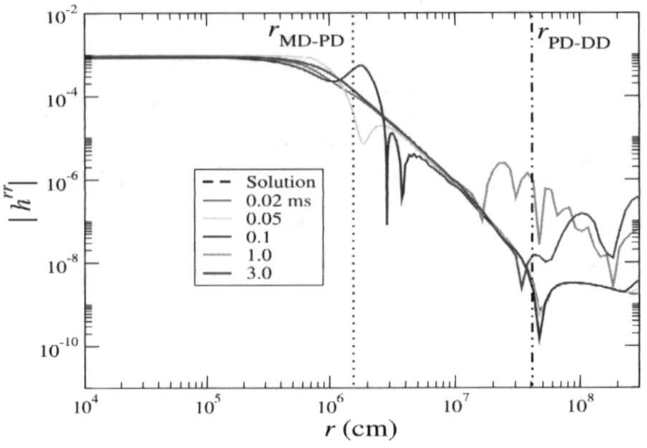


**Figure 3.** Radial profile of $|h^{rr}|$. Dashed line corresponds to the stationary solution (as reference). Continuous lines correspond to the evolution of the initial data at 0.02, 0.05, 0.1, 1.0 and 3.0 ms respectively. Vertical lines denote radius between domains.

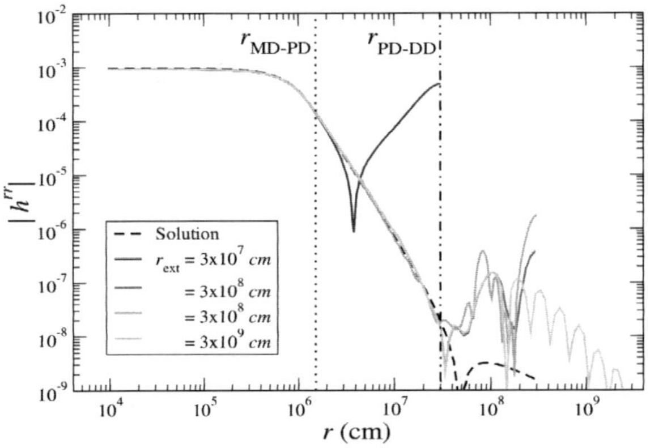


**Figure 4.** Radial profile of $|h^{rr}|$ at time $t = 3$ ms. The dashed line corresponds to the stationary solution (as reference). Continuous lines correspond to the numerical evolutions whose outer boundary are placed at $3 \times 10^7$, $3 \times 10^8$ and $3 \times 10^9$ cm. Vertical lines denote radii between domains.

radial grid points in the MD, 80 points in the PD and around 50 in the DD.

Once the outer boundary radius has been fixed, we can use the stationary initial data for both $h^{ij}$ and $\hat{A}^{ij}$ tensors. In Fig. 5 we show the absolute value of the component $h^{rr}$, in terms of the radius, for different times. No initial artificial wave is introduced. The noise comes from the outer boundary condition. Again, second order of convergence is obtained.

### *3.3. Perturbed equilibrium configuration of a rotating neutron star*

We consider a perturbed neutron star. The perturbation is proportional to $\sin(\pi\ r/r_*)$. Since the star is rotating, $r_*$ depends on $\theta$, and the perturbation is not spherically symmetric; hence, gravitational radiation is expected to be generated. The hydrodynamic equations governing the matter evolution are solved. We keep fixed $N$, $\beta^i$ and $\psi$, assuming that they do not change very

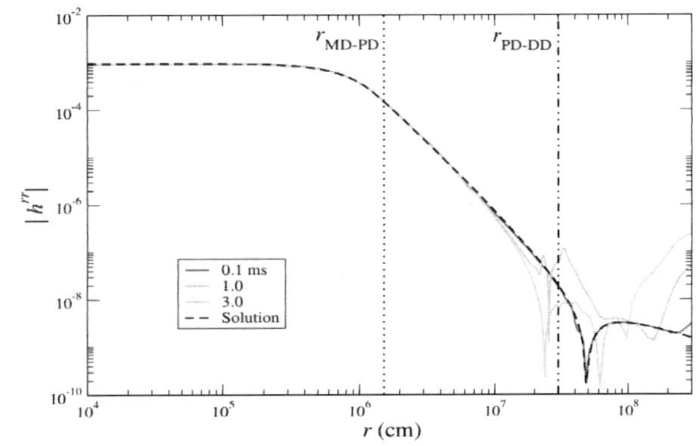


**Figure 5.** Radial profile of $|h^{rr}|$. The dashed line corresponds to the stationary solution (as reference). Continuous lines correspond to the evolution of the initial data at 0.1, 1.0 and 3.0 ms. Vertical lines denote radii between domains.

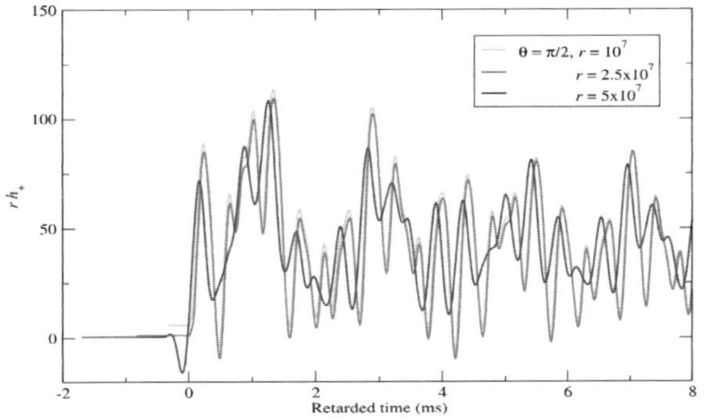


**Figure 6.** $rh_+$ versus the retarded time. The corresponding wave is extracted at $r = 10^7$, $r = 2.5 \times 10^7$ and $r = 5 \times 10^7$cm, at the equator. $r_{\rm PD-DD}$ is placed at $3 \times 10^7$cm.

much during the evolution. This test will provide where to place $r_{\rm PD-DD}$, and, consequently, where to extract the gravitational radiation.

In Fig. 6, an approximation of the real part of the Weyl scalar $\Psi_4$, $h_+$, (scaled with $r$) is plotted in terms of the retarded time, for different values of the extraction radius, at the equator. This approximation is accurate enough for the objectives of this test. $r_{\rm PD-DD}$ is placed at $3 \times 10^7$cm (around 30 stellar radii). The waves correspond to the physical gravitational waves coming from the evolution of the perturbed star. This evolution is governed by the coupled hydrodynamic and the Einstein equations (but keeping fixed $N$, $\beta^i$ and $\psi$). The different curves in Fig. 6 refer to different radii where the gravitational radiation has been extracted. From this figure, we can deduce that the speed of the waves is the light velocity and they decay as $1/r$. We also see that the wave must be extracted far from the source but inside the PD. Regarding the resolution of the logarithmic grid, we have noticed that it is enough to cover a wavelength with five points at the end of the PD in order to obtain the wave accurately.

Furthermore, it will be important to know where to place $r_{\rm PD-DD}$. In Fig. 7, $rh_+$ is plotted in terms of the retarded time in simulations with different $r_{\rm PD-DD}$. The wave is extracted close to $r_{\rm PD-DD}$ in all the cases. From the figure, it is reasonable to place $r_{\rm PD-DD}$ at $3 \times 10^7$cm (around 30 stellar radii) or further.

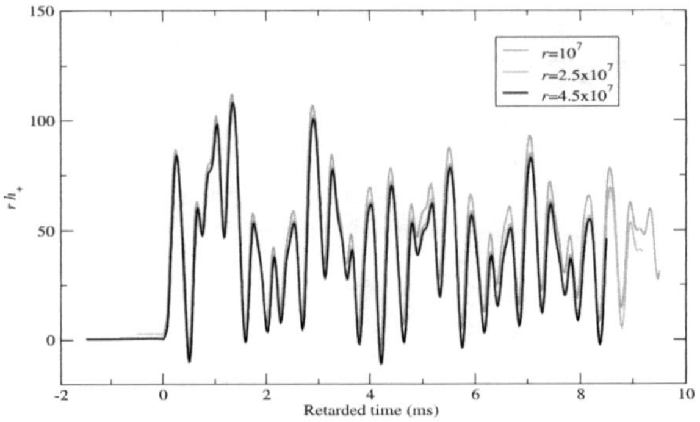


**Figure 7.** $rh_+$ versus the retarded time, at the equator. Radii are expressed in cm. The corresponding wave is extracted at $r = 10^7$ ($r_{\rm PD-DD} = 1.5 \times 10^7$), $r = 2.5 \times 10^7$ ($r_{\rm PD-DD} = 3 \times 10^7$) and $r = 4.5 \times 10^7$ ($r_{\rm PD-DD} = 5 \times 10^7$).

## 4. Conclusions

The first version of CoCoNuT's code [11] was designed to evolve matter fields of a perfect fluid in the dynamical spacetime of the CFC approximation. The current version of CoCoNuT incorporates magnetic fields, i.e., it is a general-relativistic magneto-hydrodynamic code which evolves matter in the dynamical spacetime of CFC. The present contribution to this conference, together with previous works in [3, 4], is a step towards the upgrading of the metric evolution of CoCoNuT's code by substituting the CFC approximation of the Einstein equations by the FCF of [2]. The hyperbolic sector to evolve $h^{ij}$ has already been included into CoCoNuT's code. In this contribution we have shown that this sector works properly. We have presented the numerical evolution of Teukolsky waves in a flat background. We have analyzed the behavior of the code under the Sommerfeld boundary condition, and detected the outer radius, in the case of spacetimes generated by equilibrium configurations of rotating neutron stars. Finally, we have extracted the gravitational waveforms in the case of dynamical spacetimes generated by perturbed rotating neutron stars.

## Acknowledgments

I. C.-C. acknowledges support from the Spanish Ministerio de Educación y Ciencia (AP2005-2857). This work has been supported by the Ministerio de Educación y Ciencia through Grant No. AYA2007-67626-C03-01 and by the Collaborative Research Center on Gravitational Wave Astronomy of the Deutsche Forschungsgemeinschaft (DFG SFB/Transregio 7).

## References

[1] Alcubierre M 2008 *Introduction to 3+1 numerical relativity* (Oxford: Oxford University Press)
[2] Bonazzola S, Gourgoulhon E, Grandclément P and Novak J 2004, *Phys. Rev. D* **70** 104007
[3] Cordero-Carrión I, Ibáñez J M, Gourgoulhon E, Jaramillo J L and Novak J 2008 *Phys. Rev. D* **77** 084007
[4] Cordero-Carrión I, Cerdá-Durán P, Dimmelmeier H, Jaramillo J L, Novak J and Gourgoulhon E 2009 *Phys. Rev. D* **79** 024017
[5] Isenberg J A 2008 *Int. J. Mod. Phys. D* **17** 265
[6] Wilson J R and Mathews G J 1989 *Relativistic hydrodynamics* (Frontiers in numerical relativity) (Cambridge: Cambridge University Press)
[7] Cordero-Carrión I 2009 *PhD Thesis* (University of Valencia, Spain)
[8] http://www.mpa-garching.mpg.de/hydro/COCONUT
[9] Teukolsky S A 1982 *Phys. Rev. D* **26** 745
[10] http://www.lorene.obspm.fr/
[11] Dimmelmeier H, Font J A and Mueller E 2002 *Astron. Astrophys.* **393** 523

# Numerical studies of the efficiency of resonant relaxation around a massive black hole

**Gábor Kupi, Ehud Eilon[1] and Tal Alexander**

Faculty of Physics, Weizmann Institute of Science, Rehovot 76100, Israel

E-mail: gabor.kupi@weizmann.ac.il

**Abstract.** Resonant relaxation (RR) is a rapid relaxation process that operates in the nearly-Keplerian potential near a massive black hole (MBH). RR dominates the dynamics of compact remnants that inspiral into a MBH and emit gravitational waves (extreme mass ratio inspiral events, EMRIs). RR can either increase the EMRI rate, or strongly suppress it, depending on its still poorly-determined efficiency. We use small-scale Newtonian $N$-body simulations to measure the RR efficiency and to explore its possible dependence on the stellar number density profile around the MBH, and the mass-ratio between the MBH and a star (a single-mass stellar population is assumed). We present a suite of simulations with a range of stellar density profiles and mass-ratios, and measure the mean RR efficiency in the near-Keplerian limit. We do not find a strong dependence on the density profile or the mass-ratio. Our numerical determination of the RR efficiency in the Newtonian, single-mass population approximations, suggests that RR will likely enhance the EMRI rate by a factor of a few over the rates predicted assuming only slow stochastic two-body relaxation.

## 1. Introduction

Dynamical relaxation processes near MBHs in galactic centers affect the rates of strong stellar interactions with the MBH, such as tidal disruption, tidal dissipation, or gravitational wave (GW) emission (e.g. [1]). These relaxation processes may also be reflected by the dynamical properties of the different stellar populations there [6], as observed in the Galactic Center [5, 7]. Of particular importance, in anticipation of the planned Laser Interferometer Space Antenna (LISA) GW detector, is to understand the role of relaxation in regulating the rate of GW emission events from compact remnants undergoing quasi-periodic EMRI into MBHs.

Two-body relaxation, or non-coherent relaxation (NR), is inherent to any discrete large-$N$ system, due to the cumulative effect of uncorrelated two-body encounters. These cause the orbital energy $E$ and the angular momentum $J$ to change in a random-walk fashion ($\propto \sqrt{t}$) on the typically long NR timescale $T_{NR}$. In contrast, when the gravitational potential has approximate symmetries that restrict orbital evolution (e.g. fixed ellipses in a Keplerian potential; fixed orbital planes in a spherical potential), the perturbations on a test star are no longer random, but correlated, leading to coherent ($\propto t$) torquing of $J$ on short timescales, while the symmetries hold. Over longer times, this results in resonant relaxation (RR) [10, 9], a rapid random walk of $J$ on the typically short RR timescale $T_{RR} \ll T_{NR}$. RR in a near-Keplerian potential can change both the direction and magnitude of **J** ("scalar RR"), thereby driving stars to near-radial

[1] Present address: Israel Aerospace Industries, Ben Gurion International Airport, 70100, Israel

© 2010 IOP Publishing Ltd

orbits that interact strongly with the MBH. RR in a near-spherical potential can only change the direction of **J** ("vector RR").

RR is particularly relevant in the potential near a MBH, where compact EMRI candidates originate. Hopman and Alexander [6] show that RR dominates EMRI source dynamics. Depending on its still poorly-determined efficiency, RR can either increase the EMRI rate over that predicted assuming NR only, or if too efficient, it can strongly suppress the EMRI rate by throwing the compact remnants into infall (plunge) orbits that emit a single, non-periodic and hard to detect GW burst. The still open questions about the implications of RR for EMRI rates and orbital properties provide a prime motivation for the systematic numerical investigation of RR efficiency presented here. A more detailed treatment of this subject can be found in [4].

## 2. Theory

### *2.1. Non-coherent Relaxation (NR)*

The NR time for $E$-relaxation, $T_{\rm NR}^{E}$, corresponds to the time it takes non-coherent 2-body interactions to change the stellar orbital energy by order of itself, $|\Delta E|\sim E$ (by stellar dynamical definition convention, $E>0$ for a bound orbit). Similarly, the NR time for $J$-relaxation, $T_{\rm NR}^{J}$, corresponds to the time it takes the stellar orbital angular momentum to change by order of the circular angular momentum $|\Delta J|\sim J_c$, where near a MBH of mass $M$, $J_c=GM/\sqrt{2E}$. The $E$-relaxation timescale can be estimated by considering the rate of gravitational collisions in a system of size $R$ between a test star and $N$ field stars: $T_{\rm NR}^{E}\sim(M/m)^2P/(N\ln\Lambda)$, where $m$ is the mass of stars, $P=2\pi\sqrt{R^3/GM}$ is the Keplerian period and $\Lambda$ is the Coulomb logarithm $\ln\Lambda\sim\ln(R/r_{min})$, $r_{min}$ is the minimal impact parameter where the small angle deflection assumption still holds. Near the MBH $\ln\Lambda\sim\ln Q$, where $Q\equiv M/m$.

Rauch and Tremaine [10] (RT96) parametrized the uncertainties in the evolution of the orbital $E$, $J$ and $\mathbf{J}$ by a set of numeric coefficients ($\alpha_\Lambda$, $\eta_{s,v}$, $\beta_{s,v}$), whose operational definition is tied to the procedure by which they are estimated (left unspecified by RT96). We adopt this notation and estimate these coefficients by the rms change over the stellar population. The NR changes in $E$, $J$ and $\mathbf{J}$ (here generally denoted by $x$) over the dimensionless time-lag $\tau\equiv(t_2-t_1)/P_1$ are thus formally defined as

$$\delta x(\tau)\;\equiv\;\left\langle\left(|x_2-x_1|/x_0\right)^2\right\rangle^{1/2}=K(\sqrt{N}/Q)\sqrt{\tau}\tag{1}$$

where $x_0$ is $E_1$ for $E$ and $J_{c,1}$ (angular momentum of circular orbit) for $J$ and $\mathbf{J}$. The dimensionless, order-unity constants $K$ are $\alpha\sqrt{\ln\Lambda}$, $\eta_s\sqrt{\ln\Lambda}$ and $\eta_v\sqrt{\ln\Lambda}$ respectively. The exact determination of these constants, which are system-dependent, requires detailed calculations or simulations. The corresponding NR timescales are related to these coefficients by $T_{\rm NR}^{E}=(M/m)^2P/(N\alpha_\Lambda^2)$, $T_{\rm NR}^{J}=(M/m)^2P/(N\eta_{s\Lambda}^2)$ and $T_{\rm NR}^{\mathbf{J}}=(M/m)^2P/(N\eta_{v\Lambda}^2)$.

### *2.2. Resonant Relaxation (RR)*

When the gravitational potential has symmetries that restrict the orbital evolution, for example to fixed ellipses in the potential of a point mass, or to planar annuli in a spherical potential, the perturbations on a test star are no longer random, but correlated. This leads to a coherent changes in $\mathbf{J}$ on times $P\ll t<t_\omega$, $\Delta\mathbf{J}=\mathbf{T}t$, by the residual torque $|\mathbf{T}|\sim\sqrt{N}Gm/R$ exerted by the $N$ randomly oriented, orbit-averaged mass distributions of the surrounding stars (mass wires for elliptical orbits in a Kepler potential, mass annuli for rosette-like orbits in a spherical potential). The coherence time $t_\omega$ is set by deviations from the true symmetry, which lead to a gradual orbital drift and to the randomization of $\mathbf{T}$. For example, the enclosed stellar mass leads to non-Keplerian retrograde precession; General Relativity leads to prograde precession. Ultimately, the coherent torques themselves randomize the orbits. The effective coherence time is set by the

shortest de-coherence (quenching) process. The accumulated change over $t_\omega$, $|\Delta\mathbf{J}_\omega|\sim|\mathbf{T}t_\omega|$, then becomes the basic step-size, or mean free path in $\mathbf{J}$-space, for the long-term ($t\gg t_\omega$) random-walk phase ($\propto\sqrt{t}$) of $\mathbf{J}$ by RR . Since this step-size is large, RR can be much faster than NR. The RR timescale $T_{\mathrm{RR}}$ is then defined by $|\Delta J|/J_c=(|\Delta J_\omega|/J_c)\sqrt{t/t_\omega}\equiv\sqrt{t/T_{\mathrm{RR}}}$. Note that the relaxation of $E$ is not affected by RR because the potential of the system is stationary on the coherence timescale, and so $E$ changes incoherently on all time scales. The torques exerted by elliptical mass wires in a Kepler potential can change both the direction and magnitude of $\mathbf{J}$. In contrast, the torques exerted by planar annuli can only change the direction of $\mathbf{J}$. The long-term random walk $\sqrt{\tau}$ growth continues until $|\Delta\mathbf{J}|/J_c\sim O(1)$, at which point $|\Delta J|/J_c$ and $|\Delta\mathbf{J}|/J_c$ can no longer grow, but $\mathbf{J}$ does continue to change its value randomly.

Here we consider only Newtonian dynamics. The coherence timescale for scalar RR is determined by the time it takes for the orbital apsis to precess by angle $\sim\pi$ due to the potential of the enclosed stellar mass ("mass precession"),

$$t_\omega=t_M=A_M\,(Q/N)\,P\,, \tag{2}$$

where $A_M$ is an $O(1)$ factor reflecting the approximations in this estimate. The coherence timescale for vector RR is determined by the time it takes for the coherent torques to change $|\Delta\mathbf{J}|\sim J_c$

$$t_\omega=t_\phi=A_\phi\left(\sqrt{N}/\mu\right)P\simeq\left[A_\phi Q\left/\sqrt{N}\right.\right]P\,, \tag{3}$$

where $A_\phi$ is an $O(1)$ factor, $\mu=Nm/(M+Nm)$ and where the approximate equality is for the Keplerian limit $Nm\ll M$. The RR changes in $J$ and $\mathbf{J}$ during the coherent phase ($\tau<\tau_\omega$) are defined in terms of the rms change in analogy with Eq. (1)

$$\delta J(\tau)=\beta_s\frac{\sqrt{N}}{Q}\tau\,,\quad \delta\mathbf{J}(\tau)=\beta_v\frac{\sqrt{N}}{Q}\tau\,, \tag{4}$$

where the dimensionless coefficients $\beta_s$ and $\beta_v$ depend on the parameters of the system and reflect the uncertainties introduced by the various approximations and simplification of this analysis. Accurate determination of their values requires detailed calculations or simulations.

The scalar RR change in $J$ on time-lags $\tau\gg\tau_M$ and the scalar RR timescale are then

$$\delta J(\tau)=\beta_s\sqrt{A_M/Q}\sqrt{\tau}\,,\quad T^J_{\mathrm{RR}}=\left[Q/A_M\beta_s^2\right]P\,. \tag{5}$$

The RR efficiency factor $\chi=(\beta_s/\beta_{s,\mathrm{RT96}})^2$ defined by [6] for an assumed value $A_M=1$, expresses how much shorter $T^J_{\mathrm{RR}}$ is relative to the value estimated by RT96. Scalar RR is faster than NR by a factor $T^J_{\mathrm{NR}}/T^J_{\mathrm{RR}}\propto(M/m)/N\ln\Lambda$. Similarly, the vector RR change in $\mathbf{J}$ on time-lags $\tau\gg\tau_\phi$ in the Keplerian limit and the vector RR timescale are

$$\delta\mathbf{J}(\tau)=\beta_v\sqrt{A_\phi\sqrt{N}/Q}\sqrt{\tau}\,,\quad T^{\mathbf{J}}_{\mathrm{RR}}=\left[Q\left/\sqrt{N}A_\phi\beta_v^2\right.\right]P\,. \tag{6}$$

## 3. Simulations

Our $N$-body code uses a 5th order Runge-Kutta integrator without gravity softening. The time steps were chosen to conserve total energy at the level of $|\Delta E_{\mathrm{tot}}|/E_{\mathrm{tot}}\sim O(10^{-5})-O(10^{-8})$, well below the NR energy changes expected in the simulations.

To simulate RR reliably, the $N$-body code must maximize angular momentum conservation and minimize numerical RR quenching, which can be caused by numerical precession of the apsis, which changes the torque, or to a lesser degree by a drift in the orbital period (equivalently, the

doi:10.1088/1742-6596/228/1/012056

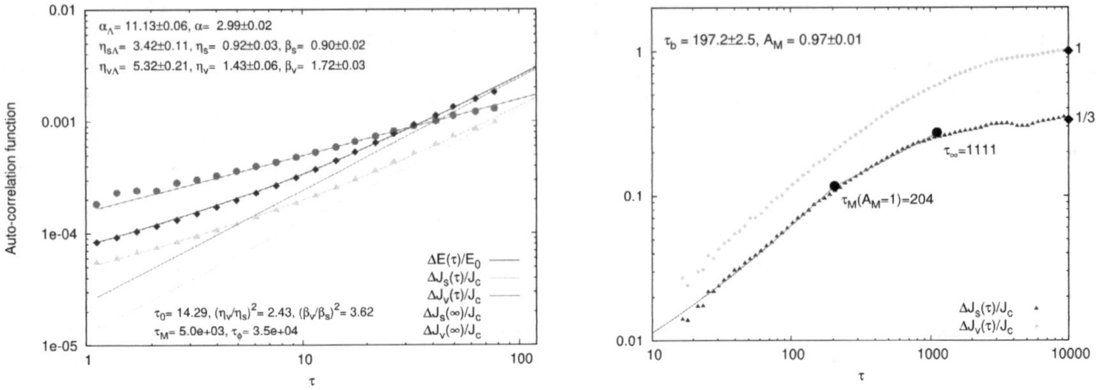


**Figure 1. Left**: The measured (points) and fitted (lines) correlation curves for $\delta E$ (top line), $\delta J$ (bottom line) and $\delta \mathbf{J}$ (middle line) in a $Q=10^6$, $\gamma=1.5$ simulation with $N=200$ particles. The points are the results of the $N$-body simulation, the thick lines are the predicted theoretical curves; the thin straight lines show the asymptotic linear behavior. The best fit coefficients and related quantities are also listed. **Right**: The long-term scalar RR (bottom curve) and vector RR (top curve) in a $Q=10^4$ simulation with $N=49$ particles and $\gamma=1$ density profile. The mass-precession timescale $\tau_M$ is calculated from Eq. (2) with the assumption that $A_M=1$ (verified here, see text). The saturation timescale of scalar RR, $\tau_\infty$, is defined by $\left(\tau_\infty/\tau_{RR}^J\right)^{1/2}=1/3$.

energy), which changes the moment of inertia of the "mass wire". We estimated these numerical drifts by integrating, over many orbits, a highly eccentric 2-body system ($e=0.99995$, $Q=3\times10^6$), since most of the numerical errors accumulate at periapse, where the acceleration is largest. We tracked the change in the period, $\Delta P=(P-P_0)/P_0$, in the angular momentum $\Delta\mathbf{J}=|\mathbf{J}-\mathbf{J}_0|/J_0$ and in the direction of the apsis $\Delta\hat{\mathbf{e}}=|\hat{\mathbf{e}}-\hat{\mathbf{e}}_0|/|\hat{\mathbf{e}}_0|$, where $\hat{\mathbf{e}}$ is the normalized Runge-Lenz vector. We find that after $\tau_{\rm sim}=4.5\times10^4$, $\Delta P\sim8\times10^{-5}$, $\Delta\mathbf{J}\sim3\times10^{-4}$ and $\Delta\hat{\mathbf{e}}\sim9\times10^{-8}$, which correspond to drift timescales of $\tau_{\Delta P}\sim(\pi/\Delta P)\tau_{\rm sim}=2\times10^9$, $\tau_{\Delta\mathbf{J}}\sim5\times10^8$ and $\tau_{\Delta\hat{\mathbf{e}}}\sim3\times10^{12}$. These are many orders of magnitude longer than our longest simulations, and can therefore be safely neglected.

We carried out two sets of Newtonian simulations. The first, designed to study the short-term behavior of RR ($\tau_{\rm sim}<\tau_M$) and to measure NR and RR parameters $\alpha$, $\eta_{s,v}$ and $\beta_{s,v}$, consisted of 200 particles (including the MBH as a fixed particle in the origin of the coordinate system). The initial orbital semi-major axes were randomly drawn from a $\rho(a)\mathrm{d}a\propto a^{2-\gamma}\mathrm{d}a$ distribution for $\gamma=1$, 1.5, 1.75, for $a$ in the range $(0,1/2)$, with eccentricities drawn from a thermal $\rho(e)\mathrm{d}e=2e\mathrm{d}e$ distribution, with random phases, orbital orientations and isotropic velocities. This distribution corresponds approximately to a $r^{-\gamma}$ number density distribution with an outer cutoff at radius $r=R=1$ from the MBH (in dimensionless units where $G=1$, $M+Nm=1$). We made $n_{\rm sim}=114$ simulations in this set.

The second set of simulations was designed to study the long term behavior of RR ($\tau\gg\tau_\phi$), the transition from the coherent RR phase to the random-walk phase, and the asymptotic saturation limits. Due to the high computational load, this set ($n_{\rm sim}=7$) was limited to a relatively small mass ratios, $Q\le10^4$, smaller number of particles, ($50\le N\le200$) and flat cusps with $\gamma=1$, where time-consuming close star-star interactions are rarer.

## 4. Results

The results of a typical Keplerian simulation are shown in Fig. 1L. Note that in such a potential, vector RR measures the combined change in the magnitude *and* direction of the

**Table 1.** Measured NR and RR coefficients [a].

| $\gamma$ | $Q$ | $n_{\rm sim}$ | $\bar{\alpha}_\Lambda$ | $\bar{\eta}_{s\Lambda}$ | $\bar{\eta}_{v\Lambda}$ | $\bar{\beta}_s$ | $\bar{\beta}_v$ |
|---|---|---|---|---|---|---|---|
| 1 | $10^6$ | 12 | 10.48±0.43 | 4.58±0.38 | 7.31±0.61 | 1.14±0.07 | 1.95±0.08 |
| 1 | $10^7$ | 12 | 9.37±0.32 | 4.18±0.26 | 6.69±0.46 | 1.13±0.07 | 1.98±0.10 |
| 1 | $10^8$ | 12 | 10.27±0.36 | 4.26±0.22 | 7.02±0.47 | 1.13±0.06 | 1.89±0.08 |
| 1.5 | $10^6$ | 13 | 12.61±0.90 | 4.12±0.48 | 6.65±0.61 | 1.00±0.05 | 1.77±0.07 |
| 1.5 | $10^7$ | 12 | 11.35±0.46 | 4.09±0.35 | 6.58±0.57 | 1.05±0.06 | 1.86±0.07 |
| 1.5 | $10^8$ | 12 | 10.51 ± 0.51 | 3.99±0.21 | 6.57±0.49 | 1.09±0.07 | 1.92±0.09 |
| 1.75 | $10^6$ | 14 | 12.06±0.54 | 3.97±0.35 | 6.50±0.48 | 0.95±0.06 | 1.68±0.06 |
| 1.75 | $10^7$ | 12 | 12.29±0.43 | 3.39 ± 0.14 | 5.29±0.18 | 0.99±0.07 | 1.75±0.09 |
| 1.75 | $10^8$ | 15 | 11.90±0.43 | 3.89±0.39 | 6.20±0.52 | 1.01±0.05 | 1.73±0.05 |
| Grand average[b] | | | $\langle\alpha\rangle$ | $\langle\eta_s\rangle$ | $\langle\eta_v\rangle$ | $\langle\beta_s\rangle$ | $\langle\beta_v\rangle$ |
| | | 114 | 2.82±0.05 | 1.01±0.03 | 1.64±0.04 | 1.05±0.02 | 1.83±0.03 |

[a] The quoted errors are the errors on the mean (the sample rms is $\sqrt{n_{\rm sim}}$ times larger).
[b] $\langle\alpha\rangle = \left\langle\alpha_\Lambda/\sqrt{\ln\Lambda}\right\rangle$, $\langle\eta_{s,v}\rangle = \left\langle\eta_{s,v\Lambda}/\sqrt{\ln\Lambda}\right\rangle$ over all simulations for $\Lambda = Q$.

angular momentum vector. The numerical prefactors $\eta_{s\Lambda}$ and $\beta_s$ are obtained by fitting the $\delta J$ curve to the function $\delta J = (\sqrt{N}/Q)\sqrt{\eta_{s\Lambda}^2\tau + \beta_s^2\tau^2}$. Note that NR dominates the evolution at early times, $\tau < \tau_0 \equiv \eta_{s,v\Lambda}^2/\beta_{s,v}^2$. A similar procedure is used to obtain $\eta_{v,\Lambda}$ and $\beta_v$.

We summarize our results in table (1). The coefficients $\alpha_\Lambda$ and $\eta_{s,v\Lambda}$ do not directly express the intrinsic properties of NR, since they should vary as $\sqrt{\ln\Lambda} \simeq \sqrt{\ln Q} \simeq 4(1\pm0.07)$ over the $Q$-range of our models. This small fractional difference and the relatively low statistics of our simulation suite may explain why we do not detect a clear $Q$-dependence in these quantities. Since we do not see a dependence on $Q$, and the $\gamma$ dependence seems weak ($\alpha_\Lambda$ shows hints of an increase with $\gamma$, and $\eta$ and $\beta$ of a decrease with $\gamma$), we adopt as the best fit estimates of the values of $\alpha$, $\eta_{s,v}$ and $\beta_{s,v}$ their grand average over all the simulations.

To explore the long-term behavior of the system, verify that the short-timescale coherent torquing phase leads to long-term relaxation and to estimate the coefficients that enter the coherence timescales, we also conducted several very long simulations ($\tau \sim 10^4 \gg \tau_\phi$). The resulting RR correlation scalar curves clearly display a first clear detection of a transition at the mass precession coherence time $\tau_M = A_M Q/N$ (Eq. 2) from the asymptotically $\propto\tau$ coherent RR phase to the $\propto\sqrt{\tau}$ RR relaxation phase (Fig 1R). The mean best-fit estimate over 7 long time-scale simulations is $A_M = 0.99\pm0.06$, where the error is dominated by the simulation to simulation scatter. The vector RR curve does not display a coherent rise beyond $\tau_M$, since by its definition it includes a scalar component, and neither a distinct random-walk phase beyond $\tau_\phi$, since the transition to the asymptotic saturation limit occurs soon after, at $\tau \gtrsim \tau_\phi$ . This non-trivial behavior makes it harder to quantify and measure $A_\phi$ by a fit to the correlation curve, and we do not attempt it here.

Our analytical calculations predict that on very long time-lags, the correlation curves saturate to $\delta J \to 1/3$ and $\delta\mathbf{J} \to 1$ [4]. These limits can be seen in Fig. (1R).

## 5. Implications for gravitational wave source dynamics

We conclude the discussion of our results by briefly considering the implications of this revised value of $\beta_s$ for EMRI rates. [6] assumed $A_M = 1$, parametrized the RR efficiency by a factor $\chi = [\beta_s/\beta_{s,\rm RT96}]^2$, and derived the $\chi$ dependence of the branching ratio of the EMRI and infall (plunge) rates (Fig. 2). The RT96 value $\chi = 1$ happens to lie very close to the maximum of the RR-accelerated EMRI rate. We confirm here that $A_M = 0.99 \pm 0.06$ and find $\beta_s/\beta_{s,\rm RT96} = 1.98 \pm 0.04$, which corresponds to a factor 4.9 increase in the EMRI rate compared to that estimated for NR only, but is a factor $\sim 1.5$ smaller than implied by the RT96 value, because the higher RR efficiency leads to a higher plunge rate at the expense of the inspiral rate.

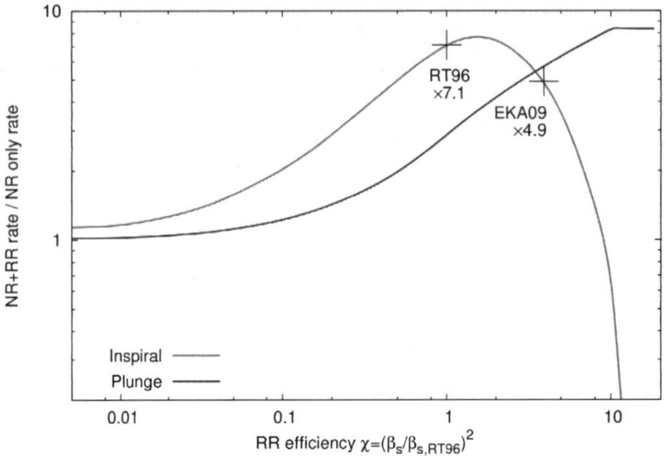


**Figure 2.** The change in the EMRI and plunge rates relative to that predicted assuming only NR, as function of the RR efficiency $\chi$ (Adapted from [6], Fig. 5). The RT96 estimate of $\beta_s$ ($\chi = 1$) predicts an increase of $\times 7.1$ in the EMRI rate due to RR, close to the maximum. Our new ([4], EKA09) higher measured RR efficiency ($\chi = 3.9$) predicts an increase of only $\times 4.9$ in the EMRI rate, and a higher plunge rate.

## 6. Summary

We characterized and measured the mean efficiency coefficients of NR ($\alpha_\Lambda, \eta_{s,\Lambda}, \eta_{v,\Lambda}$) and RR ($\beta_s, \beta_v$) in Newtonian $N$-body simulations of isotropic, thermal, near-Keplerian stellar cusps around a MBH. We measured these coefficients in a large suite of small scale $N$-body simulations with different stellar density distributions and MBH/star mass ratios. We do not find strong trends in the values of these coefficients as function of the system properties. This may require better statistics. We identified an early phase of NR-dominated relaxation that precedes the coherent RR phase. We carried out several long-term simulations and robustly detected, for the first time, the transition from the coherent to the random walk phase of RR, measured the efficiency of RR, and observed the asymptotic limits of the angular momentum correlation curves. Our measured RR efficiency suggests that RR increases the EMRI rate by a factor of $\sim 5$ above what is predicted for NR only.

Our measured RR efficiency is broadly consistent with that suggested by the analysis of the dynamical properties of the different stellar populations in the Galactic Center, which indicates an unusually rapid randomization of the stellar orbits [6, 8], and with the warping of the maser disk of NGC 4258, if interpreted as due to RR [3].

## References

[1] Alexander, T. 2005, Phys. Rep., 419, 65
[2] Bahcall, J. N. & Wolf, R. A. 1977, ApJ, 216, 883
[3] Bregman, M. & Alexander, T. 2009, ApJ, 700, L192
[4] Eilon, E., Kupi G. & Alexander, T., 2009, ApJ, 698, 641
[5] Genzel, R., Pichon, C., Eckart, A., Gerhard, O. E., & Ott, T. 2000, MNRAS, 317, 348
[6] Hopman, C. & Alexander, T. 2006, ApJ, 645, 1152
[7] Paumard, T. et al. 2006, ApJ, 643, 1011
[8] Perets, H. B., Gualandris, A., Kupi, G., Merritt, D. & Alexander, T., 2009, ApJ, 702, 884
[9] Rauch, K. P. & Ingalls, B. 1998, MNRAS, 299, 1231
[10] Rauch, K. P. & Tremaine, S. 1996, New Astronomy, 1, 149

# AUTHOR INDEX

# AUTHOR INDEX

# AUTHOR INDEX

# AUTHOR INDEX

**Institute of Physics**
Dirac House, Temple Back
Bristol BS1 6BE UK

CURRAN ASSOCIATES INC
proceedings.com